INSTRUCTOR'S SOLUTIONS MANUAL

Katy Murphy

Michael J. Sullivan, III

Michael Sullivan

ALGEBRA & TRIGONOMETRY

FOURTH EDITION

MICHAEL SULLIVAN

PRENTICE HALL, Upper Saddle River, NJ 07458

For Shannon, Patrick, and Ryan

Production Editor: *Alison Aquino*
Supplement Acquisitions Editor: *Audra Walsh*
Production Coordinator: *Joan Eurell*
Manufacturing Buyer: *Alan Fischer*

©1996 by Prentice-Hall, Inc.
Simon & Schuster / A Viacom Company
Upper Saddle River, New Jersey 07458

Printed in the United States of America

10 9 8 7 6 5 4 3 2 1

ISBN 0-13-456443-X

This manual contains detailed solutions to all the even-numbered problems in *College Algebra and Trigonometry*, Fourth Edition, by Michael Sullivan. The solutions are designed to aid the instructor in grading and verifying answers.

We wish to thank everyone who has helped with this project. The enormous job of typing was done by Brenda Dobson, and the art work was prepared by Kelly Evans. A special thank you to our families for their support during this project.

Finally, we would be grateful to hear from readers who discover any errors in this solutions manual.

Katy Murphy
Michael Sullivan, III
and
Michael Sullivan

Contents

CHAPTER 4 FUNCTIONS AND THEIR GRAPHS 101

CHAPTER 5 POLYNOMIAL AND RATIONAL FUNCTIONS 145

CHAPTER 6 EXPONENTIAL AND LOGARITHMIC FUNCTIONS 196

CHAPTER 7 TRIGONOMETRIC FUNCTIONS 217

Contents

CHAPTER 13 SEQUENCES; INDUCTION; COUNTING; PROBABILITY 451

CHAPTER 14 MISCELLANEOUS TOPICS 476

APPENDIX: GRAPHING UTILITIES 499

PRELIMINARIES

1.1 Real Numbers

2. $\dfrac{2}{5} = 0.4$ 4. $\dfrac{10}{3} = 3.333\ldots$ 6. $\dfrac{-16}{3} = -5.333\ldots$

8. $\dfrac{1}{10} = 0.1$ 10. $\dfrac{-5}{6} = -0.8333\ldots$ 12. $\dfrac{8}{11} = 0.727272\ldots$

14. $6 - 4 + 1 = 2 + 1 = 3$ 16. $8 - 4 \cdot 2 = 8 - 8 = 0$

18. $8 - 3 - 4 = 5 - 4 = 1$ 20. $2 = \dfrac{1}{2} = \dfrac{4-1}{2} = \dfrac{3}{2}$

22. $2 \cdot [8 - 3(4 + 2)] - 3 = 2 \cdot [8 - 3(6)] - 3 = 2 \cdot [8 - 18] - 3 = 2 \cdot [-10] - 3$
$= -20 - 3 = -23$

24. $1 - (4 \cdot 3 - 2 + 2) = 1 - (12 - 2 + 2) = 1 - (10 + 2) = 1 - 12 = -11$

26. $2 - 5 \cdot 4 - [6 \cdot (3 - 4)] = 2 - 5 \cdot 4 - [6 \cdot (-1)] = 2 - 5 \cdot 4 - [-6] = 2 - 5 \cdot 4 + 6$
$= 2 - 20 + 6 = -12$

28. $5 + 4 \cdot \dfrac{1}{3} = 5 + \dfrac{4}{3} = \dfrac{15 + 4}{3} = \dfrac{19}{3}$

30. $\dfrac{2}{5} - \dfrac{4}{3} = \dfrac{2 \cdot 3 - 4 \cdot 5}{5 \cdot 3} = \dfrac{6 - 20}{15} = \dfrac{-14}{15}$

32. $\dfrac{\dfrac{14}{3} + \dfrac{1}{2}}{\dfrac{3}{4}} = \dfrac{\dfrac{14 \cdot 2 + 3 \cdot 1}{3 \cdot 2}}{\dfrac{3}{4}} = \dfrac{\dfrac{28 + 3}{6}}{\dfrac{3}{4}} = \dfrac{\dfrac{31}{6}}{\dfrac{3}{4}} = \dfrac{31}{6} \cdot \dfrac{4}{3} = \dfrac{124}{18} = \dfrac{62}{9}$

34. $4(2x - 1) = 4 \cdot 2x + 4 \cdot -1 = 8x - 4$ 36. $3x(x + 5) = 3x^2 + 15x$

38. $(x + 5)(x + 1) = x^2 + x + 5x + 5 = x^2 + 6x + 5$

40. $(x - 4)(x + 1) = x^2 + x - 4x - 4 = x^2 - 3x - 4$

42. $(x - 4)(x - 2) = x^2 - 2x - 4x + 8 = x^2 - 6x + 8$

44. $(x - 3)(x + 3) = x^2 + 3x - 3x - 9 = x^2 - 9$

46. Order of operations tells us that we multiply before we add unless parentheses exist. Thus, $2 + 3 \cdot 4 = 2 + 12 = 14$ and $(2 + 3) \cdot 4 = (5)4 = 20$

48. $\dfrac{4 + 3}{2 + 5} = \dfrac{7}{7} = 1$ and $\dfrac{4}{2} + \dfrac{3}{5} = \dfrac{20}{10} + \dfrac{6}{10} = \dfrac{26}{10} = 2.6$

50. The fraction $\dfrac{2}{3} = 0.6666\ldots$, with 6 repeating. Thus $\dfrac{2}{3}$ is larger by $0.000666\ldots$.

52. No. $(a - b) - c \neq a - (b - c)$
$(4 - 3) - 2 \neq 4 - (3 - 2)$
$1 - 2 \neq 4 - 1$
$-1 \neq 3$

54. No. $(a \div b) \div c \neq a \div (b \div c)$
$a = 6, b = 3, c = 2$
$(6 \div 3) \div 2 \neq 6 \div (3 \div 2)$
$2 \div 2 \neq 6 \div 1.5$
$1 \neq 4$

56. If $x = 5$, then $x^2 + x = 30$ by the principle of substitution, since $5^2 + 5 = 30$.

58.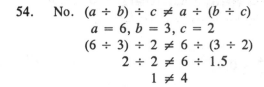

60. $z < 0$

62. $y > -5$

64. $\dfrac{1}{x} \geq 2$

66. Since $a > 0$ and $b < 0$, $a \cdot b$ is negative because the product of a positive number times a negative number is a negative number.

68. Since b is negative so is $1/b$. Thus, $a/b = a \cdot \dfrac{1}{b}$ is negative since the product of a positive number times a negative number is a negative number.

70. Since $b < 0$ and $a > 0$, then $-a < 0$ so that $b - a = b + (-a)$ is negative.

72. $|ab| > 0$ by the definition of absolute value. Note that $ab \neq 0$ since $a \neq 0$ and $b \neq 0$.

74. $a + |a|$ is positive since $a > 0$ and $|a| > 0$.

76. $|x - y| = |3 - (-2)| = |5| = 5$

78. $|x| - |y| = |3| - |-2| = 3 - 2 = 1$

80. $\dfrac{|y|}{y} = \dfrac{|-2|}{-2} = -1$

82. $|3x + 2y| = |3(3) + 2(-2)|$
$= |9 - 4| = 5$

84. $3|x| + 2|y| = 3|3| + 2|-2|$
$= 3(3) + 2(2)$
$= 9 + 4 = 13$

86. The domain of the variable x in the expression $\dfrac{-6}{x + 5}$ is $\{x \mid x \neq -5\}$ since, if $x = -5$, the denominator becomes zero, which is not defined.

88. The domain of the variable x in the expression $\dfrac{x - 2}{x - 6}$ is $\{x \mid x \neq 6\}$ since, if $x = 6$, the denominator becomes zero.

90. The domain of the variable x in the expression $\dfrac{3}{x(x-4)}$ is $\{x \mid x \neq 0, x \neq 4\}$ since, if $x = 0$ or $x = 4$, the denominator becomes zero.

92. The domain of the variable x in the expression $\dfrac{2}{x+1} + \dfrac{3}{x+2}$ is $\{x \mid x \neq -1, x \neq -2\}$ since, if $x = -1$ or $x = -2$, the denominator becomes zero.

94. The domain of the variable x in the expression $\dfrac{x-3}{x^2+6}$ is the set of all real numbers, since there is no x which causes the denominator to equal zero.

96. $P = 2(\ell + w)$
All the variables are positive real numbers.

98. $A = \dfrac{1}{2}bh$
All the variables are positive real numbers.

100. $P = 3x$; x and P are positive real numbers.

102. $S = 4\pi r^2$; r and S are positive real numbers.

104. $S = 6x^2$; x and S are positive real numbers.

106. If $F = 212°$,
$$C = \frac{5}{9}(212 - 32)$$
$$= \frac{5}{9}(180)$$
$$= 100°$$

108. If $F = -4°$
$$C = \frac{5}{9}(-4 - 32)$$
$$= \frac{5}{9}(-36)$$
$$= -20°$$

110.

Deposits	Withdrawals
$210	$-$120
$ 80	$-$ 60
$ 25	$-$ 32
$315	$-$ 5
	$-$217

Thus, the balance at the end of the month is $315 - $217 = $98.

112.
$$1 + 3 + 5 + 7 + \dots + 99 = (1 + 99) + (3 + 97) + (5 + 95) + \dots + (49 + 51)$$
$$= \underbrace{100 + 100 + 100 + \dots + 100}_{25 \text{ times}}$$
$$= 2500$$

114.
$$1 - 2 + 3 - 4 + 5 - 6 + \dots + 97 - 98 + 99$$
$$= (1 + 99) + (-2 - 98) + (3 + 97) + (-4 - 96) + \dots + (-48 - 52) + (49 + 51) - 50$$
$$= 100 - 100 + 100 - 100 + \dots + 100 - 100 + 100 - 50$$
$$= \underbrace{0 + 0 + \dots + 0}_{24 \text{ times}} + 100 - 50$$
$$= 0 + 0 + \dots + 0 + 100 - 50$$
$$= 24(0) + 100 - 50 = 50$$

116. Since $C = 2\pi r$, the circumference of a circle of radius 1 is 2π. Thus, the point P will come into contact with the line again at the coordinates $2k\pi$, where $k = 1, 2, 3, \ldots$.

1.2 Approximations; Calculators

2. (a) 25.861
   (b) 25.861

4. (a) 99.052
   (b) 99.052

6. (a) 0.054
   (b) 0.053

8. (a) 1.001
   (b) 1.000

10. $5/9 = 0.555\ldots$
   (a) 0.556
   (b) 0.555

12. $81/5 = 16.2$
   (a) 16.200
   (b) 16.200

14. Enter $\boxed{9.62}$ Press $\boxed{x^2}$

Display: 9.62   92.5444

Thus, $(9.62)^2 \approx 92.54$

16. Enter $\boxed{8.1}$ Press $\boxed{\times}$ Enter $\boxed{4.2}$ Press $\boxed{+}$

Display: 8.1   4.2   34.02

Enter $\boxed{6.1}$ Press $\boxed{=}$

Display: 6.1   40.12

Thus, $(8.1)(4.2) + 6.1 = 40.12$

18. Enter $\boxed{3.1}$ Press $\boxed{x^2}$ Press $\boxed{+}$ Enter $\boxed{9.6}$

Display: 3.1   9.61   9.6

Press $\boxed{x^2}$ Press $\boxed{=}$

Display: 92.16   101.77

Thus, $(3.1)^2 + (9.6)^2 = 101.77$

20. $9.1 - 8.2/10.2 \approx 8.296078431 \approx 8.30$

22. $\dfrac{4.73 - 2.4}{81 + 6.39}$

24. $\dfrac{21.3 - \pi}{6.1 + 8.} \approx 1.218685057 \approx 1.22$

26. $(16.5 - 7.2) \cdot (51.2) - 18.6 = 457.56$

28. $16.5 - 7.2/51.2 - 18.6 = -2.240625$
    ≈ -2.24

30. $(16.5 - 7.2)/(51.2 - 18.6) \approx 0.285276073$
    ≈ 0.29

32. $16.5 - 7.2/(51.2 - 18.6) \approx 16.2791411 \approx 16.28$

34. $\dfrac{5}{8} = 0.625 \boxed{=} 0.625$

36. $\pi = 3.1415927 \ldots \boxed{<} \dfrac{22}{7} = 3.1428571 \ldots$

38. $\dfrac{350}{1000000} = 0.00035$

40. Error message. This is not defined.

1.3 Integer Exponents

2. $-4^2 = -16$

4. $(-4)^2 = (-4) \cdot (-4) = 16$

6. $(-4)^{-2} = \dfrac{1}{(-4)^2} = \dfrac{1}{16}$

8. $(-2)^{-3} \cdot 2^0 = \dfrac{1}{(-2)^3} \cdot 2^0 = \dfrac{1}{-8} \cdot 1 = -\dfrac{1}{8}$

10. $3^{-2} + \dfrac{1}{3} = \dfrac{1}{3^2} + \dfrac{1}{3} = \dfrac{1}{9} + \dfrac{1}{3} = \dfrac{1+3}{9} = \dfrac{4}{9}$

12. $4^{-2} \cdot 4^3 = \dfrac{1}{4^2} \cdot 4^3 = \dfrac{4^3}{4^2} = 4^{3-2} = 4^1 = 4$

14. $\dfrac{(2^3)^3}{(2^2)^3} = \dfrac{2^9}{2^6} = 2^3 = 8$

16. $\left(\dfrac{3}{2}\right)^{-3} = \left(\dfrac{2}{3}\right)^3 = \dfrac{8}{27}$

18. $\dfrac{3^{-2} \cdot 5^3}{3 \cdot 5} = \dfrac{\dfrac{1}{3^2} \cdot 5^3}{3 \cdot 5} = \dfrac{\dfrac{5^3}{3^2}}{15} = \dfrac{\dfrac{125}{9}}{15} = \dfrac{125}{9} \cdot \dfrac{1}{15} = \dfrac{25}{27}$

20. $\left(\dfrac{6}{5}\right)^{-3} = \left(\dfrac{5}{6}\right)^3 = \dfrac{125}{216}$

22. $\dfrac{3^{-2}}{2} = \dfrac{\dfrac{1}{3^2}}{2} = \dfrac{\dfrac{1}{9}}{2} = \dfrac{1}{9} \cdot \dfrac{1}{2} = \dfrac{1}{18}$

24. $\dfrac{-2^{-3}}{-1} = \dfrac{-\dfrac{1}{2^3}}{-1} = \dfrac{-\dfrac{1}{8}}{1} = \dfrac{1}{8}$

26. $x^2 y = \dfrac{1}{x^1} \cdot y = \dfrac{y}{x}$

28. $x^0 y^4 = 1 \cdot y^4 = y^4$

30. $(-8x^3)^{-2} = (-8)^{-2} x^{-6} = \dfrac{1}{(-8)^2} \cdot \dfrac{1}{x^6} = \dfrac{1}{64} \cdot \dfrac{1}{x^6} = \dfrac{1}{64x^6}$

32. $(-4x)^{-1} = \dfrac{1}{(-4x)^1} = \dfrac{1}{-4x}$

34. $(3x)^0 = 3^0 \cdot x^0 = 1 \cdot 1 = 1$

36. $\dfrac{x^{-2}y}{xy^2} = \dfrac{\dfrac{1}{x^2} \cdot y}{xy^2} = \dfrac{\dfrac{y}{x^2}}{xy^2} = \dfrac{y}{x^2} \cdot \dfrac{1}{xy^2} = \dfrac{1}{x^3 y}$

38. $\dfrac{x^{-2}y^{-3}}{x} = \dfrac{\dfrac{1}{x^2} \cdot \dfrac{1}{y^3}}{x} = \dfrac{1}{x^2 y^3} \cdot \dfrac{1}{x} = \dfrac{1}{x^3 y^3}$

40. $\left[\dfrac{2x}{3}\right]^{-1} = \left[\dfrac{3}{2x}\right]^{1} = \dfrac{3}{2x}$

42. $(x^2 y)^{-2} = \dfrac{1}{(x^2 y)^2} = \dfrac{1}{x^4 y^2}$

44. $x^{-1} + y^{-1} = \dfrac{1}{x} + \dfrac{1}{y}$

46. $\dfrac{3x^{-2}yz^2}{x^4 y^{-3} z^2} = \dfrac{\dfrac{3yz^2}{x^2}}{\dfrac{x^4 z^2}{y^3}} = \dfrac{3yz^2}{x^2} \cdot \dfrac{y^3}{x^4 z^2} = \dfrac{3y^4}{x^6}$

48. $\dfrac{4x^{-2}(yz)^{-1}}{(-5)^2 x^4 y^2 z^{-2}} = \dfrac{\dfrac{4}{x^2 yz}}{\dfrac{25x^4 y^2}{z^2}} = \dfrac{4}{x^2 yz} \cdot \dfrac{z^2}{25x^4 y^2} = \dfrac{4z}{25x^6 y^3}$

50. $\dfrac{x^{-1} + y^{-1}}{x^{-1} - y^{-1}} = \dfrac{\dfrac{1}{x} + \dfrac{1}{y}}{\dfrac{1}{x} - \dfrac{1}{y}} = \dfrac{\dfrac{y+x}{xy}}{\dfrac{y-x}{xy}} = \dfrac{y+x}{xy} \cdot \dfrac{xy}{y-x} = \dfrac{y+x}{y-x}$

52. $\dfrac{\left[\dfrac{y}{x}\right]^2}{x^{-2}y} = \dfrac{\dfrac{y^2}{x^2}}{\dfrac{y}{x^2}} = \dfrac{y^2}{x^2} \cdot \dfrac{x^2}{y} = y$

54. $\left[\dfrac{5x^{-2}}{6y^{-2}}\right]^{-3} = \left[\dfrac{6y^{-2}}{5x^{-2}}\right]^{3} = \dfrac{216y^{-6}}{125x^{-6}} = \dfrac{\dfrac{216}{y^6}}{\dfrac{125}{x^6}} = \dfrac{216}{y^6} \cdot \dfrac{x^6}{125} = \dfrac{216x^6}{125y^6}$

56. $\dfrac{(3xy^{-1})^2}{(2x^{-1}y)^3} = \dfrac{9x^2 y^{-2}}{8x^{-3}y^3} = \dfrac{\dfrac{9x^2}{y^2}}{\dfrac{8y^3}{x^3}} = \dfrac{9x^2}{y^2} \cdot \dfrac{x^3}{8y^3} = \dfrac{9x^5}{8y^5}$

58. $\dfrac{\left[\dfrac{2x}{3y^2}\right]^{-1}}{\dfrac{1}{y^3}} = \dfrac{\left[\dfrac{3y^2}{2x}\right]^{1}}{\dfrac{1}{y^3}} = \dfrac{3y^2}{2x} \cdot \dfrac{y^3}{1} = \dfrac{3y^5}{2x}$

60. $\dfrac{(x^2)^{-3}y^3}{(x^3y)^{-2}} = \dfrac{x^{-6}y^3}{x^{-6}y^{-2}} = \dfrac{\dfrac{y^3}{x^6}}{\dfrac{1}{x^6y^2}} = \dfrac{y^3}{x^6} \cdot x^6y^2 = y^5$

62. $3x^3 + 2x^2 - x + 2$

   If $x = 1$, $3(1)^3 + 2(1)^2 - 1 + 2 = 3(1) + 2(1) - 1 + 2 = 3 + 2 - 1 + 2 = 6$

   If $x = 2$, $3(2)^3 + 2(2)^2 - 2 + 2 = 3(8) + 2(4) - 2 + 2 = 24 + 8 - 2 + 2 = 32$

64. $(0.1)^3(20)^3 = [0.1)(20)]^3 = 2^3 = 8$

66. $(3.7)^4 \approx 187.416$

68. $(2.2)^{-5} \approx 0.019$

70. $-(2.8)^6 \approx -481.890$

72. $-(8.11)^{-4} \approx -0.0002 = -0.000$

74. 3.214×10

76. 4.21×10^{-3}

78. 2.121×10^4

80. 5.14×10^{-2}

82. $6{,}700$

84. 0.000988

86. 411.2

88. 0.6453

90. $186{,}000x = 93{,}000{,}000$

   $x = \dfrac{93{,}000{,}000}{186{,}000} = 500$

   It will take 5.00×10^2 seconds.

1.4   Polynomials

2. $1 - 4x$ is a polynomial of degree 1

4. $-\pi$ is a polynomial of degree 0

6. $\dfrac{3}{x} + 2$ is not a polynomial due to the $3x^{-1}$ term.

8. $10z^2 + z$ is a polynomial of degree 2.

10. $\dfrac{3x^2 + 2x - 1}{x^2 + x + 1}$ is not a polynomial since it is the quotient of two polynomials and the denominator polynomial has degree greater than one.

12. $(x^3 - 3x^2 + 2) + (x^2 - x + 1) = x^3 + (-3x^2 + x^2) - x + (2 + 1) = x^3 - 2x^2 - x + 3$

14. $(x^2 - 3x - 1) - (x^3 - 2x^2 + x + 5) = x^2 - 3x - 1 - x^3 + 2x^2 - x - 5$
   $= -x^3 + (x^2 + 2x^2) + (-3x - x) + (-1 - 5)$
   $= -x^3 + 3x^2 - 4x - 6$

16. $(10x^5 - 8x^2) + (3x^3 - 2x^2 + 6) = 10x^5 + 3x^3 + (-8x^2 - 2x^2) + 6 = 10x^5 + 3x^3 - 10x^2 + 6$

18. $-2(x^2 + x + 1) + 6(-5x^2 - x + 2) = -2x^2 - 2x - 2 - 30x^2 - 6x + 12$
$$= (-2x^2 - 30x^2) + (-2x - 6x) + (-2 + 12)$$
$$= -32x^2 - 8x + 10$$

20. $8(4x^3 - 3x^2 - 1) - 6(4x^3 + 8x - 2) = 32x^3 - 24x^2 - 8 - 24x^3 - 48x + 12$
$$= (32x^3 - 24x^3) - 24x^2 - 48x + (-8 + 12)$$
$$= 8x^3 - 24x^2 - 48x + 4$$

22. $(x^2 + 1) - (4x^2 + 5) + (x^2 + x - 2) = x^2 + 1 - 4x^2 - 5 + x^2 + x - 2$
$$= (x^2 - 4x^2 + x^2) + x + (1 - 5 - 2) = -2x^2 + x - 6$$

24. $8(1 - y^3) + 2(1 + y + y^2 + y^3) = 8 - 8y^3 + 2 + 2y + 2y^2 + 2y^3$
$$= (-8y^3 + 2y^3) + 2y^2 + 2y + (8 + 2)$$
$$= -6y^3 + 2y^2 + 2y + 10$$

26. $(x - a)^2 - x^2 = x^2 - 2ax + a^2 - x^2 = a^2 - 2ax$

28. $(x - a)^3 - x^3 = x^3 - 3ax^2 + 3a^2x - a^3 - x^3 = -3ax^2 + 3a^2x - a^3$

30. $(x + 6)(x - 2) = x(x - 2) + 6(x - 2) = x^2 - 2x + 6x - 12 = x^2 + 4x - 12$

32. $(2x - 1)(x + 2) = 2x(x + 2) - 1(x + 2) = 2x^2 + 4x - x - 2 = 2x^2 + 3x - 2$

34. $(-2x - 1)(x + 1) = -2x(x + 1) - 1(x + 1) = -2x^2 - 2x - x - 1 = -2x^2 - 3x - 1$

36. $(2 - x)(2 - 3x) = 2(2 - 3x) - x(2 - 3x) = 4 - 6x - 2x + 3x^2 = 3x^2 - 8x + 4$

38. $(3x + 1)(3x - 1) = 3x(3x - 1) + 1(3x - 1) = 9x^2 - 3x + 3x - 1 = 9x^2 - 1$

40. $(3x + 1)^2 = (3x)^2 + 2(3x)(1) + (1)^2 = 9x^2 + 6x + 1$

42. $(x + 1)(x^2 - x + 1) = x(x^2 - x + 1) + 1(x^2 - x + 1)$
$$= x^3 - x^2 + x + x^2 - x + 1$$
$$= x^3 + (-x^2 + x^2) + (x - x) + 1$$
$$= x^3 + 1$$

44. $(3x - 4)(x^2 + 1) = 3x(x^2 + 1) - 4(x^2 + 1) = 3x^3 + 3x - 4x^2 - 4 = 3x^3 - 4x^2 + 3x - 4$

46. $(2x + 3)(3x^2 - 2x + 4) = 2x(3x^2 - 2x + 4) + 3(3x^2 - 2x + 4)$
$$= 6x^3 - 4x^2 + 8x + 9x^2 - 6x + 12$$
$$= 6x^3 + (-4x^2 + 9x^2) + (8x - 6x) + 12$$
$$= 6x^3 + 5x^2 + 2x + 12$$

48. $(2x + 3y)^2 = (2x)^2 + 2(2x)(3y) + (3y)^2 = 4x^2 + 12xy + 9y^2$

50. $(x - 2y)(x - y) = x(x - y) - 2y(x - y) = x^2 - xy - 2xy + 2y^2 = x^2 - 3xy + 2y^2$

52. $(x^2 - 2xy + y^2) - (x^2 - xy) = x^2 - 2xy + y^2 - x^2 + xy = (x^2 - x^2) + (-2xy + xy) + y^2$
$$= -xy + y^2$$

54. $(x - y)^2 - (x^2 + y^2) = (x^2 - 2xy + y^2) - (x^2 + y^2)$
$\qquad\qquad\qquad\qquad = x^2 - 2xy + y^2 - x^2 - y^2$
$\qquad\qquad\qquad\qquad = (x^2 - x^2) - 2xy + (y^2 - y^2)$
$\qquad\qquad\qquad\qquad = -2xy$

56. $(2x - y)^2 - (x - y)^2 = (4x^2 - 4xy + y^2) - (x^2 - 2xy + y^2)$
$\qquad\qquad\qquad\qquad = 4x^2 - 4xy + y^2 - x^2 + 2xy - y^2$
$\qquad\qquad\qquad\qquad = (4x^2 - x^2) + (-4xy + 2xy) + (y^2 - y^2)$
$\qquad\qquad\qquad\qquad = 3x^2 - 2xy$

58. $[(x - y)^2 + z^2] + [x^2 + (y - z)^2] = (x^2 - 2xy + y^2 + z^2) + (x^2 + y^2 - 2yz + z^2)$
$\qquad\qquad\qquad\qquad\qquad\qquad = (x^2 + x^2) - 2xy + (y^2 + y^2) - 2yz + (z^2 + z^2)$
$\qquad\qquad\qquad\qquad\qquad\qquad = 2x^2 - 2xy + 2y^2 - 2yz + 2z^2$

60. $(x + y)(x^2 - xy + y^2) = x(x^2 - xy + y^2) + y(x^2 - xy + y^2)$
$\qquad\qquad\qquad\qquad = x^3 - x^2y + xy^2 + x^2y - xy^2 + y^3$
$\qquad\qquad\qquad\qquad = x^3 + (-x^2y + x^2y) + (xy^2 - xy^2) + y^3$
$\qquad\qquad\qquad\qquad = x^3 + y^3$

62. $(x - 3)^2(x + 2) = (x^2 - 6x + 9)(x + 2)$
$\qquad\qquad\qquad = x^2(x + 2) - 6x(x + 2) + 9(x + 2)$
$\qquad\qquad\qquad = x^3 + 2x^2 - 6x^2 - 12x + 9x + 18$
$\qquad\qquad\qquad = x^3 + (2x^2 - 6x^2) + (-12x + 9x) + 18$
$\qquad\qquad\qquad = x^3 - 4x^2 - 3x + 18$

64. $(x + 5)^2(3x - 1) = (x^2 + 10x + 25)(3x - 1)$
$\qquad\qquad\qquad = x^2(3x - 1) + 10x(3x - 1) + 25(3x - 1)$
$\qquad\qquad\qquad = 3x^3 - x^2 + 30x^2 - 10x + 75x - 25$
$\qquad\qquad\qquad = 3x^3 + 29x^2 + 65x - 25$

66. $(x + 1)(x - 2)(x + 4) = [x(x - 2) + 1(x - 2)] (x + 4)$
$\qquad\qquad\qquad = (x^2 - 2x + x - 2)(x + 4)$
$\qquad\qquad\qquad = (x^2 - x - 2)(x + 4)$
$\qquad\qquad\qquad = x^2(x + 4) - x(x + 4) - 2(x + 4)$
$\qquad\qquad\qquad = x^3 + 4x^2 - x^2 - 4x - 2x - 8$
$\qquad\qquad\qquad = x^3 + (4x^2 - x^2) + (-4x - 2x) - 8$
$\qquad\qquad\qquad = x^3 + 3x^2 - 6x - 8$

68. $(x + 1)^3 - (x + 2)^3 = (x^3 + 3x^2 + 3x + 1) - (x^3 + 6x^2 + 12x + 8)$
$\qquad\qquad\qquad = x^3 + 3x^2 + 3x + 1 - x^3 - 6x^2 - 12x - 8$
$\qquad\qquad\qquad = (x^3 - x^3) + (3x^2 - 6x^2) + (3x - 12x) + (1 - 8)$
$\qquad\qquad\qquad = -3x^2 - 9x - 7$

70. $(x - 1)^3(x + 1) = (x^3 - 3x^2 + 3x - 1)(x + 1)$
$\qquad\qquad\qquad = x^3(x + 1) - 3x^2(x + 1) + 3x(x + 1) - 1(x + 1)$
$\qquad\qquad\qquad = x^4 + x^3 - 3x^3 - 3x^2 + 3x^2 + 3x - x - 1$
$\qquad\qquad\qquad = x^4 + (x^3 - 3x^3) + (-3x^2 + 3x^2) + (3x - x) - 1$
$\qquad\qquad\qquad = x^4 - 2x^3 + 2x - 1$

72. $(3x + 2)^3 = [(3x)^3 + 3 \cdot 2(3x)^2 + 3 \cdot 2^2(3x) + 2^3]$
$\qquad\qquad = 27x^3 + 6 \cdot 9x^2 + 12 \cdot 3x + 8$
$\qquad\qquad = 27x^3 + 54x^2 + 36x + 8$

74. $(3x - 2a)^3 = [(3x)^3 - 3 \cdot (2a)(3x)^2 + 3(2a)^2(3x) - (2a)^3]$
                  $= 27x^3 - 6a \cdot 9x^2 + 12a \cdot 3x - 8a^2$
                  $= 27x^3 - 54ax^2 + 36a^2x - 8a^3$

76. $(x + y - z)(x - y + z) = x(x - y + z) + y(x - y + z) - z(x - y + z)$
                      $= x^2 - xy + xz + xy - y^2 + yz - xz + yz - z^2$
                      $= x^2 + (-xy + xy) + (xz - xz) - y^2 + (yz + yz) - z^2$
                      $= x^2 - y^2 + 2yz - x^2$

78.
$$\begin{array}{r}
3x^2 - x + 1 \\
x\overline{)3x^3 - x^2 + x - 2} \\
\underline{3x^3} \\
-x^2 \\
\underline{-x^2} \\
x \\
\underline{x} \\
-2
\end{array}$$

The quotient is $3x^2 - x + 1$; the remainder is -2.
Check: $x(3x^2 - x + 1) - 2 = 3x^3 - x^2 + x - 2$

80.
$$\begin{array}{r}
3x^2 - 7x + 15 \\
x + 2\overline{)3x^3 - x^2 + x - 2} \\
\underline{3x^3 + 6x^2} \\
-7x^2 + x \\
\underline{-7x^2 - 14x} \\
15x - 2 \\
\underline{15x + 30} \\
-32
\end{array}$$

The quotient is $3x^2 - 7x + 15$; the remainder is -32.
Check: $(x + 2)(3x^2 - 7x + 15) - 32$
       $= 3x^3 - 7x^2 + 15x + 6x^2 - 14x + 30 - 32$
       $= 3x^3 - x^2 + x - 2$

82.
$$\begin{array}{r}
3x^2 + 11x + 45 \\
x - 4\overline{)3x^3 - x^2 + x - 2} \\
\underline{3x^3 - 12x^2} \\
-11x^2 + x \\
\underline{-11x^2 - 44x} \\
45x - 2 \\
\underline{45x - 180} \\
178
\end{array}$$

The quotient is $3x^2 + 11x + 45$; the remainder is 178.
Check: $(x - 4)(3x^2 + 11x + 45) + 178$
       $= 3x^3 + 11x^2 + 45x - 12x^2 - 44x - 180 + 178$
       $= 3x^3 - x^2 + x - 2$

84.
$$\begin{array}{r}
3x - 1 \\
x^2\overline{)3x^3 - x^2 + x - 2} \\
\underline{3x^3} \\
-x^2 \\
\underline{-x^2} \\
x - 2
\end{array}$$

The quotient is $3x - 1$; the remainder is $x - 2$.
Check: $x^2(3x - 1) + (x - 2) = 3x^3 - x^2 + x - 2$

                                                                          Chapter 1 Preliminaries

86.
$$
\begin{array}{r}
3x - 1 \\
x^2 + 2\overline{)3x^3 - x^2 + x - 2} \\
\underline{3x^3 \qquad\;\; + 6x} \\
-x^2 - 5x - 2 \\
\underline{-x^2 \qquad\; - 2} \\
-5x
\end{array}
$$

The quotient is $3x - 1$; the remainder is $-5x$.
Check: $(x^2 + 2)(3x - 1) - 5x$
$\qquad = 3x^3 - x^2 + 6x - 2 - 5x$
$\qquad = 3x^3 - x^2 + x - 2$

88.
$$
\begin{array}{r}
3 \\
x^3 - 1\overline{)3x^3 - x^2 + x - 2} \\
\underline{3x^3 \qquad\qquad\; - 3} \\
-x^2 + x + 1
\end{array}
$$

The quotient is 3; the remainder is $-x^2 + x + 1$.
Check: $3(x^3 - 1) + (-x^2 + x + 1)$
$\qquad = 3x^3 - 3 - x^2 + x + 1$
$\qquad = 3x^3 - x^2 + x - 2$

90.
$$
\begin{array}{r}
3x - 4 \\
x^2 + x + 1\overline{)3x^3 - x^2 + x - 1} \\
\underline{3x^3 + 3x^2 + 3x} \\
-4x^2 - 2x - 1 \\
\underline{-4x^2 - 4x - 4} \\
2x + 3
\end{array}
$$

The quotient is $3x - 4$; the remainder is $2x + 3$
Check: $(3x - 4)(x^2 + x + 1) + 2x + 3$
$\qquad = 3x^3 + 3x^2 + 3x - 4x^2 - 4x - 4 + 2x + 3$
$\qquad = 3x^3 - x^2 + x - 1$

92.
$$
\begin{array}{r}
3x + 8 \\
x^2 - 3x - 4\overline{)3x^3 - x^2 + x - 2} \\
\underline{3x^3 - 9x^2 - 12x} \\
8x^2 + 13x - 2 \\
\underline{8x^2 - 24x - 32} \\
37x + 30
\end{array}
$$

The quotient is $3x + 8$; the remainder is $37x + 30$.
Check: $(3x + 8)(x^2 - 3x - 4) + 37x + 30$
$\qquad = 3x^3 - 9x^2 - 12x + 8x^2 - 24x - 32 + 37x + 30$
$\qquad = 3x^3 - x^2 + x - 2$

94.
$$
\begin{array}{r}
x^3 - x^2 + x - 1 \\
x + 1\overline{)x^4 + 0x^3 + 0x^2 + 0x - 1} \\
\underline{x^4 + \;\; x^3} \\
-x^3 + 0x^2 \\
\underline{-x^3 - \;\; x^2} \\
x^2 + 0x \\
\underline{x^2 + \;\; x} \\
-x - 1 \\
\underline{-x - 1} \\
0
\end{array}
$$

The quotient is $x^3 - x^2 + x - 1$; the remainder is 0.
Check: $(x + 1)(x^3 - x^2 + x - 1)$
$\qquad = x^4 - x^3 + x^2 - x + x^3 - x^2 + x - 1$
$\qquad = x^4 - 1$

96.
$$
\begin{array}{r}
x^2 - 1 \\
x^2 + 1\overline{)x^4 + 0x^3 + 0x^2 + 0x - 1} \\
\underline{x^4 \qquad\quad + \; x^2} \\
-x^2 \qquad\;\; - 1 \\
\underline{-x^2 \qquad\;\; - 1} \\
0
\end{array}
$$

The quotient is $x^2 - 1$; the remainder is 0.
Check: $(x^2 + 1)(x^2 - 1) = x^4 - x^2 + x^2 - 1 = x^4 - 1$

98.

$$
\begin{array}{r}
-3x^3 - 3x^2 - 3x - 5 \\
x - 1 \overline{) -3x^4 + 0x^3 + 0x^2 - 2x - 1} \\
\underline{-3x^4 + 3x^3} \\
-3x^3 + 0x \\
\underline{-3x^3 + 3x^2} \\
-3x^2 - 2x \\
\underline{-3x^2 + 3x} \\
-5x - 1 \\
\underline{-5x + 5} \\
-6
\end{array}
$$

The quotient is $-3x^3 - 3x^2 - 3x - 5$; the remainder is -6.
Check: $(x - 1)(-3x^3 - 3x^2 - 3x - 5) + (-6)$
$= -3x^4 - 3x^3 - 3x^2 - 5x + 3x^3 + 3x^2 + 3x$
$ + 5 + (-6)$
$= -3x^4 - 2x - 1$

100. $\quad 1 - x^2 + x^4 = x^4 - x^2 + 1$

$$
\begin{array}{r}
x^2 + x - 1 \\
x^2 - x + 1 \overline{) x^4 + 0x^3 - x^2 + 0x + 1} \\
\underline{x^4 - x^3 + x^2} \\
x^3 - 2x^2 + 0x \\
\underline{x^3 - x^2 + x} \\
-x^2 - x + 1 \\
\underline{-x^2 + x - 1} \\
-2x + 2
\end{array}
$$

The quotient is $x^2 + x - 1$; the remainder is $-2x + 2$.
Check: $(x^2 - x + 1)(x^2 + x - 1) - 2x + 2$
$= x^4 + x^3 - x^2 - x^3 - x^2 + x + x^2$
$ + x - 1 - 2x + 2$
$= x^4 - x^2 + 1$

102. $\quad 1 + x^2 = x^2 + 1; \; 1 - x^2 + x^4 = x^4 - x^2 + 1$

$$
\begin{array}{r}
x^2 - 2 \\
x^2 + 1 \overline{) x^4 + 0x^3 - x^2 + 0x + 1} \\
\underline{x^4 + x^2} \\
-2x^2 + 0x + 1 \\
\underline{-2x^2 - 2} \\
3
\end{array}
$$

The quotient is $x^2 - 2$; the remainder is 3.
Check: $(x^2 + 1)(x^2 - 2) + 3$
$= x^4 - 2x^2 + x^2 - 2 + 3$
$= x^4 - x^2 + 1$

104.

$$
\begin{array}{r}
x^2 - ax + a^2 \\
x + a \overline{) x^3 + 0x^2 + 0x + a^3} \\
\underline{x^3 + ax^2} \\
-ax^2 + 0x \\
\underline{-ax^2 - a^2x} \\
a^2x + a^3 \\
\underline{a^2x + a^3} \\
0
\end{array}
$$

The quotient is $x^2 - ax + a^2$; the remainder is 0.
Check: $(x + a)(x^2 - ax + a^2)$
$= x^3 - ax^2 + a^2x + ax^2 - a^2x + a^3$
$= x^3 + a^3$

106.

$$\begin{array}{r} x^4 + ax^3 + a^2x^2 + a^3x + a^4 \\ \hline x - a \overline{)\, x^5 + 0x^4 + \ 0x^3 + \ 0x^2 + \ 0x - a^5} \\ \underline{x^5 - ax^4} \\ ax^4 + \ 0x^3 \\ \underline{ax^4 - a^2x^3} \\ a^2x^3 + \ 0x^2 \\ \underline{a^2x^3 - a^3x^2} \\ a^3x^2 + \ 0x \\ \underline{a^3x^2 - a^4x} \\ a^4x - a^5 \\ \underline{a^4x - a^5} \\ 0 \end{array}$$

The quotient is $x^4 + ax^3 + a^2x^2 + a^3x + a^4$; the remainder is 0.

Check: $(x - a)(x^4 + ax^3 + a^2x^2 + a^3x + a^4)$
$$= x^5 + ax^4 + a^2x^3 + a^3x^2 + a^4x$$
$$\quad - ax^4 - a^2x^3 - a^3x^2 - a^4x$$
$$\quad - a^5$$
$$= x^5 - a^5$$

108. The degree of the sum equals the degree of the leading term

If $n > m$, $(a_nx^n + a_{n-1}x^{n-1} + ... + a_1x + a_0) + (b_mx^m + b_{m-1}x^{m-1} + ... + b_1x + b_0)$
$$= a_nx^n + ... + (a_m + b_m)x^m + ... + (a_1 + b_1)x + (a_0 + b_0)$$

which is of degree n

110. (4a) $(x + a)(x + b) = x(x + b) + a(x + b) = x^2 + bx + ax + ab = x^2 + (a + b)x + ab$

(5a) $(x + a)^3 = (x + a)(x + a)^2$
$$= (x + a)(x^2 + 2ax + a^2)$$
$$= x^3 + 2ax^2 + a^2x + ax^2 + 2a^2x + a^3$$
$$= x^3 + 3an^2 + 3a^2x + a^3$$

(7) $(x + a)(x^2 - ax + a^2) = x^3 - ax^2 + a^2x + ax^2 - a^2x + a^3 = x^3 + a^3$

112. $(x + a)^5 = (x + a)(x + a)^4 = x^5 + 5ax^4 + 10a^2x^3 + 10a^3x^2 + 5a^4x + a^5$

In $(x + a)^n$, the leading term is always x^n and the last term is always x^n. The middle terms are of the form a^ix^j where $i + j = n$.

Specifically, $(x + a)^2 = x^2 + 2ax + a^2$
$$(x + a)^3 = x^3 + 3ax^2 + 3a^2x + a^3$$
$$(x + a)^4 = x^4 + 4ax^3 + 6a^2x^2 + 4a^3x + a^4$$

1.5 Factoring Polynomials

2. $7x - 14 = 7(x - 2)$

4. $ax - x = a(x - 1)$

6. $x^3 - x^2 + x = x(x^2 - x + 1)$

8. $3x^2 - 3x + 3 = 3(x^2 - x + 1)$

10. $60x^2y - 48xy^2 + 72x^3y = 12xy(5x - 4y + 6x^2)$

12. $x^2 - 4 = (x - 2)(x + 2)$

14. $1 - 9x^2 = (1 - 3x)(1 + 3x) = -(3x - 1)(3x + 1)$

16. $x^2 + 3x - 4 = (x + 4)(x - 1)$

18. $x^2 - 4x - 21 = (x - 7)(x + 3)$

20. $x^2 - 6x + 5 = (x - 5)(x - 1)$

22. $x^2 - 4x + 4 = (x - 2)^2$

24. $x^2 - 2x - 1$ prime

26. $14 + 6x + x^2$ prime

28. $x^3 + 8x^2 - 20x = x(x^2 + 8x - 20) = x(x + 10)(x - 2)$

30. $3y^3 - 18y^2 - 48y = 3y(y^2 - 6y - 16) = 3y(y - 8)(y + 2)$

32. $25x^2 + 10x + 1 = (5x + 1)^2$ 34. $9x^2 - 12x + 4 = (3x - 2)^2$

36. $bx^2 + 14b^2x + 45b^3 = b(x^2 + 14bx + 45b^2) = b(x + 5)(x + 9)$

38. $x^3 + 125 = (x + 5)(x^2 - 5x + 25)$

40. $64 - 8x^3 = 8(8 - x^3) = 8(2 - x)(4 + 2x + x^2) = -8(x - 2)(x^2 + 2x + 4)$

42. $3x^5 - 3x^2 = 3x^2(x^3 - 1) = 3x^2(x - 1)(x^2 + x + 1)$ 44. $4x^2 + 3x - 1 = (4x - 1)(x + 1)$

46. $x^4 - 1 = (x^2 - 1)(x^2 + 1) = (x - 1)(x + 1)(x^2 + 1)$ 48. $x^6 + 2x^3 + 1 = (x^3 + 1)^2$

50. $x^8 - x^5 = x^5(x^3 - 1) = x^5(x - 1)(x^2 + x + 1)$ 52. $6z^2 - z - 1 = (3z + 1)(2z - 1)$

54. $9x^2 + 24x + 16 = (3x + 4)^2$

56. $5 + 11x - 16x^2 = (1 - x)(5 + 16x) = -(x - 1)(16x + 5)$

58. $9y^2 + 9y - 4 = (3y + 4)(3y - 1)$ 60. $8x^2 - 6x - 2 = 2(4x^2 - 3x - 1) = 2(x - 1)(4x + 1)$

62. $9x^2 - 3x + 3 = 3(3x^2 - x + 1)$ 64. $x^2 + 6x + 9 = (x + 3)^2$

66. $27x^3 - 9x^2 - 6x = 3x(9x^2 - 3x - 2) = 3x(3x + 1)(3x - 2)$

68. $x^4 - 7x^2 + 9$ prime

70. $4 - 14x^2 - 8x^4 = 2(2 - 7x^2 - 4x^4) = 2(1 - 4x^2)(2 + x^2) = 2(1 - 2x)(1 + 2x)(2 + x^2)$
$$= -2(2x - 1)(2x + 1)(x^2 + 2)$$

72. $8 + 64x^3 = (2 + 4x)(4 - (2)(4x) + 16x^2) = (2 + 4x)(4 - 8x + 16x^2) = (4x + 2)(16x^2 - 8x + 4)$

74. $5(3x - 7) + x(3x - 7) = (5 + x)(3x - 7) = (x + 5)(3x - 7)$

76. $(x - 1)^2 - 2(x - 1) = (x - 1)(x - 1 - 2) = (x - 1)(x - 3)$

78. $(5x - 3)^2 - 9 = [(5x - 3) + 3][(5x - 3) - 3] = 5x(5x - 6)$

80. $(5x + 1)^3 - 1 = [(5x + 1) - 1][(5x + 1^2) + (5x + 1)(1) + 1^2]$
$$= 5x(25x^2 + 10x + 1 + 5x + 1 + 1)$$
$$= 125x^3 + 50x^2 + 25x^2 + 15x$$
$$= 125x^3 + 75x^2 + 15x$$

82. $7(x^2 - 6x + 9) + 3(x - 3) = 7(x - 3)^2 + 3(x - 3) = (x - 3)[7(x - 3) + 3] = (x - 3)(7x - 18)$

84. $x^3 - 3x^2 - x + 3 = x^2(x - 3) - 1(x - 3) = (x^2 - 1)(x - 3) = (x - 1)(x + 1)(x - 3)$

86.
$$x^4 + x^3 + x + 1 = x^3(x + 1) + 1(x + 1)$$
$$= (x^3 + 1)(x + 1)$$
$$= (x + 1)(x^2 - x + 1)(x + 1)$$
$$= (x + 1)^2(x^2 - x + 1)$$

88.
$$x^5 - x^3 + 8x^2 - 8 = x^3(x^2 - 1) + 8(x^2 - 1)$$
$$= (x^3 + 8)(x^2 - 1)$$
$$= (x^3 + 8)(x - 1)(x + 1)$$
$$= (x + 2)(x^2 - 2x + 4)(x - 1)(x + 1)$$

90. Factors of 1   1, 1   −1, −1
Sum   2   −2
The possibilities are $(x + 1)(x + 1) = x^2 + 2x + 1$ or $(x - 1)(x - 1) = x^2 - 2x + 1$, none of which equals $x^2 + x + 1$.

1.6 Rational Expressions

2. $\dfrac{4x + 8}{12x + 24} = \dfrac{4(x + 2)}{12(x + 2)} = \dfrac{1}{3}$

4. $\dfrac{15x^2 + 24x}{3x^2} = \dfrac{3x(5x + 8)}{3x(x)} = \dfrac{5x + 8}{x}$

6. $\dfrac{x^2 - 4x + 4}{x^2 - 16} = \dfrac{(x - 2)(x - 2)}{(x - 4)(x + 4)} = \dfrac{(x - 2)^2}{(x - 4)(x + 4)}$

8. $\dfrac{3y^2 - y - 2}{3y^2 + 5y + 2} = \dfrac{(3y + 2)(y - 1)}{(3y + 2)(y + 1)} = \dfrac{y - 1}{y + 1}$

10. $\dfrac{x - x^2}{x^2 + x - 2} = \dfrac{x(1 - x)}{(x - 1)(x + 2)} = \dfrac{-x(x - 1)}{(x - 1)(x + 2)} = \dfrac{-x}{x + 2}$

12. $\dfrac{x^2 + x - 6}{9 - x^2} = \dfrac{(x - 2)(x + 3)}{(3 - x)(3 + x)} = \dfrac{-(x - 2)}{(x - 3)}$

14. $\dfrac{2x^2 + 5x - 3}{1 - 2x} = \dfrac{(x + 3)(2x - 1)}{-(2x - 1)} = -(x + 3)$

16. $\dfrac{4x^4 + 2x^3 - 6x^2}{4x^4 + 26x^3 + 30x^2} = \dfrac{2x^2(2x^2 + x - 3)}{2x^2(2x^2 + 13x + 15)} = \dfrac{(x - 1)(2x + 3)}{(x + 5)(2x + 3)} = \dfrac{x - 1}{x + 5}$

18. $\dfrac{(x + 2)^2 - 8x}{(x - 2)^2} = \dfrac{x^2 + 4x + 4 - 8x}{(x - 2)^2} = \dfrac{x^2 - 4x + 4}{(x - 2)^2} = \dfrac{(x - 2)^2}{(x - 2)^2} = 1$

20. $\dfrac{3(x - 2)^2 + 17(x - 2)}{2(x - 2)^2 + 7(x - 2)} = \dfrac{(x - 2)[3(x - 2) + 17]}{(x - 2)[2(x - 2) + 7]} = \dfrac{(x - 2)(3x - 6 + 17)}{(x - 2)(2x - 4 + 7)} = \dfrac{(x - 2)(3x + 11)}{(x - 2)(2x + 3)}$
$$= \dfrac{3x + 11}{2x + 3}$$

22. $\dfrac{9x - 25}{2(x - 1)} \cdot \dfrac{(1 - x)(1 + x)}{2(3x - 5)} = \dfrac{-(9x - 25)(x - 1)(x + 1)}{4(x - 1)(3x - 5)} = \dfrac{-(x + 1)(9x - 25)}{4(3x - 5)}$

24. $\dfrac{12}{x(x - 1)} \cdot \dfrac{(x - 1)(x + 1)}{2(2x - 1)} = \dfrac{6(x + 1)}{x(2x - 1)}$ 26. $\dfrac{3(2x - 9)}{5x} \cdot \dfrac{2}{2(2x - 9)} = \dfrac{3}{5x}$

28. $\dfrac{(x + 2)(x - 3)}{(x + 1)((x - 5)} \cdot \dfrac{(x + 5)(x - 5)}{(x - 3)(x + 5)} = \dfrac{(x + 2)}{(x + 1)}$

30. $\dfrac{9x^2 - 3x - 2}{12x^2 + 5x - 2} \cdot \dfrac{8x^2 + 10x - 3}{9x^2 - 6x + 1} = \dfrac{(3x + 1)(3x - 2)}{(4x - 1)(3x + 2)} \cdot \dfrac{(4x - 1)(2x + 3)}{(3x - 1)(3x - 1)}$

$\qquad\qquad = \dfrac{(3x + 1)(3x - 2)(2x + 3)}{(3x + 2)(3x - 1)^2}$

32. $\dfrac{3x^2 + 2x - 1}{5x^2 - 9x - 2} \cdot \dfrac{10x^2 - 13x - 3}{2x^2 - x - 3} = \dfrac{(x + 1)(3x - 1)}{(x - 2)(5x + 1)} \cdot \dfrac{(2x - 3)(5x + 1)}{(2x - 3)(x + 1)} = \dfrac{3x - 1}{x - 2}$

34. $\dfrac{(x + 1)(x - 8)}{(x + 5)(x - 3)} \cdot \dfrac{(x + 2)(x + 6)}{(x - 1)(x + 7)} \cdot \dfrac{(x - 1)(x + 5)}{(x + 2)(x + 9)} = \dfrac{(x + 1)(x - 8)(x + 6)}{(x - 3)(x + 7)(x + 9)}$

36. $\dfrac{4}{x} - \dfrac{6}{x} = \dfrac{4 - 6}{x} = \dfrac{-2}{x}$ 38. $\dfrac{3x}{2x - 1} - \dfrac{4}{2x - 1} = \dfrac{3x - 4}{2x - 1}$

40. $\dfrac{(2x - 5) + (x + 4)}{(3x + 2)} = \dfrac{3x - 1}{3x + 2}$ 42. $\dfrac{(5x - 4) - (x + 1)}{3x + 4} = \dfrac{4x - 5}{3x + 4}$

44. $\dfrac{6}{x - 1} - \dfrac{x}{1 - x} = \dfrac{6}{x - 1} + \dfrac{x}{x - 1} = \dfrac{6 + x}{x - 1}$

46. $\dfrac{2(x - 5) - 3(x + 5)}{(x + 5)(x - 5)} = \dfrac{2x - 10 - 3x - 15}{(x + 5)(x - 5)} = \dfrac{-(x + 25)}{(x + 5)(x - 5)}$

48. $\dfrac{3x(x + 3) + 2x(x - 4)}{(x - 4)(x + 3)} = \dfrac{3x^2 + 9x + 2x^2 - 8x}{(x - 4)(x + 3)} = \dfrac{5x^2 + x}{(x - 4)(x + 3)} = \dfrac{x(5x + 1)}{(x - 4)(x + 3)}$

50. $\dfrac{(2x - 1)(x + 1) - (2x + 1)(x - 1)}{(x - 1)(x + 1)} = \dfrac{(2x^2 + x - 1) - (2x^2 - x - 1)}{(x - 1)(x + 1)} = \dfrac{2x}{(x - 1)(x + 1)}$

52. $\dfrac{(x - 1)(x^2 + 1) + x^4}{x^3(x^2 + 1)} = \dfrac{x^3 - x^2 + x - 1 + x^4}{x^3(x^2 + 1)} = \dfrac{x^4 + x^3 - x^2 + x - 1}{x^3(x^2 + 1)}$

54. $\dfrac{3x^2(x^2 - 1) - 4x^3}{4(x^2 - 1)} = \dfrac{3x^4 - 3x^2 - 4x^3}{4(x^2 - 1)} = \dfrac{x^2(3x^2 - 4x - 3)}{4(x - 1)(x + 1)}$

56. $\dfrac{(3x + 1)(x^2 - 1) + x(x)(x + 1) - \left[2x(x)(x - 1)\right]}{x(x - 1)(x + 1)} = \dfrac{3x^3 + x^2 - 3x - 1 + x^3 + x^2 - 2x^3 + 2x^2}{x(x - 1)(x + 1)}$

$\qquad\qquad = \dfrac{2x^3 + 4x^2 - 3x - 1}{x(x - 1)(x + 1)}$

58. $\dfrac{(x-1)(x-2) - x(x-2) - x(x-1)}{x(x-1)(x-2)} = \dfrac{x^2 - 3x + 2 - x^2 + 2x - x^2 + x}{x(x-1)(x-2)} = \dfrac{-x^2 + 2}{x(x-1)(x-2)}$

60. $(x+3)(x-4),\ (x-4)(x-4)$
$\text{LCM} = (x+3)(x-4)^2$

62. $3(x-3)(x+3),\ (x-3)(2x+5)$
$\text{LCM} = 3(x+3)(2x+5)(x-3)$

64. $x-3,\ x(x+3),\ x(x-3)(x+3)$
$\text{LCM} = x(x+3)(x-3)$

66. $(x+2)(x+2),\ x^2(x+2),$
$(x+2)(x+2)(x+2)$
$\text{LCM} = x^2(x+2)^3$

68. $\dfrac{x}{x-3} - \dfrac{x+1}{(x-3)(x+8)} = \dfrac{x(x+8) - (x+1)}{(x-3)(x+8)} = \dfrac{x^2 + 8x - x - 1}{(x-3)(x+8)} = \dfrac{x^2 + 7x - 1}{(x-3)(x+8)}$

70. $\dfrac{3}{x-1} - \dfrac{x-4}{(x-1)^2} = \dfrac{3(x-1) - (x-4)}{(x-1)^2} = \dfrac{3x - 3 - x + 4}{(x-1)^2} = \dfrac{2x+1}{(x-1)^2}$

72. $\dfrac{2(x-1) - 6(x+2)}{(x+2)^2(x-1)^2} = \dfrac{2x - 2 - 6x - 12}{(x+2)^2(x-1)^2} = \dfrac{-4x - 14}{(x+2)^2(x-1)^2} = \dfrac{-2(2x+7)}{(x+2)^2(x-1)^2}$

74. $\dfrac{2x-3}{(x+1)(x+7)} - \dfrac{x-2}{(x+1)^2} = \dfrac{(2x-3)(x+1) - (x-2)(x+7)}{(x+1)^2(x+7)}$

$= \dfrac{(2x^2 - x - 3) - (x^2 + 5x - 14)}{(x+1)^2(x+7)} = \dfrac{x^2 - 6x + 11}{(x+1)^2(x+7)}$

76. $\dfrac{x}{(x-1)^2} + \dfrac{2}{x} - \dfrac{x+1}{x^2(x-1)} = \dfrac{x^3 + 2x(x-1)^2 - (x+1)(x-1)}{x^2(x-1)^2}$

$= \dfrac{x^3 + 2x(x^2 - 2x + 1) - (x^2 - 1)}{x^2(x-1)^2} = \dfrac{x^3 + 2x^3 - 4x^2 + 2x - x^2 + 1}{x^2(x-1)^2}$

$= \dfrac{3x^3 - 5x^2 + 2x + 1}{x^2(x-1)^2}$

78. $\dfrac{1}{h}\left[\dfrac{x^2 - (x+h)^2}{x^2(x+h)^2}\right] = \dfrac{1}{h}\left[\dfrac{x^2 - x^2 - 2hx - h^2}{x^2(x+h)^2}\right] = \dfrac{-(2x+h)}{x^2(x+h)^2}$

80. $\dfrac{4 + \dfrac{1}{x^2}}{3 - \dfrac{1}{x^2}} = \dfrac{\dfrac{4x^2 + 1}{x^2}}{\dfrac{3x^2 - 1}{x^2}} = \dfrac{4x^2 + 1}{3x^2 - 1}$

82. $\dfrac{x + 1 - x}{x + 1} \cdot \dfrac{x}{2x - x + 1} = \dfrac{x}{(x+1)^2}$

84. $\dfrac{(x-2)^2 - x(x+1)}{(x+1)(x-2)} \cdot \dfrac{1}{x+3} = \dfrac{x^2 - 4x + 4 - x^2 - x}{(x+1)(x+3)(x-2)} = \dfrac{-5x + 4}{(x+1)(x+3)(x-2)}$

86.

$$\frac{(2x + 5)(x - 3) - x^2}{x(x - 3)} \cdot \frac{(x - 3)(x + 3)}{x^2(x + 3) - (x + 1)^2(x - 3)}$$

$$= \frac{2x^2 - x - 15 - x^2}{x} \cdot \frac{(x + 3)}{x^3 + 3x^2 - \left[(x^2 + 2x + 1)(x - 3)\right]}$$

$$= \frac{x^2 - x - 15}{x} \cdot \frac{(x + 3)}{x^3 + 3x^2 - (x^3 - x^2 - 5x - 3)} = \frac{(x^2 - x - 15)(x + 3)}{x(4x^2 + 5x + 3)}$$

88.

$$1 - \frac{1}{1 - \dfrac{1}{1 - x}} = 1 - \frac{1}{\dfrac{1 - x - 1}{1 - x}} = 1 - \frac{1 - x}{-x} = \frac{-x - (1 - x)}{-x} = \frac{-1}{-x} = \frac{1}{x}$$

90.

$$\frac{\dfrac{x + h + 1}{x + h - 1} - \dfrac{x + 1}{x - 1}}{h} = \frac{\dfrac{(x + h + 1)(x - 1) - (x + 1)(x + h - 1)}{(x + h - 1)(x - 1)}}{h}$$

$$= \frac{(x^2 + hx - h - 1) - (x^2 + hx + h - 1)}{h(x + h - 1)(x - 1)} = \frac{-2h}{h(x + h - 1)(x - 1)}$$

$$= \frac{-2}{(x + h - 1)(x - 1)}$$

1.7 Square Roots; Radicals

2. $\sqrt{81} = 9$

4. $\sqrt[3]{125} = 5$

6. $\sqrt[3]{-8} = -2$

8. $\sqrt[3]{\dfrac{27}{8}} = \dfrac{3}{2}$

10. $\sqrt[3]{64x^6} = 4x^2$

12. $\sqrt{4(x + 4)^2} = 2(x + 4)$

14. $\sqrt[4]{32} = 2\sqrt{2}$

16. $\sqrt{27x^3} = 3x\sqrt{3x}$

18. $\sqrt{\sqrt{x^6}} = \sqrt{x^3} = x\sqrt{x}$

20. $\sqrt[3]{\dfrac{x}{8x^4}} = \dfrac{\sqrt[3]{x}}{2x\sqrt[3]{x}} = \dfrac{1}{2x}$

22. $\sqrt[5]{x^{10}y^5} = x^2 y$

24. $\sqrt[3]{\dfrac{3xy^5}{81x^4y^2}} = \dfrac{y\sqrt{3xy}}{3x\sqrt[3]{3xy^2}} = \dfrac{y}{3x}$

26. $\sqrt{9x^5} = 3x^2\sqrt{x}$

28. $\sqrt{5x}\,\sqrt{20x^3} = \sqrt{5x} \cdot 2x\sqrt{5x} = 10x^2$

30. $\dfrac{\sqrt[3]{x^2y}\ \sqrt[3]{125x^3}}{\sqrt[3]{8x^3y^4}} = \dfrac{\sqrt[3]{x^2y}\ \cdot\ 5x}{2xy\sqrt[3]{y}} = \dfrac{5\sqrt[3]{x^2}}{2y}$ 32. $\sqrt{\dfrac{9}{16x^4y^6}} = \dfrac{\sqrt{9}}{\sqrt{16(x^2)^2(y^3)^2}} = \dfrac{3}{4x^2y^3}$

34. $\left(\sqrt[3]{3}\sqrt{10}\right)^4 = \sqrt[3]{3^4}\ \sqrt{10^4} = 3\sqrt[3]{3}\ \cdot\ 100 = 300\sqrt[3]{3}$

36. $\sqrt[3]{\dfrac{x-1}{x^2+2x+1}}\ \sqrt[3]{\dfrac{(x-1)^2}{x+1}} = \dfrac{x-1}{x+1}$ 38. $\sqrt[4]{\sqrt[3]{x}} = \sqrt[12]{x}$

40. $\sqrt{\dfrac{x^2+y^2}{xy(x^2-y^2)}}\ \sqrt{\dfrac{x^2y}{x^4-y^4}} = \sqrt{\dfrac{(x^2+y^2)(x^2y)}{xy(x^2-y^2)(x^4-y^4)}} = \sqrt{\dfrac{(x^2+y^2)x}{(x^2-y^2)(x^2-y^2)(x^2+y^2)}}$

$= \sqrt{\dfrac{x}{(x^2-y^2)^2}} = \dfrac{\sqrt{x}}{x^2-y^2}$

42. $6\sqrt{5} + \sqrt{5} - 4\sqrt{5} = (6 + 1 - 4)\sqrt{5} = 3\sqrt{5}$

44. $5\sqrt{3} + 2\sqrt{12} - 3\sqrt{27} = 5\sqrt{3} + 4\sqrt{3} - 9\sqrt{3} = (5 + 4 - 9)\sqrt{3} = 0\sqrt{3} = 0$

46. $9\sqrt[3]{24} - \sqrt[3]{81} = 18\sqrt[3]{3} - 3\sqrt[3]{3} = (18 - 3)\sqrt[3]{3} = 15\sqrt[3]{3}$

48. $\sqrt{x^2y} - 3x\sqrt{9y} + 4\sqrt{25y} = x\sqrt{y} - 9x\sqrt{y} + 20\sqrt{y} = (x - 9x + 20)\sqrt{y} = (20 - 8x)\sqrt{y}$

50. $8xy - \sqrt{25x^2y^2} + \sqrt[3]{8x^3y^3} = 8xy - 5xy + 2xy = 5xy$

52. $\left(5\sqrt{8}\right)\left(-3\sqrt{3}\right) = -15\sqrt{24} = -30\sqrt{6}$

54. $\left(\sqrt{5} - 2\right)\left(\sqrt{5} + 6\right) = 5 + 4\sqrt{5} - 12 = -7 + 4\sqrt{5}$

56. $\left(2\sqrt{6} + 3\right)^2 = 24 + 12\sqrt{6} + 9 = 33 + 12\sqrt{6}$ 58. $\left(\sqrt{x} + \sqrt{5}\right)^2 = x + 2\sqrt{5x} + 5$

60. $\left(\sqrt[3]{x} + \sqrt[3]{2}\right)^3 = \left(\sqrt[3]{x}\right)^3 + 3\sqrt[3]{2}\left(\sqrt[3]{x}\right)^2 + 3\left(\sqrt[3]{2}\right)^2\sqrt[3]{x} + \left(\sqrt[3]{2}\right)^3 = x + 3\sqrt[3]{2x^2} + 3\sqrt[3]{4x} + 2$

62. $\left(4\sqrt{x} - \sqrt{y}\right)\left(\sqrt{x} + 3\sqrt{y}\right) = 4x + 12\sqrt{xy} - \sqrt{xy} - 3y = 4x + 11\sqrt{xy} - 3y$

64. $\sqrt{1-x^2} + \dfrac{x^2}{\sqrt{1-x^2}} = \dfrac{1-x^2+x^2}{\sqrt{1-x^2}} = \dfrac{1}{\sqrt{1-x^2}} = \dfrac{\sqrt{1-x^2}}{1-x^2}$

66. $\dfrac{\sqrt{3}}{\sqrt{5}} = \dfrac{\sqrt{3}}{\sqrt{5}}\ \cdot\ \dfrac{\sqrt{5}}{\sqrt{5}} = \dfrac{\sqrt{15}}{5}$ 68. $\dfrac{5}{\sqrt{10}} = \dfrac{5}{\sqrt{10}}\ \cdot\ \dfrac{\sqrt{10}}{\sqrt{10}} = \dfrac{5\sqrt{10}}{10} = \dfrac{\sqrt{10}}{2}$

70. $$\frac{x}{\sqrt{x^2 + 4}} = \frac{x\sqrt{x^2 + 4}}{\sqrt{x^2 + 4}\sqrt{x^2 + 4}} = \frac{x\sqrt{x^2 + 4}}{x^2 + 4}$$

72. $$\frac{2}{\sqrt{7} - 2} = \frac{2}{\sqrt{7} - 2} \cdot \frac{\sqrt{7} + 2}{\sqrt{7} + 2} = \frac{4 + 2\sqrt{7}}{7 - 4} = \frac{4 + 2\sqrt{7}}{3}$$

74. $$\frac{10}{4 - \sqrt{2}} = \frac{10}{4 - \sqrt{2}} \cdot \frac{4 + \sqrt{2}}{4 + \sqrt{2}} = \frac{10(4 + \sqrt{2})}{16 - 2} = \frac{10(4 + \sqrt{2})}{14} = \frac{5(4 + \sqrt{2})}{7} = \frac{20 + 5\sqrt{2}}{7}$$

76. $$\frac{\sqrt{3}}{2\sqrt{3} + 3} \cdot \frac{2\sqrt{3} - 3}{2\sqrt{3} - 3} = \frac{6 - 3\sqrt{3}}{12 - 9} = \frac{6 - 3\sqrt{3}}{3} = 2 - \sqrt{3}$$

78. $$\frac{\sqrt{5} + \sqrt{3}}{\sqrt{5} - 3} \cdot \frac{\sqrt{5} + \sqrt{3}}{\sqrt{5} + \sqrt{3}} = \frac{5 + 2\sqrt{15} + 3}{5 - 3} = \frac{8 + 2\sqrt{15}}{2} = \frac{2(4 + 1\sqrt{15})}{2} = 4 + \sqrt{15}$$

80. $$\frac{1}{\sqrt{x} - 3} = \frac{1}{\sqrt{x} - 3} \cdot \frac{\sqrt{x} + 3}{\sqrt{x} + 3} = \frac{\sqrt{x} + 3}{\sqrt{x^2} - 9} = \frac{\sqrt{x} + 3}{x - 9}$$

82. $$\frac{\sqrt{x + h} + \sqrt{x - h}}{\sqrt{x + h} - \sqrt{x - h}} \cdot \frac{\sqrt{x + h} + \sqrt{x - h}}{\sqrt{x + h} + \sqrt{x - h}} = \frac{x + h + 2\sqrt{(x - h)(x + h)} + x - h}{x + h - (x - h)}$$

$$= \frac{2(x + \sqrt{(x - h)(x + h)})}{2h} = \frac{x + \sqrt{(x - h)(x + h)}}{h}$$

84. $$\frac{4 + \sqrt{3}}{2 + 3\sqrt{3}} = \frac{(4 + \sqrt{3})(4 - \sqrt{3})}{(2 + 3\sqrt{3})(4 - \sqrt{3})} = \frac{16 - 3}{8 + 12\sqrt{3} - 2\sqrt{3} - 9} = \frac{13}{-1 + 10\sqrt{3}} = \frac{-13}{1 - 10\sqrt{3}}$$

86. $$\frac{\frac{1}{\sqrt{x + h}} - \frac{1}{\sqrt{x}}}{h} = \frac{\sqrt{x} - \sqrt{x + h}}{h\sqrt{x + h}\sqrt{x}} = \frac{\sqrt{x} - \sqrt{x + h}}{h\sqrt{x + h}\sqrt{x}} \cdot \frac{\sqrt{x} + \sqrt{x + h}}{\sqrt{x} + \sqrt{x + h}}$$

$$= \frac{x - (x + h)}{xh\sqrt{x + h} + h\sqrt{x}(x + h)} = \frac{-h}{h(x\sqrt{x + h} + (x + h)\sqrt{x})}$$

$$= \frac{-1}{x\sqrt{x + h} + (x + h)\sqrt{x}}$$

88. $\sqrt{7} \approx 2.65$

90. $\sqrt[3]{-5} \approx -1.71$

Keystrokes: 5 | ± | SHIFT | x^y | 3 | =

92. $\dfrac{\sqrt{5} - 2}{\sqrt{2} + 4} \approx 0.04$

Keystrokes: $\boxed{5}\ \boxed{\sqrt{}}\ \boxed{-}\ \boxed{2}\ \boxed{=}\ \boxed{\div}\ \boxed{(}\ \boxed{2}\ \boxed{\sqrt{}}\ \boxed{+}\ \boxed{4}\ \boxed{)}\ \boxed{=}$

94. $\dfrac{2\sqrt{3} - \sqrt[3]{4}}{\sqrt{2}} \approx 1.33$

Keystrokes: $\boxed{2}\ \boxed{\times}\ \boxed{3}\ \boxed{\sqrt{}}\ \boxed{-}\ \boxed{4}\ \boxed{\text{SHIFT}}\ \boxed{x^y}\ \boxed{3}$

$\boxed{=}\ \boxed{\div}\ \boxed{2}\ \boxed{\sqrt{}}\ \boxed{=}$

1.8 Rational Exponents

2. $4^{3/2} = 2^3 = 8$

4. $(-64)^{4/3} = (-4)^4 = 256$

6. $(-8)^{-5/3} = -2^{-5} = \dfrac{-1}{32}$

8. $25^{-5/2} = 5^{-5} = \dfrac{1}{3125} = 0.00032$

10. $\left[\dfrac{27}{8}\right]^{2/3} = \left[\dfrac{3}{2}\right]^2 = \dfrac{9}{4}$

12. $\left[\dfrac{8}{27}\right]^{-2/3} = \left[\dfrac{2}{3}\right]^{-2} = \left[\dfrac{3}{2}\right]^2 = \dfrac{9}{4}$

14. $16^{-1.5} = 16^{-3/2} = 4^{-3} = \dfrac{1}{64}$

16. $\left[\dfrac{1}{9}\right]^{1.5} = \left[\dfrac{1}{9}\right]^{3/2} = \left[\dfrac{1}{3}\right]^3 = \dfrac{1}{27}$

18. $\left(\sqrt[3]{4}\right)^6 = 4^{6/3} = 4^2 = 16$

20. $\left(\sqrt[4]{5}\right)^{-8} = 5^{-8/4} = 5^{-2} = \dfrac{1}{25}$

22. $x^{4/3}x^{1/2}x^{-1/4} = x^{16/12+6/12-3/12} = x^{19/12}$

24. $(x^4 y^8)^{5/4} = x^5 y^{10}$

26. $(xy)^{1/4}(x^2 y^2)^{1/2} = x^{1/4}y^{1/4}xy = x^{5/4}y^{5/4}$

28. $\left(4x^{-1}y^{1/3}\right)^{3/2} = 8x^{-3/2}y^{1/2} = \dfrac{8y^{1/2}}{x^{3/2}}$

30. $\left[\dfrac{x^{1/2}}{y^2}\right]^4 \left[\dfrac{y^{1/3}}{x^{-2/3}}\right]^3 = \dfrac{x^2 y}{x^{-2}y^8} = \dfrac{x^4}{y^7}$

32. $\dfrac{1 + x}{2x^{1/2}} + x^{1/2} = \dfrac{1 + x + 2x}{2x^{1/2}} = \dfrac{1 + 3x}{2x^{1/2}}$

34. $(x + 1)^{1/3} + x \cdot \dfrac{1}{3}(x + 1)^{-2/3} = (x + 1)^{-2/3}\left[(x + 1) + \dfrac{1}{3}x\right] = (x + 1)^{-2/3}\left[\dfrac{4}{3}x + 1\right]$

$$= \dfrac{4x + 3}{3(x + 1)^{2/3}}$$

36. $\dfrac{\sqrt[3]{8x+1}}{3\sqrt[3]{(x-2)^2}} + \dfrac{\sqrt[3]{x-2}}{24\sqrt[3]{(8x+1)^2}} = \dfrac{8\sqrt[3]{(8x+1)^3} + \sqrt[3]{(x-2)^3}}{24\sqrt[3]{(x-2)^2(8x+1)^2}} = \dfrac{8(8x+1)+(x-2)}{24\sqrt[3]{(x-2)^2(8x+1)^2}}$

$$= \dfrac{65x+6}{24\sqrt[3]{(x-2)^2(8x+1)^2}}$$

38. $\dfrac{\sqrt{x^2+1} - x \cdot \dfrac{2x}{2\sqrt{x^2+1}}}{x^2+1} = \dfrac{2(x^2+1)-2x^2}{2\sqrt{x^2+1}\,(x^2+1)} = \dfrac{2x^2+2-2x^2}{2\sqrt{x^2+1}\,(x^2+1)} = \dfrac{2}{2(x^2+1)^{3/2}}$

$$= \dfrac{1}{(x^2+1)^{3/2}}$$

40. $\dfrac{(9-x^2)^{1/2} + x^2(9-x^2)^{-1/2}}{9-x^2} = \dfrac{(9-x^2)^{-1/2}[(9-x^2)+x^2]}{9-x^2} = \dfrac{9(9-x^2)^{-1/2}}{9-x^2} = \dfrac{9}{(9-x^2)^{3/2}}$

or

$\dfrac{(9-x^2)^{1/2} + x^2(9-x^2)^{-1/2}}{9-x^2} = \dfrac{(9-x^2)^{1/2} + \dfrac{x^2}{(9-x^2)^{1/2}}}{9-x^2} = \dfrac{(9-x^2)+x^2}{(9-x^2)(9-x^2)^{1/2}} = \dfrac{9}{(9-x^2)^{3/2}}$

42. $\dfrac{(x^2+4)^{1/2} - x^2(x^2+4)^{-1/2}}{x^2+4} = \dfrac{(x^2+4)^{-1/2}[(x^2+4)-x^2]}{(x^2+4)} = \dfrac{4}{(x^2+4)^{3/2}}$

44. $\dfrac{2x(1-x^2)^{1/3} + \dfrac{2}{3}x^3(1-x^2)^{-2/3}}{(1-x^2)^{2/3}} = \dfrac{x(1-x^2)^{-2/3}\left[2(1-x^2) + \dfrac{2}{3}x^2\right]}{(1-x^2)^{2/3}}$

$$= \dfrac{x(1-x^2)^{-2/3}\left[2 - \dfrac{4}{3}x^2\right]}{(1-x^2)^{2/3}} = \dfrac{x\left[2 - \dfrac{4}{3}x^2\right]}{(1-x^2)^{4/3}} = \dfrac{6x-4x^3}{3(1-x^2)^{4/3}}$$

46. $(x^2+4)^{4/3} + x \cdot \dfrac{4}{3}(x^2+4)^{1/3} \cdot 2x = (x^2+4)^{1/3}\left[x^2+4+\dfrac{8}{3}x^2\right] = (x^2+4)^{1/3}\left[\dfrac{11}{3}x^2+4\right]$

48. $6x^{1/2}(2x+3) + x^{3/2} \cdot 8 = x^{1/2}[6(2x+3)+8x] = x^{1/2}(12x+18+8x)$

$$= x^{1/2}(20x+18) = 2x^{1/2}(10x+9)$$

or

$6x^{1/2}(2x+3) + x^{3/2} \cdot 8 = 2x^{1/2}[3(2x+3)+4x] = 2x^{1/2}(10x+9)$

50. $2x(3x+4)^{4/3} + x^2 \cdot 4(3x+4)^{1/3} = 2x(3x+4)^{1/3}[3x+4+2x] = 2x(3x+4)^{1/3}(5x+4)$

52. $6(6x+1)^{1/3}(4x-3)^{3/2} + 6(6x+1)^{4/3}(4x-3)^{1/2}$
$= 6(6x+1)^{1/3}(4x-3)^{1/2}[4x-3+6x+1]$
$= 6(6x+1)^{1/3}(4x-3)^{1/2}(10x-2)$
$= 12(6x+1)^{1/3}(4x-3)^{1/2}(5x-1)$

54. If $u = 2\sqrt{3}\, x\sqrt{3x^2 + 1}$, then

$$
\begin{aligned}
1 + u^2 &= 1 + \left(2\sqrt{3}\, x\sqrt{3x^2 + 1}\right)^2 = 1 + 4 \cdot 3x^2(3x^2 + 1) \\
&= 1 + 12x^2(3x^2 + 1) = 1 + 36x^4 + 12x^2 \\
&= 36x^4 + 12x^2 + 1 = (6x^2 + 1)(6x^2 + 1) \\
&= \left(6x^2 + 1\right)^2
\end{aligned}
$$

1.9 Complex Numbers

2. $(4 - 5i) + (-8 + 2i) = -4 - 3i$

4. $(3 - 4i) - (-3 - 4i) = 6 + 0i = 6$

6. $(-8 + 4i) - (2 - 2i) = -10 + 6i$

8. $-4(2 + 8i) = -8 - 32i$

10. $5i(-3 + 4i) = -15i + 20i^2 = -20 - 15i$

12. $(5 + 3i)(2 - i) = 10 + i - 3i^2 = 13 + i$

14. $(-3 + i)(3 + i) = -9 + i^2 = -10$

16. $\dfrac{13}{5 - 12i} \cdot \dfrac{5 + 12i}{5 + 12i} = \dfrac{65 + 156i}{25 - 144i^2} = \dfrac{65 + 156i}{169} = \dfrac{5 + 12i}{13} = \dfrac{5}{13} + \dfrac{12}{13}i$

18. $\dfrac{3 - i}{-2i} \cdot \dfrac{2i}{2i} = \dfrac{6i - 2i^2}{-4i^2} = \dfrac{2 + 6i}{4} = \dfrac{2}{4} + \dfrac{6}{4}i = \dfrac{1}{2} + \dfrac{3}{2}i$

20. $\dfrac{5i}{4 - 3i} = \dfrac{5i(4 + 3i)}{(4 - 3i)(4 + 3i)} = \dfrac{20i + 15i^2}{16 - 9i^2} = \dfrac{-15 + 20i}{25} = \dfrac{-3}{5} + \dfrac{4}{5}i$

22. $\dfrac{3 - 4i}{3 + 4i} = \dfrac{(3 - 4i)(3 - 4i)}{(3 + 4i)(3 - 4i)} = \dfrac{9 - 24i + 16i^2}{9 - 16i^2} = \dfrac{-7 - 24i}{25} = \dfrac{-7}{25} - \dfrac{24}{25}i$

24. $\dfrac{4 + 3i}{1 - i} \cdot \dfrac{1 + i}{1 + i} = \dfrac{4 + 7i + 3i^2}{1 - i^2} = \dfrac{1 + 7i}{2} = \dfrac{1}{2} + \dfrac{7}{2}i$

26. $\left[\dfrac{\sqrt{3}}{2} - \dfrac{1}{2}i\right]\left[\dfrac{\sqrt{3}}{2} - \dfrac{1}{2}i\right] = \dfrac{3}{4} - \dfrac{2\sqrt{3}}{4}i + \dfrac{1}{4}i^2 = \dfrac{1}{2} - \dfrac{\sqrt{3}}{2}i$

28. $(1 - i)(1 - i) = 1 - 2i + i^2 = 0 - 2i$

30. $\dfrac{1 + 2i}{(1 - 2i)(1 - 2i)} = \dfrac{1 + 2i}{1 - 4i + 4i^2} = \dfrac{1 + 2i}{-3 - 4i} \cdot \dfrac{-3 + 4i}{-3 + 4i} = \dfrac{-3 - 2i + 8i^2}{9 - 16i^2}$

$$= \dfrac{-11 - 2i}{25} = \dfrac{-11}{25} - \dfrac{2i}{25}$$

32. $\dfrac{3}{(2 + 3i)(2 - 3i)} = \dfrac{3}{4 - 9i^2} = \dfrac{3}{13}$ **34.** $i^{13} = (i^4)^3 \cdot i = 1 \cdot i = i$

36. $1^{-23} = \dfrac{1}{i^{23}} = \dfrac{1}{(i^4)^5 \cdot i^3} = \dfrac{1}{i^2 \cdot i} = \dfrac{1}{-i} = \dfrac{-1}{i} \cdot \dfrac{i}{i} = \dfrac{-i}{-1} = i$

38. $3 + i^3 = 3 - i$ **40.** $4i^3 - 2i^2 + 1 = -4i + 2 + 1 = 3 - 4i$

42. $(3i)^4 + 1 = 81 + 1 = 82$ **44.** $2i^4(1 + i^2) = 2i^4(1 - 1) = 0$

46. $\dfrac{1 - i + i^2 - i^3 + i^4}{1 - i} = \dfrac{1 - i - 1 + i + 1}{1 - i} = \dfrac{1}{1 - i} \cdot \dfrac{1 + i}{1 + i} = \dfrac{1 + i}{2} = \dfrac{1}{2} + \dfrac{1}{2}i$

48. $i^7 + i^5 + i^3 + i = -i + i - i + i = 0$ **50.** $(8 + 3i) - (8 - 3i) = 0 + 6i = 6i$

52. $\overline{(3 - 4i) - (8 + 3i)} = \overline{-5 - 7i} = -5 + 7i$ **54.** $(3 + 4i)(8 - 3i) = 24 + 23i - 12i^2$
$= 36 + 23i$

56. $\overline{(3 + 4i)(8 + 3i)} = \overline{(3 - 4i)(8 - 3i)} = \overline{24 - 41i + 12i^2} = \overline{12 - 41i} = 12 + 41i$

58. $\overline{(\bar{z})} = \overline{\overline{(a + bi)}} = \overline{(a - bi)} = a + bi = z$

60. $\overline{z \cdot w} = \overline{(a + bi)(c + di)} = \overline{(ac - bd) + (ad + bc)i} = (ac - bd) - (ad + bc)i$
$\bar{z} \cdot \bar{w} = (a - bi)(c - di) = (ac - bd) + (-ad - bc)i = (ac - bd) - (ad + bc)i$

1.10   Geometry Topics

2. $6^2 + 8^2 = c^2$
$10 = c$

4. $4^2 + 3^2 = c^2$
$5 = c$

6. $14^2 + 48^2 = c^2$
$2500 = c^2$
$50 = c$

8. $6^2 + 8^2 = 10^2$
Right triangle
$c = 10$

10. $2^2 + 2^2 \neq 3^2$
$8 \neq 9$
Not a right triangle

12. $10^2 + 24^2 = 26^2$
Right triangle
$c = 26$

14. $5^2 + 4^2 \neq 7^2$
$41 \neq 49$
Not a right triangle

16. $A = \ell \cdot w = 6 \cdot 2 = 12;$
$P = 2(\ell + w) = 2(6 + 2) = 16$

18. $A = \ell \cdot w = \left[\dfrac{3}{4} \cdot \dfrac{4}{3}\right] = 1;$
$P = 2\left[\dfrac{3}{4} + \dfrac{4}{3}\right] = \dfrac{25}{6}$

20. $A = \dfrac{1}{2}bh = \dfrac{1}{2}(3 \cdot 5) = 7.5$

22. $A = \frac{1}{2}bh = \frac{1}{2}\left(\frac{3}{4} \cdot \frac{4}{3}\right) = \frac{1}{2}$

24. $A = \pi r^2 = 2^2 \cdot \pi = 4\pi \approx 12.56$
$C = 2\pi r = 2\pi \cdot 2 = 4\pi \approx 12.56$

26. $A = \pi r^2 = \left(\frac{5}{4}\right)^2 \cdot \pi = \frac{25}{16}\pi \approx 4.91$; $C = 2\pi r = 2\pi \cdot \frac{5}{4} = \frac{5}{2}\pi \approx 7.85$

28. $V = \ell wh = (5)(3)(4) = 60$

30. $V = \ell wh = \left(\frac{3}{4}\right)\left(\frac{1}{2}\right)\left(\frac{5}{3}\right) = \frac{5}{8}$

32. A circular disk rolls one revolution in a distance equal to its circumference. The circumference of the disk is

$C = 2\pi r = 2\pi\left(\frac{3}{2}\right) = 3\pi$ feet.

revolutions $= \frac{20}{3\pi} \approx 2.12$

34. Area of $CGF = \frac{1}{2}CG \times FG = \frac{1}{2}[BC - BG] \times FG$

$= \frac{1}{2}[10 - 4] \times 4 = 12$ square feet

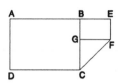

36. 20 feet $= 20$ feet $\cdot \dfrac{1\text{ mile}}{5280\text{ feet}} \approx 0.00379$ miles
$d^2 = (3960 + .00379)^2 - (3960)^2 = 30$
$d \approx 5.478$ miles

38. 100 feet ≈ 0.0189 miles; $d^2 = (3960 + 0.0189)^2 - (3960)^2 \approx 149.7$, $d \approx 12.23$ miles
150 feet ≈ 0.0284 miles; $d^2 = (3960 + 0.0284)^2 - (3960)^2 = 225$, $d = 15$ miles

1 Chapter Review

2. $8 + 8 - 6 = 10$

4. $\dfrac{9}{8} - \dfrac{6}{8} = \dfrac{3}{8}$

6. $\dfrac{\frac{18+3}{4}}{\frac{4}{3}} = \dfrac{21}{4} \cdot \dfrac{3}{4} = \dfrac{63}{16}$

8. $8 + 9 - 1 = 16$

10. $\dfrac{3^{-2} - 4^1}{2^{-2}} = \dfrac{\frac{1}{9} - 4}{\frac{1}{4}} = \dfrac{\frac{-35}{9}}{\frac{1}{4}} = \dfrac{-35}{9} \cdot 4 = \dfrac{-140}{9}$

12. $\left(2\sqrt{3} + \sqrt{2}\right)\left(2\sqrt{3} - \sqrt{2}\right) = 12 - 2 = 10$

14. $\left[\dfrac{4}{27}\right]^{-3/2} = \left[\dfrac{27}{4}\right]^{3/2} = \dfrac{\sqrt{27^3}}{8} = \dfrac{\left(3\sqrt{3}\right)^3}{8} = \dfrac{81\sqrt{3}}{8}$

16. $\dfrac{1}{\left(4\sqrt{32}\right)^{1/2}} = \dfrac{1}{2\left(4\sqrt{2}\right)^{1/2}} = \dfrac{1}{4\sqrt[4]{2}}$

18. $\left|25^{1/2} - 27^{2/3}\right| = \left|\sqrt{25} - \left(\sqrt[3]{27}\right)^2\right| = \left|5 - 3^2\right| = \left|5 - 9\right| = \left|-4\right| = 4$

20. $\sqrt[3]{-\left|4^2 - 2^3\right|} = \sqrt[3]{-\left|16 - 8\right|} = \sqrt[3]{-\left|8\right|} = \sqrt[3]{-8} = -2$

22. $\left[\dfrac{x^{-1}}{y^{-3}}\right]^2 = \dfrac{x^{-2}}{y^{-6}} = \dfrac{x^{1/2}}{\dfrac{1}{y^6}} = \dfrac{1}{x^2} \cdot \dfrac{y^6}{1} = \dfrac{y^6}{x^2}$ 24. $\dfrac{\dfrac{x^2}{y^2}}{\dfrac{x^{-1}}{y^{-1}}} = \dfrac{\dfrac{x^2}{y^2}}{\dfrac{y}{x}} = \dfrac{x^2}{y^2} \cdot \dfrac{x}{y} = \dfrac{x^3}{y^3}$

26. $\left(16^{-2/3}y^{4/3}\right)^{-3/2} = \dfrac{1}{64}xy^{-2} = \dfrac{x}{64y^2}$ 28. $\left[\dfrac{8x^{-3/2}}{y^{-3}}\right]^{-2/3} = \dfrac{\frac{1}{4}x}{y^2} = \dfrac{x}{4y^2}$

30. $(3x + 4)(2x^2 - 8x - 2) = 6x^3 - 16x^2 - 38x - 8$

32. $8(1 - x^2 + x^3) - 4(1 + 2x^2 - 4x^4) = 8 - 8x^2 + 8x^3 - 4 - 8x^2 + 16x^4 = 16x^4 + 8x^3 - 16x^2 + 4$

34. $(1 - 2x)(1 - 4x) = 1 - 6x + 8x^2 = 8x^2 - 6x + 1$

36. $(x^2 + 1)(x + 1) = x^3 + x^2 + x + 1$

38. $(x - y)^2(x + y)^2 = (x^2 - 2xy + y^2)(x^2 + 2xy + y^2)$
$= x^4 + 2x^3y + x^2y^2 - 2x^3y - 4x^2y^2 - 2xy^3 + x^2y^2 + 2xy^3 + y^4$
$= x^4 - 2x^2y^2 + y^4$

40.
$$
\begin{array}{r}
2x^2 + x + 3 \\
x - 2\,\overline{)\,2x^3 - 3x^2 + x + 1} \\
\underline{2x^3 - 4x^2} \\
x^2 + x \\
\underline{x^2 - 2x} \\
3x + 1 \\
\underline{3x - 6} \\
7
\end{array}
$$

The quotient is $2x^2 + x + 3$; the remainder is 7.
Check: $(x - 2)(2x^2 + x + 3) + 7$
$= 2x^3 + x^2 + 3x - 4x^2 - 2x - 6 + 7$
$= 2x^3 - 3x^2 + x + 1$

42.
$$
\begin{array}{r}
-4x + 1 \\
x^2 - 1\,\overline{)\,-4x^3 + x^2 + 0x - 2} \\
\underline{-4x^3 + 4x} \\
x^2 - 4x - 2 \\
\underline{x^2 - 1} \\
-4x - 1
\end{array}
$$

The quotient is $-4x + 1$; the remainder is $-4x - 1$.
Check: $(x^2 - 1)(-4x + 1) - 4x - 1$
$= -4x^3 + x^2 + 4x - 1 - 4x - 1$
$= -4x^3 + x^2 - 2$

44.

$$
\begin{array}{r}
3x^2 - 10x + 36 \\
x^2 + 3x - 2\,\overline{\smash{\big)}\,3x^4 -\ x^3 +\ 0x^2 -\ \ 8x +\ 4} \\
\underline{3x^4 + 9x^3 -\ 6x^2} \\
-10x^3 +\ 6x^2 -\ \ 8x \\
\underline{-10x^3 - 30x^2 +\ 20x} \\
36x^2 -\ 28x +\ 4 \\
\underline{36x^2 + 108x -\ 72} \\
-136x + 76
\end{array}
$$

The quotient is $3x^2 - 10x + 36$;
the remainder is $-136x + 76$.
Check:

$$
\begin{aligned}
(x^2 + 3x - 2)&(3x^2 - 10x + 36) - 136x + 76 \\
&= 3x^4 - 10x^3 + 36x^2 + 9x^3 - 30x^2 \\
&\quad + 108x - 6x^2 + 20x - 72 \\
&\quad - 136x + 76 \\
&= 3x^4 - x^3 - 8x + 4
\end{aligned}
$$

46.

$$
\begin{array}{r}
x^4 + x^3 + x^2 + x + 1 \\
x - 1\,\overline{\smash{\big)}\,x^5 + 0x^4 + 0x^3 + 0x^2 + 0x - 1} \\
\underline{x^5 -\ x^4} \\
x^4 + 0x^3 \\
\underline{x^4 -\ x^3} \\
x^3 + 0x^2 \\
\underline{x^3 -\ x^2} \\
x^2 + 0x \\
\underline{x^2 -\ x} \\
x - 1 \\
\underline{x - 1} \\
0
\end{array}
$$

The quotient is $x^4 + x^3 + x^2 + x + 1$; the
remainder is 0.
Check: $\ (x - 1)(x^4 + x^3 + x^2 + x + 1)$

$$
\begin{aligned}
&= x^5 + x^4 + x^3 + x^2 + x - x^4 \\
&\quad\ - x^3 - x^2 - x - 1 \\
&= x^5 - 1
\end{aligned}
$$

48.

$$
\begin{array}{r}
3x^4 - 2x^2 + 1 \\
2x - 1\,\overline{\smash{\big)}\,6x^5 - 3x^4 - 4x^3 + 2x^2 + 2x - 1} \\
\underline{6x^5 - 3x^4} \\
-4x^3 + 2x^2 \\
\underline{-4x^3 + 2x^2} \\
2x - 1 \\
\underline{2x - 1} \\
0
\end{array}
$$

The quotient is $3x^4 - 2x^2 + 1$; the remainder is 0.
Check: $\ (2x - 1)(3x^4 - 2x^2 + 1)$

$$
\begin{aligned}
&= 6x^5 - 4x^3 + 2x - 3x^4 + 2x^2 - 1 \\
&= 6x^5 - 3x^4 - 4x^3 + 2x^2 + 2x - 1
\end{aligned}
$$

50. $\ (x - 14)(x - 1)$

52. $\ (3x + 2)(2x - 1)$

54. $\ 2x(x^2 + 9x + 14) = 2x(x + 2)(x + 7)$

56. $\ (3x - 2)(9x^2 + 6x + 4)$

58. $\ x^2(2x + 3) + 1(2x + 3) = (x^2 + 1)(2x + 3)$

60. $\ 16x^2 - 1 = (4x - 1)(4x + 1)$

62. prime

64. $\ \dfrac{(x + 2)(x - 7)}{(2 - x)(2 + x)} = \dfrac{x - 7}{2 - x} = -\dfrac{(x - 7)}{x - 2}$

66. $\ \dfrac{(x - 5)(x + 5) \cdot x(x + 1)}{(x(x + 1)(x - 5)(1 - x)(1 + x)} = \dfrac{x + 5}{(1 - x)(1 + x)} = \dfrac{-(x + 5)}{(x - 1)(x + 1)}$

68. $\ \dfrac{x(x + 2) - 2x(x + 1)}{(x + 1)(x + 2)} = \dfrac{x^2 + 2x - 2x^2 - 2x}{(x + 1)(x + 2)} = \dfrac{-x^2}{(x + 1)(x + 2)}$

70. $$\frac{x^2}{(2x-1)(x+3)} + \frac{x^2}{(x-2)(2x-1)} = \frac{x^2(x-2) + x^2(x+3)}{(x+3)(x-2)(2x-1)} = \frac{x^3 - 2x^2 + x^3 + 3x^2}{(x+3)(x-2)(2x-1)}$$
$$= \frac{2x^3 + x^2}{(x+3)(x-2)(2x-1)} = \frac{x^2(2x+1)}{(x+3)(x-2)(2x-1)}$$

72. $$1 - \frac{1}{\dfrac{x+1}{x}} = 1 - \frac{x}{x+1} = \frac{x+1-x}{x+1} = \frac{1}{x+1}$$

74. $$\frac{-2}{\sqrt{5}} \cdot \frac{\sqrt{5}}{\sqrt{5}} = \frac{-2\sqrt{5}}{5}$$

76. $$\frac{-4}{1+\sqrt{3}} \cdot \frac{1-\sqrt{3}}{1-\sqrt{3}} = \frac{-4 + 4\sqrt{3}}{1-3} = \frac{-4(1-\sqrt{3})}{-2} = 2(1 - \sqrt{3})$$

78. $$\frac{4\sqrt{3} + 2}{2\sqrt{3} + 1} \cdot \frac{2\sqrt{3} - 1}{2\sqrt{3} - 1} = \frac{24 - 2}{12 - 1} = \frac{22}{11} = 2$$

80. $$(x^2 + 4)^{-1/3}\left[x^2 + 4 + \frac{4}{3}x^2\right] = (x^2 + 4)^{-1/3}\left(\frac{7}{3}x^2 + 4\right) = \frac{\frac{7}{3}x^2 + 4}{(x^2 + 4)^{1/3}} = \frac{7x^2 + 12}{3(x^2 + 4)^{1/3}}$$

82. $$\frac{2x(x^2 + 4)^{-1/2}\left[x^2 + 4 - \dfrac{x^2}{2}\right]}{x^2 + 4} = \frac{2x\left[\dfrac{1}{2}x^2 + 4\right]}{(x^2 + 4)^{3/2}} = \frac{x^3 + 8x}{(x^2 + 4)^{3/2}}$$

84. $2 - i$

86. $2 + 2i - 6 + 9i = -4 + 11i$

88. $$\frac{4}{2-i} \cdot \frac{2+i}{2+i} = \frac{8 + 4i}{4 - i^2} = \frac{8}{5} + \frac{4}{5}i$$

90. $i^{29} = (i^4)^7 \cdot i = 1 \cdot i = i$

92. $(3 - 2i)^2(3 - 2i) = (9 - 12i + 4i^2)(3 - 2i) = (5 - 12i)(3 - 2i) = 15 - 46i + 24i^2$
$$= -9 - 46i$$

94. $$C = 3000 + 6x - \frac{x^2}{1000}$$

(a) If $x = 1000$,
$$C = 3000 + 6(1000) - \frac{(1000)^2}{1000}$$
$$= 3000 + 6000 - 1000$$
$$= 8000$$

(b) If $x = 3000$,
$$C = 3000 + 6(3000) - \frac{(3000)^2}{1000}$$
$$= 3000 + 18000 - 9000$$
$$= 12000$$

96. Yes. He could see about 229 miles.
35,000 feet = 6.629 miles
$(3966.629)^2 - (3960)^2 = (229.228)^2$

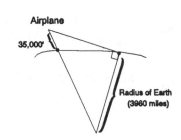

ients of $333.58
nents of $962.96
nents of $609.15
nents of $349.75

ck: $12,008.88   Amount of interest: $2008.88
ck: $11,555.52   Amount of interest: $2055.52
ick: $10,964.70   Amount of interest: $1964.70
ick: $12,591.00   Amount of interest: $2091.00

car from Dealer #3 is cheaper over all, but they might prefer spacing out the
l, so as not to have to "save up" the larger payments. And all things being
the color of the car!

Chapter 2

EQUATIONS AND INEQUALITIES

2.1 Equations

2. $3x = -24$
   $x = -8$

4. $3x + 18 = 0$
   $x = -6$

6. $3x + 4 = -8$
   $x = -4$

8. $\frac{2}{3}x = \frac{9}{2}$

   $x = \frac{27}{4}$

10. $2x + 7 = 3x + 5$
    $2x = 3x - 2$
    $x = 2$

12. $5x + 6 = -18 - y$
    $6y = -24$
    $y = -4$

14. $3 - 2x = 2 - x$
    $-2x = -1 - x$
    $x = 1$

16. $3 - 2m = 3m + 1$
    $-5m = -2$
    $m = \frac{2}{5}$

18. $3(2 - x) = 2x - 1$
    $6 - 3x = 2x - 1$
    $-5x = -7$
    $x = \frac{7}{5}$

20. $5 - (2x - 1) = 10$
    $5 - 2x + 1 = 10$
    $-2x = 4$
    $x = -2$

22. $\frac{1}{3}x = 2 - \frac{2}{3}x$

    $x = 3\left[2 - \frac{2}{3}x\right]$

    $x = 6 - 2x$
    $3x = 6$
    $x = 2$

24. $1 - \frac{1}{2}x = 6$

    $2\left[1 - \frac{1}{2}x\right] = 12$

    $2 - x = 12$
    $x = -10$

26. $\frac{1}{2} - \frac{1}{3}p = \frac{4}{3}$

    $6\left[\frac{1}{2} - \frac{1}{3}p\right] = 6\left[\frac{4}{3}\right]$

    $3 - 2p = 8$
    $-2p = 5$
    $p = \frac{-5}{2}$

28. $0.9t = 1 + t$
    $-0.1t = 1$
    $t = -10$

30.
$$\frac{2x + 1}{3} + 16 = 3x$$
$$3\left[\frac{2x + 1}{3} + 16\right] = 3(3x)$$
$$2x + 1 + 48 = 9x$$
$$-7x = -49$$
$$x = 7$$

32.
$$\frac{4}{y} - 5 = \frac{5}{2y}$$
$$2y\left[\frac{4}{y} - 5\right] = 2y\left[\frac{5}{2y}\right]$$
$$8 - 10y = 5$$
$$-10y = -3$$
$$y = \frac{3}{10}$$

Because we multiplied both sides of the equation by an expression containing a variable, we must check the apparent solution. On checking $y = \dfrac{3}{10}$, we find that it is, in fact, a solution of the original equation.

34.
$$\frac{3}{x} - \frac{1}{3} = \frac{1}{4}$$
$$12x\left[\frac{3}{x} - \frac{1}{3}\right] = 12x\left[\frac{1}{4}\right]$$
$$36 - 4x = 3x$$
$$36 = 7x$$
$$x = \frac{36}{7}$$

36.
$$(x + 2)(x - 3) = (x - 3)^2$$
$$x^2 - x - 6 = x^2 - 6x + 9$$
$$-x - 6 = -6x + 9$$
$$5x = 15$$
$$x = 3$$

38.
$$x(1 + 2x) = (2x - 1)(x - 2)$$
$$x + 2x^2 = 2x^2 - 5x + 2$$
$$x = -5x + 2$$
$$6x = 2$$
$$x = \frac{1}{3}$$

40.
$$w(4 - w^2) = 8 - w^3$$
$$4w - w^3 = 8 - w^3$$
$$4w = 8$$
$$w = 2$$

42.
$$\frac{3x}{x + 2} = \frac{-6}{x + 2} - 2$$
Note that $x + 2 \neq 0$, so $x \neq -2$ is not in the domain of the variable.
$$(x + 2) \cdot \frac{3x}{x + 2} = (x + 2)\left[\frac{-6}{x + 2} - 2\right]$$
$$3x = -6 - 2(x + 2)$$
$$3x = -6 - 2x - 4$$
$$5x = -10$$
$$x = -2$$
But $x = -2$ is not in the domain of the variable. The equation has no solution.

44.
$$x^3 = x^2$$
$$x^3 - x^2 = 0$$
$$x^2(x - 1) = 0$$
$$x^2 = 0 \qquad x - 1 = 0$$
$$x = 0 \qquad\quad x = 1$$
The solution set is $\{0, 1\}$.

46.
$$4z^3 - 8z^2 = 0$$
$$4z^2(z - 2) = 0$$
$$z^2 = 0 \quad z - 2 = 0$$
$$z = 0 \qquad z = 2$$
The solution set is $\{0, 2\}$.

48.
$$\frac{x}{x^2 - 9} + \frac{1}{x + 3} = \frac{3}{x^2 - 9}$$

Note that $x^2 - 9 \neq 0$ and $x + 3 \neq 0$, so $x = -3$ and $x = 3$ are not in the domain of the variable.

$$(x^2 - 9)\left[\frac{x}{x^2 - 9} + \frac{1}{x + 3}\right] = (x^2 - 9)\frac{3}{x^2 - 9}$$

$$x + \frac{x^2 - 9}{x + 3} = 3$$

$$x + \frac{(x - 3)(x + 3)}{x + 3} = 3$$

$$x + x - 3 = 3$$

$$2x = 6$$

$$x = 3$$

But $x = 3$ is not in the domain of the variable. The equation has no solution.

50.
$$\frac{3x}{x - 1} = 2, \; x \neq 1$$

On checking $x = -2$, we find that it is, in fact, a solution of the original equation.

$$(x - 1)\left[\frac{3x}{x - 1}\right] = (x - 1)(2)$$

$$3x = 2x - 2$$

$$x = -2$$

52.
$$\frac{-2}{x + 4} = \frac{-3}{x + 6}, \; x \neq -4, -1$$

$$(x + 6)(x + 4)\left[\frac{-2}{x + 4}\right] = (x + 6)(x + 4)\left[\frac{-3}{x + 6}\right]$$

$$-2(x + 6) = -3(x + 4)$$

$$-2x - 12 = -3x - 12$$

$$x = 0$$

On checking $x = 0$, we find that it is, in fact, a solution of the original equation.

54.
$$\frac{8w + 5}{10w - 7} = \frac{4w - 3}{5w + 7}, \; w \neq \frac{-7}{5}, \frac{7}{10}$$

$$(10w - 7)(5w + 7)\left[\frac{8w + 5}{10w - 7}\right] = (10w - 7)(5w + 7)\left[\frac{4w - 3}{5w + 7}\right]$$

$$(5w + 7)(8w + 5) = (10w - 7)(4w - 3)$$

$$40w^2 + 81w + 35 = 40w^2 - 58w + 21$$

$$81w + 35 = -58w + 21$$

$$139w = -14$$

$$w = \frac{-14}{139}$$

On checking, we find that $w = \dfrac{-14}{139}$ is, in fact, a solution of the original equation.

56.
$$\frac{-4}{2x + 3} + \frac{1}{x - 1} = \frac{1}{(2x + 3)(x - 1)}, \; x \neq \frac{-3}{2}, 1$$

$$(2x + 3)(x - 1)\left[\frac{-4}{2x + 3} + \frac{1}{x - 1}\right] = (2x + 3)(x - 1)\left[\frac{1}{(2x + 3)(x - 1)}\right]$$

$$-4(x - 1) + (2x + 3) = 1$$

$$-2x + 7 = 1$$

$$2x = 6$$

$$x = 3$$

On checking, we find that $x = 3$ is, in fact, a solution of the original equation.

58.

$$\frac{5}{5z - 11} + \frac{4}{2z - 3} = \frac{-3}{5 - z}, \quad z \neq \frac{3}{2}, \frac{11}{5}, 5$$

$$(5z - 11)(2z - 3)(5 - z)\left[\frac{5}{5z - 11} + \frac{4}{2z - 3}\right] = (5z - 11)(2z - 3)(5 - z)\left[\frac{-3}{5 - z}\right]$$

$$5(2z - 3)(5 - z) + 4(5z - 11)(5 - z) = -3(5z - 11)(2z - 3)$$

$$5(-2z^2 + 13z - 15) + 4(-5z^2 + 36z - 55) = -3(10z^2 - 37z + 33)$$

$$-10z^2 + 65z - 75 - 20z^2 + 144z - 220 = -30z^2 + 111z - 99$$

$$-30z^2 + 209z - 295 = -30z^2 + 111z - 99$$

$$98z = 196$$

$$z = 2$$

On checking $z = 2$, we find that it is in fact, a solution of the original equation.

60.

$$\frac{x + 1}{x^2 + 2x} - \frac{x + 4}{x^2 + x} = \frac{-3}{x^2 + 3x + 2}, \quad x \neq -1, x \neq -2$$

$$\frac{x + 1}{x(x + 2)} - \frac{x + 4}{x(x + 1)} = \frac{-3}{(x + 1)(x + 2)}$$

The domain is $\{x \mid x \neq -2, x \neq -1, x \neq 0\}$.

$$x(x + 1)(x + 2)\left[\frac{x + 1}{x(x + 2)} - \frac{x + 4}{x(x + 1)}\right] = \left[\frac{-3}{(x + 1)(x + 2)}\right]x(x + 1)(x + 2)$$

$$(x + 1)(x + 1) - (x + 4)(x + 2) = -3x$$

$$x^2 + 2x + 1 - (x^2 + 6x + 8) = -3x$$

$$-4x - 7 = -3x$$

$$-x = 7$$

$$x = -7$$

On checking, we find that $x = -7$ is, in fact, a solution of the original equation.

62.

$$6.2x - \frac{19.1}{83.72} = 0.195$$

$$83.72\left[6.2x - \frac{19.1}{83.72}\right] = 83.72(0.195)$$

$$519.064x - 19.1 = 16.3254$$

$$519.064x = 35.4254$$

$$x \approx 0.07$$

64.

$$18.63x - \frac{21.2}{2.6} = \frac{14x}{2.32} - 20$$

$$6.032\left[18.63x - \frac{21.2}{2.6}\right] = 6.032\left[\frac{14x}{2.32} - 20\right]$$

$$112.38x - 49.184 = 36.4x - 120.64$$

$$75.98x = -71.456$$

$$x \approx -0.94$$

66.

$$1 - ax = b, \quad a \neq 0$$

$$-ax = b - 1$$

$$x = \frac{b - 1}{-a} = \frac{1 - b}{a}$$

68.

$$\frac{a}{x} + \frac{b}{x} = c$$

$$\frac{a + b}{x} = c$$

$$a + b = cx$$

$$x = \frac{a + b}{c}$$

Letting $x = \dfrac{a + b}{c}$ in the original equation, we get:

$$\frac{a}{\frac{a + b}{c}} + \frac{b}{\frac{a + b}{c}} = \frac{a + b}{\frac{a + b}{c}} = c$$

Therefore, $x = \dfrac{a + b}{c}$ is a solution.

70.

$$\frac{b + c}{x + a} = \frac{b - c}{x - a}$$

$$(x + a)(x - a)\left[\frac{b + c}{x + a}\right] = (x + a)(x - a)\left[\frac{b - c}{x - a}\right]$$

$$(x - a)(b + c) = (x + a)(b - c)$$

$$xb + xc - ab - ac = xb - xc + ab - ac$$

$$xc - ab = -xc + ab$$

$$2xc = 2ab$$

$$x = \frac{ab}{c}$$

Letting $x = \dfrac{ab}{c}$ in the original equation, we get:

$$\frac{b + c}{\dfrac{ab}{c} + a} \stackrel{?}{=} \frac{b - c}{\dfrac{ab}{c} - a}$$

$$\frac{b + c}{\dfrac{ab + ac}{c}} \stackrel{?}{=} \frac{b - c}{\dfrac{ab - ac}{c}}$$

$$(b + c)\left[a\frac{ab - ac}{c}\right] \stackrel{?}{=} (b - c)\left[\frac{ab + ac}{c}\right]$$

$$\frac{ab^2 - ac^2}{c} = \frac{ab^2 - ac^2}{c}$$

Therefore, $x = \dfrac{ab}{c}$ is a solution.

72.

$$x + 2b = x - 4 + 2bx$$
$$2 + 2b = 2 - 4 + 4b$$
$$2 + 2b = -2 + 4b$$
$$-2b = -4$$
$$b = 2$$

74.

$$A = P(1 + rt)$$
$$A = P + Prt$$
$$Prt = A - P$$
$$r = \frac{A - P}{Pt}$$

76.

$$PV = nRT$$
$$nRT = PV$$
$$T = \frac{PV}{nR}$$

78.

$$v = -gt + v_0$$
$$gt = v_0 - v$$
$$t = \frac{v_0 - v}{g}$$

80. (b)

2.2 Setting Up Equations: Applications

2. If C is the circumference, and r the radius, then $C = 2\pi r$

4. If P is the perimeter, and s the length of a side, then $P = 4s$

6. If P is pressure, F the force, and A the unit area, then $P = \dfrac{F}{A}$

8. If K is kinetic energy, M the mass, and V the velocity, then $K = \dfrac{1}{2}MV^2$

10. R = total revenue; $R = 250x$

12.

Katy	Mike	Total
x	$x - 2000$	10000

$$x + (x - 2000) = 10000$$
$$2x = 12000$$
$$x = 6000$$

Katy will receive \$6000 and Mike will receive \$4000.

14.

Mike	Colleen	Cost
x	$\frac{2}{3}x$	18

$$x + \frac{2}{3}x = 18$$
$$\left(1 + \frac{2}{3}\right)x = 18$$
$$\frac{5}{3}x = 18$$
$$x = \frac{3}{5}(18)$$
$$x = \frac{54}{5} = \$10.80$$

Mike pays \$10.80 and Colleen pays \$7.20.

16.

	Hourly Wage	Salary
Regular hours, 40	x	$40x$
Overtime Hours, 10:		
Regular O.T. 6	$1.5x$	$6(1.5x) = 9x$
Sunday O.T. 4	$2x$	$4(2x) = 8x$

$$40x + 9x + 8x = \$342$$
$$57x = 342$$
$$x = 6$$

Colleen's hourly wage is \$6.00 per hour.

18.

		Total Points
# of Field Goals	x	$2x$
# of Free Throws	$\frac{1}{3}x$	$\frac{1}{3}x$

$$2x + \frac{1}{3}x = 70$$
$$\frac{7}{3}x = 70$$
$$7x = 210$$
$$x = 30$$

The team had 30 field goals (and $\frac{1}{3}(30) = 10$ free throws).

20.
$$2\ell + 2w = 42$$
$$\ell = 2w$$
$$2(2w + 2w) = 42$$
$$4w + 2w = 42$$
$$6w = 42$$
$$w = 7$$
$$\ell = 14$$

The length is 14 meters and width is 7 meters.

22.

	Amount	Concentration of Acid	Pure Acid
Pure Acid	x	$100\% = 1.0$	x
30% Solution	20	$30\% = 0.3$	$.3(20) = 6$
50% Solution	$x + 20$	$50\% = 0.5$	$.5(x + 20)$

$$x + 6 = .5(x + 20)$$
$$x + 6 = .5x + 10$$
$$.5x = 4$$
$$x = 8$$

Add 8 cubic centimeters of pure acid.

24.

	Principal (P)	Rate (r)	Time (t)	Interest (I)
Bonds	x	0.15	1	$0.15x$
Certificate	$50,000 - x$	0.07	1	$0.07(50,000 - x)$
Total	50,000		1	7,000

$$I = Prt$$
$$0.15x + 0.07(50000 - x) = 7000$$
$$0.15x + 3500 - 0.07x = 7000$$
$$0.08x = 3500$$
$$x = 43750$$

Thus, the investor should place \$43,750 in the bonds and \$6,250(50,000 − 43,750) in the certificate.

26.

	Principal (\$)	Rate	Time (yr)	Interest (\$)
Amount loaned at 16%	x	0.16	1	$0.16x$
Amount loaned at 19%	$1,000,000 - x$	0.19	1	$0.19(1,000,000 - x)$
Total	1,000,000	0.18	1	$0.18(1,000,000)$

$$0.16x + 0.19(1,000,000 - x) = 0.18(1,000,000)$$
$$0.16x + 190,000 - 0.19x = 180,000$$
$$-0.03x = -10,000$$
$$x = 333,333.33$$

The loan officer should lend \$333,333.33 at the 16% rate.

28.

	Amount	Cost/Unit	Total Cost
Caramels	x	0.25	$0.25x$
Cremes	$30 - x$	0.45	$0.45(30 - x)$
Total	30		9.50

$$0.25x + 0.45(30 - x) = 9.50$$
$$0.25x + 13.5 - 0.45x = 9.50$$
$$-0.2x = -4$$
$$x = 20$$

There should be 20 caramels and 10(30 − 20) cremes in a box.

30.

	Amount	Cost/Unit	Total Cost
Coffee I	x	2.75	$2.75x$
Coffee II	$100 - x$	5.00	$5.00(100 - x)$
Total	100	3.90	$3.90(100) = 290$

$$2.75x + 5.00(100 - x) = 390$$
$$2.75x + 500 - 5.00x = 390$$
$$-2.25x = -110$$
$$x \approx 48.9$$

≈ 49 pounds of Coffee I and ≈ 51 pounds $(100 - 49)$ of Coffee II should be blended.

32. Let x be the original price,
then
$$x - 0.3x = 399$$
$$0.7x = 399$$
$$x = \$570.00$$

34. Let x be the list price,
then
$$x - 0.15x = 8000$$
$$0.85x = 8000$$
$$x = \$9,411.76$$
The amount saved is
$$\$9,411.76 - \$8,000 = \$1,411.76$$

36. Let x be the dealer's cost of a new car,
then $x = (.85)(12,000)$
$$= \$10,200$$
If the dealer is willing to accept \$100 over cost, then I will pay $\$10,200 + \$100 = \$10,300$.

38.

	Hours to do job	Part of job done in 1 hour
Painter	10	$\dfrac{1}{10}$
Helper	x	$\dfrac{1}{x}$
Together	6	$\dfrac{1}{6}$

$$\frac{1}{10} + \frac{1}{x} = \frac{1}{6}$$

$$\frac{x + 10}{10x} = \frac{1}{6}$$
$$6(x + 10) = 10x$$
$$6x + 60 = 10x$$
$$4x = 60$$
$$x = 15$$

It will take the helper 15 hours.

40. Let $8x$ represent the $\frac{2}{3}$ remaining in the final grade since 4 scores represent $\frac{1}{3}$

To get a B, Dan needs...

$$\frac{86 + 80 + 84 + 90 + 8x}{12} = 80$$

$$\frac{340 + 8x}{12} = 80$$

$$340 + 8x = 960$$

$$8x = 620$$

$$x = 77.5 \approx 78$$

To get an A, Dan needs...

$$\frac{86 + 80 + 84 + 90 + 8x}{12} = 90$$

$$340 + 8x = 1080$$

$$8x = 740$$

$$x = 92.5 \approx 93$$

To earn a "B", Dan needs 78% on the final exam; and to earn an "A", Dan needs 93% on the final exam.

42. Let x be the number of miles of highway travel, then

$$\frac{1 \text{ gallon}}{40 \text{ miles}} \cdot x \text{ miles} + \frac{1 \text{ gallon}}{25 \text{ miles}} \cdot (30,000 - x) \text{miles} = 900 \text{ gallons}$$

$$\frac{x}{40} + \frac{30,000 - x}{25} = 900$$

$$\frac{5x + 8(30,000 - x)}{200} = 900$$

$$5x + 240,000 - 8x = 180,000$$

$$-3x = -60,000$$

$$x = 20,000 \text{ miles}$$

20,000 miles should be allowed as a business expense.

44. Let x be the number of liters drained and replaced by pure antifreeze. Then $(0.4)x$ is the amount removed and $(1)x(100\%)$ is the amount replaced. Then,

$$0.4(15) + 0.4x + x = 0.6(15)$$

$$0.6x = 3$$

$$x = \frac{3}{0.6} = 5$$

46.

	Amount	% Gold	Pure Gold
12 karat	x	$\frac{12}{24} = 50\% = 0.5$	$.5x$
Pure	$60 - x$	$\frac{24}{24} = 100\% = 1.0$	$60 - x$
16 karat	60	$\frac{16}{24} = 66\frac{2}{3}\% = 0.67$	$.67(60)$

$$.5x + (60 - x) = .67(60)$$

$$.5x + 60 - x = 40$$

$$-.5x = -20$$

$$x = 40$$

Mix 40 grams of 12 karat gold with the pure gold.

48.

	Amount	Concentration of Water	Pure Water
Water	x	$100\% = 1.0$	x
3% Salt	240	$97\% = .97$	$.97(240)$
5% Salt	$240 - x$	$95\% = .95$	$.95(240 - x)$

$$x + .95(240 - x) = .97(240)$$
$$x + 228 - .95x = 232.8$$
$$.05x = 4.8$$
$$x = 96$$

Evaporate 96 ounces of water from the 3% salt solution.

50. Let $h =$ atoms of hydrogen in a sugar molecule
Let $x =$ atoms of oxygen in a sugar molecule
Let $c =$ atoms of carbon in a sugar molecule

$$h = 2x$$
$$c = x + 1$$
$$45 = h + c + x$$
$$45 = 2x + x + 1 + x$$
$$45 = 4x + 1$$
$$44 = 4x$$
$$11 = x$$
$$h = 22$$
$$c = 12$$

There are 11 atoms of oxygen and 22 atoms of hydrogen in a 45-atom sugar molecule.

52. Let m be the speed of the motor boat.

	Velocity (mph)	Time (hrs)	Distance
Upstream	$m - 3$	5	$5(m - 3)$
Return	$m + 3$	2.5	$2.5(m + 3)$

Distance Upstream = Distance to Return
The two distances are equal, so
$$5(m - 3) = 2.5(m + 3)$$
$$5m - 15 = 2.5m + 7.5$$
$$2.5m = 19.5$$
$$m = 7.8 \text{ miles per hour}$$

The speed of the motorboat is 7.8 miles per hour.

54. Let x be the average speed of the slower car and $x + 10$ be the average speed of the faster car. Let's set up a table.

	Velocity	Time	Distance
Slower Car	x	3.5	$3.5x$
Faster Car	$x + 10$	3	$3(x + 10)$

$$3.5x = 3(x + 10)$$
$$3.5x = 3x + 30$$
$$.5x = 30$$
$$x = 60$$

Thus, the slower car travels at a speed of 60 miles per hour and the faster car travels at a speed of 70 miles per hour. After 3-1/2 hours, the slower car, traveling at an average speed of 60 mph, goes a distance of $(3.5)(60) = 310$ miles. The faster car travels $(70)(3)$ or 210 miles.

56.

Amount	% Cement	% Sand
20	.25	.75
x	1.00	0
$20 + x$	.40	.60

$$.25(20) + x = .40(20 + x)$$
$$5 + x = 8 + .4x$$
$$.6x = 3$$
$$x = 5$$

Five pounds must be added to produce a cement mix that is 40% cement.

58.

	Velocity (mph)	Time (hrs)	Distance
Without wind	300	5	1500
With wind	$300 - 30$	5	$(270)5$

$$(270)5 = 1350$$

It can fly 1350 miles.

60.

	Distance	Velocity	Time
Lewis	100	$\dfrac{100}{9.99}$	9.99
Burke	$\dfrac{100}{12}(9.99)$	$\dfrac{100}{12}$	9.99

In 9.99 seconds, Burke ran a distance of $\dfrac{100}{12}(9.99) = 83.25$ meters. Lewis would win by 16.75 meters.

2.3 Quadratic Equations

2.
$$x^2 + 4x = 0$$
$$x(x + 4) = 0$$
$x = 0$ or $x = -4$
The solution set is $\{0, -4\}$.

4.
$$x^2 - 9 = 0$$
$$(x + 3)(x - 3) = 0$$
$x = -3$ or $x = 3$
The solution set is $\{-3, 3\}$.

6.
$$v^2 + 7v + 12 = 0$$
$$(v + 3)(v + 4) = 0$$
$v = -3$ or $v = -4$
The solution set is $\{-3, -4\}$.

8.
$$3x^2 + 5x + 2 = 0$$
$$(3x + 2)(x + 1) = 0$$
$$x = -\frac{2}{3} \text{ or } x = -1$$

The solution set is $\left\{-\dfrac{2}{3}, -1\right\}$.

10.
$$2y^2 - 50 = 0$$
$$2(y^2 - 25) = 0$$
$$2(y - 5)(y + 5) = 0$$
The solution set is $\{-5, 5\}$.

12.
$$x(x + 1) = 12$$
$$x^2 + x - 12 = 0$$
$$(x + 4)(x - 3) = 0$$
$x = -4$ or $x = 3$
The solution set is $\{-4, 3\}$.

14.
$$25x^2 + 16 = 40x$$
$$25x^2 - 40x + 16 = 0$$
$$(5x - 4)(5x - 4) = 0$$
$$x = \frac{4}{5} \text{ or } x = \frac{4}{5}$$

The repeated solution is $\frac{4}{5}$.

16.
$$2(2u^2 - 4u) + 3 = 0$$
$$4u^2 - 8u + 3 = 0$$
$$(2u - 3)(2u - 1) = 0$$
$$u = \frac{3}{2} \text{ or } u = \frac{1}{2}$$

The solution set is $\left\{\frac{3}{2}, \frac{1}{2}\right\}$.

18.
$$x + \frac{12}{x} = 7$$
$$x^2 + 12 = 7x$$
$$x^2 - 7x + 12 = 0$$
$$(x - 3)(x - 4) = 0$$
$$x = 3 \text{ or } x = 4$$
The solution set is $\{3, 4\}$.

20.
$$\frac{5}{x + 4} = 4 + \frac{3}{x - 2}$$
$$5(x - 2) = 4(x + 4)(x - 2) + 3(x + 4)$$
$$5x - 10 = 4(x^2 + 2x - 8) + 3(x + 4)$$
$$5x - 10 = 4x^2 + 8x - 32 + 3x + 12$$
$$5x - 10 = 4x^2 + 11x - 20$$
$$4x^2 + 6x - 10 = 0$$
$$2(2x^2 + 3x - 5) = 0$$
$$2(2x + 5)(x - 1) = 0$$
$$x = -\frac{5}{2} \text{ or } x = 1$$

The solution set is $\left\{-\frac{5}{2}, 1\right\}$.

22. $x^2 + 4x + 2 = 0$
$$x = \frac{-4 \pm \sqrt{16 - 8}}{2}$$
$$= \frac{-4 \pm \sqrt{8}}{2}$$
$$= \frac{-4 \pm 2\sqrt{2}}{2}$$
$$= \frac{2\left(-2 \pm \sqrt{2}\right)}{2}$$
$$= -2 \pm \sqrt{2}$$

The solution set is $\left\{-2 + \sqrt{2}, -2 - \sqrt{2}\right\}$.

24. $x^2 + 6x + 1 = 0$
$$x = \frac{-6 \pm \sqrt{36 - 4}}{2}$$
$$x = \frac{-6 \pm 4\sqrt{2}}{2}$$

The solution set is $\left\{-3 - 2\sqrt{2}, -3 + 2\sqrt{2}\right\}$.

26. $2x^2 + 5x + 3 = 0$
$$x = \frac{-5 \pm \sqrt{25 - 24}}{4}$$
$$x = \frac{-5 \pm 1}{4}$$

The solution set is $\left\{-1, -\frac{3}{2}\right\}$.

28. $4t^2 + t + 1 = 0$
$$t = \frac{-1 \pm \sqrt{1 - 16}}{8}$$
Since $b^2 - 4ac = 1 - 16$
$$= -15 < 0$$
The equation has no real solution.

30. $2x^2 + 2x - 1 = 0$

$$x = \frac{-2 \pm \sqrt{4 + 8}}{4}$$

$$x = \frac{-2 \pm 2\sqrt{3}}{4}$$

$$x = \frac{2(-1 \pm \sqrt{3})}{4}$$

$$x = \frac{-1 \pm \sqrt{3}}{2}$$

The solution set is $\left\{ \dfrac{-1 + \sqrt{3}}{2}, \dfrac{-1 - \sqrt{3}}{2} \right\}$.

32. $$5x = 4x^2$$
$$4x^2 - 5x = 0$$

$$x = \frac{5 \pm \sqrt{25}}{8}$$

$$x = \frac{5 \pm 5}{8}$$

The solution set is $\left\{ 0, \dfrac{5}{4} \right\}$.

34. $4u^2 - 6u + 9 = 0$

$$u = \frac{6 \pm \sqrt{36 - 144}}{8}$$

Since $b^2 - 4ac = 36 - 144$
$$= -108 < 0,$$
the equation has no real solution.

36. $2x^2 - 3x - 1 = 0$

$$x = \frac{3 \pm \sqrt{9 + 8}}{4}$$

$$x = \frac{3 \pm \sqrt{17}}{4}$$

The solution set is

$$\left\{ \frac{3 + \sqrt{17}}{4}, \frac{3 - \sqrt{17}}{4} \right\}.$$

38. $4 + \dfrac{1}{x} - \dfrac{1}{x^2} = 0$

$$4x^2 + x - 1 = 0$$

$$x = \frac{-1 \pm \sqrt{1 + 16}}{8}$$

$$x = \frac{-1 \pm \sqrt{17}}{8}$$

The solution set is

$$\left\{ \frac{-1 + \sqrt{17}}{8}, \frac{-1 - \sqrt{17}}{8} \right\}.$$

40. $x = 1 - \dfrac{4}{x}$

$$x^2 - x + 4 = 0$$

$$x = \frac{1 \pm \sqrt{1 - 16}}{2}$$

Since $b^2 - 4ac = 1 - 16$
$$= -15 < 0,$$
the equation has no real solution.

42. $x^2 + 4x + 2 = 0$

$$x = \frac{-4 \pm \sqrt{16 - 8}}{2}$$

$$x = \frac{-4 \pm \sqrt{8}}{2}$$

The solution set is $\{-0.59, -3.41\}$.

44. $x^2 + \sqrt{2}x - 2 = 0$

$$x = \frac{-\sqrt{2} \pm \sqrt{2 + 8}}{2}$$

$$x = \frac{-\sqrt{2} \pm \sqrt{10}}{2}$$

The solution set is $\{0.87, -2.29\}$.

46. $\pi x^2 + \pi x - 2 = 0$

$$x = \frac{-\pi \pm \sqrt{\pi^2 + 8\pi}}{2\pi}$$

The solution set is $\{0.44, -1.44\}$.

48. $\pi x^2 - 15\sqrt{2}x + 20 = 0$

$$x = \frac{15\sqrt{2} \pm \sqrt{450 - 80\pi}}{2\pi}$$

$$x = \frac{15\sqrt{2} \pm \sqrt{198.673}}{6.283}$$

The solution set is $\{1.13, 5.62\}$.

50. $x^2 - 8 = 0$

$x^2 = 8$

$x = \pm 2\sqrt{2}$

The solution set is $\left\{-2\sqrt{2}, 2\sqrt{2}\right\}$.

52. $9x^2 - 6x + 1 = 0$

$$x = \frac{6 \pm \sqrt{36 - 36}}{18}$$

$$x = \frac{1}{3}$$

The repeated solution is $\dfrac{1}{3}$.

54. $6x^2 + 7x - 20 = 0$

$(2x + 5)(3x - 4) = 0$

$x = -\dfrac{5}{2}$ or $x = \dfrac{4}{3}$

The solution set is $\left\{-\dfrac{5}{2}, \dfrac{4}{3}\right\}$.

56. $6y^2 + y - 2 = 0$

$$y = \frac{-1 \pm \sqrt{1 + 48}}{12}$$

$$y = \frac{-1 \pm 7}{12}$$

The solution set is $\left\{-\dfrac{2}{3}, \dfrac{1}{2}\right\}$.

58. $\dfrac{1}{2}x^2 - \sqrt{2}x - 1 = 0$

$$x = \frac{\sqrt{2} + \sqrt{2 + 2}}{1}$$

$$x = \sqrt{2} \pm 2$$

The solution set is $\left\{\sqrt{2} + 2, \sqrt{2} - 2\right\}$.

60. $x^2 + x - 1 = 0$

$$x = \frac{-1 \pm \sqrt{1 + 4}}{2}$$

$$x = \frac{-1 \pm \sqrt{5}}{2}$$

The solution set is $\left\{\dfrac{-1 + \sqrt{5}}{2}, \dfrac{-1 - \sqrt{5}}{2}\right\}$.

62. $x^2 + 4x + 7 = 0$

Since $b^2 - 4ac = 16 - 28 = -8 < 0$, there is no real solution.

64. $25x^2 - 20x + 4 = 0$

Since $b^2 - 4ac = 400 - 400 = 0$, there is a repeated real solution.

66. $2x^2 - 3x - 4 = 0$

Since $b^2 - 4ac = 9 + 32$
$= 41 > 0$,

there are two unequal real solutions.

68. $x^2 + 8x$

To complete the square,

add $\left[\dfrac{1}{2} \cdot 8\right]^2 = 16$

70. $x^2 - \dfrac{1}{3}x$

To complete the square,

add $\left[\dfrac{1}{2} \cdot -\dfrac{1}{3}\right]^2 = \dfrac{1}{36}$

72. $x^2 - \dfrac{2}{5}x$

To complete the square,

add $\left[\dfrac{1}{2} \cdot -\dfrac{2}{5}\right]^2 = \dfrac{1}{25}$

74.

$$x^2 - 6x = 13$$
$$x^2 - 6x + 9 = 13 + 9$$
$$(x - 3)^2 = 22$$
$$x - 3 = \pm\sqrt{22}$$
$$x = 3 \pm \sqrt{22}$$

The solution set is $\left\{3 + \sqrt{22},\ 3 - \sqrt{22}\right\}$.

76.

$$x^2 + \frac{2}{3}x = \frac{1}{3}$$
$$x^2 + \frac{2}{3} + \frac{1}{9} = \frac{1}{3} + \frac{1}{9}$$
$$\left(x + \frac{1}{3}\right)^2 = \frac{4}{9}$$
$$x + \frac{1}{3} = \pm\sqrt{\frac{4}{9}}$$
$$x = -\frac{1}{3} \pm \frac{2}{3}$$

The solution set is $\left\{\frac{1}{3},\ -1\right\}$.

78.

$$2x^2 - 3x = 1$$
$$2\left[x^2 - \frac{3}{2}x\right] = 1$$
$$2\left[x^2 - \frac{3}{2}x + \frac{9}{16}\right] = 1 + \frac{9}{8}$$
$$2\left[x - \frac{3}{4}\right]^2 = \frac{17}{8}$$
$$\left[x - \frac{3}{4}\right]^2 = \frac{17}{16}$$
$$x - \frac{3}{4} = \pm\sqrt{\frac{17}{16}}$$
$$x = \frac{3}{4} \pm \frac{\sqrt{17}}{4}$$

The solution set is $\left\{\dfrac{3 + \sqrt{17}}{4},\ \dfrac{3 - \sqrt{17}}{4}\right\}$.

80.

$$w = \text{width of opening}$$
$$w + 1 = \text{length of opening}$$
$$\text{Area of opening} =$$
$$w(w + 1) = 306$$
$$w^2 + w = 306$$
$$w^2 + w - 306 = 0$$
$$(w + 18)(w - 17) = 0$$
$$w + 18 = 0 \qquad w - 17 = 0$$
$$w = -18 \qquad w = 17$$

Since the width must be positive, the dimensions are 17 centimeters by 18 centimeters.

82. Let s be the length of a side of the 1250 square foot field, so that $s^2 = 1250$.

Let x be the least distance setting for the water distribution, which equals the radius of the circle. Then $2x$ equals the diameter of the circle which equals the diagonal of the square. By the Pythagorean Theorem,

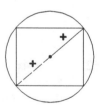

$$(2x)^2 = s^2 + s^2$$
$$4x^2 = 2(1250) = 2500$$
$$x^2 = 625$$
$$x = 25$$

The least distance to set the device is 25 feet.

84. Let x be the width of sheet metal. Then $2x$ is the length. The resulting box has length $2x - 2$, width $x - 2$, and height 1.

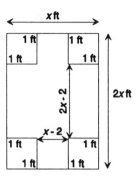

Thus, the volume of the box
(length $\times$ width $\times$ height) is
$(2x - 2)(x - 2)(1) = 4$ cubic feet
$$2x^2 - 6x = 0$$
$$2x(x - 3) = 0$$
$$x = 0 \text{ or } x = 3$$
We discard the solution $x = 0$ because the dimension must be greater than zero. Thus, $x = 3$. So the dimensions of the sheet metal are 6 feet by 3 feet.

86. $A = 2\pi r^2 + 2\pi rh$.

Since $A = 188.5$ square inches and $h = 7$ inches,
$$188.5 = 2\pi r^2 + 2\pi r(7)$$
$$2\pi r^2 + 14\pi r - 188.5 = 0$$

$$r = \frac{-14\pi + \sqrt{(14\pi)^2 + 8\pi(188.5)}}{4\pi}$$

$$r \approx 3 \text{ or } r \approx -10$$

We discard the solution -10 because the radius must be positive. Thus, the radius is 3 inches.

88.

	Hours to do job	Part of job done in one hour
Smaller pump	x	$\dfrac{1}{x}$
Larger pump	y	$\dfrac{1}{y}$
Together	5	$\dfrac{1}{5}$

Since $y = x - 4$
$$\frac{1}{x} + \frac{1}{y} = \frac{1}{5}$$
$$\frac{1}{x} + \frac{1}{x - 4} = \frac{1}{5}$$
$$\frac{x - 4 + x}{(x)(x - 4)} = \frac{1}{5}$$
$$5(2x - 4) = x^2 - 4x$$
$$x^2 - 14x + 20 = 0$$
$$x = \frac{14 + \sqrt{116}}{2} \quad \text{or} \quad x = \frac{14 - \sqrt{116}}{2} \quad \text{(impossible)}$$
$$x = 12.39 \text{ hours or 12 hours and 23 minutes}$$
It will take the smaller pump 12 hours and 23 minutes to do the job alone.

90.　ℓ = length; w = width; h = height

$\ell = 2w$, $h = 4$ in $= \dfrac{1}{3}$ ft $\Rightarrow \ell w h = 2w \cdot w \cdot \dfrac{1}{3} = 216$ cubic feet

$$\frac{2}{3}w^2 = 216$$
$$2w^2 - 648 = 0$$
$$w^2 = \frac{648}{2} = 324$$
$$w = \pm 18$$

We discard the solution $w = -18$ because the width is always positive. Thus, $w = 18$ feet $= 216$ inches and $\ell = 36$ feet $= 432$ inches. The dimensions of the patio will be $432'' \times 216'' \times 4''$.

92.　$s = -4.9t^2 + 20t$

(a)　$-4.9t^2 + 20t = 15$
$$-4.9t^2 + 20t - 15 = 0$$
$$t = \frac{-20 \pm \sqrt{400 - 294}}{-9.8}$$
$$t = \frac{-20 \pm \sqrt{106}}{-9.8}$$
$$t = 0.99 \text{ or } t = 3.09$$

Thus, the object will be 15 meters above the ground after 0.99 seconds and after 3.09 seconds.

(b)　$-4.9t^2 + 20t = 0$
$$t(-4.9t + 20) = 0$$
$$t = 0 \text{ or } t = \frac{20}{4.9} = 4.08$$

Thus, it will strike the ground in 4.08 seconds.

(c)　$-4.9t^2 + 20t = 100$
$$-4.9t^2 + 20t - 100 = 0$$
$$t = \frac{-20 \pm \sqrt{400 - 1960}}{2(-4.9)}$$

Since there is so solution when the object is 100 meters, it will never reach that height.

(d)　Since it will strike the ground in 4.08 seconds, it will reach its maximum height at the halfway point, i.e., 2.04 seconds. Thus, the maximum height is
$$-4.9(2.04)^2 + 20(2.04) = 20.41 \text{ meters}$$

94.　Volume of old candy bar $= \ell w h = 12 \cdot 7 \cdot 3 = 252$ c.c. The volume is reduced by 20% so $(0.2)(252) = 50.4$. Thus, the new volume is $252 - 50.4 = 201.6$. Let x be the amount of reduction of the length and the width. Then,
$$(12 - x)(7 - x)(3) = 201.6$$
$$(84 - 19x + x^2)(3) = 201.6$$
$$252 - 57x + 3x^2 = 201.6$$
$$3x^2 - 57x + 50.4 = 0$$
$$3(x^2 - 19x + 16.8) = 0$$
$$x = \frac{19 \pm \sqrt{361 - 67.2}}{2}$$
$$x = 18.1 \text{ or } x = 0.93$$

We disregard the solution $x = 18.0$ because this would make the length and width negative. Therefore, at $x = 0.93$, length $= 12 - 0.93 = 11.07$ cm; width $= 7 - 0.93 = 6.07$ cm; height $= 3$ cm.

96.　Let x be the width of the border.

$$\text{Area of border} = \pi(5 + x)^2 - 25\pi$$
$$= 10\pi x + \pi x^2$$

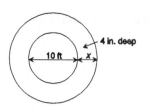

Since 4 in. $= \dfrac{1}{3}$ ft. and 1 cubic yard $= 27$ cubic feet,

$$\left(\pi x^2 + 10\pi x\right)\frac{1}{3} = 27$$

$$\frac{\pi}{3}x^2 + \frac{10\pi}{3}x - 27 = 0$$

$$x = \frac{-\dfrac{10\pi}{3} + \sqrt{\left[\dfrac{10\pi}{3}\right]^2 + 36\pi}}{\dfrac{2\pi}{3}}$$

$$x = 2.13 \text{ ft.}$$

98.
$$(x)^2 + (17 - x)^2 = 13^2$$
$$x^2 + 289 - 34x + x^2 = 169$$
$$2x^2 - 34x + 120 = 0$$
$$2(x^2 - 17x + 60) = 0$$
$$2(x - 5)(x - 12) = 0$$
$$x = 5 \text{ or } x = 12$$

One leg is 5 centimeters; the other leg is 12 centimeters.

100.　(a)　The object will strike the ground in 8 seconds, because the distance is 0 when it strikes the ground

$$0 = 1280 - 32t - 16t^2$$
$$0 = -16(t^2 + 2t - 80)$$
$$0 = -16(t + 10)(t - 8)$$
$$t = 8 \text{ seconds}$$

(b)
$$s = 1280 - 32(4) - 16(4)^2$$
$$s = 1280 - 128 - 256$$
$$s = 896 \text{ feet}$$
The height is 896 feet.

102.
$$\left[\frac{-b + \sqrt{b^2 - 4ac}}{2a}\right]\left[\frac{-b - \sqrt{b^2 - 4ac}}{2a}\right] = \frac{b^2 - (b^2 - 4ac)}{4a^2} = \frac{4ac}{4a^2} = \frac{c}{a}$$

104.　$x^2 - kx + 4 = 0$ has a repeated real solution if $b^2 - 4ac = 0$. Thus,
$$k^2 - 16 = 0$$
$$k^2 = 16$$
$$k = 4 \text{ or } k = -4$$

106.　$ax^2 + bx + c = 0 \quad x_1 = \dfrac{-b + \sqrt{b^2 - 4ac}}{2a}, \ x_2 = \dfrac{-b - \sqrt{b^2 - 4ac}}{2a}$

　　$cx^2 + bx + a = 0 \quad x_3 = \dfrac{-b + \sqrt{b^2 - 4ac}}{2c}, \ x_4 = \dfrac{-b - \sqrt{b^2 - 4ac}}{2c}$

$$x_1 x_4 = \frac{b^2 - (b^2 - 4ac)}{4ac} = 1; \ x_2 x_3 = \frac{b^2 - (b^2 - 4ac)}{4ac} = 1$$

108.
$$\frac{1}{2}n(n - 3) = 65$$
$$n(n - 3) = 130$$
$$n^2 - 3n - 130 = 0$$
$$n = \frac{3 \pm \sqrt{529}}{2}$$
$$n = \frac{3 \pm 23}{2}$$
$$n = 13, \, n > 0$$
Polygon with 65 diagonals has 13 sides.
$$\frac{1}{2}n(n - 3) = 80$$
$$n(n - 3) = 160$$
$$n^2 - 3n - 160 = 0$$
$$n = \frac{3 \pm \sqrt{9 - 4(-160)}}{2}$$
There is no polygon with 80 diagonals.

110. Since $d = rt$, the time traveled with the tail wind was
$$919 = 550t$$
$$t = \frac{919}{550} \approx 1.671 \text{ hr.}$$
Since the plane was 20 minutes early, the time in still air would have been
$$1.671 + 20 \text{ min or}$$
$$1.671 + 0.333 \approx 2 \text{ hr.}$$
Thus, the rate in still air is
$$919 = r(2)$$
$$r \approx 460$$
Hence the tail wind was
$$550 - 460 = 60 \text{ nautical miles per hour.}$$

2.4 Quadratic Equations with a Negative Discriminant

2. $\sqrt{-9} = 3i$

4. $\sqrt{-64} = 8i$

6. $\sqrt{(4 + 3i)(3i - 4)} = \sqrt{-16 + 9i^2} = \sqrt{-25} = 5i$

8. $\left(\sqrt{-4} + 3\right)\left(\sqrt{-4} + 3\right) = (2i + 3)(2i + 3) = 4i^2 + 12i + 9 = -4 + 12i + 9 = 12i + 5$

10. $\left(8 + \sqrt{-4}\right)\left(2 - \sqrt{-9}\right) = 16 - 8\sqrt{-9} + 2\sqrt{-4} - 6i^2 = 22 - 8 \cdot 3i + 2 \cdot 2i = 22 - 20i$

12.
$$x^2 + 64 = 0$$
$$x^2 = -64$$
$$x = \pm 8i$$
The solution set is $\{-8i, 8i\}$.

14.
$$x^3 + 25x = 0$$
$$x(x^2 + 25) = 0$$
$$x = 0 \quad x^2 = -25 \quad x = \pm 5i$$
The solution set is $\{-5i, 0, 5i\}$.

16. $x^2 + 4x + 8 = 0$
$$x = \frac{-4 \pm \sqrt{16 - 32}}{2}$$
$$x = \frac{-4 \pm \sqrt{-16}}{2} = \frac{-4 \pm 4i}{2} = -2 \pm 2i$$
The solution set is $\{-2 + 2i, -2 - 2i\}$.

18. $x^2 - 2x + 5 = 0$
$$x = \frac{2 \pm \sqrt{4 - 20}}{2}$$
$$x = \frac{2 \pm \sqrt{-16}}{2} = \frac{2 \pm 4i}{2} = 1 \pm 2i$$
The solution set is $\{1 + 2i, 1 - 2i\}$.

20. $10x^2 + 6x + 1 = 0$

$$x = \frac{-6 \pm \sqrt{36 - 40}}{20}$$

$$x = \frac{-6 \pm \sqrt{-4}}{20} = \frac{-6 \pm 2i}{20} = \frac{-3 \pm i}{10}$$

The solution set is

$$\left\{ \frac{-3}{10} + \frac{1}{10}i, \frac{-3}{10} - \frac{1}{10}i \right\}.$$

22. $13x^2 + 6x + 1 = 0$

$$x = \frac{-6 \pm \sqrt{36 - 52}}{26}$$

$$x = \frac{-6 \pm \sqrt{-16}}{26} = \frac{-6 \pm 4i}{26} = \frac{-3 \pm 2i}{13}$$

The solution set is

$$\left\{ \frac{-3}{13} + \frac{2}{13}i, \frac{-3}{13} - \frac{2}{13}i \right\}.$$

24. $x^2 - x + 1 = 0$

$$x = \frac{1 \pm \sqrt{1 - 4}}{2} = \frac{1 \pm \sqrt{3}i}{2}$$

The solution set is $\left\{ \frac{1}{2} + \frac{\sqrt{3}}{2}i, \frac{1}{2} - \frac{\sqrt{3}}{2}i \right\}.$

26. $x^3 + 27 = (x + 3)(x^2 - 3x + 9) = 0$

$x + 3 = 0 \quad x^2 - 3x + 9 = 0$

$$x = -3 \quad \text{or} \quad x = \frac{3 \pm \sqrt{9 - 36}}{2}$$

$$= \frac{3 \pm \sqrt{27}\,i}{2}$$

$$= \frac{3 \pm 3\sqrt{3}\,i}{2}$$

The solution set is

$$\left\{ -3, \frac{3 - 3\sqrt{3}\,i}{2}, \frac{3 + 3\sqrt{3}\,i}{2} \right\}.$$

28. $x^4 - 1 = (x^2 - 1)(x^2 + 1) = (x - 1)(x + 1)(x^2 + 1) = 0$

$x - 1 = 0 \quad \text{or} \quad x + 1 = 0 \quad \text{or} \quad x^2 + 1 = 0$

$x = 1 \quad \text{or} \quad x = -1 \quad \text{or} \quad x^2 = -1$

$$x = \pm i$$

The solution set is $\{-1, 1, -i, i\}$.

30. $x^4 + 5x^2 + 4 = 0$

$(x^2 + 4)(x^2 + 1) = 0$

$x^2 + 4 = 0 \quad \text{or} \quad x^2 + 1 = 0$

$x^2 = -4 \quad \text{or} \quad x^2 = -1$

$x = \pm 2i \quad \text{or} \quad x = \pm i$

The solution set is $\{-2i, -i, i, 2i\}$.

32. $2x^2 - 4x + 1 = 0$

$b^2 - 4ac = 16 - 8 = 8$

so the solutions are two unequal real numbers.

34. $x^2 + 2x + 6 = 0$

$b^2 - 4ac = 9 - 4(6) = 9 - 24 = -15$

so the solutions are complex numbers that are conjugates of each other.

36. $4x^2 + 12x + 9 = 0$

$b^2 - 4ac = 144 - 144 = 0$

so the solution is a repeated real number.

38. The other solution is the conjugate of $4 - i$, which is $4 + i$.

2.
$$\sqrt{3t + 4} = 2$$
$$3t + 4 = 4$$
$$3t = 0$$
$$t = 0$$

4.
$$\sqrt{5t + 4} = -3$$

No real solution.

6.
$$\sqrt[3]{1 - 2x} - 1 = 0$$
$$\sqrt[3]{1 - 2x} = 1$$
$$1 - 2x = 1$$
$$x = 0$$

8.
$$\sqrt{12 - x} = x$$
$$12 - x = x^2$$
$$x^2 + x - 12 = 0$$
$$(x - 3)(x + 4) = 0$$
$$x = 3 \text{ or } x = -4$$

Since $\sqrt{16} \neq -4$, $x = 3$ is the only solution.

10.
$$x = 2\sqrt{-x - 1}$$
$$x^2 = 4(-x - 1)$$
$$x^2 + 4x + 4 = 0$$
$$(x + 2)(x + 2) = 0$$
$$x = -2$$

Since $-2 \neq 2\sqrt{-(-2) - 1}$, there is no real solution.

12.
$$\sqrt{3 - x + x^2} = x - 2$$
$$3 - x + x^2 = x^2 - 4x + 4$$
$$3x - 1 = 0$$
$$x = \frac{1}{3}$$

Since $\frac{1}{3}$ is not a solution, there is no real solution.

14.
$$2 + \sqrt{12 - 2x} = x$$
$$\sqrt{12 - 2x} = x - 2$$
$$12 - 2x = x^2 - 4x + 4$$
$$0 = x^2 - 2x - 8 = (x - 4)(x + 2)$$

$x = 4$ or $x = -2$
Upon checking, we find the equation has only one real solution, $x = 4$; the solution $x = -2$ is extraneous.

16.
$$\sqrt{3x + 7} + \sqrt{x + 2} = 1$$
$$\sqrt{3x + 7} = 1 - \sqrt{x + 2}$$
$$3x + 7 = x + 3 - 2\sqrt{x + 2}$$
$$2x + 4 = -2\sqrt{x + 2}$$
$$4x^2 + 16x + 16 = 4(x + 2)$$
$$4x^2 + 12x + 8 = 0$$
$$4(x^2 + 3x + 2) = 0$$
$$4(x + 1)(x + 2) = 0$$
$$x = -1 \text{ or } x = -2$$

Upon checking, we find the equation has only one real solution, $x = -2$; the solution $x = -1$ is extraneous.

18. $\sqrt{3x - 5} - \sqrt{x + 7} = 2$

$\sqrt{3x - 5} = 2 + \sqrt{x + 7}$

$3x - 5 = 4 + 4\sqrt{x + 7} + x + 7$

$2x - 16 = 4\sqrt{x + 7}$

$4x^2 - 64x + 256 = 16(x + 7)$

$4x^2 - 80x + 144 = 0$

$4(x^2 - 20x + 36) = 0$

$4(x - 18)(x - 2) = 0$

$x = 18 \quad \text{or} \quad x = 2$

Upon checking, we find the equation has only one real solution, 18; the solution $x = 2$ is extraneous.

20. $\sqrt{10 + 3\sqrt{x}} = \sqrt{x}$

$10 + 3\sqrt{x} = x$

$3\sqrt{x} = x - 10$

$9x = x^2 - 20x + 100$

$0 = x^2 - 29x + 100$

$0 = (x - 25)(x - 4)$

$0 = x - 25 \quad \text{or} \quad 0 = x - 4$

$x = 25 \quad \text{or} \quad x = 4$

Does not check.

The solution is 25.

22. $(3x - 5)^{1/2} = 2$

$3x - 5 = 4$

$3x = 9$

$x = 3$

The solution is 3.

24. $(2x + 1)^{1/3} = -2$

$2x + 1 = -8$

$2x = -9$

$x = \dfrac{-9}{2}$

The solution is $\dfrac{-9}{2}$.

26. $(x^2 - 16)^{1/2} = 9$

$x^2 - 16 = 81$

$x^2 = 97$

$x = \pm\sqrt{97}$

The solution set is $\left\{-\sqrt{97}, \sqrt{97}\right\}$.

28. $(2x + 3)^2 - (2x + 3) - 6 = 0$

Let $u = 2x + 3$

$u^2 - u - 6 = 0$

$(u - 3)(u + 2) = 0$

$u - 3 = 0 \quad u + 2 = 0$

$u = 3 \qquad u = -2$

$2x + 3 = 3 \qquad 2x + 3 = -2$

$2x = 0 \qquad\qquad 2x = -5$

$x = 0 \qquad\qquad x = \dfrac{-5}{2}$

The solution set is $\left\{\dfrac{-5}{2}, 0\right\}$.

30. $(2 - x)^2 + (2 - x) - 20 = 0$

Let $u = 2 - x$

$u^2 + u - 20 = 0$

$(u + 5)(u - 4) = 0$

$u + 5 = 0 \quad u - 4 = 0$

$u = -5 \qquad u = 4$

$2 - x = -5 \qquad 2 - x = 4$

$-x = -7 \qquad\quad -x = 2$

$x = 7 \qquad\qquad x = -2$

The solution is $\{-2, 7\}$.

32.　$3(1 - y)^2 + 5(1 - y) + 2 = 0$
Let $u = 1 - y$, then
$$u^2 = (1 - y)^2$$
$$3u^2 + 5u + 2 = 0$$
$$(3u + 2)(u + 1) = 0$$
$$u = -\frac{2}{3} \text{ or } u = -1$$

Thus, $1 - y = -\frac{2}{3}$
or $1 - y = -1$
The solution set is $\left\{\frac{5}{3}, 2\right\}$.

34.　$x + 8\sqrt{x} = 0$
Let $u = \sqrt{x}$, then $u^2 = x$
$$u^2 + 8u = 0$$
$$u(u + 8) = 0$$
$$u = 0 \text{ or } u = -8$$
Thus, $\sqrt{x} = 0$ or $\sqrt{x} = -8$ (impossible)
The solution is 0.

36.　$x + 7\sqrt{x} + 6 = 0$
Let $u = \sqrt{x}$
$$u^2 + 7u + 6 = 0$$
$$(u + 1)(u + 6) = 0$$
$$u + 1 = 0 \text{ or } u + 6 = 0$$
$$u = -1 \qquad u = -6$$
Since $u = \sqrt{x} > 0$, there is no real solution.

38.　$z^{1/2} + 2z^{1/4} + 1 = 0$
Let $u = z^{1/4}$, then $u^2 = z^{1/2}$
$$u^2 + 2u + 1 = 0$$
$$(u + 1)(u + 1) = 0$$
$$u = -1$$
Thus, $z^{1/4} = -1$. This is impossible.
Hence, there is no real solution.

40.　$x^{1/2} - 3x^{1/4} + 2 = 0$
Let $u = x^{1/4}$, then $u^2 = x^{1/2}$
$$u^2 - 3u + 2 = 0$$
$$(u - 1)(u - 2) = 0$$
$$u = 1 \text{ or } u = 2$$
Thus, $x^{1/4} = 1$ or $x^{1/4} = 2$
The solution set is $\{1, 16\}$.

42.　
$$\sqrt[4]{4 - 5x^2} = x$$
$$4 - 5x^2 = x^4$$
$$x^4 + 5x^2 - 4 = 0$$
Let $u = x^2$, then $u^2 = x^4$
$$u^2 + 5u - 4 = 0$$
$$u = \frac{-5 \pm \sqrt{41}}{2}$$
Thus, $x^2 = \frac{-5 \pm \sqrt{41}}{2}$
$$x = \sqrt{\frac{-5 \pm \sqrt{41}}{2}},$$
(since x cannot be negative)
The solution is $\sqrt{\dfrac{-5 + \sqrt{41}}{2}}$.

44. $x^2 - 3x - \sqrt{x^2 - 3x} = 2$

Let $u = \sqrt{x^2 - 3x}$, then $u^2 = x^2 - 3x$
$$u^2 - u - 2 = 0$$
$$(u - 2)(u + 1) = 0$$
$$u = 2 \text{ or } u = -1$$

Thus, $\sqrt{x^2 - 3x} = 2$

or $\sqrt{x^2 - 3x} = -1 < 0$ (impossible)
$$x^2 - 3x = 4$$
$$x^2 - 3x - 4 = 4$$
$$(x - 4)(x + 1) = 0$$
$$x = 4 \text{ or } x = -1$$
The solution set is $\{4, -1\}$.

46. $\dfrac{1}{(x - 1)^2} + \dfrac{1}{x - 1} = 12$

Let $u = \dfrac{1}{x - 1}$, then $u^2 = \dfrac{1}{(x - 1)^2}$
$$u^2 + u - 12 = 0$$
$$(u + 4)(u - 3) = 0$$
$$u = -4 \text{ or } u = 3$$

Thus, $\dfrac{1}{x - 1} = -4 \qquad$ or $\qquad \dfrac{1}{x - 1} = 3$

$x - 1 = -\dfrac{1}{4} \qquad$ or $\qquad x - 1 = \dfrac{1}{3}$

$x = \dfrac{3}{4} \qquad$ or $\qquad x = \dfrac{4}{3}$

The solution set is $\left\{ \dfrac{3}{4}, \dfrac{4}{3} \right\}$.

48. $2x^{-2} - 3x^{-1} - 4 = 0$

Let $u = x^{-1}$, then $u^2 = x^{-2}$

$$u = \frac{3 \pm \sqrt{9 - 4(-8)}}{4}$$

$$u = \frac{3 \pm \sqrt{41}}{4}$$

$$x^{-1} = \frac{3 \pm \sqrt{41}}{4}$$

$$x = \frac{4}{3 \pm \sqrt{41}}$$

The solution set is

$$\left\{ \frac{4}{3 - \sqrt{41}}, \frac{4}{3 + \sqrt{41}} \right\}.$$

50. $3x^{4/3} + 5x^{2/3} + 2 = 0$

Let $x^{2/3} = u$, then $x^{4/3} = u^2$
$$3u^2 + 5u + 2 = 0$$
$$(3u + 2)(u + 1) = 0$$
$$3u + 2 = 0 \quad u + 1 = 0$$

$u = \dfrac{-2}{3} \qquad$ or $\qquad u = -1$

$x^{2/3} = \dfrac{-2}{3} \qquad$ or $\qquad x^{2/3} = -1$

$x = \left[\dfrac{-2}{3} \right]^{3/2} \qquad$ or $\qquad x = (-1)^{3/2}$

$$x = \sqrt[2]{-1} \Rightarrow \text{Not defined}$$

$x = \sqrt{\dfrac{-8}{27}} \Rightarrow$ Not defined

There is no solution.

52. $\left[\dfrac{y}{y - 1} \right]^2 = 6 \left[\dfrac{y}{y - 1} \right] + 7$

Let $u = \left[\dfrac{y}{y - 1} \right]$, then $u^2 = \left[\dfrac{y}{y - 1} \right]^2$
$$u^2 - 6u - 7 = 0$$
$$(u - 7)(u + 1) = 0$$
$$u = 7 \text{ or } u = -1$$

Thus, $\dfrac{y}{y - 1} = 7 \qquad$ or $\qquad \dfrac{y}{y - 1} = -1$

$y = 7y - 7 \quad$ or $\quad y = -y + 1$

$-6y = -7 \quad$ or $\quad 2y = 1$

$y = \dfrac{7}{6} \quad$ or $\quad y = \dfrac{1}{2}$

The solution set is $\left\{ \dfrac{7}{6}, \dfrac{1}{2} \right\}$.

54. $x^4 = 4x^3$
$$x^4 - 4x^3 = 0$$
$$x^3(x - 4) = 0$$
$$x^3 = 0, \quad x - 4 = 0$$
$$x = 0, \qquad x = 4$$
$$x = 0, \qquad x = 4$$
$$\{0, 4\}$$

56.
$$x^4 - 10x^2 + 25 = 0$$
$$(x^2 - 5)^2 = 0$$
$$x^2 - 5 = 0$$
$$x^2 = 5$$
$$x = -\sqrt{5}, \quad x = \sqrt{5}$$
$$\left\{-\sqrt{5}, \sqrt{5}\right\}$$

58.
$$2x^4 - 5x^2 - 12 = 0$$
$$(2x^2 + 3)(x^2 - 4) = 0$$
$$2x^2 + 3 = 0 \qquad \cdot x^2 - 4 = 0$$
$$2x^2 = -3 \qquad x^2 = 4$$
No real solution $x = -2, x = 2$
$$\{-2, 2\}$$

60.
$$x^6 - 7x^3 - 8 = 0$$
$$(x^3 - 8)(x^3 + 1) = 0$$
$$x^3 = 8, \ x^3 = -1$$
$$x = 2, \ x = -1$$
$$\{-1, 2\}$$

62.
$$x = 4\sqrt{x}$$
$$x - 4\sqrt{x} = 0$$
$$\sqrt{x}\left(\sqrt{x} - 4\right) = 0$$
$$\sqrt{x} = 0 \quad \text{or} \quad \sqrt{x} - 4 = 0$$
$$x = 0 \quad \text{or} \qquad x = 16$$
The solution set is $\{0, 36\}$.

64.
$$x^{3/4} - 4x^{1/4} = 0$$
$$x^{1/4}(x^{1/2} - 4) = 0$$
$$x^{1/4} = 0 \quad \text{or} \quad x^{1/2} - 4 = 0$$
$$x = 0 \quad \text{or} \qquad x = 16$$
The solution set is $\{0, 16\}$.

66.
$$x^3 + 6x^2 - 7x = 0$$
$$x(x^2 + 6x - 7) = 0$$
$$x(x + 7)(x - 1) = 0$$
$$x = 0 \text{ or } x = -7 \text{ or } x = 1$$
The solution set is $\{-7, 0, 1\}$.

68.
$$x^3 + x^2 - x - 1 = 0$$
$$x^2(x + 1) - (x + 1) = 0$$
$$(x + 1)(x^2 - 1) = 0$$
$$(x + 1)(x + 1)(x - 1) = 0$$
$$x = -1 \text{ or } x = 1$$
The solution set is $\{-1, 1\}$.

70.
$$x^3 - 3x^2 - x + 3 = 0$$
$$x^2(x - 3) - (x - 3) = 0$$
$$(x - 3)(x^2 - 1) = 0$$
$$(x - 3)(x + 1)(x - 1) = 0$$
$$x = 3 \text{ or } x = -1 \text{ or } x = 1$$
The solution set is $\{3, -1, 1\}$.

72.
$$y^6 - 4y^4 - y^2 + 4 = 0$$
$$y^4(y^2 - 4) - (y^2 - 4) = 0$$
$$(y^2 - 4)(y^4 - 1) = 0$$
$$(y - 2)(y + 2)(y^2 - 1)(y^2 + 1) = 0$$
$$(y - 2)(y + 2)(y - 1)(y + 1)(y^2 + 1) = 0$$
$$y = 2 \text{ or } y = -2 \text{ or } y = 1 \text{ or } y = -1$$
The solution set is $\{2, -2, 1, -1\}$.

74.
$$x^{2/3} + 4x^{1/3} + 2 = 0$$
Let $u = x^{1/3}$, then $u^2 = x^{2/3}$
$$u^2 + 4u + 2 = 0$$
$$u = \frac{-4 \pm \sqrt{8}}{2}$$
$$u = -2 \pm \sqrt{2}$$
$$x^{1/3} = -2 + \sqrt{2} \quad \text{or} \quad x^{1/3} = 2 - \sqrt{2}$$
$$x = -0.20 \qquad \text{or} \qquad x = -39.80$$
The solution set is $\{-0.20, -39.80\}$.

76. $x^4 + \sqrt{2}\,x^2 - 2 = 0$
Let $u = x^2$, then $u^2 = x^4$

$$u^2 + \sqrt{2}\,u - 2 = 0$$

$$u = \frac{-\sqrt{2} \pm \sqrt{10}}{2}$$

$$x^2 = \frac{-\sqrt{2} + \sqrt{10}}{2} \quad \text{or}$$

$$x^2 = \frac{-\sqrt{2} - \sqrt{10}}{2} < 0 \text{ (impossible)}$$

$$x = \pm \sqrt{\frac{-\sqrt{2} + \sqrt{10}}{2}}$$

$x = 0.93$ or $x = -0.93$
The solution set is $\{0.93, -0.93\}$.

78. $\pi(1 + r)^2 = 2 + \pi(1 + r)$
$\pi(1 + r)^2 - \pi(1 + r) - 2 = 0$
Let $u = (1 + r)$, then $u^2 = (1 + r)^2$
$\pi u^2 - \pi u - 2 = 0$

$$u = \frac{\pi \pm \sqrt{\pi^2 + 8\pi}}{2\pi}$$

$u = 1.44$ or $u = -0.44$
$1 + r = 1.44$ or $1 + r = -0.44$
$r = 0.44$ or $r = -1.44$
The solution set is $\{0.44, -1.44\}$.

80.
$$k^2 - 3k = 28$$
$$k^2 - 3k - 28 = 0$$
$$(k - 7)(k + 4) = 0$$

$k - 7 = 0$	$k + 4 = 0$
$k = 7$	$k = -4$
$\dfrac{x + 3}{x - 4} = 7$	$\dfrac{x + 3}{x - 4} = -4$
$x + 3 = 7x - 28$	$x + 3 = -4x + 16$
$31 = 6x$	$5x = 13$
$x = \dfrac{31}{6}$	$x = \dfrac{13}{5}$

$$\left\{ \frac{13}{5}, \frac{31}{6} \right\}$$

2.6 Inequalities

2. $[2, \infty); 2 \le x < \infty$

4. $(-\infty, 0]; -\infty < x \le 0$

6. $(-\infty, 0]$ or $(1, \infty);$
$-\infty < x \le 0$ or $1 < x < \infty$

8. $(-1, 1]; -1 < x \le 1$

10. $<$

12. $>$

14. $<$

16. $<$

18.

$x < 4$

20.

$x > 3$ and $x \le 7$

22.

$x > 0$ or $x \ge 5$

24.

$x \geq 4$ and $x < -2$

26.

$x \geq 4$ or $x < -2$

28. $(-1, 5)$

30. $(-2, 0)$

32. $(-\infty, 5]$

34. $(1, \infty)$

36. $1 < x < 2$

38. $0 \leq x < 1$

40. $x \leq 2$

42. $x > -8$

44. If $a \leq b$ and $c < 0$, show that $ac \geq bc$

$a - b \leq 0$ and $c < 0$

$c(a - b) \geq c(0)$ Reverse signs when multiplying by a negative number.

$ca - cb \geq 0$

$\qquad ca \geq cb$

$\qquad ac \geq bc$

46. $\dfrac{a + b}{2} - a = \dfrac{a + b - 2a}{2} = \dfrac{b - a}{2}$

$b - \dfrac{a + b}{2} = \dfrac{2b - a - b}{2} = \dfrac{b - a}{2}$

$\therefore \dfrac{a + b}{2}$ is equidistant from a and from b.

48. We want to show that

$$\sqrt{ab} < \dfrac{a + b}{2}$$

$$\dfrac{a + b}{2} - \sqrt{ab} = \dfrac{1}{2}\left[a - 2\sqrt{ab} + b\right]$$

$$= \dfrac{1}{2}\left(\sqrt{a} - \sqrt{b}\right)^2 > 0$$

since $a > b$

Therefore, $\sqrt{ab} < \dfrac{a + b}{2}$

50. We want to show that

$$h = \frac{(\text{geometric mean})^2}{\text{arithmetic mean}} = \frac{\left(\sqrt{ab}\right)^2}{\frac{1}{2}(a+b)}$$

Now, $\dfrac{1}{h} = \dfrac{1}{2}\left[\dfrac{1}{a} + \dfrac{1}{b}\right]$

$$\frac{2}{h} = \frac{1}{a} + \frac{1}{b} = \frac{a+b}{ab}$$

$$\frac{h}{2} = \frac{ab}{a+b}$$

$$h = 2\frac{ab}{a+b} = \frac{(ab)^2}{\frac{1}{2}(a+b)}$$

52. $40 \le \text{Age} < 60$

2.7 Linear Inequalities

2. $2 > 1$

(a) $\quad -3 + 2 > -3 + 1$
$\qquad -1 > -2$

(b) $\quad 2 - 5 > 1 - 5$
$\qquad -3 > -4$

(c) $\quad (3)2 > (3)1$
$\qquad 6 > 3$

(d) $\quad (-2)2 < (-2)1$
$\qquad -4 < -2$

4. $1 - 2x > 5$

(a) $\quad -3 + (1 - 2x) > -3 + 5$
$\qquad -2 - 2x > 2$

(b) $\quad (1 - 2x) - 5 > 5 - 5$
$\qquad -4 - 2x > 0$

(c) $\quad (3)(1 - 2x) > (3)5$
$\qquad 3 - 6x > 15$

(d) $\quad (-2)(1 - 2x) < (-2)5$
$\qquad -2 + 4x < -10$

6. $x - 6 < 1$
$\qquad x < 7$

$\{x \mid x < 7\}$ or $(7, \infty)$

8. $2 - 3x \le 5$
$\qquad -3x \le 3$
$\qquad x \ge -1$

$\{x \mid x \ge -1\}$ or $[-1, \infty)$

10. $2x + 5 > 1$
$\qquad 2x > -4$
$\qquad x > -2$

$\{x \mid x > -2\}$ or $(-2, \infty)$

12. $2x - 2 \ge 3 + x$
$\qquad x \ge 5$

$\{x \mid x \ge 5\}$ or $[5, \infty)$

14. $-3(1 - x) < 12$

$-3 + 3x < 12$

$3x < 15$

$x < 5$

$\{x \mid x < 5\}$ or $(-\infty, 5)$

16. $8 - 4(2 - x) \le -2x$

$8 - 8 + 4x \le -2x$

$6x \le 0$

$x \le 0$

$\{x \mid x \le 0\}$ or $(-\infty, 0]$

18. $3x + 4 > \dfrac{1}{3}(x - 2)$

$3x + 4 > \dfrac{1}{3}x - \dfrac{2}{3}$

$\dfrac{8}{3}x > -\dfrac{14}{3}$

$x > -\dfrac{7}{4}$

$\left\{x \mid x > -\dfrac{7}{4}\right\}$ or $\left[-\dfrac{7}{4}, \infty\right]$

20. $\dfrac{x}{3} \ge 2 + \dfrac{x}{6}$

$2x \ge 12 + x$

$x \ge 12$

$\{x \mid x \ge 12\}$ or $[12, \infty)$

22. $4 \le 2x + 2 \le 10$

$2 \le 2x \le 8$

$1 \le x \le 4$

$\{x \mid 1 \le x \le 4\}$ or $[1, 4]$

24. $-3 \le 3 - 2x \le 9$

$-6 \le -2x \le 6$

$3 \ge x \ge -3$

$-3 \le x \le 3$

$\{x \mid -3 \le x \le 3\}$ or $[-3, 3]$

26. $0 < \dfrac{3x + 2}{2} < 4$

$0 < 3x + 2 < 8$

$-2 < 3x < 6$

$-\dfrac{2}{3} < x < 2$

$\left\{x \mid \dfrac{-2}{3} < x < 2\right\}$ or $\left[\dfrac{-2}{3}, 2\right]$

28. $0 < 1 - \dfrac{1}{3}x < 1$

$-1 < -\dfrac{1}{3}x < 0$

$3 > x > 0$

$0 < x < 3$

$\{x \mid 0 < x < 3\}$ or $(0, 3)$

30. $(x - 1)(x + 1) > (x - 3)(x + 4)$

$x^2 - 1 > x^2 + x - 12$

$-1 > x - 12$

$11 > x$

$x < 11$

$\{x \mid x < 11\}$ or $(-\infty, 11)$

32. $x(9x - 5) \le (3x - 1)^2$

$9x^2 - 5x \le 9x^2 - 6x + 1$

$-5x \le -6x + 1$

$x \le 1$

$\{x \mid x \le 1\}$ or $(-\infty, 1]$

34.
$$\frac{1}{3} < \frac{x+1}{2} \le \frac{2}{3}$$
$$\frac{2}{3} < x+1 \le \frac{4}{3}$$
$$-\frac{1}{3} < x \le \frac{1}{3}$$

$$\left\{ x \,\middle|\, -\frac{1}{3} < x \le \frac{1}{3} \right\} \text{ or } \left[-\frac{1}{3}, \frac{1}{3} \right]$$

36.
$$(2x-1)^{-1} > 0$$
$$\frac{1}{2x-1} > 0$$
$$2x-1 > 0$$
$$2x > 1$$
$$x > \frac{1}{2}$$

$$\left\{ x \,\middle|\, x > \frac{1}{2} \right\} \text{ or } \left[\frac{1}{2}, \infty \right]$$

38.
$$0 < \frac{1}{x} < \frac{1}{3}$$
$$x > 3$$

$$\{ x \mid x > 3 \} \text{ or } (3, \infty)$$

40.
$$0 < (3x+6)^{-1} < \frac{1}{3}$$
$$0 < \frac{1}{3x+6} < \frac{1}{3}$$
$$3x+6 > 3$$
$$3x > -3$$
$$x > -1$$

$$\{ x \mid x > -1 \} \text{ or } (-1, \infty)$$

42.
$$-3 < x < 2$$
$$-3-6 < x-6 < 2-6$$
$$-9 < x-6 < -4$$
$$a = -9, b = -4$$

44.
$$-4 < x < 0$$
$$\frac{1}{2}(-4) < \frac{1}{2}x < \frac{1}{2}(0)$$
$$-2 < \frac{1}{2}x < 0$$
$$a = -2, b = 0$$

46.
$$-3 < x < 3$$
$$-2(-3) > -2x > -2(3)$$
$$6 > -2x > -6$$
$$-6 < -2x < 6$$
$$1-6 < 1-2x < 1+6$$
$$-5 < 1-2x < 7$$
$$a = -5, b = 7$$

48.
$$2 < x < 4$$
$$2-6 < x-6 < 4-6$$
$$-4 < x-6 < -2$$
$$\frac{1}{-4} > \frac{1}{x-6} > \frac{1}{-2}$$
$$-\frac{1}{2} < \frac{1}{x-6} < -\frac{1}{4}$$
$$a = -\frac{1}{2}, b = -\frac{1}{4}$$

50.
$$0 < 2x < 6$$
$$\frac{1}{2}(0) < \frac{1}{2}(2x) < \frac{1}{2}(6)$$
$$0 < x < 3$$
$$0^2 < x^2 < 3^2$$
$$0 < x^2 < 9$$
$$a = 0, b = 9$$

52. $x > 3$
Since $x > 3$, then $x - 3 > 0$
$$\frac{x}{x-3} > 2$$
$$(x-3)\frac{x}{x-3} > (x-3)(2)$$
$$x > 2x-6$$
$$-x > -6$$
$$x < 6$$
Here, $x < 6$ or $x > 3$.
$$\{ x \mid x < 6 \text{ or } x > 3 \}$$

54. $\sqrt{8 + 2x}$ is defined when

$$8 + 2x \geq 0$$
$$2x \geq -8$$
$$x \geq -4$$

$$\{x \mid x \geq -4\}$$

56. I = Interest, P = Principal, r = Rate, t = Time

Simple Interest = $I = Prt$

$$90 \leq I \leq 110$$
$$90 \leq Prt \leq 110$$
$$90 \leq (1000)r(1) < 110$$
$$\frac{90}{1000} \leq r \leq \frac{110}{1000}$$
$$\frac{9}{100} \leq r \leq \frac{11}{100}$$

The interest rates needed are between 9% and 11%, inclusive.

58. C = Commission

$$25 + .4(20) \leq C \leq 25 + .4(250)$$
$$33 \leq C \leq 125$$

The commission will vary between \$33 and \$125.

60. A = Amount withheld $\quad A = 0.28(600 - 476) + 63.90$

W = Weekly wage $\quad A = 0.28(700 - 476) + 63.90$

$A = 0.28(W - 476) + 63.90$

$600 \leq$	A	≤ 700
$(0.28)(600) \leq$	$.28W$	$\leq (0.28)(700)$
$168 \leq$	$.28W$	≤ 196
$168 - 69.38 \leq$	$0.28W - 69.38$	$\leq 196 - 69.38$
$98.62 \leq$	A	≤ 126.62

The amount withheld varies from \$98.62 to \$126.62, inclusive.

62. W = Water usage (number of 1000 gallons in excess of 12,000)

$$\frac{28.40 - 21.60}{1.70} \leq W \leq \frac{65.75 - 21.60}{1.70}$$
$$4 \leq W \leq 26$$

The water usage varied from $4000 + 12000 = 16000$ gallons to $26000 + 12000 = 38000$ gallons.

64. T = Test scores of people in the top 2.5%

$$T > 1.95(12) + 100$$
$$T > 123.4$$

The people in the top 2.5% have test scores greater than 123.4.

66.
$$80 \leq \frac{68 + 82 + 87 + 89 + 2x}{6} < 90$$
$$80 \leq \frac{326 + 2x}{6} < 90$$
$$480 \leq 326 + 2x < 540$$
$$154 \leq 2x < 214$$
$$77 \leq x < 107$$

You will need a score greater than or equal to 77 to get a B.

68. x = Number of gallons of gasoline in the tank at the start of the trip

$$25x \leq 250$$
$$x \leq 10$$

There was 10 gallons or less of gasoline at the start of the trip.

2. $(x - 5)(x + 2) > 0$

    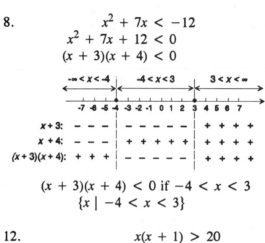

    $(x - 5)(x + 2) > 0$ if $-\infty < x < -2$
    or $5 < x < \infty$
    $\{x \mid -\infty < x < -2 \text{ or } 5 < x < \infty\}$

4. $x^2 + 8x > 0$
    $x(x + 8) > 0$

    $x(x + 8) > 0$ if $-\infty < x < -8$
    or $0 < x < \infty$
    $\{x \mid -\infty < x < -8 \text{ or } 0 < x < \infty\}$

6. $x^2 - 1 < 0$
    $(x - 1)(x + 1) < 0$

    $(x - 1)(x + 1) < 0$ if $-1 < x < 1$
    $\{x \mid -1 < x < 1\}$

8. $x^2 + 7x < -12$
    $x^2 + 7x + 12 < 0$
    $(x + 3)(x + 4) < 0$

    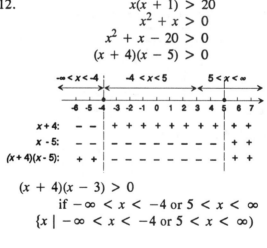

    $(x + 3)(x + 4) < 0$ if $-4 < x < 3$
    $\{x \mid -4 < x < 3\}$

10. $6x^2 < 6 + 5x$
    $6x^2 - 5x - 6 < 0$
    $(3x + 2)(2x - 3) < 0$

    $(3x + 2)(2x - 3) < 0$ if
    $-\dfrac{2}{3} < x < \dfrac{3}{2}$
    $\left\{x \mid -\dfrac{2}{3} < x < \dfrac{3}{2}\right\}$

12. $x(x + 1) > 20$
    $x^2 + x > 0$
    $x^2 + x - 20 > 0$
    $(x + 4)(x - 5) > 0$

    $(x + 4)(x - 3) > 0$
    if $-\infty < x < -4$ or $5 < x < \infty$
    $\{x \mid -\infty < x < -4 \text{ or } 5 < x < \infty)$

14.

$$25x^2 + 16 < 40x$$
$$25x^2 - 40x + 16 < 0$$
$$(5x - 4)(5x - 4) < 0$$
$$(5x - 4)^2 < 0$$

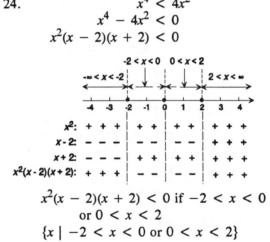

$-\infty < x < \frac{4}{5}$ $\frac{4}{5} < x < \infty$

$5x - 4$: $- - - - - - - | + + + +$
$5x - 4$: $- - - - - - - | + + + +$
$(5x - 4)^2$: $+ + + + + + + | + + + +$

Since the square of a quantity is always positive, there is no solution.

16.

$$2(2x^2 - 3x) > -9$$
$$4x^2 - 6x + 9 > 0$$

Since $b^2 - 4ac = 36 - 144 = -108 < 0$, $4x^2 - 6x + 9$ has no real solution. Therefore, its value is always positive or always negative. Because $4(0)^2 - 6(0) + 9 = 9$ is positive, we conclude that $4x^2 - 6x + 9 > 0$ for all x.

18. $(x + 2)(x^2 - x + 1) > 0$

Since the discriminant of $x^2 - x + 1$ is $1 - 4 = -3$, it has no real solution. Testing 0, we conclude $x^2 - x + 1 > 0$ for all x. Therefore, $x + 2 > 0 \Rightarrow x > -2$.

$$\{x \mid -2 < x < \infty\}$$

20. $(x + 1)(x + 2)(x + 3) < 0$

$-\infty < x < -3$ $-3 < x < -2$ $-2 < x < -1$ $-1 < x < \infty$

$x + 1$: $- - - \mid - \mid - \mid + + + + +$
$x + 2$: $- - - \mid - \mid + \mid + + + + +$
$x + 3$: $- - - \mid + \mid + \mid + + + + +$
$(x + 1)(x + 2)(x + 3)$: $- - - \mid + \mid - \mid + + + + +$

$(x + 1)(x + 2)(x + 3) < 0$
   if $-\infty < x < -3$ or $-2 < x < -1$.
$\{x \mid -\infty < x < -3$ or $-2 < x < -1\}$

22.

$$x^3 + 2x^2 - 3x > 0$$
$$x(x^2 + 2x - 3) > 0$$
$$x(x - 1)(x - 2) > 0$$

$-\infty < x < 0$ $0 < x < 1$ $1 < x < 2$ $2 < x < \infty$

x: $- - \mid + + \mid + + \mid + +$
$x - 1$: $- - \mid - - \mid + + \mid + +$
$x - 2$: $- - \mid - - \mid - - \mid + +$
$x(x - 1)(x - 2)$: $- - \mid + + \mid - - \mid + +$

$x(x + 4)(x - 2) > 0$
   if $0 < x < 1$ or $2 < x < \infty$
$\{x \mid 0 < x < 1$ or $2 < x < \infty\}$

24.

$$x^4 < 4x^2$$
$$x^4 - 4x^2 < 0$$
$$x^2(x - 2)(x + 2) < 0$$

$-\infty < x < -2$ $-2 < x < 0$ $0 < x < 2$ $2 < x < \infty$

x^2: $+ + + \mid + + \mid + + \mid + + +$
$x - 2$: $- - - \mid - - \mid - - \mid + + +$
$x + 2$: $- - - \mid + + \mid + + \mid + + +$
$x^2(x - 2)(x + 2)$: $+ + + \mid - - \mid - - \mid + + +$

$x^2(x - 2)(x + 2) < 0$ if $-2 < x < 0$
   or $0 < x < 2$
$\{x \mid -2 < x < 0$ or $0 < x < 2\}$

26.

$$x^3 < 3x^2$$
$$x^3 - 3x^2 < 0$$
$$x^2(x - 3) < 0$$

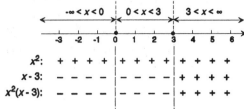

$$x^2(x - 3) < 0 \text{ if } -3 < x < 0 \text{ or } 0 < x < 3$$
$$\{x \mid -3 < x < 0 \text{ or } 0 < x < 3\}$$

28.

$$x^3 > 1$$
$$x^3 - 1 > 0$$
$$(x - 1)(x^2 + x + 1) > 0$$

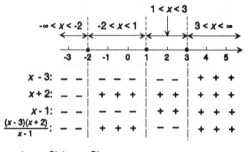

$$(x - 1)(x^2 + x + 1) > 0 \text{ if } 1 < x < \infty$$
$$\{x \mid 1 < x < \infty\}$$

30. $\dfrac{x - 3}{x + 1} > 0$

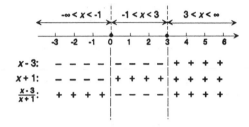

$$\frac{x - 3}{x + 1} > 0 \text{ if } -\infty < x < -1$$
$$\text{or } 3 < x < \infty$$
$$\{x \mid -\infty < x < -1 \text{ or } 3 < x < \infty\}$$

32. $\dfrac{(x - 3)(x + 2)}{x - 1} < 0$

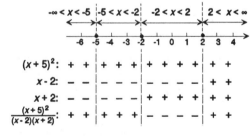

$$\frac{(x - 3)(x + 2)}{x - 1} < 0 \text{ if } -\infty < x < -2$$
$$\text{or } 1 < x < 3$$
$$\{x \mid -\infty < x < -2 \text{ or } 1 < x < 3\}$$

34. $\dfrac{(x + 5)^2}{x^2 - 4} \geq 0$

$$\frac{(x + 5)^2}{(x - 2)(x + 2)} \geq 0$$

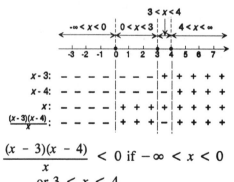

$$\frac{(x + 5)^2}{(x - 2)(x + 2)} \geq 0 \text{ if } -\infty < x < -2$$
$$\text{or } 2 < x < \infty$$
$$\{x \mid -\infty < x < -2 \text{ or } 2 < x < \infty\}$$

36. $x + \dfrac{12}{x} < 7$

$$\frac{x^2 - 7x + 12}{x} < 0$$
$$\frac{(x - 3)(x - 4)}{x} < 0$$

$$\frac{(x - 3)(x - 4)}{x} < 0 \text{ if } -\infty < x < 0$$
$$\text{or } 3 < x < 4$$
$$\{x \mid -\infty < x < 0 \text{ or } 3 < x < 4\}$$

38.
$$\frac{x+2}{x-4} \geq 1$$
$$\frac{x+2-(x-4)}{x-4} \geq 0$$
$$\frac{6}{x-4} \geq 0$$
$$x - 4 > 0$$
$$x > 4$$
$$\{x \mid 4 < x < \infty\}$$

40.
$$\frac{x-4}{2x+4} \geq 1$$
$$\frac{x-4-(2x+4)}{2x+4} \geq 0$$
$$\frac{-x-8}{2(x+2)} \geq 0$$
$$\frac{-(x+8)}{2(x+2)} \geq 0$$

	$-\infty < x < -8$	$-8 < x < -2$	$-2 < x < \infty$
$-\frac{1}{2}$:	$-\ -$	$-\ -\ -\ -$	$-\ -\ -\ -$
$x+8$:	$-\ -$	$+\ +\ +\ +$	$+\ +\ +\ +$
$x+2$:	$-\ -$	$-\ -\ -\ -$	$+\ +\ +\ +$
$\frac{-(x+8)}{2(x+2)}$:	$-\ -$	$+\ +\ +\ +$	$-\ -\ -\ -$

$$\frac{-(x+8)}{2(x+2)} \geq 0 \text{ if } -8 \leq x < -2$$
$$\{x \mid -8 \leq x < -2\}$$

42.
$$\frac{5}{x-3} > \frac{3}{x+1}$$
$$\frac{5}{x-3} - \frac{3}{x+1} > 0$$
$$\frac{5(x+1)-3(x+1)}{(x-3)(x+1)} > 0$$
$$\frac{2x+2}{(x-3)(x+1)} > 0$$
$$\frac{2(x+1)}{(x-3)(x+1)} > 0$$
$$\frac{2}{x-3} > 0$$

	$-\infty < x < 3$	$3 < x < \infty$
$x - 3$:	$-\ -\ -\ -\ -$	$+\ +\ +\ +\ +$

$$\frac{2}{x-3} > 0 \text{ if } 3 < x < \infty$$
$$\{x \mid 3 < x < \infty\}$$

44.
$$\frac{1}{x+2} > \frac{3}{x+1}$$
$$\frac{1}{x+2} - \frac{3}{x+1} > 0$$
$$\frac{x+1-3(x+2)}{(x+2)(x+1)} > 0$$
$$\frac{-2x-5}{(x+2)(x+1)} > 0$$

	$-\infty < x < -\frac{5}{2}$	$-\frac{5}{2} < x < -2$	$-2 < x < -1$	$-1 < x < \infty$
$-2x-5$:	$+\ +$	$-$	$-\ -$	$-\ -\ -\ -\ -\ -$
$x+2$:	$-\ -$	$-$	$+\ +$	$+\ +\ +\ +\ +\ +$
$x+1$:	$-\ -$	$-$	$-\ -$	$+\ +\ +\ +\ +\ +$
$\frac{-2x-5}{(x+2)(x+1)}$:	$+\ +$	$-$	$+\ +$	$-\ -\ -\ -\ -\ -$

$$\frac{-2x-5}{(x+2)(x+1)} > 0 \text{ if } -\infty < x < \frac{-5}{2}$$
$$\text{or } -2 < x < -1$$
$$\left\{ x \mid -\infty < x < -\frac{5}{2} \text{ or } -2 < x < -1 \right\}$$

46. $\dfrac{x(x^2 + 1)(x - 2)}{(x - 1)(x + 1)} > 0$

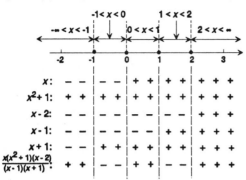

$\dfrac{x(x^2 + 1)(x - 2)}{(x - 1)(x + 1)} > 0$ if $-\infty < x < -1$
  or $0 < x < 1$ or $2 < x < \infty$

48.
$$x^2 > 2x, \; x > 0$$
$$x^2 - 2x > 0$$
$$x(x - 2) > 0$$
$$x > 0 \text{ and } x > 2$$
Thus, $\{x \mid 2 < x < \infty\}$

50. $x^3 - 3x^2 \geq 0$
$x^2(x - 3) \geq 0$

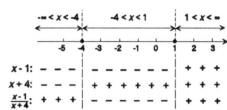

$x^2(x - 3) \geq 0$ if $x = 0$ or $3 < x < \infty$
$\{x \mid x = 0 \text{ or } 3 < x < \infty\}$

52. $\dfrac{x - 1}{x + 4} \geq 0$

$\dfrac{x - 1}{x + 4} \geq 0$ if $-\infty < x < -4$
  or $1 \leq x < \infty$
$\{x \mid -\infty < x < -4 \text{ or } 1 \leq x < \infty\}$

54.
$$80t - 16t^2 < 64$$
$$-16t^2 + 80t - 64 < 0$$
$$-t^2 + 5t - 4 < 0$$
$$t^2 - 5t + 4 > 0$$
$$(t - 4)(t - 1) > 0$$
$t > 4$ and $t > 1$ or $t < 4$ and $t < 1$
$\{x \mid -\infty < x < 1 \text{ or } 4 < x < \infty\}$

Thus, the ball is less than 64 feet above the ground when $t > 4$ or $t < 1$.

56.
$$x(5 - 0.05x - 1.50x \geq 60$$
$$5x - 0.05x^2 - 1.50x - 60 \geq 0$$
$$-0.05x + 3.5x - 60 \geq 0$$
$$x^2 - 70x + 1200 \leq 0$$
$$(x - 30)(x - 40) \leq 0$$
$x \leq 30$ and $x \geq 40$ (impossible) or
$x \geq 30$ and $x \leq 40$

Thus, $30 \leq x \leq 40$. Thus, between 30 and 40, inclusive, boxes of candy must be sold.

58. The equation $x^2 + kx + 1 = 0$ has no real solution when the discriminant $b^2 - 4ac < 0$. In this equation, $a = 1$, $b = k$, $c = 1$; therefore, $b^2 - 4ac = k^2 - 4(1)(1) = k^2 - 4 = (k - 2)(k + 2)$
$(k - 2)(k + 2) < 0$ when

	Test Number	$k - 2$	$k + 2$	$(k - 2)(k + 2)$
$k < -2$	-3	$-$	$-$	$+$
$-2 < k < 2$	0	$-$	$+$	$-$
$k > 2$	3	$+$	$+$	$+$

The solution set is $\{k \mid -2 < k < 2\}$.

2. $|3x| = 15$

$\qquad 3x = 15 \text{ or } 3x = -15$

$\qquad\quad x = 5 \text{ or } x = -5$

The solution set is $\{5, -5\}$.

4. $|3x - 1| = 2$

$\qquad 3x - 1 = 2 \text{ or } 3x - 1 = -2$

$\qquad\qquad 3x = 3 \text{ or } \qquad 3x = -1$

$\qquad\qquad\quad x = 1 \text{ or } \qquad\quad x = -\dfrac{1}{3}$

The solution set is $\left\{1, -\dfrac{1}{3}\right\}$.

6. $|1 - 2z| = 3$

$\qquad 1 - 2z = 3 \text{ or } 1 - 2z = -3$

$\qquad\quad -2z = 2 \text{ or } \qquad -2z = -4$

$\qquad\qquad z = -1 \text{ or } \qquad\quad z = 2$

The solution set is $\{-1, 2\}$.

8. $|-x| = 1$

$\qquad -x = 1 \text{ or } \qquad -x = -1$

$\qquad\quad x = -1 \text{ or } \qquad\quad x = 1$

The solution set is $\{-1, 1\}$.

10. $|3|x = 9$

$\qquad 3x = 9$

$\qquad\ x = 3$

The solution is 3.

12. $\dfrac{3}{4}|x| = 9$

$\qquad |x| = 12$

$\qquad x = 12 \text{ or } x = -12$

The solution set is $\{12, -12\}$.

14. $\left|\dfrac{x}{2} - \dfrac{1}{3}\right| = 1$

$\qquad \dfrac{x}{2} - \dfrac{1}{3} = 1 \quad\text{or}\quad \dfrac{x}{2} - \dfrac{1}{3} = -1$

$\qquad\qquad \dfrac{x}{2} = \dfrac{4}{3} \quad\text{or}\quad \dfrac{x}{2} = \dfrac{-2}{3}$

$\qquad\qquad\ x = \dfrac{8}{3} \quad\text{or}\quad x = \dfrac{-4}{3}$

The solution set is $\left\{\dfrac{8}{3}, \dfrac{-4}{3}\right\}$.

16. $|2 - v| = -1$

No solution. The absolute value is always nonnegative.

18. $|x^2 - 4| = 0$

$\qquad\quad x^2 - 4 = 0$

$\quad (x - 2)(x + 2) = 0$

$\qquad\qquad x = 2 \text{ or } x = -2$

The solution set is $\{2, -2\}$.

20. $|x^2 + x| = 12$

$\qquad x^2 + x = 12 \quad\text{or}\qquad x^2 + x = -12$

$x^2 + x - 12 = 0 \quad\text{or}\quad x^2 + x + 12 = 0 \quad \text{(no real solution)}$

$\quad (x + 4)(x - 3) = 0$

$\qquad\qquad\qquad x = -4 \text{ or } \qquad x = 3$

The solution set is $\{-4, 3\}$.

22. $|x^2 + 3x - 2| = 2$

$x^2 + 3x - 2 = 2$ or $x^2 + 3x - 2 = -2$

$x^2 + 3x - 4 = 0$ or $x^2 + 3x = 0$

$(x + 4)(x - 1) = 0$ or $x(x + 3) = 0$

$x = -4$ or $x = 1$ or $x = 0$ or $x = -3$

The solution set is $\{-4, 1, 0, -3\}$

24. $|3x| < 15$

$-15 < 3x < 15$

$-5 < x < 5$

The solution set consists of all numbers x for which $-5 < x < 5$.

$(-5, 5)$

26. $|2x| > 6$

$2x < -6$ or $2x > 6$

$x < -3$ or $x > 3$

The solution set consists of all numbers x for which $x < -3$ or $x > 3$.

$(-\infty, -3)$ or $(3, \infty)$

28. $|x + 4| < 2$

$-2 < x + 4 < 2$

$-6 < x < -2$

The solution set consists of all numbers x for which $-6 < x < -2$.

$(-6, -2)$

30. $|2u + 5| \leq 7$

$-7 \leq 2u + 5 \leq 7$

$-12 \leq 2u \leq 2$

$-6 \leq u \leq 1$

The solution set consists of all number u for which $-6 \leq u \leq 1$.

$[-6, 1]$

32. $|x + 4| \geq 2$

$x + 4 \leq -2$ or $x + 4 \geq 2$

$x \leq -6$ or $x \geq -2$

The solution set consists of all numbers x for which $x \leq -6$ or $x \geq -2$.

$(-\infty, -6]$ or $[-2, \infty)$

34. $|1 - 2x| < 3$

$-3 < 1 - 2x < 3$

$-4 < -2x < 2$

$2 > x > -1$

$-1 < x < 2$

The solution set consists of all numbers x for which $-1 < x < 2$.

$(-1, 2)$

36. $|2 - 3x| > 1$

$2 - 3x < -1$ or $2 - 3x > 1$

$-3x < -3$ or $-3x > -1$

$x > 1$ or $x < \dfrac{1}{3}$

The solution set consists of all numbers x

for which $x < \dfrac{1}{3}$ or $x > 1$.

$\left[-\infty, \dfrac{1}{3}\right]$ or $(1, \infty)$

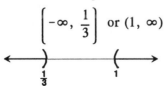

38. $|3 - x| > 0$

This is true for all x, except $x = 3$, since the absolute value is always nonnegative.

$(-\infty, 3)$ or $(3, \infty)$

40. $|2 - x| > -2$

This is true for all x since the absolute value is always nonnegative.

$$(-\infty, \infty)$$

42.
$$|3x - 2| < 0.02$$
$$-0.02 < 3x - 2 < 0.02$$
$$1.98 < 3x < 2.02$$
$$0.66 < x < 0.673$$

The solution set consists of all number x for which $0.66 < x < 0.673$.

$$(0.66, 0.673)$$

44.
$$|x + 2| < 5$$
$$-5 < x + 2 < 5$$
$$-7 < x < 3$$
$$-7 - 2 < x - 2 < 3 - 2$$
$$-9 < x - 2 < 1$$
$$a = -9, b = 1$$

46.
$$|x - 3| \leq 1$$
$$-1 \leq x - 3 \leq 1$$
$$2 \leq x \leq 4$$
$$6 \leq 3x \leq 12$$
$$7 \leq 3x + 1 \leq 13$$
$$a = 7, b = 13$$

48.
$$|x + 1| \leq 3$$
$$-3 \leq x + 1 \leq 3$$
$$1 \leq x + 5 \leq 7$$
$$\frac{1}{7} \leq \frac{1}{x + 5} \leq 1$$
$$a = \frac{1}{7}, b = 1$$

50. $a \leq |a|$ since if $a \geq 0$, then $a = |a|$ and if $a < 0$, then $-a = |a|$.

52. $|a| = |(a - b) + b| \leq |a - b| + |b|$. Thus, $|a| - |b| \leq |a - b|$ or $|a - b| \geq |a| - |b|$.

54.
$$|x - (-4)| < 1$$
$$-1 < x + 4 < 1$$
$$-5 < x < -3$$

56.
$$|x - 2| > 3$$
$$x - 2 < -3 \quad \text{or} \quad x - 2 > 3$$
$$x < -1 \quad \text{or} \quad x > 5$$

58.
$$|x - 115| \leq 5$$
$$-5 \leq x - 115 \leq 5$$
$$110 \leq x < 120$$

60. $x^2 > a \Rightarrow x^2 - a > 0$
$$\Rightarrow \left(x - \sqrt{a}\right)\left(x + \sqrt{a}\right) > 0$$
Therefore, $x > \sqrt{a}$
or $x < -\sqrt{a}$.

62.
$$x^2 < 4$$
$$-2 < x < 2$$

64.
$$x^2 \geq 1$$
$$x \geq 1 \quad \text{or} \quad x \leq -1$$

66.
$$x^2 \leq 9$$
$$-3 \leq x \leq 3$$

68.
$$x^2 > 16$$
$$x > 4 \quad \text{or} \quad x < -4$$

70. $x + |3x - 2| = -2$
     $|3x - 2| = -2 - x$
     $3x - 2 = -(-2 - x)$ $3x - 2 = -2 - x$
     $3x - 2 = 2 + x$ $4x = 0$
     $2x = 4$ $x = 0$
     $x = 2$
Does not check
     $x + |3x - 2| = 2$
     $|3x - 2| = 2 - x$
            $3x - 2 = (-(2 - x))$ $3x - 2 = 2 - x$
            $3x - 2 = -2 + x$ $4x = 4$
            $2x = 0$ $x = 1$
            $x = 0$
The solutions are $x = 0$ and $x = 1$.

2 Chapter Review

2. $\dfrac{x}{4} - 2 = 6$

    $\dfrac{x}{4} = 8$

    $x = 32$

4. $(6 - 3x) - 2(1 + x) = 6x$
     $6 - 3x - 2 - 2x - 6x = 0$
                  $-11x + 4 = 0$

                  $x = \dfrac{4}{11}$

6. $\dfrac{4 - 2x}{3} + \dfrac{1}{6} = 2x$

     $2(4 - 2x) + 1 = 12x$
     $8 - 4x + 1 - 12x = 0$
                $-16x = -9$

                $x = \dfrac{9}{16}$

8. $\dfrac{4x - 5}{3 - 7x} = 4$

     $4x - 5 = 4(3 - 7x)$
     $4x - 5 = 12 - 28x$
     $32x = 17$

     $x = \dfrac{17}{32}$

10. $x(1 + x) = 6$
      $x + x^2 - 6 = 0$
      $x^2 + x - 6 = 0$
      $(x + 3)(x - 2) = 0$
      $x = -3$ or $x = 2$
      $\{-3, 2\}$

12. $\dfrac{1 - 3x}{4} = \dfrac{x + 6}{3} + \dfrac{1}{2}$

      $3(1 - 3x) = 4(x + 6) + 6$
      $3 - 9x = 4x + 24 + 6$
      $-13x = 27$

      $x = -\dfrac{27}{13}$

14.

$$x(2 - x) = 3(x - 4)$$
$$2x - x^2 = 3x - 12$$
$$x^2 + x - 12 = 0$$
$$(x + 4)(x - 3) = 0$$
$$x = -4 \quad \text{or} \quad x = 3$$
$$\{-4, 3\}$$

16.

$$1 + 6x = 4x^2$$
$$4x^2 - 6x - 1 = 0$$
$$x = \frac{6 \pm \sqrt{36 + 16}}{8}$$
$$x = \frac{6 \pm \sqrt{52}}{8}$$
$$x = \frac{6 \pm 2\sqrt{13}}{8}$$
$$x = \frac{3 \pm \sqrt{13}}{4}$$
$$\left\{ \frac{3 - \sqrt{13}}{4}, \frac{3 + \sqrt{13}}{4} \right\}$$

18.

$$\sqrt{1 + x^3} = 3$$
$$1 + x^3 = 9$$
$$x^3 = 8$$
$$x = 2$$

20. $3x^2 - x + 1 = 0$
Since $b^2 - 4ac = 1 - 12 = -11 < 0$, there is no real solution.

22. $3x^4 + 4x^2 + 1 = 0$
Let $u = x^2$, then $u^2 = x^4$
$$3u^2 + 4u + 1 = 0$$
$$(3u + 1)(u + 1) = 0$$
$$u = -\frac{1}{3} \quad \text{or } u = -1$$
$$x^2 = -\frac{1}{3} \quad \text{or } x^2 = -1$$

These are both impossible, so there is no real solution.

24.

$$\sqrt{2x - 1} = x - 2$$
$$2x - 1 = x^2 - 4x + 4$$
$$x^2 - 6x + 5 = 0$$
$$(x - 5)(x - 1) = 0$$
$$x = 5 \quad \text{or} \quad x = 1$$

Since $1 \neq -1$, $x = 1$ is not a solution.
Thus, the only solution is $x = 5$.

26.

$$x^{2/3} + x = 0$$
$$x^{2/3}(1 + x^{1/3}) = 0$$
$$x^{2/3} = 0 \quad \text{or} \quad 1 + x^{1/3} = 0$$
$$x = 0 \quad \text{or} \qquad x = -1$$
$$\{-1, 0\}$$

28.

$$\sqrt{2x - 1} - \sqrt{x - 5} = 3$$
$$\sqrt{2x - 1} = 3 + \sqrt{x - 5}$$
$$2x - 1 = 9 + 6\sqrt{x - 5} + (x - 5)$$
$$x - 5 = 6\sqrt{x - 5}$$
$$x^2 - 10x + 25 = 36(x - 5)$$
$$x^2 - 46x + 205 = 0$$
$$(x - 41)(x - 5) = 0$$
$$x = 41 \quad \text{or} \quad x = 5$$
$$\{5, 41\}$$

30.

$$4\sqrt[3]{x^2} = 1$$
$$4x^{2/3} = 1$$
$$4x^{2/3} - 1 = 0$$

Let $x^{1/3} = u$, then $x^{2/3} = u^2$

$$4u^2 - 1 = 0$$
$$(2u - 1)(2u + 1) = 0$$

$$u = \frac{1}{2} \quad \text{or} \quad u = \frac{-1}{2}$$

$$x^{1/3} = \frac{1}{2} \quad \text{or} \quad x^{1/3} = \frac{-1}{2}$$

$$x = \left[\frac{1}{2}\right]^3 \quad \text{or} \quad x = \left[\frac{-1}{2}\right]^3$$

$$x = \frac{1}{8} \quad \text{or} \quad x = \frac{-1}{8}$$

$$\left\{-\frac{1}{8}, \frac{1}{8}\right\}$$

32.

$$6x^{-1} - 5x^{-1/2} + 1 = 0$$

Let $u = x^{-1/2}$, then $u^2 = x^{-1}$

$$6u^2 - 5u + 1 = 0$$
$$(3u - 1)(2u - 1) = 0$$

$$u = \frac{1}{3} \quad \text{or} \quad u = \frac{1}{2}$$

$$x^{-1/2} = \frac{1}{3} \quad \text{or} \quad x^{-1/2} = \frac{1}{2}$$

$$\frac{1}{\sqrt{x}} = \frac{1}{3} \quad \text{or} \quad \frac{1}{\sqrt{x}} = \frac{1}{2}$$

$$\sqrt{x} = 3 \quad \text{or} \quad \sqrt{x} = 2$$
$$x = 9 \quad \text{or} \quad x = 4$$

$$\{4, 9\}$$

34.

$$b^2x^2 + 2ax = x^2 + a^2$$
$$(b^2 - 1)x^2 + 2ax - a^2 = 0$$

$$x = \frac{-2a \pm \sqrt{4a^2 + (4b^2 - 4)a^2}}{2(b^2 - 1)}$$

$$x = \frac{-2a \pm \sqrt{4a^2 + 4b^2a^2 - 4a^2}}{2(b^2 - 1)}$$

$$x = \frac{-2a \pm 2ba}{2(b^2 - 1)}$$

$$x = \frac{-a \pm ba}{b^2 - 1}$$

$$x = \frac{-a(1 - b)}{b^2 - 1} \quad \text{or} \quad x = \frac{-a(1 + b)}{b^2 - 1}$$

$$x = \frac{a(b - 1)}{(b - 1)(b + 1)} \quad \text{or} \quad x = \frac{-a(1 + b)}{(b - 1)(b + 1)}$$

$$x = \frac{a}{b + 1} \quad \text{or} \quad x = \frac{-a}{b - 1}$$

$$\left\{-\frac{a}{b - 1}, \frac{a}{b + 1}\right\}$$

36.

$$\frac{1}{x - m} + \frac{1}{x - n} = \frac{2}{x}$$
$$x(x - n) + x(x - m) = 2(x - m)(x - n)$$
$$x^2 - xn + x^2 - xm = 2(x^2 - xn - xm + mn)$$
$$2x^2 - 2x^2 - xn + 2xn - xm + 2xm - 2mn = 0$$
$$xn + xm - 2mn = 0$$
$$x(n + m) = 2mn$$
$$x = \frac{2mn}{(n + m)}$$

38. $\sqrt{x^2 + 3x + 7} - \sqrt{x^2 - 3x + 9} = 2$

$$\sqrt{x^2 + 3x + 7} = \sqrt{x^2 + 3x + 9} + 2$$

$$x^2 + 3x + 7 = x^2 + 3x + 9 + 4\sqrt{x^2 + 3x + 9} + 4$$

$$-6 = 4\sqrt{x^2 + 3x + 9}$$

$$36 = 16(x^2 + 3x + 9)$$

$$16x^2 + 48x + 108 = 0$$

$$4x^2 + 12x + 27 = 0$$

Since $b^2 - 4ac = 144 - 432 = -288 < 0$, there is no real solution.

40. $\dfrac{5 - x}{3} \leq 6x - 4$

$$5 - x \leq 18x - 12$$

$$-19x \leq -17$$

$$x \geq \frac{17}{19}$$

$$\left\{ x \,\middle|\, \frac{17}{19} \leq x < \infty \right\}$$

42. $-4 < \dfrac{2x - 2}{3} < 6$

$$-12 < 2x - 2 < 18$$

$$-10 < 2x < 20$$

$$-5 < x < 10$$

$$\{ x \mid -5 < x < 10 \}$$

44. $6 > \dfrac{5 - 3x}{2} \geq -3$

$$12 > 5 - 3x \geq -6$$

$$7 > -3x \geq -11$$

$$-\frac{7}{3} < x \leq \frac{11}{3}$$

$$\left\{ x \,\middle|\, -\frac{7}{3} < x \leq \frac{11}{3} \right\}$$

46.
$$3x^2 - 2x - 1 \geq 0$$

$$(3x + 1)(x - 1) \geq 0$$

$(3x + 1) \geq 0 \quad \text{and} (x - 1) \geq 0 \quad \text{or} \quad (3x + 1) \leq 0 \quad \text{and } (x - 1) \leq 0$

$x \geq -\dfrac{1}{3} \text{ and} \qquad x \geq 1 \text{ or} \qquad x \leq -\dfrac{1}{3} \text{ and} \qquad x \leq 1$

$$x \geq 1 \text{ or} \qquad x \leq -\frac{1}{3}$$

$$\left\{ x \,\middle|\, -\infty < x \leq -\frac{1}{3} \text{ or } 1 \leq x < \infty \right\}$$

48.
$$\frac{-2}{1 - 3x} < 1, \quad x \neq \frac{1}{3}$$

$$\frac{-2 - (1 - 3x)}{1 - 3x} < 0$$

$$\frac{-3 + 3x}{1 - 3x} < 0$$

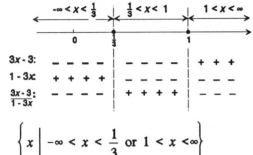

$$\left\{ x \mid -\infty < x < \frac{1}{3} \text{ or } 1 < x < \infty \right\}$$

50.
$$\frac{3 - 2x}{2x + 5} \geq 2$$

$$\frac{3 - 2x - 2(2x + 5)}{2x + 5} \geq 0$$

$$\frac{-6x - 7}{2x + 5} \geq 0$$

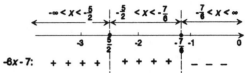

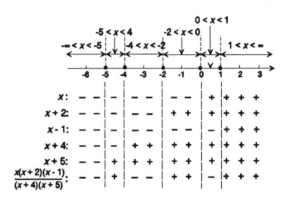

$$\left\{ x \mid -\frac{5}{2} < x \leq -\frac{7}{6} \right\}$$

52.
$$\frac{x + 1}{x(x - 5)} \leq 0, \quad x \neq 0, 5$$

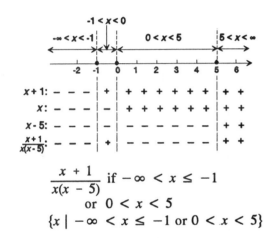

$$\frac{x + 1}{x(x - 5)} \text{ if } -\infty < x \leq -1$$
$$\text{or } 0 < x < 5$$
$$\{x \mid -\infty < x \leq -1 \text{ or } 0 < x < 5\}$$

54.
$$\frac{x(x^2 + x - 2)}{x^2 + 9x + 20} \leq 0, \quad x \neq -5, \ x \neq -4$$

$$\frac{x(x + 2)(x - 1)}{(x + 4)(x + 5)} \leq 0$$

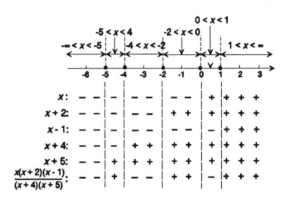

$$\frac{x(x + 2)(x - 1)}{(x + 4)(x + 5)} \leq 0 \text{ if } -\infty < x < -5$$
$$\text{or } -4 < x \leq -2 \text{ or } 0 < x \leq 1$$
$$\{x \mid -\infty < x < -5 \text{ or } -4 < x \leq -2$$
$$\text{or } 0 < x \leq 1\}$$

56. $|1 - 2x| < \dfrac{1}{3}$

$$-\dfrac{1}{3} < 1 - 2x < \dfrac{1}{3}$$

$$-\dfrac{4}{3} < -2x < -\dfrac{2}{3}$$

$$\dfrac{2}{3} > x > \dfrac{1}{3}$$

$$\dfrac{1}{3} < x < \dfrac{2}{3}$$

$$\left\{ x \;\middle|\; \dfrac{1}{3} < x < \dfrac{2}{3} \right\}$$

58. $|3x + 1| \geq 10$

$3x + 1 \leq -10 \quad$ or $\quad 3x + 1 \geq 10$

$3x \leq -11 \qquad$ or $\qquad 3x \geq 9$

$x \leq \dfrac{-11}{3} \qquad$ or $\qquad x \geq 3$

$$\left\{ x \;\middle|\; \dfrac{-11}{3} \leq x < \infty \text{ or } 3 \leq x < \infty \right\}$$

60. $x^2 - x + 1 = 0$

$$x = \dfrac{1 \pm \sqrt{-3}}{2}$$

$$x = \dfrac{1}{2} + \dfrac{\sqrt{3}}{2}i \quad \text{or} \quad x = \dfrac{1}{2} - \dfrac{\sqrt{3}}{2}i$$

$$\left\{ \dfrac{1 + \sqrt{3}\,i}{2}, \dfrac{1 - \sqrt{3}\,i}{2} \right\}$$

62. $3x^2 - 2x - 1 = 0$

$(3x + 1)(x - 1) = 0$

$$x = -\dfrac{1}{3} \quad \text{or} \quad x = 1$$

$$\left\{ -\dfrac{1}{3}, 1 \right\}$$

64. $2x^2 + 1 = 2x$

$2x^2 - 2x + 1 = 0$

$$x = \dfrac{2 \pm \sqrt{-4}}{4}$$

$$x = \dfrac{2 \pm 2i}{4}$$

$$x = \dfrac{1}{2} + \dfrac{1}{2}i \quad \text{or} \quad x = \dfrac{1}{2} - \dfrac{1}{2}i$$

$$\left\{ \dfrac{1}{2} - \dfrac{1}{2}i, \dfrac{1}{2} + \dfrac{1}{2}i \right\}$$

66. $x(1 + x) = 2$

$x^2 - x - 2 = 0$

$(x + 2)(x - 1) = 0$

$x = -2$ or $x = 1$

$\{-2, 1\}$

68. $x^4 - 8x^2 - 9 = 0$

Let $u = x^2$, then $u^2 = x^4$

$u^2 - 8u - 9 = 0$

$(u - 9)(u + 1) = 0$

$u = 9$ or $u = -1$

$x^2 = 9$ or $x^2 = -1$

$x = 3$ or $x = -3$

or $x = i$ or $x = -i$

$\{-3, 3, -i, i\}$

70.

$$1600 \leq I \leq 3600$$

$$1600 \leq \dfrac{900}{x^2} \leq 3600$$

$$1600 \leq \dfrac{900}{x^2} \text{ and } \dfrac{900}{x^2} \leq 3600$$

$$1600x^2 \leq 900 \text{ and } 900 \leq 3600x^2$$

$$x^2 \leq 0.5625 \text{ and } 0.25 \leq x^2$$

$$x \leq 0.75 \text{ and } 0.5 \leq x$$

$$0.5 \leq x \leq 0.75$$

The range of distances is from 0.5 meters to 0.75 meters.

72.
$$(250 + 30)t = (250 - 30)(2 - t)$$
$$280t = 220(2 - t)$$
$$280t = 440 - 220t$$
$$500t = 440$$
$$t = 0.88$$
$$d = 280(0.88) = 246.4$$

The plane can extend its search 246.4 miles.

74.
$$\frac{d}{3} = \frac{(150 - d)}{5} = \text{time needed to meet}$$
$$5d = 450 - 3d$$
$$8d = 450$$
$$d = 56.25$$
$$t = \frac{56.25}{3} = 18.75 \text{ seconds}$$

The bees meet for the first time in 18.75 seconds.
$$18.75 \times 2 = 37.5$$
The bees meet for the second time in 37.5 seconds.

76.
$$\ell^2 = 4(\ell + 4) \quad (\text{Note: } n = 4)$$
$$\ell^2 - 4\ell - 16 = 0$$
$$\ell = \frac{4 + \sqrt{80}}{2} \quad (\ell > 0)$$
$$\ell = 4.472 \text{ feet}$$

The length should be cut by 4.528 feet to be a length of 4.472 feet with width remaining 4 feet.

78.

	Hours To Do Job	Part of Job Done in One Minute
New Copier	$x - 1$	$\dfrac{1}{x - 1}$
Old Copier	x	$\dfrac{1}{x}$
Together	$\dfrac{72}{60}$	$\dfrac{60}{72}$

$$\frac{1}{x - 1} + \frac{1}{x} = \frac{60}{72}$$
$$\frac{x + x - 1}{x(x - 1)} = \frac{5}{6}$$
$$6(2x - 1) = 5x(x - 1)$$
$$12x - 6 = 5x^2 - 5x$$
$$5x^2 - 17x + 6 = 0$$
$$(5x - 2)(x - 3) = 0$$
$$x = \frac{2}{5}, \ x = 3$$

$x = \dfrac{2}{5}$ is not possible since together it takes 72 minutes. The older copier takes 3 hours to do the job itself.

Mission Possible

1. (Sketch of candy bar)

2. 252 cu. cm.

3. 226.8 cu. cum.

4. 183.708 cu. cm.

5. (a) 11.384 × 6.641 × 3 (changing length and width but not thickness)
   (b) 12 × 6.641 × 2.846 (changing width and thickness but not length)
   (c) 11.384 × 7 × 2.846 (changing length and thickness but not width)

6. 11.586 × 6.758 × 2.896

7. (Answers will vary. I would lean toward leaving the thickness unchanged because the "feel" of the candy bar would seem the same.)

8. (Answers will vary, but probably—Yes. Yes. Read the labels or read the back page of *Consumer Reports*.)

Chapter 3

GRAPHS

3.1 Rectangular Coordinates

2. (a) Quadrant I
    (b) Quadrant III
    (c) Quadrant II
    (d) Quadrant I
    (e) positive y-axis
    (f)   negative x-axis

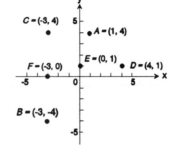

4. The points will be on a horizontal line that is three units above the x-axis.

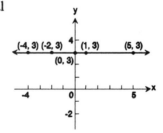

6. $d(P_1, P_2) = \sqrt{(-2 - 0)^2 + (1 - 0)^2} = \sqrt{4 + 1} = \sqrt{5}$

8. $d(P_1, P_2) = \sqrt{(2 - (-1))^2 + (2 - 1)^2} = \sqrt{9 + 1} = \sqrt{10}$

10. $d(P_1, P_2) = \sqrt{(3)^2 + (4)^2} = \sqrt{25} = 5$ 12. $d(P_1, P_2) = \sqrt{(2)^2 + (5)^2} = \sqrt{29}$

14. $d(P_1, P_2) = \sqrt{(10)^2 + (5)^2} = \sqrt{125} = 5\sqrt{5}$

16. $d(P_1, P_2) = \sqrt{(-1.5)^2 + (-1.2)^2} = \sqrt{3.69} \approx 1.92$

18. $d(P_1, P_2) = \sqrt{(-a)^2 + (-a)^2} = \sqrt{2a^2} = \sqrt{2}\,|a|$

20.
$$d(A, C) = \sqrt{(12)^2 + (-16)^2} = \sqrt{400} = 20$$
$$d(A, B) = \sqrt{(14)^2 + (-2)^2} = \sqrt{200} = 10\sqrt{2}$$
$$d(B, C) = \sqrt{(-2)^2 + (-14)^2} = \sqrt{200} = 10\sqrt{2}$$
$$[d(A, C)]^2 = [d(A, B)]^2 + [d(B, C)]^2$$
$$20^2 = \left(10\sqrt{2}\right)^2 + \left(10\sqrt{2}\right)^2$$
$$400 = 200 + 200$$
$$400 = 400$$
$$A = \frac{1}{2}bh$$
$$A = \frac{1}{2}\left(10\sqrt{2}\right)\left(10\sqrt{2}\right)$$
Area = 100 square units

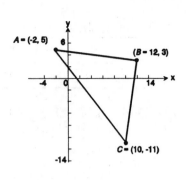

22.
$$d(A, B) = \sqrt{(9)^2 + (-8)^2} = \sqrt{145}$$
$$d(A, C) = \sqrt{(5)^2 + (2)^2} = \sqrt{29}$$
$$d(C, B) = \sqrt{(-4)^2 + (10)^2} = \sqrt{116}$$
$$[d(A, B)]^2 = [d(A, C)]^2 + [d(C, B)]^2$$
$$145 = 29 + 116$$
$$145 = 145$$
$$A = \frac{1}{2}bh$$
$$A = \frac{1}{2}\left(\sqrt{29}\right)\left(\sqrt{116}\right)$$
Area = 29 square units

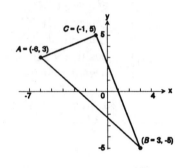

24.
$$d(A, B) = \sqrt{(0)^2 + (4)^2} = 4$$
$$d(A, C) = \sqrt{(-2)^2 + (4)^2} = \sqrt{20}$$
$$= 2\sqrt{5}$$
$$d(C, B) = \sqrt{(2)^2 + (0)^2} = 2$$
$$[d(A, C)]^2 = [d(A, B)]^2 + [d(B, C)]^2$$
$$\left(2\sqrt{5}\right)^2 = 4^2 + 2^2$$
$$20 = 16 + 4$$
$$20 = 20$$
$$A = \frac{1}{2}bh$$
$$A = \frac{1}{2}(4)(2)$$
Area = 4 square units

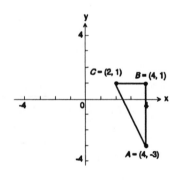

26. $d[(1, 2),(x, -3)] = 13$

$\sqrt{(x - 1)^2 + (-5)^2} = 13$

$\sqrt{(x - 1)^2 + 25} = 13$

$(x - 1)^2 + 25 = 169$

$x^2 - 2x + 1 + 25 = 169$

$x^2 - 2x - 143 = 0$

$(x - 13)(x + 11) = 0$

$x - 13 = 0 \text{ or } x + 11 = 0$

$x = 13 \quad \text{or} \quad x = -11$

Thus, $(13, -3)$ or $(-11, -3)$ are 13 units from $(1, 2)$.

28. $d[(4, 4),(0, y)] = 5$

$\sqrt{(4)^2 + (y - 4)^2} = 5$

$\sqrt{16 + (y - 4)^2} = 5$

$16 + (y - 4)^2 = 25$

$(y - 4)^2 = 9$

$y - 4 = \pm 3$

$y = \pm 3 + 4$

$y = 7 \text{ or } y = 1$

Thus, $(0, 1)$ and $(0, 7)$ are 5 units from $(-4, 4)$.

30. $x = \dfrac{-1 + 2}{2} \quad y = \dfrac{0 + 4}{2}$

$x = \dfrac{1}{2} \quad\quad y = \dfrac{4}{2}$

Midpoint $= \left(\dfrac{1}{2}, 2\right)$

32. $x = \dfrac{2 + 4}{2} \quad y = \dfrac{-3 + 2}{2}$

$x = 3 \quad\quad y = -\dfrac{1}{2}$

Midpoint $= \left(3, -\dfrac{1}{2}\right)$

34. $x = \dfrac{-4 + 2}{2} \quad y = \dfrac{-3 + 2}{2}$

$x = -1 \quad\quad y = -\dfrac{1}{2}$

Midpoint $= \left(-1, -\dfrac{1}{2}\right)$

36. $x = \dfrac{1.2 - 0.3}{2} \quad y = \dfrac{2.3 + 1.1}{2}$

$x = 0.45 \quad\quad y = 1.7$

Midpoint $= (0.45, 1.7)$

38. $x = \dfrac{a + 0}{2} \quad y = \dfrac{a + 0}{2}$

$x = \dfrac{a}{2} \quad\quad y = \dfrac{a}{2}$

Midpoint $= \left(\dfrac{a}{2}, \dfrac{a}{2}\right)$

40. Let $P_1 = (0, 4)$, $P_2 = (0, 0)$, $P = (x, y)$

$d(P_1, P_2) = 4; \; d(P_2, P) = \sqrt{x^2 + y^2} = 4$

$d(P_1, P) = \sqrt{x^2 + (y - 4)^2} = 4$

$x^2 + y^2 = 16$

$x^2 + (y - 4)^2 = 16$

$x^2 + y^2 - 8y + 16 = 16$

$16 - 8y + 16 = 16$

$-8y = -16$

$y = 2$

$x^2 + 2^2 = 16$

$x^2 = 12$

$x = \pm 2\sqrt{3}$

Two triangles are possible.

The third vertex is $\left(2\sqrt{3}, 2\right)$ or $\left(-2\sqrt{3}, 2\right)$

42. $d(P_1, P_2) = \sqrt{(7)^2 + (-2)^2} = \sqrt{53}$

$\quad\quad d(P_2, P_3) = \sqrt{(-2)^2 + (-7)^2} = \sqrt{53}$

$\quad\quad d(P_1, P_3) = \sqrt{(5)^2 + (-9)^2} = \sqrt{106}$

$\quad\quad [d(P_1, P_3)]^2 = [d(P_2, P_3)]^2 + [d(P_1, P_2)]^2$

$\quad\quad\quad\quad\quad 106 = 53 + 53$

$\quad\quad\quad\quad\quad 106 = 106$

Isosceles and right triangle

44. $d(P_1, P_2) = \sqrt{(-11)^2 + (-2)^2} = \sqrt{125} = 5\sqrt{5}$

$\quad\quad d(P_2, P_3) = \sqrt{(8)^2 + (6)^2} = \sqrt{100} = 10$

$\quad\quad d(P_1, P_3) = \sqrt{(-3)^2 + (4)^2} = \sqrt{25} = 5$

$\quad\quad [d(P_1, P_2)]^2 = [d(P_2, P_3)]^2 + [d(P_1, P_3)]^2$

$\quad\quad\quad\quad\quad 125 = 100 + 25$

$\quad\quad\quad\quad\quad 125 = 125$

Right triangle

46. $\quad\quad P_1 = (x_1, y_1) \quad\quad P_2 = (x_2, y_2)$

$$M = \left[\frac{x_1 + x_2}{2}, \frac{y_1 + y_2}{2}\right]$$

$$= \left[\frac{1}{2}(x_1 + x_2), \frac{1}{2}(y_1 + y_2)\right]$$

$$d(P_1, M) = \sqrt{\left[\frac{x_2 - x_1}{2}\right]^2 + \left[\frac{y_2 - y_1}{2}\right]^2}$$

$$d(P_1, P_2) = \sqrt{(x_2 - x_1)^2 + (y_2 - y_1)^2}$$

$$\frac{d(P_1 M)}{d(P_1, P_2)} = \frac{\sqrt{\left[\frac{x_2 - x_1}{2}\right]^2 + \left[\frac{y_2 - y_1}{2}\right]^2}}{\sqrt{(x_2 - x_1)^2 + (y_2 - y_1)^2}} = \frac{1}{2}$$

48. $x = (1 - 0)x_1 + 0x_2$

$\quad\quad x = x_1$

$\quad\quad P = (x_1, y_1) = P_1$

$\quad\quad y = (1 - 0)y_1 + 0y_2$

$\quad\quad y = y_1$

50. $d(P_2, P_1) = \sqrt{(-4)^2 + (-3)^2} = \sqrt{25} = 5$

Using the result in Problem 47 and letting $P = (x, y)$, $r = 3$, we have

$\quad\quad x = (1 - 3) \cdot 0 + 3(-4) = -12$

$\quad\quad y = (1 - 3) \cdot 4 + 3(1) = -5$

Thus, $P = (-12, -5)$. On checking, $d(P, P_1) = \sqrt{(-12)^2 + (-9)^2} = \sqrt{225} = 15 = 3d(P_2, P_1)$

52. $\quad\quad 60^2 + 60^2 = c^2$

$\quad\quad 3600 + 3600 = c^2$

$\quad\quad\quad\quad 7200 = c^2$

$\quad\quad\quad\quad 84.9' = c$

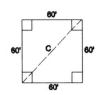

54. (a)

(b) $d = \sqrt{(180 - 60)^2 + (20 - 60)^2}$

$\quad\quad = \sqrt{(120)^2 + (40)^2}$

$\quad\quad \approx 126.5$ feet

(c) $d = \sqrt{(220 - 0)^2 + (220 - 60)^2}$

$\quad\quad = \sqrt{(220)^2 + (160)^2}$

$\quad\quad = 272.0$ feet

56. Let $P = (a, 0)$, $Q = \left[\dfrac{a}{2}, \dfrac{\sqrt{3}\,a}{2}\right]$

$$d(0, P) = \sqrt{a^2} = |a|$$

$$d(P, Q) = \sqrt{\left[-\dfrac{a}{2}\right]^2 + \left[\dfrac{\sqrt{3}\,a}{2}\right]^2} = \sqrt{\dfrac{a^2}{4} + \dfrac{3a^2}{4}} = \sqrt{\dfrac{4a^2}{4}} = |a|$$

$$d(0, Q) = \sqrt{\left[\dfrac{a}{2}\right]^2 + \left[\dfrac{\sqrt{3}\,a}{2}\right]^2} = \sqrt{\dfrac{a^2}{4} + \dfrac{3a^2}{4}} = \sqrt{\dfrac{4a^2}{4}} = |a|$$

The midpoint of O and P is $M_1 = \left[\dfrac{a}{2}, 0\right]$

The midpoint of O and Q is $M_2 = \left[\dfrac{a}{4}, \dfrac{\sqrt{3}\,a}{4}\right]$

The midpoint of P and Q is $M_3 = \left[\dfrac{3a}{4}, \dfrac{\sqrt{3}\,a}{4}\right]$

$$d(M_1, M_2) = \sqrt{\left[\dfrac{a}{4} - \dfrac{2a}{4}\right]^2 + \left[\dfrac{\sqrt{3}\,a}{4}\right]^2} = \sqrt{\dfrac{a^2}{16} + \dfrac{3a^2}{16}} = \sqrt{\dfrac{4a^2}{16}} = \dfrac{|a|}{2}$$

$$d(M_2, M_3) = \sqrt{\left[\dfrac{3a}{4} - \dfrac{a}{4}\right]^2 + \left[\dfrac{\sqrt{3}\,a}{4} - \dfrac{\sqrt{3}\,a}{4}\right]^2} = \sqrt{\left[\dfrac{2a}{4}\right]^2} = \sqrt{\dfrac{a^2}{4}} = \dfrac{|a|}{2}$$

$$d(M_1, M_3) = \sqrt{\left[\dfrac{3a}{4} - \dfrac{2a}{4}\right]^2 + \left[\dfrac{\sqrt{3}\,a}{4}\right]^2} = \sqrt{\dfrac{a^2}{16} + \dfrac{3a^2}{16}} = \sqrt{\dfrac{4a^2}{16}} = \dfrac{|a|}{2}$$

58. $d = \sqrt{(100)^2 + \left[15 \cdot \dfrac{5280}{1} \cdot \dfrac{1}{3600}t\right]^2} = \sqrt{10000 + 484t^2}$

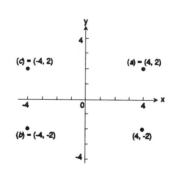

60. $(3, 4)$

62. $(-3, -4)$

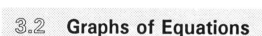

2.

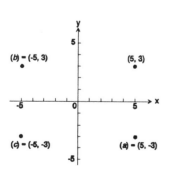

4.

6.

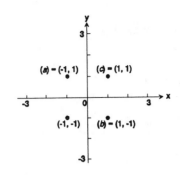

8.

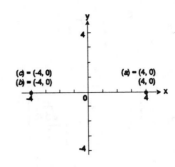

10.

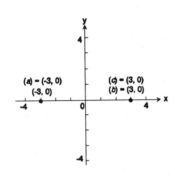

12. (a) $(0, 1)$
    (b) none

14. (a) $(0, 0)$
    (b) origin

16. (a) $(-2, 0)$, $(2, 0)$,
       $(0, -2) (0, 2)$
    (b) x-axis, y-axis, origin

18. (a) $(0, 0)$
    (b) none

20. (a) $(-3, 0)$, $(0, 2)$, $(2, 0)$
    (b) none

22. (a) $(0, 1)$, $(2, 0)$
    (b) none

24. (a) $(0, 0)$
    (b) origin

26. (a) none
    (b) x-axis

28. $(0, 0)$: $0^3 - 2\sqrt{0} = 0$; yes

    $(1, 1)$: $1^3 - 2\sqrt{1} = 1 - 2 = -1 \neq 1$;
               no

    $(1, -1)$: $1^3 - 2\sqrt{1} = 1 - 2 = -1$; yes

30. $(1, 2)$: $2^3 = 8$ $(0, 1)$: $1^3 = 1$ $(-1, 0)$: $0^3 = 0$
           $1 + 1 = 2$ $0 + 1 = 1$ $-1 + 1 = 0$
           No   Yes   Yes

32. $(0, 1)$: $0^2 + 4.1^2 = 4$; yes
    $(2, 0)$: $2^2 + 4.0^2 = 4$; yes

    $\left[2, \dfrac{1}{2}\right]$: $2^2 + 4 \cdot \left[\dfrac{1}{2}\right]^2 = 4 + 1 = 5 \neq 4$; no

34. $y = x^2 + 3x$
$b = 2^2 + 3(2)$
$b = 10$

36. $y = mx + b$
$0 = m(2) + b$ and
$5 = m(0) + b$
$5 = b$
$0 = m(2) + 5$
$-2m = 5$
$m = -\dfrac{5}{2}$

38.

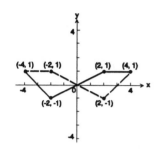

40.

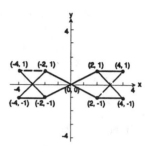

42.

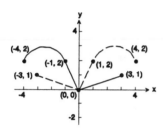

44.

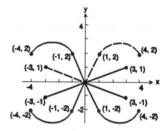

46. $y^2 = x$ Intercept: (0, 0)

x-axis: $(-y)^2 = x$
              $y^2 = x$ Thus, symmetric with respect to x-axis

y-axis: $y^2 = (-x)$
              $y^2 = -x$ Thus, no symmetry with respect to y-axis

origin: $(-y^2) = (-x)$
              $y^2 = -x$ Thus, no symmetry with respect to origin

48. $y = -5x$ Intercept: (0, 0)

x-axis: $(-y) = -5x$
              $-y = -5x$ Thus, no symmetry with respect to y-axis

y-axis: $y = -5(-x)$
              $y = 5x$ Thus, no symmetry with respect to y-axis

origin: $(-y) = -5(-x)$
              $-y = 5x$
              $y = -5x$ Thus, symmetric with respect to origin

50. $y^2 - x - 4 = 0$ Intercepts: $(-4, 0)$, $(0, -2)$, $(0, 2)$

x-axis: $(-y)^2 - x - 4 = 0$
              $y^2 - x - 4 = 0$ Thus, symmetric with respect to x-axis

y-axis: $y^2 - (-x) - 4 = 0$
              $y^2 + x - 4 = 0$ Thus, no symmetry with respect to y-axis

origin: $(-y^2) - (-x) - 4 = 0$
              $y^2 + x - 4 = 0$ Thus, no symmetry with respect to origin

52. $4x^2 + y^2 = 4$ Intercepts: $(-1, 0), (1, 0), (0, -2), (0, 2)$

     x-axis:

$$4x^2 + (-y)^2 = 4$$
$$4x^2 + y^2 = 4$$

Thus, symmetric with respect to *x*-axis

     y-axis:

$$4(-x)^2 + y^2 = 4$$
$$4x^2 + y^2 = 4$$

Thus, symmetric with respect to *y*-axis

     origin:

$$4(-x^2) - (-y)^2 = 4$$
$$4x^2 + y^2 = 4$$

Thus, symmetric with respect to origin

54. $y = x^4 - 1$ Intercepts: $(-1, 0), (1, 0), (0, -1)$

     x-axis: $-y = x^4 - 1$ Thus, no symmetry with respect to *x*-axis

     y-axis:

$$y = (-x)^4 - 1$$
$$y = x^4 - 1$$

Thus, symmetric with respect to *y*-axis

     origin:

$$-y = (-x)^4 - 1$$
$$-y = x^4 - 1$$

Thus, no symmetry with respect to origin

56. $y = x^2 + 4$ Intercept: $(0, 4)$

     x-axis: $-y = x^2 + 4$ Thus, no symmetry with respect to *x*-axis

     y-axis:

$$y = (-x)^2 + 4$$
$$y = x^2 + 4$$

Thus, symmetric with respect to *y*-axis

     origin:

$$-y = (-x)^4 + 4$$
$$-y = x^2 + 4$$

Thus, no symmetry with respect to origin

58. $y = \dfrac{x^2 - 4}{x}$ Intercepts: $(2, 0), (-2, 0)$

     x-axis:

$$-y = \frac{x^2 - 4}{x}$$

Thus, no symmetry with respect to *x*-axis

     y-axis:

$$y = \frac{(-x)^2 - 4}{-x}$$

$$y = \frac{x^2 - 4}{-x}$$

Thus, no symmetry with respect of *y*-axis

     origin:

$$-y = \frac{(-x)^2 - 4}{-x}$$

$$-y = \frac{x^2 - 4}{-x}$$

$$y = \frac{x^2 - 4}{x}$$

Thus, symmetric with respect to origin

3.3   The Straight Line

2.   slope $= \dfrac{1 - 0}{-2 - 0} = -\dfrac{1}{2}$ **4.**   slope $= \dfrac{2 - 1}{2 - (-1)} = \dfrac{1}{3}$

6. slope $= \dfrac{4 - 2}{3 - 4} = \dfrac{2}{-1} = -2$

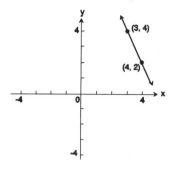

8. slope $= \dfrac{3 - 1}{2 - (-1)} = \dfrac{2}{3}$

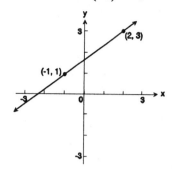

10. slope $= \dfrac{2 - 2}{-5 - 4} = 0$

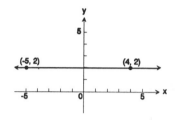

12. slope $= \dfrac{2 - 0}{2 - 2}$

slope undefined

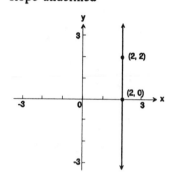

14. slope $= \dfrac{\sqrt{5} - 0}{4 - \left(-2\sqrt{2}\right)} = \dfrac{\sqrt{5}}{4 + 2\sqrt{2}}$

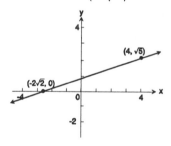

16.

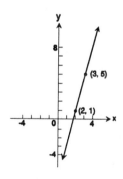

18.

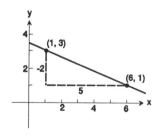

20.

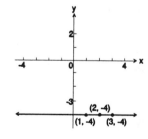

22.

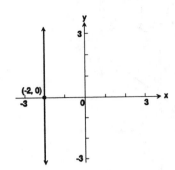

24. $\text{slope} = -\dfrac{1}{2},\ b = 0$

$$y = -\dfrac{1}{2}x$$

$$\dfrac{1}{2}x + y = 0 \text{ or } y = -\dfrac{1}{2}x$$

26. $\text{slope} = \dfrac{1}{3}$

$$y - 1 = \dfrac{1}{3}(x + 1)$$

$$y - 1 = \dfrac{1}{3}x + \dfrac{1}{3}$$

$$\dfrac{1}{3}x - y + \dfrac{4}{3} = 0 \text{ or } y = \dfrac{1}{3}x + \dfrac{4}{3}$$

28.
$$y + 3 = 2(x - 4)$$
$$y + 3 = 2x - 8$$
$$2x - y - 11 = 0 \text{ or } y = 2x - 11$$

30.
$$y - 1 = \dfrac{1}{2}(x - 3)$$
$$y - 1 = \dfrac{1}{2}x - \dfrac{3}{2}$$
$$\dfrac{1}{2}x - y - \dfrac{1}{2} = 0 \text{ or } y = \dfrac{1}{2}x - \dfrac{1}{2}$$

32.
$$m = \dfrac{1}{5}$$
$$y - 5 = \dfrac{1}{5}(x - 2)$$
$$y - 5 = \dfrac{1}{5}x - \dfrac{2}{5}$$
$$\dfrac{1}{5}x - y + \dfrac{23}{5} = 0 \text{ or } y = \dfrac{1}{5}x + \dfrac{23}{5}$$

34.
$$y = -2x - 2$$
$$2x + y + 2 = 0 \text{ or } y = -2x - 2$$

36. $(-4, 0)$ and $(0, 4)$
$$\text{slope} = \dfrac{4}{4} = 1$$
$$y = x + 4$$
$$x - y + 4 = 0 \text{ or } y = x + 4$$

38. $(3, 8)$, slope undefined
$$x = 3$$
$$x - 3 = 0;\ \text{no slope-intercept form}$$

40. $y = -3x + 4$
$\text{slope} = -3,\ y\text{-intercept} = 4$

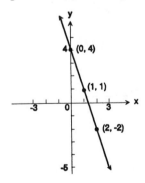

42. $y = -\dfrac{1}{3}x + 2$

$\text{slope} = -\dfrac{1}{3},\ y\text{-intercept} = 2$

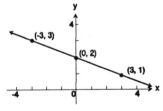

44. $y = 2x + \dfrac{1}{2}$

slope $= 2$, y-intercept $= \dfrac{1}{2}$

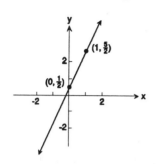

46. $-x + 3y = 6$

$3y = x + 6$

$y = \dfrac{1}{3}x + 2$

slope $= \dfrac{1}{3}$, y-intercept $= 2$

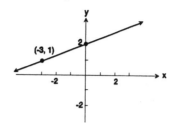

48. $2y = -3x + 6$

$y = -\dfrac{3}{2}x + 3$

slope $= -\dfrac{3}{2}$, y-intercept $= 3$

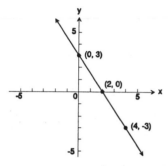

50. $-y = -x + 2$

$y = x - 2$

slope $= 1$, y-intercept $= -2$

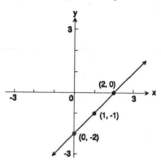

52. $y = -1$

slope $= 0$, y-intercept $= -1$

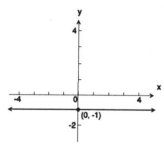

54.

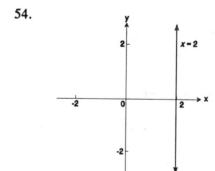

Slope is undefined, no y-intercept

56. $x + y = 0$
$y = -x$

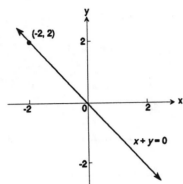

slope $= -1$, y-intercept $= 0$

58. $3x + 2y = 0$
$2y = -3x$
$y = -\dfrac{3}{2}x$

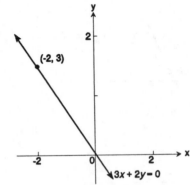

slope $= -\dfrac{3}{2}$, y-intercept $= 0$

60. $x = 0$

62. (a) $K = °C + 273$

(b) $°C = \dfrac{5}{9}(°F - 32)$

$K = \dfrac{5}{9}(°F - 32) + 273$

$K = \dfrac{5}{9}°F - \dfrac{160}{9} + \dfrac{2457}{9}$

$K = \dfrac{5}{9}°F + \dfrac{2297}{9}$

64. (a) $P = 0.55x - 125$

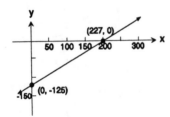

(b) $P = 0.55(1000) - 125$
$= \$425$

(c) $P = 0.55(5000) - 125$
$= \$2625$

66. $(a, 0)$ and $(0, b)$ are points on the line.

slope $= \dfrac{b}{-a}$

$y - 0 = \dfrac{-b}{a}(x - a)$

$y = \dfrac{-bx}{a} + b$

$\dfrac{bx}{a} + y = b$

$\dfrac{x}{a} + \dfrac{y}{b} = 1$

68. (c)

70. (a)

72. $x + y - 1 = 0$ or $y = -x + 1$

74. $x + 2y + 2 = 0$ or $y = -\dfrac{1}{2}x - 1$

76. $2x - 3y = 0$ or $y = \dfrac{2}{3}x$

78. No

80. They are the same line

82. No

2. $y = -5x$
   (a) -5
   (b) $\dfrac{1}{5}$

4. $y = \dfrac{2}{3}x - 1$
   (a) $\dfrac{2}{3}$
   (b) $\dfrac{-3}{2}$

6. $3x + y = 4$
   $\quad\quad y = -3x + 4$
   (a) -3
   (b) $\dfrac{1}{3}$

8. $4x - 3y + 7 = 0$
   $\quad\quad 3y = 4x + 7$
   $\quad\quad y = \dfrac{4}{3} + \dfrac{7}{3}$
   (a) $\dfrac{4}{3}$
   (b) $\dfrac{-3}{4}$

10. $y = 5$ is a horizontal line; slope = 0
    (a) 0
    (b) Since $y = 5$ is horizontal, a line perpendicular would be vertical; slope is undefined.

12. $y - y_1 = m(x - x_1), m = -1$
    $\quad y - 2 = -1(x - 1)$
    $\quad y - 2 = -x + 1$
    $\quad x + y - 3 = 0$ or $y = -x + 3$

14. $y - y_1 = m(x - x_1), m = 1$
    $\quad y - 1 = 1(x + 1)$
    $\quad y - 1 = x + 1$
    $x - y + 2 = 0$ or $y = x + 2$

16. slope = -3
    $y - 2 = -3(x + 1)$
    $y - 2 = -3x - 3$
    $3x + y + 1 = 0$ or $y = -3x - 1$

18. $-2y = -x - 5$
    $\quad y = \dfrac{1}{2}x + \dfrac{5}{2}$
    slope = $\dfrac{1}{2}$
    $y - 0 = \dfrac{1}{2}(x - 0)$
    $\quad y = \dfrac{1}{2}x$
    $\dfrac{1}{2}x - y = 0$ or $y = \dfrac{1}{2}x$

20. slope = 0
    $y - 2 = 0(x - 4)$
    $y - 2 = 0$ or $y = 2$

22. slope = $-\dfrac{1}{2}$
    $y + 2 = -\dfrac{1}{2}(x - 1)$
    $y + 2 = -\dfrac{1}{2}x + \dfrac{1}{2}$
    $\dfrac{1}{2}x + y + \dfrac{3}{2} = 0$ or $y = -\dfrac{1}{2}x - \dfrac{3}{2}$

24. $-2y = -x - 5$
    $\quad y = \dfrac{1}{2}x + \dfrac{5}{2}$
    slope = -2
    $y - 4 = -2(x - 0)$
    $y - 4 = -2x$
    $2x + y - 4 = 0$ or $y = -2x + 4$

26. slope undefined
    $x = 3$
    $x - 3 = 0$; no slope-intercept form

28. Center (1, 2)
    Radius = Distance from (1, 2) to (1, 0) = 2
    $(x - 1)^2 + (y - 2)^2 = 4$

30. Radius = $\frac{1}{2}\sqrt{(2 - 0)^2 + (3 - 1)^2} = \frac{1}{2}\sqrt{8} = \sqrt{2}$

    Center = (1, 2)
    $(x - 1)^2 + (y - 2)^2 = 2$

32. $(x + 2)^2 + (y - 1)^2 = 4$
    $x^2 + 4x + 4 + y^2 - 2y + 1 = 4$
    $x^2 + y^2 + 4x - 2y + 1 = 0$

34. $(x - 1)^2 + y^2 = 9$
    $x^2 - 2x + 1 + y^2 = 9$
    $x^2 + y^2 - 2x - 8 = 0$

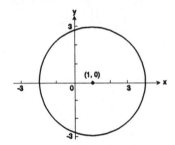

36. $(x - 2)^2 + (y + 3)^2 = 16$
    $x^2 - 4x + 4 + y^2 + 6y + 9 = 16$
    $x^2 + y^2 - 4x + 6y - 3 = 0$

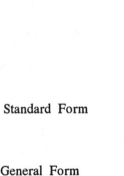

38. $x^2 + y^2 = 9$
    $x^2 + y^2 - 9 = 0$

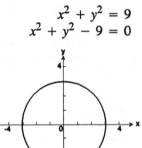

40. $r = \frac{1}{2}$; $(h, k) = \left[0, -\frac{1}{2}\right]$

    $(x - 0)^2 + \left[y + \frac{1}{2}\right]^2 = \frac{1}{4}$ Standard Form

    $x^2 + y^2 + y + \frac{1}{4} = \frac{1}{4}$

    $x^2 + y^2 + y = 0$ General Form

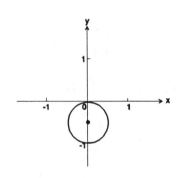

42. $r = 1$; $(h, k) = (0, 1)$

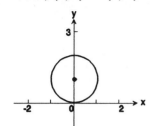

44. $r = \sqrt{2}$; $(h, k) = (-1, 1)$

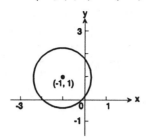

46.
$$x^2 - 6x + y^2 + 2y = -9$$
$$(x^2 - 6x + 9) + (y^2 + 2y + 1) = -9 + 9 + 1$$
$$(x - 3)^2 + (y + 1)^2 = 1$$
$$r = 1;\ (h, k) = (3, -1)$$

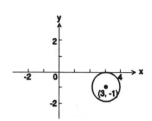

48.
$$x^2 + x + y^2 + y = \frac{1}{2}$$
$$\left[x^2 + x + \frac{1}{4}\right] + \left[y^2 + y + \frac{1}{4}\right] = \frac{1}{2} + \frac{1}{4} + \frac{1}{4}$$
$$\left[x + \frac{1}{2}\right]^2 + \left[y + \frac{1}{2}\right]^2 = 1$$
$$r = 1;\ (h, k) = \left[-\frac{1}{2},\ -\frac{1}{2}\right]$$

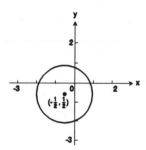

50.
$$2x^2 + 8x + 2y^2 = -7$$
$$x^2 + 4x + y^2 = -\frac{7}{2}$$
$$x^2 + 4x + 4 + y^2 = -\frac{7}{2} + 4$$
$$(x + 2)^2 + y^2 = \frac{1}{2}$$
$$r = \sqrt{\frac{1}{2}}\ ;\ (h, k) = (-2, 0)$$

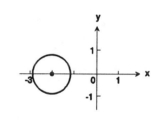

52.
$$(x - 1)^2 + y^2 = r^2$$
$$(-3 - 1)^2 + 2^2 = r^2$$
$$16 + 4 = r^2$$
$$r^2 = 20$$
$$x^2 - 2x + 1 + y^2 = 20$$
$$x^2 + y^2 - 2x - 19 = 0$$

54.
$$(x + 3)^2 + (y - 1)^2 = r^2$$
$$\text{touches } y\text{-axis at } (0, 1)$$
$$(3)^2 + 0 = r^2$$
$$3 = r$$
$$x^2 + 6x + 9 + y^2 - 2y + 1 = 9$$
$$x^2 + y^2 + 6x - 2y + 1 = 0$$

56. $d[(4, 3),\ (0, 1)] = \sqrt{(-4)^2 + (-2)^2}\ = \sqrt{16 + 4}\ = \sqrt{20}\ = 2\sqrt{5}\ = $ diameter
$r = \sqrt{5}$
Center $(2, 2)$
$$(x - 2)^2 + (y - 2)^2 = 5$$
$$x^2 - 4x + 4 + y^2 - 4y + 4 - 5 = 0$$
$$x^2 + y^2 - 4x - 4y + 3 = 0$$

58. $P_1 = (1, -1)$, $P_3 = (2, 2)$, $m = \dfrac{3}{1} = 3$

$P_2 = (4, 1)$, $P_4 = (5, 4)$, $m = \dfrac{3}{1} = 3$

$P_3 = (2, 2)$, $P_4 = (5, 4)$, $m = \dfrac{2}{3}$

$P_1 = (1, -1)$, $P_2 = (4, 1)$, $m = \dfrac{2}{3}$

Opposite sides are parallel.

60. $P_1 = (0, 0)$, $P_2 = (1, 3)$, $m = 3$
$P_3 = (4, 2)$, $P_4 = (3, -1)$, $m = 3$

$P_2 = (1, 3)$, $P_3 = (4, 2)$, $m = \dfrac{-1}{3}$

$P_1 = (0, 0)$, $P_4 = (3, -1)$, $m = \dfrac{-1}{3}$

Opposite sides are parallel; adjacent sides are perpendicular

$d(P_1, P_2) = \sqrt{1^2 + 3^2} = \sqrt{10}$

$d(P_2, P_3) = \sqrt{(4 - 1)^2 + (2 - 3)^2} = \sqrt{3^2 + (-1)^2} = \sqrt{10}$

$d(P_3, P_4) = \sqrt{(3 - 4)^2 + (-1 - 2)^2} = \sqrt{(-1)^2 + (-3)^2} = \sqrt{10}$

$d(P_4, P_1) = \sqrt{3^2 + (-1)^2} = \sqrt{10}$

All sides have equal lengths. Thus, the quadrilateral is a square.

62. (d)

64. (a)

66. $(x - 4)^2 + (y + 2)^2 = 9$

68. $(x - 1)^2 + (y - 3)^2 = 4$

70. $b = 3$, $a = 2$
$A = 3 \cdot 2 = 6$ square units

72. $b = 1$, $a = 2$
$A = 1 \cdot 2 = 2$ square units

74. $b = \sqrt{5}$, $a = \sqrt{5}$
$A = \sqrt{5} \cdot \sqrt{5} = 5$ square units

76. Earth:
$$x^2 + y^2 + 2x + 4y - 4091 = 0$$
$$(x^2 + 2x \qquad) + (y^2 + 4y \qquad) = 4091$$
$$(x^2 + 2x + 1) + (y^2 + 4y + 4) = 4091 + 1 + 4$$
$$(x + 1)^2 + (y + 2)^2 = 4096$$
$$(h, k) = (-1, -2)$$
$$r = \sqrt{4096}$$
$$= 64$$

Orbit of satellite has center at $(-1, -2)$
$r = 0.6 + 64$
$$[x - (-1)]^2 + [y - (-2)]^2 = (64.6)^2$$
$$(x + 1)^2 + (y + 2)^2 = 4173.16$$
$$x^2 + 2x + 1 + y^2 + 4y + 4 = 4173.16$$
$$x^2 + y^2 + 2x + 4y - 4168.16 = 0$$

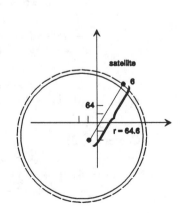

78. $x^2 + y^2 = 9$

Center $(0, 0)$

Slope from center to

$\left(1, 2\sqrt{2}\right)$ is $\dfrac{2\sqrt{2}}{1} = 2\sqrt{2}$

Slope of tangent line is $\dfrac{-1}{2\sqrt{2}} = \dfrac{-\sqrt{2}}{4}$

Thus, the equation of the tangent line is

$$y - 2\sqrt{2} = \dfrac{-\sqrt{2}}{4}(x - 1)$$

$$\sqrt{2}x + 4y - 9\sqrt{2} = 0$$

80. Let (h, k) be the center

$$x - 2y + 4 = 0$$
$$2y = x + 4$$
$$y = \dfrac{1}{2}x + 2$$

Slope of tangent line is $\dfrac{1}{2}$

Slope from (h, k) to $(0, 2)$ is -2

$$\dfrac{2 - k}{-h} = -2$$
$$2 - k = 2h$$
$$1 - \dfrac{1}{2}k = h$$

$$y = 2x - 7$$

Slope of tangent line is 2

Slope from (h, k) to $(3, -1)$ is $\dfrac{-1}{2}$

$$\dfrac{-1 - k}{3 - h} = \dfrac{-1}{2}$$

$$\dfrac{-1 - k}{3 - \left[1 - \dfrac{1}{2}k\right]} = -\dfrac{1}{2}$$

$$\dfrac{-1 - k}{2 + \dfrac{1}{2}k} = \dfrac{-1}{2}$$

$$-2 - \dfrac{1}{2}k = -2 - 2k$$

$$\dfrac{-1}{2}k = -2k$$

$$k = 0$$

$$h = 1 - \dfrac{1}{2}k = 1 - 0 = 1$$

Thus, $(h, k) = (1, 0)$

82. Slope from (a, b) to (b, a) is $\dfrac{a - b}{b - a} = -1$

Slope of the line $y = x$ is 1

Since $-1 \cdot 1 = -1$, the line containing the points (a, b) and (b, a) is perpendicular to the line $y = x$.

The midpoint of (a, b) and $(b, a) = \left[\dfrac{a + b}{2}, \dfrac{b + a}{2}\right]$

Since $\dfrac{b + a}{2} = \dfrac{a + b}{2}$, the midpoint lies on the line $y = x$.

84. The family of lines $Cx + y + 4 = 0$ intersect at the point $(0, -4)$.

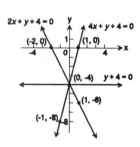

2.
$$v = kt$$
$$16 = k(2)$$
$$8 = k$$
$$v = 8t$$

4.
$$V = kx^3$$
$$36\pi = k(3)^3$$
$$36\pi = 27k$$
$$\frac{4\pi}{3} = k$$
$$V = \frac{4\pi}{3}x^3$$

6.
$$y = \frac{k}{\sqrt{x}}$$
$$4 = \frac{k}{\sqrt{9}}$$
$$4 = \frac{1}{3}k$$
$$12 = k$$
$$y = \frac{12}{\sqrt{x}}$$

8.
$$T = k\sqrt[3]{x} \; d^2$$
$$18 = k\sqrt[3]{8} \cdot 3^2$$
$$18 = 18k$$
$$1 = k$$
$$T = d^2 \sqrt[3]{x}$$

10.
$$z = k(x^3 + y^2)$$
$$1 = k(2^3 + 3^2)$$
$$1 = 17k$$
$$\frac{1}{17} = k$$
$$z = \frac{1}{17}(x^3 + y^2)$$

12.
$$z^3 = k(x^2 + y^2)$$
$$2^3 = k(9^2 + 4^2)$$
$$8 = k(97)$$
$$\frac{8}{97} = k$$
$$z^3 = \frac{8}{97}(x^2 + y^2)$$

14. Letting c denote the hypotenuse and a, b denote the legs of a right triangle,
$$c^2 = 1(a^2 + b^2)$$
$$c^2 = a^2 + b^2$$

16. Letting p denote perimeter and ℓ and w denote the lengths of the sides of a rectangle,
$$p = 2(\ell + w)$$

18. Letting V denote volume, r denote radius, and h denote height of a right circular cone,
$$V = \frac{\pi}{3}r^2h$$

20. Letting T denote period and ℓ denote length of a simple pendulum,
$$T = \frac{2\pi}{\sqrt{32}} \sqrt{\ell}$$

22. Letting v denote the velocity of a falling object and t denote the time of the fall,
$$v = kt$$
$$\frac{64 \text{ ft}}{\text{sec}} = k(2 \text{ sec})$$
$$k = \frac{32 \text{ ft}}{\text{sec}^2}$$
$$v = \frac{32 \text{ ft}}{\text{sec}^2} \cdot 3 \text{ sec}$$
$$v = \frac{96 \text{ ft}}{\text{sec}}$$

24. Letting v denote the rate of vibration of a string and ℓ denote the length of the string,
$$v = \frac{k}{\ell}$$
$$\frac{256 \text{ times}}{\text{second}} = \frac{k}{48 \text{ inches}}$$
$$12{,}288 = k$$
$$576 = \frac{12{,}288}{\ell}$$
$$576\ell = 12{,}288$$
$$\ell = \frac{64}{3} \text{ inches}$$

26. Letting f denote the force exerted by the wind, a denote area, and v denote the velocity of the wind,

$$f = kav^2$$
$$11 = 20(22)^2(k)$$
$$11 = 9680\,k$$
$$k \approx 0.0011364$$
$$f = (47.125)(36.5)^2(0.0011364)$$
$$f = 71.346$$

28. Letting w denote the weight of a body and d denote the distance from the center of the earth,

$$w = \frac{k}{d^2}$$
$$200 = \frac{k}{4000^2}$$
$$32 \cdot 10^8 = k$$
$$w = \frac{32 \cdot 10^8}{4001^2}$$
$$w = 199.9 \text{ lbs.}$$

30. Letting r denote the electrical resistance of a wire, ℓ denote the length and d the diameter of the wire,

$$r = \frac{k\ell}{d^2}$$
$$1.24 = \frac{k(432)}{4^2}$$
$$1.24 = 27k$$
$$\frac{31}{675} = k$$
$$1.44 = \frac{31}{675} \cdot \ell \cdot \frac{1}{3^2}$$
$$1.44 = \frac{31\ell}{6075}$$
$$\ell = \frac{8748}{31} \approx 282.2$$

32. Letting s denote the maximum safe load for a horizontal rectangular beam, and w denote the width, t the thickness, and ℓ the length of the beam,

$$s = \frac{kwt^2}{\ell}$$
$$750 = \frac{k \cdot 2 \cdot 4^2}{8}$$
$$750 = \frac{32k}{8}$$
$$k = 187.5$$
$$s = \frac{187.5 \cdot 2 \cdot 6}{10}$$
$$s = 1350$$

34.
$$V = \frac{kt}{P}$$
$$100 = k\frac{300}{15}$$
$$k = \frac{15}{300}(100) = 5$$
$$V = \frac{ST}{P}$$
$$80 = \frac{5(310)}{P}$$
$$80P = 1550$$
$$P = 19.4 \text{ atmospheres}$$

36.
$$v = \sqrt{gr}$$
$$v = \sqrt{79,036(4,000 + 500)}$$
$$v = \sqrt{79,036(4,500)}$$
$$v = \sqrt{355,662,000}$$
$$v \approx 18,859 \text{ miles per hour}$$

38. Letting s denote the distance of a satellite from the surface of the earth,

$$18,630 = \sqrt{79,036(4000 + s)}$$
$$18,630 = \sqrt{316,144,000 + 79,036s}$$
$$347,076,900 = 316,144,000 + 79,036s$$
$$30,932,900 = 79,036s$$
$$391.4 \text{ miles} = s$$

40. $V = \sqrt{gr}$ where $r = 4000 + h$ feet and $V = \dfrac{\text{circumference}}{\text{time}}$

$$\frac{2\pi(4000 + h)}{24 \cdot 3600} = \sqrt{g}\ \sqrt{4000 + h} \quad \text{(Change to seconds)}$$

$$\sqrt{4000} + h = 12\sqrt{g}\,\pi \cdot 3600$$

$$4000 + h = 12\sqrt{32}\,\pi \cdot 3600$$

$$h = 763730 \text{ feet} = 144.6 \text{ miles}$$

42. $F = \dfrac{mv^2}{r}$

$F = \dfrac{150(120000)^2}{100}$

$F = 2.16 \times 10^{10}$ newtons

44. $2.16 \times 10^{10} = \dfrac{150v^2}{50}$

$2.16 \times 10^{10} = 3v^2$

84.8 kilometers per hour $= v$

46. $F = \dfrac{mv^2}{2r}$

$2rF = mv^2$

More; twice as much.

3 Chapter Review

2. $y - 4 = 0(x + 5)$
$y - 4 = 0$ or $y = 4$

4. $(2, 0)$ and $(4, -5)$

slope $= -\dfrac{5}{2}$

$y = -\dfrac{5}{2}(x - 2)$

$y = -\dfrac{5}{2}x + 5$

$\dfrac{5}{2}x + y - 5 = 0$ or $y = -\dfrac{5}{2}x + 5$

6. slope $= -\dfrac{5}{1} = -5$

$y + 4 = -5(x - 3)$

$y + 4 = -5x + 15$

$5x + y - 11 = 0$ or $y = -5x + 11$

8. $y = -x + 2$

slope $= -1$

$y + 3 = -(x - 1)$

$y + 3 = -x + 1$

$x + y + 2 = 0$ or $y = -x - 2$

10. $3x - y + 4 = 0$
$$-y = -3x - 4$$
$$y = 3x + 4$$
$$\text{slope} = 3$$

Slope of perpendicular line is $-\dfrac{1}{3}$

$$y - 4 = -\frac{1}{3}(x + 2)$$
$$y - 4 = -\frac{1}{3}x - \frac{2}{3}$$
$$\frac{1}{3}x + y - \frac{10}{3} = 0 \text{ or } y = -\frac{1}{3}x + \frac{10}{3}$$

12. $3x + 4y - 12 = 0$
x-intercept: $(4, 0)$
y-intercept: $(0, 3)$

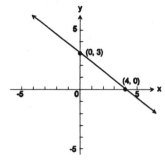

14. $\dfrac{-3}{4}x + \dfrac{1}{2}y = 0$
intercept: $(0, 0)$

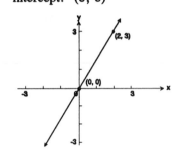

16. $\dfrac{x}{3} + \dfrac{y}{4} = 1$
x-intercept: $(3, 0)$
y-intercept: $(0, 4)$

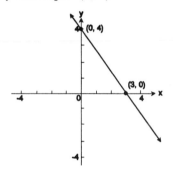

18. $$x^2 + 4x + y^2 - 4y = 1$$
$$x^2 + 4x + 4 + y^2 - 4y + 4 = 1 + 4 + 4$$
$$(x + 2)^2 + (y - 2)^2 = 9$$
$$(h, k) = (-2, 2), \text{ radius} = 3$$

20. $$2x^2 - 4x + 2y^2 = 0$$
$$x^2 - 2x + y^2 = 0$$
$$x^2 - 2x + 1 + y^2 = 0 + 1$$
$$(x - 1)^2 + y^2 = 1$$
$$(h, k) = (1, 0), \text{ radius} = 1$$

22. $(2, 5)$ and $(6, -3)$
$$\text{slope} = \frac{-3 - 5}{6 - 2} = \frac{-8}{4} = -2$$
$$\text{distance} = \sqrt{(2 - 6)^2 + [5 - (-3)]^2} = \sqrt{(-4)^2 + 8^2}$$
$$= \sqrt{16 + 64} = \sqrt{80} = 4\sqrt{5}$$
$$\text{midpoint:} \quad x = \frac{2 + 6}{2} = \frac{8}{2} = 4, \quad y = \frac{5 - 3}{2} = \frac{2}{2} = 1$$

Midpoint is $(4, 1)$

24. The lines are perpendicular so their slopes have a product of -1. The x-intercepts are both positive; the y-intercepts are opposite in sign.
   (a)   The slopes do not have a product equal to -1.
   (b)   The slopes have a product of -1, but the x-intercept of each one is 0.
   (c)   The slopes do not have a product equal to -1.
   (d)   These might be the lines.
   (e)   These lines are perpendicular; each has a negative y-intercept though.
   The only possibility is (d).

26. $y = 5x$ Intercept: (0, 0)
   x-axis: $-y = 5x$ Thus, not symmetric with respect to x-axis
   y-axis: $y = -5x$ Thus, not symmetric with respect to y-axis
   origin: $-y = -5x$
                   $y = 5x$ Thus, symmetric with respect to origin

28. $9x^2 - y^2 = 9$ Intercepts: $(1, 0)$, $(-1, 0)$
   x-axis: $9x^2 - (-y)^2 = 9$
                       $9x^2 - y^2 = 9$ Thus, symmetric with respect to x-axis
   y-axis: $9(-x)^2 - y^2 = 9$
                       $9x^2 - y^2 = 9$ Thus, symmetric with respect to y-axis
   origin: $9(-x)^2 - (-y)^2 = 9$
                       $9x^2 - y^2 = 9$ Thus, symmetric with respect to origin

30. $y = x^3 - x$ Intercepts: $(-1, 0)$, $(0, 0)$, $(1, 0)$
   x-axis: $-y = x^3 - x$ Thus, not symmetric with respect to x-axis
   y-axis: $y = (-x)^3 - (-x)$
                   $y = -x^3 + x$ Thus, not symmetric with respect to y-axis
   origin: $-y = -x^3 + x$
                   $y = x^3 - x$ Thus, symmetric with respect to origin

32. $x^2 + 4x + y^2 - 2y = 0$ Intercepts: $(-4, 0)$, $(0, 0)$, $(0, 2)$
   x-axis: $x^2 + 4x + (-y)^2 - 2(-y) = 0$
                       $x^2 + 4x + y^2 + 2y = 0$ Thus, not symmetric with respect to x-axis
   y-axis: $(-x)^2 + 4(-x) + y^2 - 2y = 0$
                       $x^2 - 4x + y^2 - 2y = 0$ Thus, not symmetric with respect to y-axis
   origin: $(-x)^2 - 4(-x) + (-y)^2 - 2(-y) = 0$
                       $x^2 + 4x + y^2 + 2y = 0$ Thus, not symmetric with respect to origin

34.   Letting p denote pitch in a vibrating string and t denote the tension of the string,

$$p = k\sqrt{t}$$
$$300 = k\sqrt{9}$$
$$300 = 3k$$
$$100 = k$$
$$400 = 100\sqrt{t}$$
$$4 = \sqrt{t}$$
$$t = 16 \text{ pounds}$$

36.   Letting T denote the period of revolution of a planet and a denote the mean distance of the planet from the sun,

$$T^2 = ka^3$$
$$T^2 = \frac{a^3}{8.04 \cdot 10^{23}}$$
$$\left(5\sqrt{5}\right)^2 = \frac{a^3}{8.04 \cdot 10^{23}}$$
$$a^3 = 1.005 \cdot 10^{26}$$
$$a = 4.65 \cdot 10^8$$

Jupiter is 465 million miles from the sun.

38. We need to show that $\overline{M_1M_2}$ and $\overline{BA}$ are parallel.

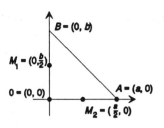

$$\text{slope of } \overline{M_1M_2} = \frac{\dfrac{-b}{2}}{\dfrac{a}{2}} = \frac{-b}{a}$$

$$\text{slope of } \overline{BA} = \frac{-b}{a}$$

Because the two lines have equal slopes, they are parallel.

40. $A = (-2, 0)$, $B = (-4, 4)$ and $C = (8, 5)$

(a) $d(A, B) = \sqrt{[-4-(-2)]^2 + (4 - 0)^2} = \sqrt{4 + 16} = \sqrt{20} = 2\sqrt{5}$

$d(B, C) = \sqrt{[8 - (-4)]^2 + (5 - 4)^2} = \sqrt{144 + 1} = \sqrt{145}$

$d(A, C) = \sqrt{[8 - (-2)]^2 + (5 - 0)^2} = \sqrt{100 + 25} = \sqrt{125} = 5\sqrt{5}$

Since $d[A, B]^2 + d[A, C]^2 = d[B, C]^2$

$$\sqrt{20}^2 + \sqrt{125}^2 = \sqrt{145}^2$$
$$20 + 125 = 145$$

Then by the converse of the Pythagorean Theorem, the points A, B, and C are vertices of a right triangle.

(b) Slope of $AB = \dfrac{4}{-2} = -2$

Slope of $BC = \dfrac{1}{12}$

Slope of $AC = \dfrac{5}{10} = \dfrac{1}{2}$

Since $(-2)\left[\dfrac{1}{2}\right] = -1$, the lines AB and AC are perpendicular and hence form a right angle.

42. $A = (1, 5)$, $B = (2, 4)$ and $C = (-3, 5)$, $(h, k) = (-1, 2)$
$(x - h)^2 + (y - k)^2 = r^2$ or $(x + 1)^2 + (y - 2)^2 = r^2$
If $(1, 5)$ lies on the circle, then
$$(1 + 1)^2 + (5 - 2)^2 = r^2$$
$$2^2 + 3^2 = r^2$$
$$13 = r^2$$

The radius is $\sqrt{13}$
$$(2 + 1)^2 + (4 - 2)^2 = 3^2 + 2^2 = 13$$
Thus, $B = (2, 4)$ lies on the circle.
$$(-3 + 1)^2 + (5 - 2)^2 = 2^2 + 3^2 = 13$$
Thus, $C = (-3, 5)$ lies on the circle.

44.
$$d[(-3, 2), (5, y)] = 10$$
$$\sqrt{[5 - (-3)]^2 + (y - 2)^2} = 10$$
$$\sqrt{64 + (y - 2)^2} = 10$$
$$64 + (y - 2)^2 = 100$$
$$y^2 - 4y + 68 = 100$$
$$y^2 - 4y - 32 = 0$$
$$(y - 8)(y + 4) = 0$$
$$y - 8 = 0 \quad \text{or} \quad y + 4 = 0$$
$$y = 8 \quad \text{or} \quad y = -4$$

46. (a) y-axis
     (b) x-axis
     (c)   line through origin, slope -1
     (d)   the x-axis or the y-axis or both
     (e)   the origin

Mission Possible

1.   (Graph of data points)

2.   (On graph) Perhaps they tried to keep an equal number of points above and below the line. Perhaps they chose to ignore certain "anomalous" points.

3.   Approximate equations: Men's: $y = 20.8 - .027x$. Women's: $y = 24.2 - .071x$.

4.   The two lines appear to cross near the year 2024. Solving the two equations given above algebraically gives an intersection at $y = 77.28$ (which needs to be added to the 1948 at the y-axis to give a year of 2025.28) and $x = 18.71$ seconds.

5.   The graphing calculator should match the results in #4.

6.   Answers may vary. Mention might be made of the fact that women's times may have decreased more sharply over the last few decades because more women are competing and their training has intensified; this might have caused an anomalous decrease. Also, doctors of sports medicine say that men and women can compete equally if they are carrying the same amount of muscle mass; this is different from matching weight since women generally have a higher percentage of fat to muscle. In other words, a woman must weigh more than a man in order to have the possibility of equal muscle mass. Anatomically, women tend to be shorter than men and to have wider hips; these facts can also affect running times because of the physics of leg length and center of gravity.

FUNCTIONS AND THEIR GRAPHS

4.1 Functions

2. $f(x) = 2x^2 + x - 1$
   (a) $f(0) = 2(0)^2 + 0 - 1 = -1$
   (b) $f(1) = 2(1)^2 + 1 - 1 = 2$
   (c) $f(-1) = 2(-1)^2 - 1 - 1 = 0$
   (d) $f(3) = 2(3)^2 + 3 - 1 = 20$

4. $f(x) = \dfrac{x^2 - 1}{x + 4}$
   (a) $f(0) = \dfrac{(0)^2 - 1}{0 + 4} = \dfrac{-1}{4}$
   (b) $f(1) = \dfrac{(1)^2 - 1}{1 + 4} = \dfrac{0}{5} = 0$
   (c) $f(-1) = \dfrac{(-1)^2 - 1}{(-1) + 4} = \dfrac{0}{3} = 0$
   (d) $f(3) = \dfrac{(3)^2 - 1}{3 + 4} = \dfrac{8}{7}$

6. $f(x) = \sqrt{x^2 + x}$
   (a) $f(0) = \sqrt{(0)^2 + 0} = 0$
   (b) $f(1) = \sqrt{(1)^2 + 1} = \sqrt{2}$
   (c) $f(-1) = \sqrt{(-1)^2 - 1} = 0$
   (d) $f(3) = \sqrt{(3)^2 + 3} = \sqrt{12} = 2\sqrt{3}$

8. $f(x) = 1 - \dfrac{1}{(x + 2)^2}$
   (a) $f(0) = 1 - \dfrac{1}{(0 + 2)^2} = 1 - \dfrac{1}{4} = \dfrac{3}{4}$
   (b) $f(1) = 1 - \dfrac{1}{(1 + 2)^2} = 1 - \dfrac{1}{9} = \dfrac{8}{9}$
   (c) $f(-1) = 1 - \dfrac{1}{(-1 + 2)^2} = 1 - 1 = 0$
   (d) $f(3) = 1 - \dfrac{1}{(3 + 2)^2} = 1 - \dfrac{1}{25} = \dfrac{24}{25}$

10. $f(6) = 0$ and $f(11) = 1$

12. $f(8)$ is negative

14. $f(x) > 0$ when $-3 < x < 6$ and $10 < x \le 11$

16. The range of f is $\{y \mid -3 \le y \le 4\}$

18. 3

20. The line $y = 3$ intersects the graph twice.

22. $f(x) = \dfrac{x^2 + 2}{x + 4}$

    (a) $\dfrac{3}{5} \overset{?}{=} \dfrac{1^2 + 2}{1 + 4}$

         $\dfrac{3}{5} = \dfrac{3}{5}$

        Yes, $\left(1, \dfrac{3}{5}\right)$ is on the graph of f.

    (b) $f(0) = \dfrac{0^2 + 2}{0 + 4}$

        $f(0) = \dfrac{1}{2}$

    (c) $\dfrac{1}{2} = \dfrac{x^2 + 2}{x + 4}$

        $x + 4 = 2x^2 + 4$

        $2x^2 - x = 0$

        $x(2x - 1) = 0$

        $x = 0$ or $2x - 1 = 0$

        $x = 0$ or $x = \dfrac{1}{2}$

    (d) Domain of $f = \{x \mid x \neq -4\}$

24. $f(x) = \dfrac{2x}{x - 2}$

    (a) $\dfrac{-2}{3} \overset{?}{=} \dfrac{2\left[\frac{1}{2}\right]}{\frac{1}{2} - 2}$

        $\dfrac{-2}{3} \overset{?}{=} \dfrac{1}{\frac{-3}{2}}$

        $\dfrac{-2}{3} = \dfrac{-2}{3}$

        Yes, $\left(\dfrac{1}{2}, \dfrac{-2}{3}\right)$ is on the graph of f.

    (b) $f(4) = \dfrac{2 \cdot 4}{4 - 2} = \dfrac{8}{2}$

        $f(4) = 4$

    (c) $1 = \dfrac{2x}{x - 2}$

        $x - 2 = 2x$

        $-2 = x$

    (d) Domain of $f = \{x \mid x \neq 2\}$

26. Function   (a) Domain: all real numbers; Range: $\{y \mid 0 < y < \infty\}$
                 (b) (0, 1)
                 (c) None

28. Function   (a) Domain: $\left\{x \mid -\dfrac{\pi}{2} < x < \dfrac{\pi}{2}\right\}$; Range: all real numbers
                 (b) (0, 0)
                 (c) Origin

30. Not a function

32. Function   (a) Domain: $\{x \mid 0 \leq x \leq 4\}$; Range: $\{y \mid 0 \leq y \leq 4\}$
                 (b) (0, 0)
                 (c) None

34. Function   (a) Domain: $\{x \mid -3 \leq x < \infty\}$; Range: $\{y \mid 0 \leq y < \infty\}$
                 (b) $(-3, 0)$, $(0, 2)$, $(2, 0)$
                 (c) None

36. Function   (a) Domain: $\{x \mid x \neq -1\}$; Range: $\{y \mid 0 < y < \infty\}$
                 (b) (0, 2)
                 (c) None

38. $f(x) = 5x^2 + 2$
    The domain is the set of all real numbers.

40. $f(x) = \dfrac{x^2}{x^2 + 1}$
    The domain is the set of all real numbers.

42. $h(x) = \dfrac{x}{x - 1}$

The domain is the set of all real numbers x except $x = 1$, or $\{x \mid x \neq 1\}$.

44. $G(x) = \dfrac{x + 4}{x^3 - 4x}$

$x^3 - 4x \neq 0$

$x(x^2 - 4) \neq 0$

$x \neq 0$ or $x^2 - 4 \neq 0$

                $(x - 2)(x + 2) \neq 0$

                    $x \neq 2$ or $x \neq -2$

The domain is the set of all real numbers x except $x = 0$, $x = -2$, or $x = 2$, or $\{x \mid x \neq 0, x \neq -2, x \neq 2\}$.

46. $G(x) = \sqrt{1 - x}$

We require that $1 - x \geq 0$

            $-x \geq -1$

             $x \leq 1$

The domain is $(-\infty, 1]$.

48. $f(x) = \dfrac{1}{\sqrt{x^2 - 4}}$

We require that $x^2 - 4 > 0$

           $x^2 > 4$

       $x < -2$ or $x > 2$

The domain is $(-\infty, -2)$ or $(2, \infty)$.

50. $g(x) = \sqrt{x^2 - x - 2}$

We require that $x^2 - x - 2 \geq 0$

        $(x - 2)(x + 1) \geq 0$

Thus, $x - 2 \geq 0$ and $x + 1 \geq 0$ or $x - 2 \leq 0$ and $x + 1 \leq 0$

               $x \geq 2$ and $x \geq -1$ or $x \leq 2$ and $x \leq -1$

                      $x \geq 2$ or $x \leq -1$

The domain is $(-\infty, -1]$ or $[2, \infty)$.

52. $f(x) = 3x^2 - Bx + 4$

$f(-1) = 3(-1)^2 - B(-1) + 4 = 12$

               $3 + B + 4 = 12$

                   $7 + B = 12$

                      $B = 5$

54. $f(x) = \dfrac{(2x - B)}{3x + 4}$

$f(2) = \dfrac{4 - B}{6 + 4} = \dfrac{1}{2}$

        $\dfrac{4 - B}{10} = \dfrac{1}{2}$

      $8 - 2B = 10$

         $-2B = 2$

            $B = -1$

56. $f(x) = \dfrac{(x - B)}{(x - A)}$

$f(2) = \dfrac{2 - B}{2 - A} = 0$ and

$f(1) = \dfrac{1 - B}{1 - A}$ is undefined.

Since $f(1)$ is undefined, $A = 1$.

$\dfrac{2 - B}{2 - 1} = 0 \Rightarrow 2 - B = 0 \Rightarrow B = 2$

58. $H(x) = 20 - 13x^2$

(a) $H(1) = 20 - 13(1)^2 = 7$ meters

       $H(1.1) = 20 - 13(1.1)^2 = 4.27$ meters

       $H(1.2) = 20 - 13(1.2)^2 = 1.28$ meters

(b) $H(x) = 20 - 13x^2 = 0$

              $-13x^2 = -20$

                  $x^2 = \dfrac{20}{13}$

                    $x = \sqrt{\dfrac{20}{13}} \approx 1.24$ seconds

60. $A(x) = \dfrac{1}{2}x^2$

62. $G(x) = 100 + 10x$

64.   The time is least for $x = 1.50$ mi.

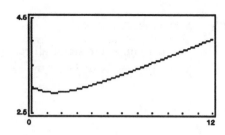

66.   (a) $T(x) = \dfrac{\sqrt{9 + x^2}}{12} + \dfrac{20 - x}{5}$

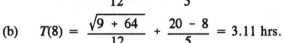

       (b) $T(8) = \dfrac{\sqrt{9 + 64}}{12} + \dfrac{20 - 8}{5} = 3.11$ hrs.

       (c) $T(12) = \dfrac{\sqrt{9 + 144}}{12} + \dfrac{20 - 12}{5} = 2.63$ hrs.

       (d)   The time is least for $x = 20$ miles

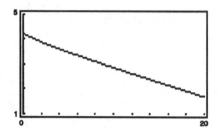

68. $C(x) = 100 + \dfrac{x}{10} + \dfrac{36,000}{x}$

       (a) $C(500) = 100 + \dfrac{500}{10} + \dfrac{36,000}{500} = 100 + 50 + 72 = \$222$
              The cost per passenger is \$222.

       (b) $C(450) = 100 + \dfrac{450}{10} + \dfrac{36,000}{450} = 100 + 45 + 80 = \$225$
              The cost per passenger is \$225.

       (c) $C(600) = 100 + \dfrac{600}{10} + \dfrac{36,000}{600} = 100 + 60 + 60 = \$220$
              The cost per passenger is \$220.

       (d) $C(400) = 100 + \dfrac{400}{10} + \dfrac{36,000}{400} = 100 + 60 + 60 = \$220$
              The cost per passenger is \$230.

       (e)   The cost is least for $x = 600$ mi/hr

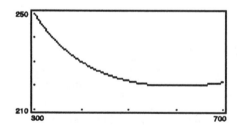

70. $W(h) = m\left[\dfrac{4000}{4000 + h}\right]^2$

       If $m = 120$, and $h = 14,110$ feet $= 2.67$ miles, then

       $$W(2.67) = 120\left[\dfrac{4000}{4000 + 2.67}\right]^2 \approx 119.8$$

       On Pike's Peak, 14,110 feet above sea level, a 120 pound woman weighs above 119.8 pounds.

72. $y = x^3 - 3x$ is a function since no two ordered pairs have the same first element.

74. $y = \dfrac{3}{x} - 3$ is a function since no two ordered pairs have the same first element.

76. $y = \pm\sqrt{1 - 2x}$
Since $(0, 1)$ and $(0, -1)$ are both ordered pairs defined by this equation, it is not a function.

78. $x + 2y^2 = 1$
Since $(-1, 1)$ and $(-1, -1)$ are both ordered pairs defined by this equation, it is not a function.

84. Yes

4.2 More about Functions

2. A   **4.** G   **6.** D   **8.** H

10. (a) Domain: $\{x \mid -2 \leq x \leq 3\}$; Range: $\{y \mid -1 \leq y \leq 3\}$
    (b) In interval notation,
         increasing on $[-2, 0]$ and on $[1, 3]$
         decreasing on $[0, 1]$.
         In inequality notation,
         increasing on $-2 \leq x \leq 0$ and on $1 \leq x \leq 3$
         decreasing on $0 \leq x \leq 1$
    (c) Neither
    (d) $(-2, 0)$, $(0, 3)$, $(1, 0)$

12. (a) Domain: $\{x \mid 0 < x < \infty\}$;
         Range: all real numbers
    (b) In interval notation, increasing on
         $(0, \infty)$
         In inequality notation, increasing on
         $0 < x < \infty$.
    (c) Neither
    (d) $(1, 0)$

14. (a) Domain: $\left\{x \mid -\dfrac{\pi}{2} < x < \dfrac{\pi}{2}\right\}$;
         Range: all real numbers
    (b) In interval notation, increasing on
         $\left[-\dfrac{\pi}{2}, \dfrac{\pi}{2}\right]$
         In inequality notation, increasing on
         $-\dfrac{\pi}{2} < x < \dfrac{\pi}{2}$
    (c) Odd
    (d) $(0, 0)$

16. (a) Domain: $\{x \mid x \neq -1\}$;
         Range: $\{y \mid 0 < y < \infty\}$
    (b) In interval notation, increasing on
         $(-\infty, -1)$ and decreasing on
         $(-1, \infty)$
         In inequality notation, increasing on
         $-\infty < x < -1$ and decreasing
         on $-1 < x < \infty$
    (c) Neither
    (d) $(0, 2)$

18. (a) Domain: $\{x \mid x \neq 0\}$;
         Range: $\{y \mid -\infty < y < -2$ or
         $2 < y < \infty\}$
    (b) In interval notation, increasing on
         $[-1, 0)$ and on $(0, 1]$, and
         decreasing on $(-\infty, -1]$ and
         on $[1, \infty)$.
         In inequality notation, increasing on
         $-1 \leq x < 0$ and on $0 < x \leq 1$,
         and decreasing on
         $-\infty < x \leq -1$ and on
         $1 \leq x < \infty$
    (c) Odd
    (d) None

20. (a) Domain: $\{x \mid x \neq -1, x \neq 1\}$;
    Range: all real numbers
    (b) In interval notation, increasing on
        $(-\infty, -1)$ and on $(-1, 1)$ and
        on $(1, \infty)$
        In inequality notation, increasing on
        $-\infty < x < -1$ and on
        $-1 < x < 1$ and on $1 < x < \infty$
    (c) Odd
    (d) $(0, 0)$

22. (a) Domain: $\{x \mid -4 \leq x \leq 4$;
    Range: $\{y \mid -2 \leq y \leq 2\}$
    (b) Increasing on $[-2, 2]$ or $-2 \leq x \leq 2$
        Decreasing on $[-4, -2]$ and on $[2, 4]$
        $-4 \leq x \leq -2$ or $2 \leq x \leq 4$
    (c) Odd
    (d) $(0, 0)$

24. (a) 0
    (b) 0
    (c) -1

26. (a) -1
    (b) 2
    (c) 5

28. $f(x) = 3 - x$
    (a) $f(-x) = 3 - (-x) = 3 + x$
    (b) $-f(x) = -(3 - x) = x - 3$
    (c) $f(2x) = 3 - 2x$
    (d) $f(x - 3) = 3 - (x - 3) = 6 - x$
    (e) $f\left[\dfrac{1}{x}\right] = 3 - \dfrac{1}{x}$
    (f) $\dfrac{1}{f(x)} = \dfrac{1}{3 - x}$

30. $f(x) = x^3 + 1$
    (a) $f(-x) = (-x)^3 + 1 = -x^3 + 1$
    (b) $-f(x) = -(x^3 + 1) = -x^3 - 1$
    (c) $f(2x) = (2x)^3 + 1 = 8x^3 + 1$
    (d) $f(x - 3)^3 + 1$
        $= x^3 - 9x^2 + 27x - 27 + 1$
        $= x^3 - 9x^2 + 27x - 26$
    (e) $f\left[\dfrac{1}{x}\right] = \left[\dfrac{1}{x}\right]^3 + 1 = \dfrac{1}{x^3} + 1$
    (f) $\dfrac{1}{f(x)} = \dfrac{1}{x^3 + 1}$

32. $f(x) = x^2 + x$
    (a) $f(-x) = (-x)^2 + x = x^2 + x$
    (b) $-f(x) = -(x^2 + x) = -x^2 - x$
    (c) $f(2x) = (2x)^2 + 2x = 4x^2 + 2x$
    (d) $f(x - 3) = (x - 3)^2 + (x - 3)$
        $= x^2 - 6x + 9 + x - 3$
        $= x^2 - 5x + 6$

    (e) $f\left[\dfrac{1}{x}\right] = \left[\dfrac{1}{x}\right]^2 + \dfrac{1}{x} = \dfrac{1}{x^2} + \dfrac{1}{x}$
    (f) $\dfrac{1}{f(x)} = \dfrac{1}{x^2 + x}$

34. $f(x) = \dfrac{x^2}{x^2 + 1}$

    (a) $f(-x) = \dfrac{(-x)^2}{(-x)^2 + 1} = \dfrac{x^2}{x^2 + 1}$

    (b) $-f(x) = -\left[\dfrac{x^2}{x^2 + 1}\right] = \dfrac{-x^2}{x^2 + 1}$

    (c) $f(2x) = \dfrac{(2x)^2}{(2x)^2 + 1} = \dfrac{4x^2}{4x^2 + 1}$

    (d) $f(x - 3) = \dfrac{x^2 - 6x + 9}{x^2 - 6x + 10}$
        $= \dfrac{x^2 - 6x + 9}{x^2 - 6x + 10}$

    (e) $f\left[\dfrac{1}{x}\right] = \dfrac{\left[\dfrac{1}{x}\right]^2}{\left[\dfrac{1}{x}\right]^2 + 1} = \dfrac{\dfrac{1}{x^2}}{\dfrac{1 + x^2}{x^2}}$
        $= \dfrac{x^2}{x^2(1 + x^2)} = \dfrac{1}{1 + x^2}$

    (f) $\dfrac{1}{f(x)} = \dfrac{1}{\dfrac{x^2}{x^2 + 1}} = \dfrac{x^2 + 1}{x^2}$

36. $f(x) = \dfrac{1}{x}$

    (a) $f(-x) = \dfrac{1}{-x} = \dfrac{-1}{x}$

    (b) $-f(x) = \dfrac{-1}{x}$

    (c) $f(2x) = \dfrac{1}{2x}$

    (d) $f(x - 3) = \dfrac{1}{x - 3}$

    (e) $f\left(\dfrac{1}{x}\right) = \dfrac{1}{\frac{1}{x}} = x$

    (f) $\dfrac{1}{f(x)} = \dfrac{1}{\frac{1}{x}} = x$

38.   (a) $f(-x) = 4 + \dfrac{2}{-x} = 4 - \dfrac{2}{x}$

    (b) $-f(x) = -\left(4 + \dfrac{2}{x}\right) = -4 - \dfrac{2}{x}$

    (c) $f(2x) = 4 + \dfrac{2}{2x} = 4 + \dfrac{1}{x}$

    (d) $f(x - 3) = 4 + \dfrac{2}{x - 3}$

    (e) $f\left(\dfrac{1}{x}\right) = 4 + \dfrac{2}{\frac{1}{x}} = 4 + 2x$

    (f) $\dfrac{1}{f(x)} = \dfrac{1}{4 + \frac{2}{x}} = \dfrac{x}{4x + 2}$

40. $\dfrac{f(x) - f(1)}{x - 1} = \dfrac{-2x + 2}{x - 1} = \dfrac{-2(x - 1)}{x - 1} = -2$

42. $\dfrac{f(x) - f(1)}{x - 1} = \dfrac{(x^2 + 1) - (2)}{x - 1} = \dfrac{x^2 - 1}{x - 1} = \dfrac{(x - 1)(x + 1)}{x - 1} = x + 1$

44. $\dfrac{f(x) - f(1)}{x - 1} = \dfrac{(4x - 2x^2) - (2)}{x - 1} = \dfrac{-2x^2 + 4x - 2}{x - 1} = \dfrac{-2(x^2 - 2x + 1)}{x - 1} = \dfrac{-2(x - 1)^2}{x - 1} = -2(x - 1)$

46. $\dfrac{f(x) - f(1)}{x - 1} = \dfrac{(x^3 + x) - (2)}{x - 1} = \dfrac{x^3 + x - 2}{x - 1} = \dfrac{(x - 1)(x^2 + x + 2)}{x - 1} = x^2 + x + 2$

48. $\dfrac{f(x) - f(1)}{x - 1} = \dfrac{\frac{1}{x^2} - 1}{x - 1} = \dfrac{\frac{1 - x^2}{x^2}}{x - 1} = \dfrac{1 - x^2}{x^2(x - 1)} = \dfrac{(1 - x)(1 + x)}{x^2(x - 1)} = \dfrac{-(1 + x)}{x^2}$

50. $\dfrac{f(x) - f(1)}{x - 1} = \dfrac{\sqrt{x + 3} - 2}{x - 1} = \dfrac{\sqrt{x + 3} - 2}{x - 1} \cdot \dfrac{\sqrt{x + 3} + 2}{\sqrt{x + 3} + 2} = \dfrac{(x + 3) - 4}{(x - 1)\left[\sqrt{x + 3} + 2\right]}$

                   $= \dfrac{x - 1}{(x - 1)\left[\sqrt{x + 3} + 2\right]} = \dfrac{1}{\sqrt{x + 3} + 2}$

52. $f(x) = 2x^4 - x^2$

    Since $f(-x) = 2(-x)^4 - (-x)^2$
    $= 2x^4 - x^2 = f(x)$, f is even.

54. $h(x) = 3x^3 + 2$

    Since $h(-x) = -3x^3 + 2 \neq h(x)$ or $-h(x)$,
    h is neither even nor odd.

56. $G(x) = \sqrt{x}$

    Since $G(-x) = \sqrt{-x}$ is undefined, G is
    neither even nor odd.

58. $f(x) = \sqrt[3]{2x^2 + 1}$

    Since $f(-x) = \sqrt[3]{2(-x)^2 + 1} = \sqrt[3]{2x^2 + 1}$
    $= f(x)$, f is even.

60. $h(x) = \dfrac{x}{x^2 - 1}$

Since $h(-x) = \dfrac{-x}{(-x)^2 - 1} = \dfrac{-x}{x^2 - 1}$
$\qquad\qquad = -h(x)$, h is odd.

62. $F(x) = \dfrac{x}{|x|}$

Since $F(-x) = \dfrac{-x}{|-x|} = \dfrac{-x}{|x|} = -F(x)$,

F is odd.

64. A function can have at most one y-intercept. If there were more than one y-intercept, a vertical line would contain more than one point.

66. $f(x) = 4 - 2x$

(a) The domain is all real numbers.

(b) x-intercept(s): $\qquad$ y-intercept(s):

$\quad 0 = 4 - 2x \qquad\qquad y = 4 - 2(0)$
$\quad 2x = 4 \qquad\qquad\qquad y = 4$
$\quad\; x = 2$

The intercepts are $(2, 0)$, $(0, 4)$

(c)

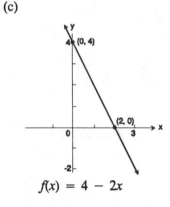

$f(x) = 4 - 2x$

(d) The range is all real numbers.

68. $g(x) = x^2 + 4$

(a) The domain is all real numbers.

(b) x-intercept(s): $\qquad$ y-intercept:

$\quad 0 = x^2 + 4 \qquad\qquad y = 0^2 + 4$
$\quad -4 = x^2 \qquad\qquad\quad y = 4$

No solution.

The intercept is $(0, 4)$.

(c)

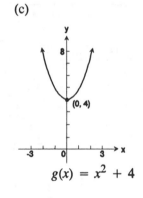

$g(x) = x^2 + 4$

(d) The range is $\{y \mid 4 \le y < \infty\}$.

70. $F(x) = 2x^2$

(a) The domain is all real numbers.

(b) x-intercept(s): $\qquad\qquad$ y-intercept:

$\quad 0 = 2x^2 \qquad\qquad\qquad y = 2(0)^2$
$\quad 0 = x \qquad\qquad\qquad\quad\; y = 0$

The intercept is $(0, 0)$.

(c)

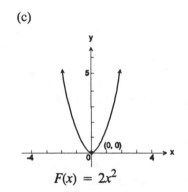

$F(x) = 2x^2$

(d) The range is $\{y \mid 0 \le y < \infty\}$.

72. $g(x) = \sqrt{x} + 2$
    (a)   The domain is $\{y \mid 0 \leq x < \infty\}$
    (b) x-intercept(s): y-intercept:

$$0 = \sqrt{x} + 2 \qquad y = \sqrt{0} + 2$$
$$-2 = \sqrt{x} \qquad\quad y = 2$$

No solution.
The intercept is $(0, 2)$.

(c)

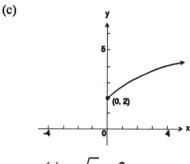

$$g(x) = \sqrt{x} + 2$$

(d)   The range is $\{y \mid 2 \leq y < \infty\}$

74. $F(x) = -\sqrt{x}$
    (a)   The domain is $\{x \mid 0 \leq x < \infty\}$
    (b) x-intercept(s): y-intercept:

$$0 = -\sqrt{x} \qquad y = -\sqrt{0}$$
$$0 = x \qquad\quad\ y = 0$$

The intercept is $(0, 0)$.

(c)

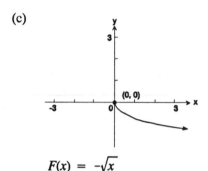

$$F(x) = -\sqrt{x}$$

(d)   The range is $\{y \mid -\infty < y \leq 0\}$.

76. $g(x) = |x + 3|$
    (a)   The domain is all real numbers.
    (b) x-intercept(s): y-intercept:

$$0 = |x + 3| \qquad y = |0 + 3|$$
$$0 = x + 3 \qquad\ y = 3$$
$$-3 = x$$

The intercepts are $(-3, 0)$, $(0, 3)$.

(c)

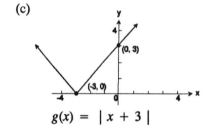

$$g(x) = |x + 3|$$

(d)   The range is $\{y \mid 0 \leq y < \infty\}$.

78. $F(x) = |3 - x|$
    (a)   The domain is all real numbers.
    (b) x-intercept(s): y-intercept:

$$0 = |3 - x| \qquad y = |3 - 0|$$
$$0 = 3 - x \qquad\ y = 3$$
$$x = 3$$

The intercepts are $(3, 0)$, $(0, 3)$.

(c)

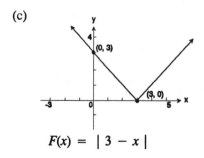

$$F(x) = |3 - x|$$

(d)   The range is $\{y \mid 0 \leq y < \infty\}$.

80. $f(x) = \begin{cases} 3x & \text{if} \quad x \neq 0 \\ 4 & \text{if} \quad x = 0 \end{cases}$

(a)   The domain is all real numbers.

(b) x-intercept(s): y-intercept:

   $0 = 3x$ $y = 4$ if $x = 0$

   $0 = x$

   The intercept is $(0, 4)$.

(c)

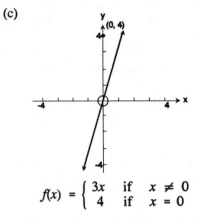

$f(x) = \begin{cases} 3x & \text{if} \quad x \neq 0 \\ 4 & \text{if} \quad x = 0 \end{cases}$

(d)   The range is all real numbers.

82. $f(x) = \begin{cases} \dfrac{1}{x} & \text{if} \quad x < 0 \\ \sqrt{x} & \text{if} \quad x \geq 0 \end{cases}$

(a)   The domain is all real numbers.

(b) x-intercept(s): y-intercept:

   $0 = \sqrt{x}$ $y = \sqrt{0}$

   $0 = x$ $y = 0$

   The intercept is $(0, 0)$.

(c)

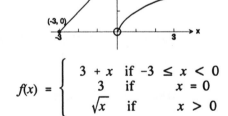

$f(x) = \begin{cases} \dfrac{1}{x} & \text{if} \quad x < 0 \\ \sqrt{x} & \text{if} \quad x \geq 0 \end{cases}$

(d)   The range is all real numbers.

84. $f(x) = \begin{cases} 3 + x & \text{if} \quad -3 \leq x < 0 \\ 3 & \text{if} \quad x = 0 \\ \sqrt{x} & \text{if} \quad x > 0 \end{cases}$

(a)   The domain is $\{x \mid -3 \leq x < \infty\}$

(b) x-intercept(s): y-intercept:

   $0 = 3 + x$ $y = 3$ if $x = 0$

   $x = -3$

   The intercepts are $(-3, 0)$, $(0, 3)$.

(c)

$f(x) = \begin{cases} 3 + x & \text{if} \quad -3 \leq x < 0 \\ 3 & \text{if} \quad x = 0 \\ \sqrt{x} & \text{if} \quad x > 0 \end{cases}$

(d)   The range is $\{y \mid 0 \leq y < \infty\}$

86. $g(x) = \begin{cases} x & \text{if } x \geq 1 \\ 1 & \text{if } x < 1 \end{cases}$

   (a) The domain is all real numbers.

   (b) x-intercept(s):   y-intercept:
       None $y = 1$ if $x = 0$
       The intercept is (0, 1).

   (c)

   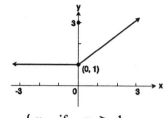

   $g(x) = \begin{cases} x & \text{if } x \geq 1 \\ 1 & \text{if } x < 1 \end{cases}$

   (d) The range is $\{y \mid 1 \leq y < \infty\}$

88. $f(x) = [\![2x]\!]$

   (a) The domain is all real numbers.

   (b) x-intercept(s):   y-intercept:
       $0 = [\![2x]\!]$ $y = [\![2(0)]\!]$

       $0 \leq x < \dfrac{1}{2}$ $y = 0$

       The intercepts are all ordered pairs $(x, 0)$ where

       $0 \leq x < \dfrac{1}{2}$.

   (c)

   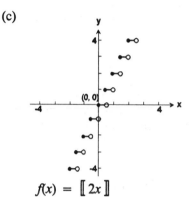

   $f(x) = [\![2x]\!]$

   (d) The range is the set of integers.

90. $G(x) = |x^2 - 4|$

   (a) The domain is all real numbers

   (b) x-intercepts: y-intercept:
       $0 = |x^2 - 4|$ $y = |0^2 - 4| = 4$
       $x^2 - 4 = 0$
       $x = -2, x = 2$
       The intercepts are $(-2, 0), (2, 0), (0, 4)$

   (d) The range is $\{y \mid 0 \leq y < \infty\}$.

   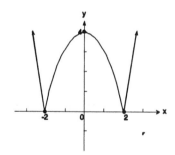

92. $f(x) = -3x + 2$

   $\dfrac{f(x + h) - f(x)}{h} = \dfrac{-3(x + h) + 2 - (-3x + 2)}{h} = \dfrac{-3x - 3h + 2 + 3x - 2}{h} = \dfrac{-3h}{h} = -3$

94. $f(x) = \dfrac{1}{x}$

   $\dfrac{f(x + h) - f(x)}{h} = \dfrac{\dfrac{1}{x + h} - \dfrac{1}{x}}{h} = \dfrac{\dfrac{x - (x + h)}{(x + h)x}}{h} = \dfrac{\dfrac{-h}{(x + h)x}}{h} = \dfrac{-h}{h(x + h)x} = \dfrac{-1}{(x + h)x}$

96. $f(x) = \begin{cases} x & \text{if } -1 \leq x \leq 0 \\ 1 & \text{if } 0 < x \leq 2 \end{cases}$ 98. $f(x) = \begin{cases} 2x + 2 & \text{if } -1 \leq x \leq 0 \\ x & \text{if } x > 0 \end{cases}$

100. $f(x) = \begin{cases} x^2 + 4 & \text{if } x \neq 2 \\ 5 & \text{if } x = 2 \end{cases}$

To prove that f is even, we need to show that $f(-x) = f(x)$ for **all** values x.

$f(-x) = (-x)^2 + 4 = x^2 + 4 = f(x)$

However, when $x = 2$,

$f(-2) = (-2)^2 + 4 = 8$, but $f(2) = 5$

Because $f(-2) \neq f(2)$, f is not even.

102. Each graph is that of $y = x^2$, but shifted vertically. If $y = (x - k)^2$ and $k > 0$, the shift is k units to the right. If $y = (x + k)^2$ and $k > 0$, the shift is k units to the left.

104. The graph of $y = -f(x)$ is the reflection about the x-axis of the graph of $y = f(x)$.

106. Shift $y = x^3$ to the right 1 unit and then up 2 units.

108. (a) For 40 therms, the charge C is $4 + 0.1402 \times 40 = \$9.61$

(b) For 202 therms, the charge C is $4 + 0.1402 \times 50 + 0.0547 = 152 + 0.2406 \times 202$
    $= \$67.93$

(c) If C is the charge, then

$$C = \begin{cases} 4 + 0.1402x + 0.2406x & \text{if } 0 \leq x \leq 50 \\ 4 + 0.1402(50) + 0.0547(x - 50) + 0.2406x & \text{if } x > 50 \end{cases}$$

$$= \begin{cases} 4 + 0.3808x & \text{if } 0 \leq x \leq 50 \\ 8.28 + 0.2953x & \text{if } x > 50 \end{cases}$$

(d)

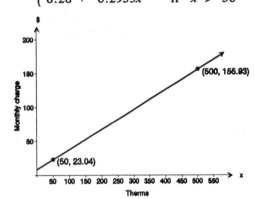

110. We are given that
$$\min(f, g)(x) = \begin{cases} f(x) & \text{if } f(x) \leq g(x) \\ g(x) & \text{if } f(x) > g(x) \end{cases}$$

$$\max(f, g)(x) = \begin{cases} g(x) & \text{if } f(x) \leq g(x) \\ f(x) & \text{if } f(x) > g(x) \end{cases}$$

If $f(x)$ is the min, then $f(x) \leq g(x)$, and

$f(x) - g(x) \leq 0$, so $|f(x) - g(x)| = g(x) - f(x)$

Let us use this to show that:

$$\min(f, g)(x) = \frac{f(x) + g(x)}{2} - \frac{|f(x) - g(x)|}{2} = \frac{f(x) + g(x)}{2} - \left[\frac{g(x) - f(x)}{2}\right] = \frac{2f(x)}{2} = f(x)$$

If $g(x)$ is the min, then $f(x) \geq g(x)$, and $f(x) - g(x) \geq 0$, so $|f(x) - g(x)| = f(x) - g(x)$

$$\min(f, g)(x) = \frac{f(x) + g(x)}{2} - \frac{|f(x) - g(x)|}{2} = \frac{f(x) + g(x)}{2} - \left[\frac{f(x) - g(x)}{2}\right] = \frac{2g(x)}{2} = g(x)$$

A similar formula is:

$$\max(f, g)(x) = \frac{f(x) + g(x)}{2} + \frac{|f(x) - g(x)|}{2}$$

To show that the formula is true, we will prove the two cases.

If $f(x) \geq g(x)$, then $|f(x) - g(x)| = f(x) - g(x)$

so $\max(f, g)(x) = \dfrac{f(x) + g(x)}{2} + \dfrac{|f(x) - g(x)|}{2} = \dfrac{2f(x)}{2} = f(x)$

If $g(x) \geq f(x)$, then $|f(x) - g(x)| = g(x) - f(x)$

so $\max(f, g)(x) = \dfrac{f(x) + g(x)}{2} + \dfrac{|f(x) - g(x)|}{2} = \dfrac{f(x) + g(x)}{2} + \dfrac{g(x) - f(x)}{2} = \dfrac{2g(x)}{2} = g(x)$

4.3   Graphing Techniques

2. E   4. D   6. A   8. C   10. J   12. K

14. $y = (x + 4)^3$ 16. $y = x^3 - 4$

18. $y = -x^3$ 20. $y = (4x)^3 = 64x^3$

22. $f(x) = x^2 + 4$ 24. $g(x) = x^3 - 1$

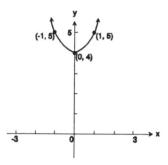   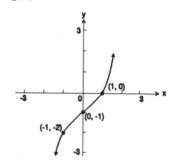

26. $h(x) = \sqrt{x + 1}$ 28. $f(x) = (x + 2)^3$

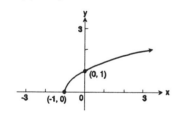   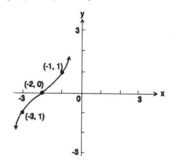

30. $g(x) = \frac{1}{2}\sqrt{x}$

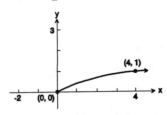

32. $h(x) = \frac{4}{x}$

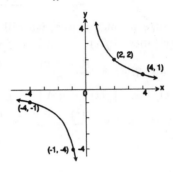

34. $f(x) = -\sqrt{x}$

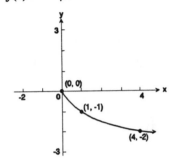

36. $g(x) = -x^3$

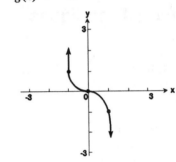

38. $h(x) = \frac{1}{-x}$

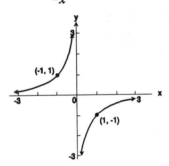

40. $f(x) = (x - 2)^2 + 1$

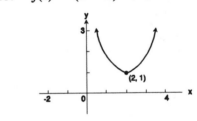

42. $g(x) = |x + 1| - 3$

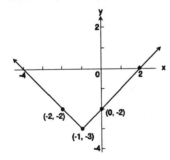

44. $h(x) = \frac{4}{x} + 2$

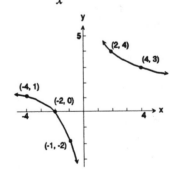

Chapter 4 Functions and Their Graphs

46. $f(x) = 4\sqrt{x - 1}$

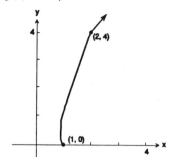

48. $g(x) = 4\sqrt{2 - x}$

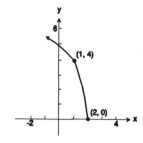

50. $h(x) = -x^3 + 2$

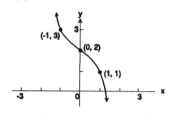

52. (a) $F(x) = f(x) + 3$

(b) $G(x) = f(x + 2)$

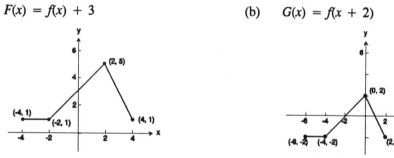

(c) $P(x) = -f(x)$

(d) $Q(x) = \frac{1}{2}f(x)$

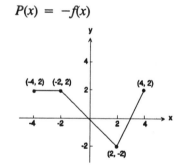

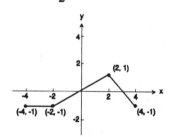

(e) $g(x) = f(-x)$

(f) $h(x) = 3f(x)$

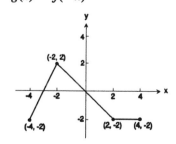

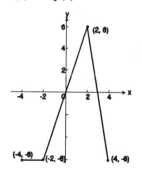

54. (a) $F(x) = f(x) + 3$

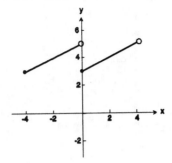

(b) $G(x) = f(x + 2)$

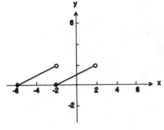

(c) $P(x) = -f(x)$

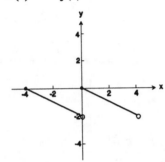

(d) $Q(x) = \frac{1}{2}f(x)$

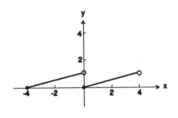

(e) $g(x) = f(-x)$

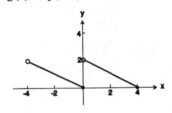

(f) $f(x) = 3f(x)$

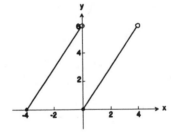

56. (a) $F(x) = f(x) + 3$

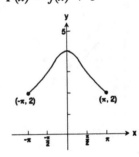

(b) $G(x) = f(x + 2)$

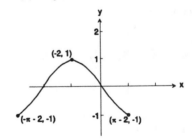

(c) $P(x) = -f(x)$

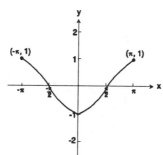

(-π, 1)   (π, 1)

(d) $Q(x) = \frac{1}{2}f(x)$

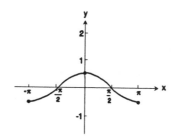

(e) $g(x) = f(-x)$

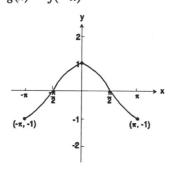

(-π, -1)   (π, -1)

(f) $f(x) = 3f(x)$

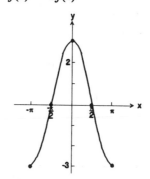

58. $f(x) = x^2 - 6x$
       $f(x) = (x^2 - 6x + 9) - 9$
       $f(x) = (x - 3)^2 - 9$

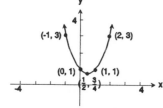

(2, -8)   (4, -8)
    (3, -9)

60. $f(x) = x^2 + 4x + 2$
       $f(x) = (x^2 + 4x + 4) + 2 - 4$
       $f(x) = (x + 2)^2 - 2$

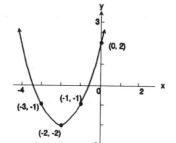

(0, 2)
(-3, -1)   (-1, -1)
    (-2, -2)

62. $f(x) = x^2 - x + 1$

$f(x) = \left[x^2 - x + \dfrac{1}{4}\right] + 1 - \dfrac{1}{4}$

$f(x) = \left[x - \dfrac{1}{2}\right]^2 + \dfrac{3}{4}$

(-1, 3)   (2, 3)

(0, 1)   (1, 1)
    $\left(\dfrac{1}{2}, \dfrac{3}{4}\right)$

64.

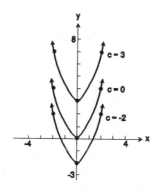

c = 3
c = 0
c = -2

Section 4.3 Graphing Techniques

117

66. An increase in ℓ increases the period T.

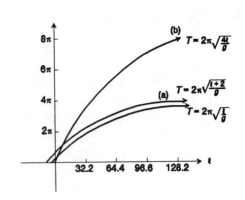

68. (a)

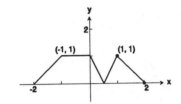

(b)

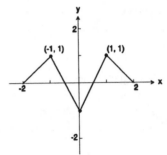

70. (a)

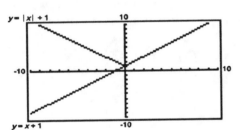

(b)

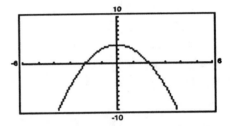

(c)

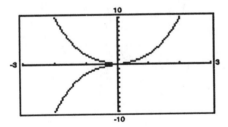

(d) For the graph of $y = f(\,|\,x\,|\,)$, the graph of $y = f(x)$ to the left of the y-axis is replaced by the reflection about the y-axis of the graph of $y = f(x)$ to the right of the y-axis. To the right of the y-axis, both graphs are the same.

4.4 Operations on Functions; Composite Functions

2. $f(x) = 2x + 1$, $g(x) = 3x - 2$

(a) $(f + g)(x) = (2x + 1) + (3x - 2) = 5x - 1$
The domain is the set of all real numbers.

(b) $(f - g)(x) = (2x + 1) - (3x - 2) = -x + 3$
The domain is the set of all real numbers.

(c) $(f \cdot g)(x) = (2x + 1)(3x - 2) = 6x^2 - x - 2$
The domain is the set of all real numbers.

(d) $\left[\dfrac{f}{g}\right](x) = \dfrac{2x + 1}{3x - 2}$ The domain is the set of all real numbers x except $x = \dfrac{2}{3}$.

4. $f(x) = 2x^2 + 3$, $g(x) = 4x^3 + 1$
    (a) $(f + g)(x) = (2x^2 + 3) + (4x^3 + 1) = 4x^3 + 2x^2 + 4$
        The domain is the set of all real numbers.
    (b) $(f - g)(x) = (2x^2 + 3) - (4x^3 + 1) = -4x^3 - 2x^2 + 2$
        The domain is the set of all real numbers.
    (c) $(f \cdot g)(x) = (2x^2 + 3)(4x^3 + 1) = 8x^5 + 12x^3 + 2x^2 + 3$
        The domain is the set of all real numbers.
    (d) $\left[\dfrac{f}{g}\right](x) = \dfrac{2x^2 + 3}{4x^3 + 1}$ The domain is the set of all real numbers x except $x = \sqrt[3]{\dfrac{-1}{4}}$

6. $f(x) = |x|$, $g(x) = x$
    (a) $(f + g)(x) = |x| + x$
        The domain is the set of all real numbers.
    (b) $(f - g)(x) = |x| - x$
        The domain is the set of all real numbers.
    (c) $(f \cdot g)(x) = |x| x$
        The domain is the set of all real numbers.
    (d) $\left[\dfrac{f}{g}\right](x) = \dfrac{|x|}{x}$
        The domain is the set of all real numbers x except $x = 0$

8. $f(x) = 2x^2 - x$, $g(x) = 2x^2 + x$
    (a) $(f + g)(x) = (2x^2 - x) + (2x^2 + x) = 4x^2$
        The domain is the set of all real numbers.
    (b) $(f - g)(x) = (2x^2 - x) - (2x^2 + x) = -2x$
        The domain is the set of all real numbers.
    (c) $(f \cdot g)(x) = (2x^2 - x)(2x^2 + x) = 4x^2 - x^2$
        $= 3x^2$
        The domain is the set of all real numbers.
    (d) $\left[\dfrac{f}{g}\right](x) = \dfrac{2x^2 - x}{2x^2 + x}$
        The domain is the set of all real numbers except all x such that $x(2x + 1) = 0$. Thus, $x = 0$ and $x = -\dfrac{1}{2}$ are excluded from the domain.

10. $f(x) = \sqrt{x + 1}$, $g(x) = \dfrac{2}{x}$

    (a) $(f + g)(x) = \sqrt{x + 1} + \dfrac{2}{x}$
        The domain is the set of all numbers for which $x \geq -1$, but $x \neq 0$.

    (b) $(f - g)(x) = \sqrt{x + 1} - \dfrac{2}{x}$
        The domain is the set of all numbers for which $x \geq -1$, but $x \neq 0$.

    (c) $(f \cdot g)(x) = \dfrac{2\sqrt{x + 1}}{x}$ The domain is the set of all numbers for which $x \geq -1$, but $x \neq 0$.

    (d) $\left[\dfrac{f}{g}\right](x) = \dfrac{\sqrt{x + 1}}{\dfrac{2}{x}} = \dfrac{x\sqrt{x + 1}}{2}$ The domain is the set of all numbers x for which $x \geq -1$.

12. If $f(x) = \dfrac{1}{x}$ and $\left[\dfrac{f}{g}\right](x) = \dfrac{x + 1}{x^2 - x}$, then

    $\dfrac{\dfrac{1}{x}}{g(x)} = \dfrac{x + 1}{x^2 - x}$

    $\dfrac{1}{x} = \dfrac{x + 1}{x^2 - x} g(x)$

    $g(x) = \dfrac{x^2 - x}{x^2 + x}$

14. $f(x) = |x| + \sqrt{x}$

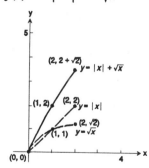

16. $f(x) = x^3 + x^2$

18. $f(x) = 3x + 2, g(x) = 2x^2 - 1$

(a) $(f \circ g)(4) = f(g(4)) = 3 \cdot 31 + 2 = 95$
$\uparrow$
$g(4) = 31$

(b) $(g \circ f)(2) = g(f(2)) = 2 \cdot 64 - 1 = 127$
$\uparrow$
$f(2) = 8$

(c) $(f \circ f)(1) = f(f(1)) = 3 \cdot 5 + 2 = 17$
$\uparrow$
$f(1) = 5$

(d) $(g \circ g)(0) = g(g(0)) = 2(1) - 1 = 1$
$\uparrow$
$g(0) = -1$

20. $f(x) = 2x^2, g(x) = 1 - 3x^2$

(a) $(f \circ g)(4) = f(g(4)) = 2(-47)^2 = 4418$
$\uparrow$
$g(4) = -47$

(b) $(g \circ f)(2) = g(f(2)) = 1 - 3(64) = -191$
$\uparrow$
$f(2) = 8$

(c) $(f \circ f)(1) = f(f(1)) = 2 \cdot (2)^2 = 8$
$\uparrow$
$f(1) = 2$

(d) $(g \circ g)(0) = g(g(0)) = 1 - (3)1 = -2$
$\uparrow$
$g(0) = 1$

22. $f(x) = \sqrt{x + 1}, g(x) = 3x$

(a) $(f \circ g)(4) = f(g(4)) = \sqrt{12 + 1} = \sqrt{13}$
$\uparrow$
$g(4) = 12$

(b) $(g \circ f)(2) = g(f(2)) = 3\sqrt{3}$
$\uparrow$
$f(2) = \sqrt{3}$

(c) $(f \circ f)(1) = f(f(1)) = \sqrt{\sqrt{2} + 1}$
$\uparrow$
$f(1) = \sqrt{2}$

(d) $(g \circ g)(0) = g(g(0)) = 3(0) = 0$
$\uparrow$
$g(0) = 0$

24. $f(x) = |x - 2|$, $g(x) = \dfrac{3}{x^2 + 2}$

    (a) $(f \circ g)(4) = f(g(4)) = \left| \dfrac{1}{6} - 2 \right| = \dfrac{11}{6}$

                    $\uparrow$

                  $g(4) = \dfrac{3}{18} = \dfrac{1}{6}$

    (b) $(g \circ f)(2) = g(f(2)) = \dfrac{3}{2}$

                   $\uparrow$

                $f(2) = 0$

    (c) $(f \circ f)(1) = f(f(1)) = |1 - 2| = 1$

                  $\uparrow$

               $f(1) = 1$

    (d) $(g \circ g)(0) = g(g(0)) = \dfrac{3}{\left(\dfrac{3}{2} \right)^2 + 2} = \dfrac{3}{\dfrac{17}{4}} = \dfrac{12}{17}$

                  $\uparrow$

                $g(0) = \dfrac{3}{2}$

26. $f(x) = x^3$, $g(x) = \dfrac{2}{x^2 + 1}$

    (a) $(f \circ g)(4) = f(g(4)) = \left(\dfrac{2}{17} \right)^3 = \dfrac{8}{4913}$

                  $\uparrow$

                $g(4) = \dfrac{2}{17}$

    (b) $(g \circ f)(2) = g(f(2)) = \dfrac{2}{65}$

                  $\uparrow$

                $f(2) = 8$

    (c) $(f \circ f)(1) = f(f(1)) = 1^3 = 1$

                  $\uparrow$

                $f(1) = 1$

    (d) $(g \circ g)(0) = g(g(0)) = \dfrac{2}{4 + 1} = \dfrac{2}{5}$

                  $\uparrow$

                $g(0) = 2$

28. $f(x) = -x$, $g(x) = 2x - 4$
    (a) $(f \circ g)(x) = f(g(x)) = f(2x - 4) = -(2x - 4) = 4 - 2x$
    (b) $(g \circ f)(x) = g(f(x)) = g(-x) = 2(-x) - 4 = -2x - 4$
    (c) $(f \circ f)(x) = f(f(x)) = f(-x) = -(-x) = x$
    (d) $(g \circ g)(x) = g(g(x)) = g(2x - 4) = 2(2x - 4) - 4 = 4x - 12$

30. $f(x) = \sqrt{x + 1}$, $g(x) = x + 4$

    (a) $(f \circ g)(x) = f(g(x)) = f(x + 4) = \sqrt{(x + 4) + 1} = \sqrt{x + 5}$

    (b) $(g \circ f)(x) = g(f(x)) = g\left(\sqrt{x + 1}\right) = \sqrt{x + 1} + 4$

    (c) $(f \circ f)(x) = f(f(x)) = f\left(\sqrt{x + 1}\right) = \sqrt{\sqrt{x + 1} + 1}$

    (d) $(g \circ g)(x) = g(g(x)) = g(x + 4) = (x + 4) + 4 = x + 8$

32. $f(x) = \sqrt{x + 1}$, $g(x) = \dfrac{1}{x^2}$

    (a) $(f \circ g)(x) = f(g(x)) = f\left(\dfrac{1}{x^2}\right) = \sqrt{\dfrac{1}{x^2} + 1}$

    (b) $(g \circ f)(x) = g(f(x)) = g\left(\sqrt{x + 1}\right) = \dfrac{1}{\left(\sqrt{x + 1}\right)^2} = \dfrac{1}{x + 1}$

    (c) $(f \circ f)(x) = f(f(x)) = f\left(\sqrt{x + 1}\right) = \sqrt{\sqrt{x + 1} + 1}$

    (d) $(g \circ g)(x) = g(g(x)) = g\left(\dfrac{1}{x^2}\right) = \dfrac{1}{\left[\dfrac{1}{x^2}\right]^2} = x^4$

34. $f(x) = x + \dfrac{1}{x}$, $g(x) = x^2$

    (a) $(f \circ g)(x) = f(g(x)) = x^2 + \dfrac{1}{x^2}$

    (b) $(g \circ f)(x) = g(f(x)) = g\left(x + \dfrac{1}{x}\right) = \left(x + \dfrac{1}{x}\right)^2$

    (c) $(f \circ f)(x) = f(f(x)) = f\left(x + \dfrac{1}{x}\right) = \left(x + \dfrac{1}{x}\right) + \dfrac{1}{x + \dfrac{1}{x}} = \dfrac{x^2 + 1}{x} + \dfrac{x}{x^2 + 1}$

    (d) $(g \circ g)(x) = g(g(x)) = (x^2)^2 = x^4$

36. $f(x) = 2x + 4$, $g(x) = \dfrac{1}{2}x - 2$

    (a) $(f \circ g)(x) = f(g(x)) = f\left(\dfrac{1}{2}x - 2\right) = 2\left(\dfrac{1}{2}x - 2\right) + 4 = x$

    (b) $(g \circ f)(x) = g(f(x)) = g(2x + 4) = \dfrac{1}{2}(2x + 4) - 2 = x$

    (c) $(f \circ f)(x) = f(f(x)) = f(2x + 4) = 2(2x + 4) + 4 = 4x + 12$

    (d) $(g \circ g)(x) = g(g(x)) = g\left(\dfrac{1}{2}x - 2\right) = \dfrac{1}{2}\left(\dfrac{1}{2}x - 2\right) - 2 = \dfrac{1}{4}x - 3$

38. $f(x) = \dfrac{x+1}{x-1}$, $g(x) = \dfrac{x-1}{x+1}$

(a) $(f \circ g)(x) = f(g(x)) = f\left(\dfrac{x-1}{x+1}\right) = \dfrac{\dfrac{x-1}{x+1} + 1}{\dfrac{x-1}{x+1} - 1} = \dfrac{\dfrac{x-1+(x+1)}{x+1}}{\dfrac{x-1-(x+1)}{x+1}} = \dfrac{2x}{-2} = -x$

(b) $(g \circ f)(x) = g(f(x)) = g\left(\dfrac{x+1}{x-1}\right) = \dfrac{\dfrac{x+1}{x-1} - 1}{\dfrac{x+1}{x-1} + 1} = \dfrac{\dfrac{x+1-(x-1)}{x-1}}{\dfrac{x+1+(x-1)}{x-1}} = \dfrac{2}{2x} = \dfrac{1}{x}$

(c) $(f \circ f)(x) = f(f(x)) = f\left(\dfrac{x+1}{x-1}\right) = \dfrac{\dfrac{x+1}{x-1} + 1}{\dfrac{x+1}{x-1} - 1} = \dfrac{\dfrac{x+1+(x-1)}{x-1}}{\dfrac{x+1-(x-1)}{x-1}} = \dfrac{2x}{2} = x$

(d) $(g \circ g)(x) = g(g(x)) = g\left(\dfrac{x-1}{x+1}\right) = \dfrac{\dfrac{x-1}{x+1} - 1}{\dfrac{x-1}{x+1} + 1} = \dfrac{\dfrac{x-1-(x+1)}{x+1}}{\dfrac{x-1-(x+1)}{x+1}} = \dfrac{-2}{2x} = \dfrac{-1}{x}$

40. $f(x) = \dfrac{ax+b}{cx+d}$, $g(x) = mx$

(a) $(f \circ g)(x) = f(g(x)) = f(mx) = \dfrac{amx+b}{cmx+d}$

(b) $(g \circ f)(x) = g(f(x)) = g\left(\dfrac{ax+b}{cx+d}\right) = m\left(\dfrac{ax+b}{cx+d}\right) = \dfrac{max+mb}{cx+d}$

(c) $(f \circ f)(x) = f(f(x)) = f\left(\dfrac{ax+b}{cx+d}\right) = \dfrac{a\left[\dfrac{ax+b}{cx+d}\right] + b}{c\left[\dfrac{ax+b}{cx+d}\right] + d}$

$= \dfrac{\dfrac{a^2x + ab + bcx + bd}{cx+d}}{\dfrac{cax + cb + dcx + d^2}{cx+d}} = \dfrac{a^2x + ab + bcx + bd}{cax + cb + dcx + d^2}$

(d) $(g \circ g)(x) = g(g(x)) = g(mx) = m(mx) = m^2x$

42. $(f \circ g)(x) = f(g(x)) = f\left[\dfrac{1}{4}x\right] = 4\left[\dfrac{1}{4}x\right] = x$

$(g \circ f)(x) = g(f(x)) = g(4x) = \dfrac{1}{4}(4x) = x$

44. $(f \circ g)(x) = f(g(x)) = f(x-5) = (x-5)+5 = x$

$(g \circ f)(x) = g(f(x)) = g(x+5) = (x+5)-5 = x$

46. $(f \circ g)(x) = f(g(x)) = f\left[\dfrac{1}{3}(4-x)\right] = 4 - 3\left[\dfrac{1}{3}(4-x)\right] = x$

$(g \circ f)(x) = g(f(x)) = g(4-3x) = \dfrac{1}{3}\big(4-(4-3x)\big) = x$

48. $(f \circ g)(x) = f(g(x)) = f\left[\dfrac{1}{x}\right] = \dfrac{1}{\dfrac{1}{x}} = x$

$(g \circ f)(x) = g(f(x)) = g\left[\dfrac{1}{x}\right] = \dfrac{1}{\dfrac{1}{x}} = x$

50. $f(x) = \dfrac{x}{x-1}$

$(f \circ f)(x) = f(f(x)) = f\left[\dfrac{x}{x-1}\right] = \dfrac{\dfrac{x}{x-1}}{\dfrac{x}{x-1} - 1} = \dfrac{\dfrac{x}{x-1}}{\dfrac{x - (x-1)}{x-1}} = x$

52. $((f \circ g) \circ h)(x) = f(g(x)) \circ h = f(\sqrt{x} + 2) \circ h = \left(\sqrt{x} + 2\right)^2 \circ h$

$= \left(\sqrt{1 - 3x} + 2\right)^2 = 5 - 3x + 4\sqrt{1 - 3x}$

54. $[(f \circ h) + (g \circ h)](x) = f(h(x)) + g(h(x)) = f(1 - 3x) + g(1 - 3x)$

$= (1 - 3x)^2 + \sqrt{1 - 3x} + 2$

$= 9x^2 - 6x + 3 + \sqrt{1 - 3x}$

56. $G(x) = 3x^2 = g(f(x)) = g \circ f$

58. $p(x) = 3\sqrt{x} + 3 = g(h(x)) = g \circ h$

60. $R(x) = 9x = g(g(x)) = g \circ g$

62. $Q(x) = \sqrt{\sqrt{x} + 1} + 1 = h(h(x)) = h \circ h$

64. $H(x) = (1 + x^2)^{3/2}$
If $f(x) = x^{3/2}$ and $g(x) = 1 + x^2$, then $f \circ g = H$

66. $H(x) = \dfrac{1}{1 + x^2}$
If $f(x) = \dfrac{1}{x}$ and $g(x) = 1 + x^2$, then $f \circ g = H$

68. $H(x) = |2x^2 + 3|$
If $f(x) = |x|$ and $g(x) = 2x^2 + 3$, then $f \circ g = H$

70. $H(x) = (4 - x^2)^{-4}$
If $f(x) = x^{-4}$ and $g(x) = 4 - x^2$, then $f \circ g = H$

72. $f(x) = 3x^2 - 7,\ g(x) = 2x + a$
$(f \circ g)(x) = f(g(x)) = f(2x + a) = 3(2x + a)^2 - 7$
When $x = 0$, then $(f \circ g)(x) = (f \circ g)(0) = 68$
$$3a^2 - 7 = 68$$
$$3a^2 = 75$$
$$a^2 = 25$$
$$a = -5, 5$$

74. $V(r) = \dfrac{4}{3}r^3,\ r(t) = \dfrac{2}{3}t^3,\ t \geq 0$

Thus, $V(r(t)) = V\left[\dfrac{2}{3}t^3\right] = \dfrac{4}{3}\left[\dfrac{2}{3}t^3\right]^3 = \dfrac{4}{3} \cdot \dfrac{8}{27}t^9 = \dfrac{32}{81}t^9$

76. $A(r) = \pi r^2$ and $r = r(t) = 200\sqrt{t}$

$A(r(t)) = \pi\left(200\sqrt{t}\right)^2 = 40000\pi t$

78. $P = -\dfrac{1}{5}x + 200$

$\dfrac{1}{5}x = 200 - p$

$x = 5(200 - p)$

$x = 5(200 - p)$

$C = \dfrac{\sqrt{x}}{10} + 400 = \dfrac{\sqrt{5(200 - p)}}{10} + 400$

80. f is odd so $f(-x) = -f(x)$; g is even so $g(-x) = g(x)$

$(f \circ g)(-x) = f(g(-x)) = f(g(x)) = (f \circ g)(x)$ so $(f \circ g)$ is even

$(g \circ f)(-x) = g(f(-x)) = g(-f(x)) = g(f(x))$ so $g \circ f$ is even

4.5 One-to-One Functions; Inverse Functions

2. Because a horizontal line will intersect the graph of f exactly once, f is one-to-one.

4. Because a horizontal line will intersect the graph of f at two points, f is not one-to-one.

6. Because a horizontal line will intersect the graph of f at infinitely many points, f is not one-to-one.

8.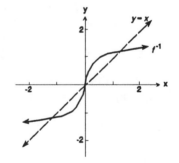

10.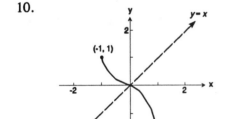

12. Reflect about the line, $y = x$.

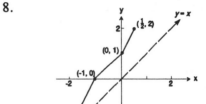

14. $f(x) = 3 - 2x$, $g(x) = -\dfrac{1}{2}(x - 3)$

$f(g(x)) = f\left[-\dfrac{1}{2}(x - 3)\right] = 3 - 2\left[-\dfrac{1}{2}(x - 3)\right] = 3 + x - 3 = x$

$g(f(x)) = g(3 - 2x) = \dfrac{-1}{2}(3 - 2x - 3) = x$

16. $f(x) = 2x + 6$, $g(x) = \frac{1}{2}x - 3$

$f(g(x)) = f\left[\frac{1}{2}x - 3\right] = 2\left[\frac{1}{2}x - 3\right] + 6 = x - 6 + 6 = x$

$g(f(x)) = g(2x + 6) = \frac{1}{2}(2x + 6) - 3 = x + 3 - 3 = x$

18. $f(x) = (x - 2)^2$, $x \geq 2$, $g(x) = \sqrt{x} + 2$, $x \geq 0$

$f(g(x)) = f(\sqrt{x} + 2) = \left[(\sqrt{x} + 2) - 2\right]^2 = (\sqrt{x})^2 = |x| = x$

$g(f(x)) = g\left[(x - 2)^2\right] = \sqrt{(x - 2)^2} + 2 = x - 2 + 2 = x, x \geq 0$

20. $f(x) = x$, $g(x) = x$
$f(g(x)) = f(x) = x$
$g(f(x)) = g(x) = x$

22. $f(x) = \frac{x - 5}{2x + 3}$, $g(x) = \frac{3x + 5}{1 - 2x}$

$f(g(x)) = f\left[\frac{3x + 5}{1 - 2x}\right] = \dfrac{\dfrac{3x + 5}{1 - 2x} - 5}{2\left[\dfrac{3x + 5}{1 - 2x}\right] + 3} = \dfrac{\dfrac{3x + 5 - 5(1 - 2x)}{1 - 2x}}{\dfrac{6x + 10 + 3(1 - 2x)}{1 - 2x}} = \frac{13x}{1 - 2x} \cdot \frac{1 - 2x}{13} = x$

$g(f(x)) = g\left[\frac{x - 5}{2x + 3}\right] = \dfrac{3\left[\dfrac{x - 5}{2x + 3}\right] + 5}{1 - 2\left[\dfrac{x - 5}{2x + 3}\right]} = \dfrac{\dfrac{3x - 15 + 10x + 15}{2x + 3}}{\dfrac{2x + 3 - 2x + 10}{2x + 3}} = \frac{13x}{13} = x$

24. $f(x) = -4x$
$y = -4x$
$x = -4y$
$y = \frac{-x}{4}$
$f^{-1}(x) = \frac{-x}{4}$

Verify: $f(f^{-1}(x)) = f\left[\frac{-x}{4}\right] = -4\left[\frac{-x}{4}\right] = x$

$f^{-1}(f(x)) = f^{-1}(-4x) = \frac{-(-4x)}{4} = x$

Domain of f = range of f^{-1} = $(-\infty, \infty)$
Range of f = domain of f^{-1} = $(-\infty, \infty)$

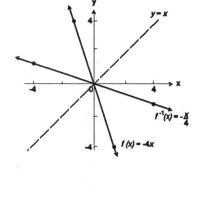

26. $f(x) = 1 - 3x$
$y = 1 - 3x$
$x = 1 - 3y$
$3y = -x + 1$
$y = \frac{-x + 1}{3}$
$f^{-1}(x) = \frac{-x + 1}{3}$

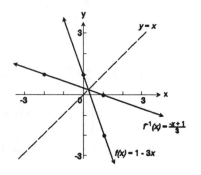

Chapter 4 Functions and Their Graphs

Verify: $f(f^{-1}(x)) = f\left[\dfrac{-x+1}{3}\right] = 1 - 3\left[\dfrac{-x+1}{3}\right] = 1 + x - 1 = x$

$f^{-1}(f(x)) = f^{-1}(1 - 3x) = \dfrac{-(1 - 3x) + 1}{3} = \dfrac{3x}{3} = x$

Domain of f = range of $f^{-1} = (-\infty, \infty)$
Range of f = domain of $f^{-1} = (-\infty, \infty)$

28. $f(x) = x^3 + 1$
$y = x^3 + 1$
$x = y^3 + 1$
$y^3 = x - 1$
$y = \sqrt[3]{x - 1}$
$f^{-1}(x) = \sqrt[3]{x - 1}$

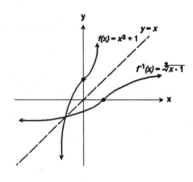

Verify: $f(f^{-1}(x)) = f\left(\sqrt[3]{x - 1}\right) = \left(\sqrt[3]{x - 1}\right)^3 + 1 = x$

$f^{-1}(f(x)) = f^{-1}(x^3 + 1) = \sqrt[3]{x^3 + 1 - 1} = x$

Domain of f = range of $f^{-1} = (-\infty, \infty)$
Range of f = domain of $f^{-1} = (-\infty, \infty)$

30. $f(x) = x^2 + 9, x \geq 0$
$y = x^2 + 9, x \geq 0$
$x = y^2 + 9$
$y^2 = x - 9$
$y = \sqrt{x - 9}$
$f^{-1}(x) = \sqrt{x - 9}$

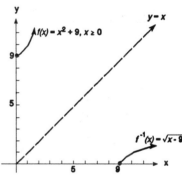

Verify: $f(f^{-1}(x)) = f\left(\sqrt{x - 9}\right) = \left(\sqrt{x - 9}\right)^2 + 9 = x$

$f^{-1}(f(x)) = f^{-1}(x^2 + 9) = \sqrt{x^2 + 9 - 9} = \sqrt{x^2} = |x| = x, x \geq 0$

Domain of f = range of $f^{-1} = [0, \infty)$
Range of f = domain of $f^{-1} = [9, \infty)$

32. $f(x) = \dfrac{-3}{x}$

$y = \dfrac{-3}{x}$

$x = \dfrac{-3}{y}$

$xy = -3$

$f^{-1}(x) = \dfrac{-3}{x}$

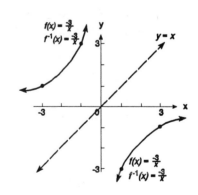

Verify: $f(f^{-1}(x)) = f\left[\dfrac{-3}{x}\right] = \dfrac{-3}{\dfrac{-3}{x}} = x$

$f^{-1}(f(x)) = f^{-1}\left[\dfrac{-3}{x}\right] = \dfrac{-3}{\dfrac{-3}{x}} = x$

Domain of f = range of f^{-1} = domain of f^{-1} = range of f = all real numbers except 0

34. $f(x) = \dfrac{4}{x + 2}$

$y = \dfrac{4}{x + 2}$

$x = \dfrac{4}{x + 2}$

$xy + 2x = 4$

$xy = 4 - 2x$

$y = \dfrac{4 - 2x}{x}$

$f^{-1}(x) = \dfrac{4 - 2x}{x}$

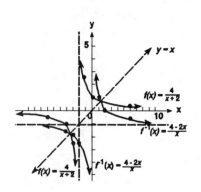

Verify: $f(f^{-1}(x)) = f\left[\dfrac{4 - 2x}{x}\right] = \dfrac{4}{\dfrac{4 - 2x}{x} + 2} = \dfrac{4}{\dfrac{4 - 2x + 2x}{x}} = x$

$f^{-1}(f(x)) = f^{-1}\left[\dfrac{4}{x + 2}\right] = \dfrac{4 - 2\left[\dfrac{4}{x + 2}\right]}{\dfrac{4}{x + 2}} = \dfrac{\dfrac{4x + 8 - 8}{x + 2}}{\dfrac{4}{x + 2}} = x$

Domain of f = range of f^{-1} = all real numbers except -2
Range of f = domain of f^{-1} = all real numbers except 0

36. $f(x) = \dfrac{4}{2 - x}$ Verify:

$y = \dfrac{4}{2 - x}$ $f(f^{-1}(x)) = f\left[\dfrac{2x - 4}{x}\right] = \dfrac{4}{2 - \left[\dfrac{2x - 4}{x}\right]} = \dfrac{4}{\dfrac{2x - 2x + 4}{x}} = x$

$x = \dfrac{4}{2 - y}$

$2x - xy = 4$

$xy = 2x - 4$ $f^{-1}(f(x)) = f^{-1}\left[\dfrac{4}{2 - x}\right] = \dfrac{2\left[\dfrac{4}{2 - x}\right] - 4}{\dfrac{4}{2 - x}} = \dfrac{\dfrac{8 - 8 + 4x}{2 - x}}{\dfrac{4}{2 - x}} = x$

$y = \dfrac{2x - 4}{x}$

$f^{-1}(x) = \dfrac{2x - 4}{x}$ Domain of f = range of f^{-1} = all real numbers except 2
Range of f = domain of f^{-1} = all real numbers except 0

38. $f(x) = (x - 1)^2, x \geq 1$
$y = (x - 1)^2, x \geq 1$ Verify: $f(f^{-1}(x)) = f(\sqrt{x} + 1) = \left(\sqrt{x} + 1 - 1\right)^2 = x$
$x = (y - 1)^2$
$\sqrt{x} = y - 1$ $f^{-1}(f(x)) = f^{-1}(x - 1)^2) = \sqrt{(x - 1)^2} + 1 = x$
$y = \sqrt{x} + 1$ Domain of f = range of f^{-1} = $[1, \infty)$
Range of f = domain of f^{-1} = $[0, \infty)$
$f^{-1}(x) = \sqrt{x} + 1$

Chapter 4 Functions and Their Graphs

40.

$$f(x) = \frac{3x + 1}{x}$$ Verify:

$$y = \frac{3x + 1}{x}$$

$$x = \frac{3y + 1}{y}$$

$$xy = 3y + 1$$

$$xy - 3y = 1$$

$$y(x - 3) = 1$$

$$y = \frac{1}{x - 3}$$

$$f^{-1}(x) = \frac{1}{x - 3}$$

$$f(f^{-1}(x)) = f\left(\frac{1}{x - 3}\right) = \frac{3\left[\dfrac{1}{x - 3}\right] + 1}{\dfrac{1}{x - 3}} = \frac{\dfrac{3 + x - 3}{x - 3}}{\dfrac{1}{x - 3}} = x$$

$$f^{-1}(f(x)) = f^{-1}\left(\frac{3x + 1}{x}\right) = \frac{1}{\dfrac{3x + 1}{x} - 3} = \frac{1}{\dfrac{3x + 1 - 3x}{x}} = x$$

Domain f = range f^{-1} = all real numbers except 0
Domain f^{-1} = range f = all real numbers except 3

42.

$$f(x) = \frac{2x - 3}{x + 4}$$

$$y = \frac{2x - 3}{x + 4}$$

$$x = \frac{2y - 3}{y + 4}$$

$$xy + 4x = 2y - 3$$

$$xy - 2y = -4x - 3$$

$$y(x - 2) = -4x - 3$$

$$y = \frac{-4x - 3}{x - 2}$$

$$f^{-1}(x) = \frac{-4x - 3}{x - 2}$$

Verify: $$f(f^{-1}(x)) = f\left(\frac{-4x - 3}{x - 2}\right) = \frac{2\left[\dfrac{-4x - 3}{x - 2}\right] - 3}{\left[\dfrac{-4x - 3}{x - 2}\right] + 4} = \frac{\dfrac{-8x - 6 - 3x + 6}{x - 2}}{\dfrac{-4x - 3 + 4x - 8}{x - 2}} = \frac{-11x}{-11} = x$$

$$f^{-1}(f(x)) = f^{-1}\left(\frac{2x - 3}{x + 4}\right) = \frac{-4\left[\dfrac{2x - 3}{x + 4}\right] - 3}{\left[\dfrac{2x - 3}{x + 4}\right] - 2} = \frac{\dfrac{-8x + 12 - 3x - 12}{x + 4}}{\dfrac{2x - 3 - 2x - 8}{x + 4}} = \frac{-11x}{-11} = x$$

Domain f = range f^{-1} = all real numbers except -4
Range f = domain f^{-1} = all real numbers except 2

44.

$$f(x) = \frac{-3x - 4}{x - 2}$$

$$y = \frac{-3x - 4}{x - 2}$$

$$x = \frac{-3y - 4}{y - 2}$$

$$xy - 2x = -3y - 4$$

$$xy + 3y = 2x - 4$$

$$y(x + 3) = 2x - 4$$

$$y = \frac{2x - 4}{x + 3}$$

$$f^{-1}(x) = \frac{2x - 4}{x + 3}$$

Verify: $f(f^{-1}(x)) = f\left[\dfrac{2x-4}{x+3}\right] = \dfrac{-3\left[\dfrac{2x-4}{x+3}\right]-4}{\left[\dfrac{2x-4}{x+3}\right]-2} = \dfrac{\dfrac{-6x+12-4x-12}{x+3}}{\dfrac{2x-4-2x-6}{x+3}} = \dfrac{-10x}{-10} = x$

$f^{-1}(f(x)) = f^{-1}\left[\dfrac{-3x-4}{x-2}\right] = \dfrac{2\left[\dfrac{-3x-4}{x-2}\right]-4}{\left[\dfrac{-3x-4}{x-2}\right]+3} = \dfrac{\dfrac{-6x-8-4x+8}{x-2}}{\dfrac{-3x-4+3x-6}{x-2}} = \dfrac{-10x}{-10} = x$

Domain f = range f^{-1} = all real numbers except 2
Range f = domain f^{-1} = all real numbers except -3

46. $f(x) = \dfrac{4}{\sqrt{x}}$

$y = \dfrac{4}{\sqrt{x}}$

$x = \dfrac{4}{\sqrt{y}}$

$\sqrt{y} = \dfrac{4}{x}$

$y = \dfrac{16}{x^2}$

$f^{-1}(x) = \dfrac{16}{x^2}$

Verify: $f(f^{-1}(x)) = f\left[\dfrac{16}{x^2}\right] = \dfrac{4}{\sqrt{\dfrac{16}{x^2}}} = \dfrac{4}{\dfrac{4}{x}} = x$

$f^{-1}(f(x)) = f^{-1}\left[\dfrac{4}{\sqrt{x}}\right] = \dfrac{16}{\left[\dfrac{4}{\sqrt{x}}\right]^2} = \dfrac{16}{\dfrac{16}{x}} = x$

Domain f = range f^{-1} = $(0, \infty)$
Range f = domain f^{-1} = all real numbers except 0

48. $f(x) = \sqrt{r^2 - x^2},\ 0 \le x \le r$

$y = \sqrt{r^2 - x^2}$

$x = \sqrt{r^2 - y^2}$
$x^2 = r^2 - y^2$
$y^2 = r^2 - x^2$

$y = \sqrt{r^2 - x^2}$

$f^{-1}(x) = \sqrt{r^2 - x^2}$

50. No

52. Quadrant II

54. $f(x) = x^4,\ x \ge 0$, is one-to-one
We have $y = x^4,\ x \ge 0$
$x = y^4$

$y = \sqrt[4]{x}$

$f^{-1}(x) = \sqrt[4]{x},\ x \ge 0$

Verify: $f(f^{-1}(x)) = f\left(\sqrt[4]{x}\right) = \left(\sqrt[4]{x}\right)^4 = x$

$f^{-1}(f(x)) = f^{-1}(x^4) = \sqrt[4]{x^4} = x$

56. $p(x) = 300 - 5x$
$p = 300 - 5x$
$5x = 300 - p$

$x = \dfrac{1}{5}(300 - p)$

$x(p) = \dfrac{1}{5}(300 - p)$

58. $f(x) = \begin{cases} x & \text{if } x \text{ is rational} \\ -x & \text{if } x \text{ is irrational} \end{cases}$

Chapter 4 Functions and Their Graphs

60. (a) $$f(x) = \frac{2x + 5}{x - 3}$$

$$x = \frac{2y + 5}{y - 3}$$
$$xy - 3x = 2y + 5$$
$$xy - 2y = 3x + 5$$
$$y(x - 2) = 3x + 5$$
$$y = \frac{3x + 5}{x - 2}$$
$$f^{-1}(x) = \frac{3x + 5}{x - 2}$$

The range of f is $\{y \mid y \neq 2\}$.

(b) $$g(x) = 4 - \frac{2}{x}$$

$$y = 4 - \frac{2}{x}$$
$$x = 4 - \frac{2}{y}$$
$$x - y = \frac{-2}{y}$$
$$xy - 4y = -2$$
$$y(x - 4) = -2$$
$$g^{-1}(x) = \frac{-2}{x - 4}$$

The range of g is $\{y \mid y \neq 4\}$.

(c) $$F(x) = \frac{3}{4 - x}$$

$$y = \frac{3}{4 - x}$$
$$x = \frac{3}{4 - y}$$
$$4x - xy = 3$$
$$xy = 4x - 3$$
$$y = \frac{4x - 3}{x}$$
$$F^{-1}(x) = \frac{4x - 3}{x}$$

The range of F is $\{y \mid y \neq 0\}$.

68. Yes. The graph of f is symmetric with respect to the line $y = x$.

4.6 Mathematical Models: Constructing Functions

2. If $V = \frac{1}{3}\pi r^2 h$ and $h = 2r$, then $V(r) = \frac{1}{3}\pi r^2 (2r) = \frac{2}{3}\pi r^3$

4. If $p = \frac{-1}{3}x + 100$ and $R = xp$, then $R(x) = x\left(\frac{-1}{3}x + 100\right) = \frac{-1}{3}x^2 + 100x$

6. If $x = -20p + 500$ and $R = xp$, then

$$p = \frac{500 - x}{20} \text{ and } R(x) = x\left(\frac{500 - x}{20}\right) = \frac{-1}{20}x^2 + 25x$$

8. (a) Let P = perimeter to be fenced, x = length, w = width.

Since only 3 sides require fencing, $P = x + 2w = 3000$, so $w = \frac{3000 - x}{2}$. The area of the

rectangle as a function of x, represented by $A(x) = xw = x\left(\frac{3000 - x}{2}\right) = \frac{-x^2}{2} + 1500x$,

$0 < x < 3000$

(b) *A* is largest when $x = 1500$ feet.

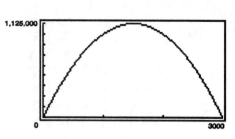

10. (a) Let P = perimeter of a square, s = side.

We know that $P = 4s$, by definition. If a wire of length x is bent into the shape of a square, then x is the perimeter, so $P = x = 4s$. Expressing the perimeter of the square as a function of x, we have $P(x) = x$.

(b) We know that $P = x = 4s$, so $s = \dfrac{x}{4}$. By definition, the area of a square is $A = s^2$.

Expressing the area of the square as a function of x, we have $A(x) = \left(\dfrac{x}{4}\right)^2 = \dfrac{x^2}{16}$

12. By definition, a triangle has area $A = \dfrac{1}{2}bh$ where b = base, h = height. Because a vertex of the triangle is at the origin, we know that $b = x$ and $h = y$. Expressing the area of the triangle as a function of x, we have $A(x) = \dfrac{1}{2}xy = \dfrac{1}{2}x(9 - x^2) = \dfrac{-1}{2}x^3 + \dfrac{9}{2}x$

14. (a) The distance d from P to the point $(0, -1)$ is $d = \sqrt{x^2 + (y + 1)^2}$. Since P is a point on the graph of $y = x^2 - 8$, we have

$$d(x) = \sqrt{x^2 + (x^2 - 8 + 1)^2}$$
$$d(x) = \sqrt{x^2 + (x^2 - 7)^2}$$
$$d(x) = \sqrt{x^4 - 13x^2 + 49}$$

(b) If $x = 0$, the distance d is $d(0) = \sqrt{49} = 7$.

(c) If $x = -1$, the distance d is $d(-1) = \sqrt{1 - 13 + 49} = \sqrt{37}$

16. The distance d from P to the origin is

$d = \sqrt{x^2 + y^2}$. Since P is a point on the

graph of $y = \dfrac{1}{x}$, we have

$$d(x) = \sqrt{x^2 + \left(\dfrac{1}{x}\right)^2}$$

$$d(x) = \sqrt{\dfrac{x^4 + 1}{x^2}}$$

18.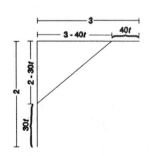

(a) $d^2 = (2 - 30t)^2 + (3 - 40t)^2$

$d(t) = \sqrt{(2 - 30t)^2 + (3 - 40t)^2}$

(b) d is smallest when $t = 0.072$ hours

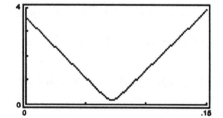

Chapter 4 Functions and Their Graphs

20. (a) A = Amount of material used
V = Volume, h = height
$V = (x)(x)(h) = x^2h$
$\quad\quad = 10$ ft^3

$$h = \frac{10}{x^2}$$

$A = x^2 + 4xh$

$$A(x) = x^2 + 4x\left[\frac{10}{x^2}\right]$$

$$A(x) = x^2 + \frac{40}{x}$$

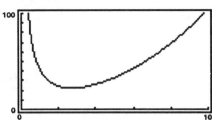

(b) A is smallest if $x = 2.71$ feet.

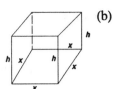

22. V = volume, r = radius, S = surface area

$$V = \frac{4}{3}\pi r^3 \text{ and } S = 4\pi r^2$$

$$\frac{S}{4\pi} = r^2$$

$$r = \sqrt{\frac{S}{4\pi}}$$

$$V(S) = \frac{4}{3}\pi\left[\sqrt{\frac{S}{4\pi}}\right]^3$$

$$V(S) = \frac{4}{3}\pi \cdot \frac{S}{4\pi}\sqrt{\frac{S}{4\pi}} = \frac{S}{3}\sqrt{\frac{S}{4\pi}}$$

$$V(S) = \frac{S}{6}\sqrt{\frac{S}{\pi}}$$

If the surface area doubles, then S becomes $2S$ and we have

$$V(2S) = \frac{2S}{6}\sqrt{\frac{2S}{\pi}} = \frac{S}{3}\sqrt{\frac{2S}{\pi}}$$

Thus, when the surface area doubles, the volume is multiplied by $2\sqrt{2}$.

24. A = Area of the rectangle
(a) $A = (2x)(y) = 2xy$
Since $x^2 + y^2 = 4$,

then $y = \sqrt{4 - x^2}$

$A(x) = 2x\sqrt{4 - x^2}$

(b) p = Perimeter of the rectangle
$p = 4x + 2y$

$$p(x) = 4x + 2\sqrt{4 - x^2}$$

(c) A is largest if $x = 1.41$.

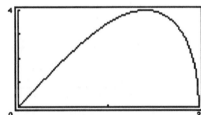

(d) p is largest if $x = 1.78$.

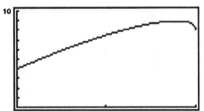

26. A = Area, p = Perimeter, r = radius
(a) $A(r) = (2r)^2 = 4r^2$
(b) $p(r) = 4(2r) = 8r$

28. A = Amount of material required, r = radius

Volume = $\pi r^2 h = 100 \Rightarrow h = \dfrac{100}{\pi r^2}$

(a) $A = 2\pi r^2 + 2\pi rh$

$$A(r) = 2\pi r^2 + 2\pi r \left[\frac{100}{\pi r^2}\right]$$

$$A(r) = 2\pi r^2 + \frac{200}{r}$$

(b)   If $r = 3$,

$$A(3) = 2\pi(3)^2 + \frac{200}{r}$$

$$= 18\pi + \frac{200}{3} \approx 123.22 \text{ ft.}$$

(c)   If $r = 4$,

$$A(4) = 2\pi(4)^2 + \frac{200}{4}$$

$$= 32\pi + \frac{200}{4} \approx 150.53 \text{ sq. ft.}$$

(d)   If $r = 5$,

$$A(5) = 2\pi(5)^2 + \frac{200}{5}$$

$$= 50\pi + \frac{200}{5} \approx 197.08 \text{ sq ft.}$$

(e) A is smallest for $r = 2.51$

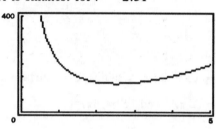

30. C = circumference, A = Area, r = radius, x = side of the equilateral triangle

$$C = 2\pi r = 10 - 3x; \ r = \frac{10 - 3x}{2\pi}$$

(a) $A = \pi r^2 + \dfrac{x^2}{4}\sqrt{3}$

$$A(x) = \pi \left[\frac{10 - 3x}{2\pi}\right]^2 + \frac{x^2}{4}\sqrt{3}$$

$$= \frac{(10 - 3x)^2}{4\pi} + \frac{\sqrt{3}}{4}x^2$$

(b)   Since all lengths must be positive, we have

$$x > 0 \text{ and } 10 - 3x > 0$$

$$-3x > -10$$

$$x < \frac{10}{3}$$

Thus, domain of $A = \left\{x \,\middle|\, 0 < x < \dfrac{10}{3}\right\}$

(c) A is smallest if $x = 2.07$ meters

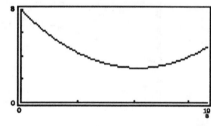

32. Let A = area of circle minus the area of the equilateral triangle

$$A = \pi r^2 - x^2 \left[\frac{\sqrt{3}}{4}\right]$$

$$A(x) = \frac{\pi x^2}{3} - x^2 \left[\frac{\sqrt{3}}{4}\right]$$

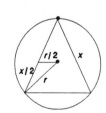

$$\left[\frac{x}{2}\right]^2 + \left[\frac{r}{2}\right]^2 = r^2$$

$$\frac{x^2}{4} = \frac{3r^2}{4}$$

$$r^2 = \frac{x^2}{3}$$

36. (a) Let x be the number of miles and let C be the cost. Then

$$C(x) = \begin{cases} 0.50x & \text{if } 0 < x \le 100 \\ 0.50(100) + 0.40(x - 100) & \text{if } 100 < x \le 400 \\ 0.50(100) + 0.40(300) + 0.25(x - 400) & \text{if } 400 < x \le 800 \\ 0.50(100) + 0.40(300) + 0.25(400) + 0(x - 800) & \text{if } 800 < x \le 960 \end{cases}$$

$$C(x) = \begin{cases} 0.50x & \text{if } 0 < x \le 100 \\ 10 + 0.40x & \text{if } 100 < x \le 400 \\ 70 + 0.25x & \text{if } 400 < x \le 800 \\ 270 & \text{if } 800 < x \le 960 \end{cases}$$

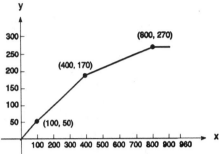

(b) For hauls between 100 and 400 miles, the cost as a function of miles is $C(x) = 10 + 0.40x$.

(c) For hauls between 400 and 800 miles, the cost as a function of miles is $C(x) = 70 + 0.25x$.

36. $$C(x) = \begin{cases} 219 & \text{if } x = 7 \\ 264 & \text{if } 7 < x \le 8 \\ 309 & \text{if } 8 < x \le 9 \\ 354 & \text{if } 9 < x \le 10 \\ 399 & \text{if } 10 < x \le 11 \\ 438 & \text{if } 11 < x \le 14 \end{cases}$$

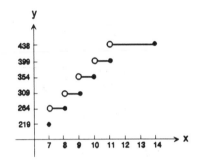

38. Schedule X: $$f(x) = \begin{cases} .15x & \text{if } 0 < x \le 22750 \\ 3412.50 + .28(x - 22750) & \text{if } 22750 < x \le 55100 \\ 12470.50 + .31(x - 55100) & \text{if } 55100 < x \le 115000 \\ 31039.50 + 36(x - 115000) & \text{if } 115000 < x \le 250000 \\ 79,639.50 + .396(x - 25000) & \text{if } x > 250000 \end{cases}$$

Schedule Y-1: $$f(x) = \begin{cases} .15x & \text{if } 0 < x \le 38000 \\ 5700 + .28(x - 38000) & \text{if } 38000 < x \le 91850 \\ 20,778 + .31(x - 91850) & \text{if } 91850 < x \le 140000 \\ 35,704.50 + .36(x - 140000) & \text{if } 140000 < x \le 250000 \\ 75,304.50 + .396(x - 250000) & \text{if } x > 250000 \end{cases}$$

4 Chapter Review

2. Using the point-slope form,
$y - y_1 = m(x - x_1)$, where $m = -4$ and
when $x = -2$, $y = 2$, we have
$$y - 2 = -4[x - (-2)]$$
$$y - 2 = -4x - 8$$
$$y = g(x) = -4x - 6$$

4. $g(x) = \dfrac{A}{x} + \dfrac{8}{x^2}$ and $g(-1) = 0$

$$0 = \frac{A}{-1} + \frac{8}{(-1)^2}$$

$$0 = \frac{A - 8}{-1}$$

$$A - 8 = 0$$

$$A = 8$$

6. (a) Domain: $\{x \mid -5 \le x \le 4\}$
        Range: $\{y \mid -3 \le y \le 1\}$
    (c) f is constant on $[-5, -1]$

    (b) f is increasing on $[3, 4]$

    (d) Intercepts: $(0, 0), (4, 0)$

8. $f(x) = \dfrac{x^2}{x + 2}$

    (a) $f(-x) = \dfrac{(-x)^2}{(-x) + 2} = \dfrac{x^2}{-x + 2}$

    (c) $f(x + 2) = \dfrac{(x + 2)^2}{(x + 2) + 2} = \dfrac{x^2 + 4x + 4}{x + 4}$

    (b) $-f(x) = \dfrac{-x^2}{x + 2}$

    (d) $f(x - 2) = \dfrac{(x - 2)^2}{(x - 2) + 2} = \dfrac{x^2 - 4x + 4}{x}$

10. $f(x) = \mid x^2 - 4 \mid$
    (a) $f(-x) = \mid (-x)^2 - 4 \mid = \mid x^2 - 4 \mid$ (b) $-f(x) = \mid x^2 - 4 \mid$
    (c) $f(x + 2) = \mid (x + 2)^2 - 4 \mid = \mid x^2 + 4x + 4 - 4 \mid = \mid x^2 + 4x \mid$
    (d) $f(x - 2) = \mid (x - 2)^2 - 4 \mid = \mid x^2 - 4x + 4 - 4 \mid = \mid x^2 - 4x \mid$

12. $f(x) = \dfrac{x^3}{x^2 - 4}$

    (a) $f(-x) = \dfrac{(-x)^3}{(-x)^2 - 4} = \dfrac{-x^3}{x^2 - 4}$

    (b) $-f(x) = \dfrac{-x^3}{x^2 - 4}$

    (c) $f(x + 2) = \dfrac{(x + 2)^3}{(x + 2)^2 - 4} = \dfrac{x^3 + 6x^2 + 12x + 8}{x^2 + 4x + 4 - 4} = \dfrac{x^3 + 6x^2 + 12x + 8}{x^2 + 4x}$

    (d) $f(x - 2) = \dfrac{(x - 2)^3}{(x - 2)^2 - 4} = \dfrac{x^3 - 6x^2 + 12x - 8}{x^2 - 4x + 4 - 4} = \dfrac{x^3 - 6x^2 + 12x - 8}{x^2 - 4x}$

14. $g(x) = \dfrac{4 + x^2}{1 + x^4}$

    $g(-x) = \dfrac{4 + (-x)^2}{1 + (-x)^4} = \dfrac{4 + x^2}{1 + x^4} = g(x)$
    Thus, g is an even function.

16. $F(x) = \sqrt{1 - x^3}$

    $F(-x) = \sqrt{1 - (-x)^3} = \sqrt{1 + x^3}$
    Because $F(-x) \ne F(x)$ and $F(-x) \ne -F(x)$, F is neither even nor odd.

18. $H(x) = 1 + x + x^2$
    $H(-x) = 1 + (-x) + (-x)^2 = 1 - x + x^2$
    Because $H(x) \ne H(-x)$ and $H(x) \ne -H(x)$, H is neither even nor odd.

20. $f(x) = \dfrac{3x^2}{x - 2}$ domain $= \{x \mid x \ne 2\}$

22. $f(x) = \sqrt{x + 2}$ domain $= \{x \mid x \ge -2\}$ or $[-2, \infty)$

24. $g(x) = \dfrac{\mid x \mid}{x}$ domain $= \{x \mid x \ne 0\}$

26. $F(x) = \dfrac{1}{x^2 - 3x - 4}$ The domain is such that $x^2 - 3x - 4 \ne 0$

$$(x - 4)(x + 1) \ne 0$$
$$x \ne 4, -1$$

domain $= \{x \mid x \ne -1, x \ne 4\}$

28. $H(x) = \begin{cases} \dfrac{1}{x} & \text{if } 0 < x < 4 \\ x - 4 & \text{if } 4 \le x \le 8 \end{cases}$ domain: $\{x \mid 0 < x \le 8\}$ or $(0, 8]$

30. $g(x) = \begin{cases} |1 - x| & \text{if } x < 1 \\ 3 & \text{if } x = 1 \\ x + 1 & \text{if } 1 < x \le 3 \end{cases}$ domain: $\{x \mid -\infty < x \le 3\}$ or $(-\infty, 3]$

32. $f(x) = |x| + 4$

(a) The domain is the set of all real numbers.

(b) x-intercept(s): y-intercept:
$$0 = |x| + 4 \qquad y = |0| + 4$$
$$|x| = -4 \qquad\quad y = 4$$
No solution.
The intercept is $(0, 4)$.

(c)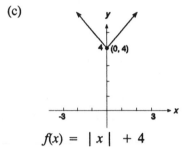

$f(x) = |x| + 4$

(d) The range is $\{y \mid 4 \le y < \infty\}$ or $[4, \infty)$.

34. $g(x) = \dfrac{1}{2} |x|$

(a) The domain is the set of all real numbers.

(b) x-intercept(s): y-intercept:
$$0 = \frac{1}{2} |x| \qquad y = \frac{1}{2} |0|$$
$$0 = |x| \qquad\qquad y = 0$$
$$x = 0$$
The intercept is $(0, 0)$.

(c)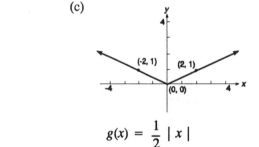

$g(x) = \dfrac{1}{2} |x|$

(d) The range is $\{y \mid 0 \le y < \infty\}$ or $[0, \infty)$.

36. $h(x) = \sqrt{x} - 1$

(a) The domain is $\{x \mid 0 \le x < \infty\}$ or $[0, \infty)$.

(b) x-intercept(s): y-intercept:
$$0 = \sqrt{x} - 1 \qquad y = \sqrt{0} - 1$$
$$\sqrt{x} = 1 \qquad\qquad y = -1$$
$$x = 1$$
The intercepts are $(1, 0)$, $(0, -1)$.

(c)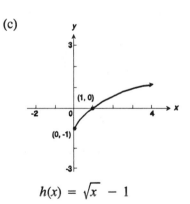

$h(x) = \sqrt{x} - 1$

(d) The range is $\{y \mid -1 \le y < \infty\}$ or $[-1, \infty)$.

38. $f(x) = -\sqrt{x}$

    (a)   The domain is $\{x \mid 0 \le x < \infty\}$ or $[0, \infty)$.

    (b) x-intercept(s): y-intercept:

$$0 = -\sqrt{x} \qquad y = -\sqrt{0}$$
$$x = 0 \qquad\qquad y = 0$$

        The intercept is $(0, 0)$.

(c)

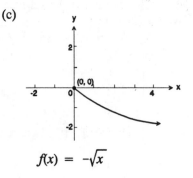

$$f(x) = -\sqrt{x}$$

    (d)   The range is $\{y \mid -\infty < y \le 0\}$ or $(-\infty, 0]$.

40. $H(x) = \begin{cases} |1 - x| & \text{if } 0 \le x \le 2 \\ |x - 1| & \text{if } x > 2 \end{cases}$

    (a)   The domain is $\{x \mid 0 \le x < \infty\}$ or $[0, \infty)$.

    (b) x-intercept(s): y-intercept:

$$0 = |1 - x| \qquad y = |1 - 0|$$
$$0 = 1 - x \qquad\quad y = 1$$
$$x = 1$$

        The intercepts are $(1, 0)$, $(0, 1)$.

(c)

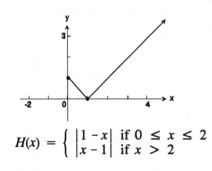

$$H(x) = \begin{cases} |1 - x| & \text{if } 0 \le x \le 2 \\ |x - 1| & \text{if } x > 2 \end{cases}$$

    (d)   The range is $\{y \mid 0 \le y < \infty\}$ or $[0, \infty)$.

42. $h(x) = (x + 2)^2 - 3$

    (a)   The domain is set of all real numbers.

    (b) x-intercept(s): y-intercept:

$$0 = (x + 2)^2 - 3 \qquad y = (0 + 2)^2 - 3$$
$$0 = x^2 + 4x + 4 - 3 \quad y = 1$$
$$0 = x^2 + 4x + 1$$

$$x = \frac{-4 \pm \sqrt{16 - 4}}{2}$$

$$x = \frac{-4 \pm \sqrt{12}}{2}$$

$$x = \frac{-4 \pm 2\sqrt{3}}{2}$$

$$x = -2 \pm \sqrt{3}$$

        The intercepts are

$$\left(-2 + \sqrt{3}, 0\right), \left(-2 - \sqrt{3}, 0\right), (0, 1).$$

(c)

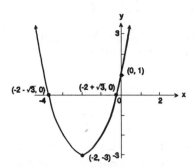

    (d)   The range is $\{y \mid -3 \le y < \infty\}$ or $[-3, \infty)$.

44. $g(x) = (x + 2)^3 - 8$

    (a)   The domain is the set of all real numbers.

    (b) x-intercept(s): y-intercept:

$$0 = (x + 2)^3 - 8 \qquad y = (0 + 2)^3 - 8$$
$$(x + 2)^3 = 8 \qquad\qquad y = 0$$
$$x + 2 = \sqrt[3]{8}$$
$$x = \sqrt[3]{8} - 2 = 2 - 2 = 0$$

        The intercept is $(0, 0)$.

(c)

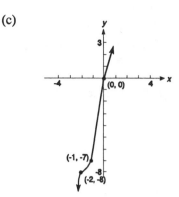

    (d)   The range is the set of all real numbers.

46. $f(x) = \begin{cases} 3|x| & \text{if } x < 0 \\ \sqrt{1 - x} & \text{if } 0 \le x \le 1 \end{cases}$

    (a)   The domain is $\{x \mid -\infty < x \le -1\}$ or $(-\infty, 1]$.

    (b) x-intercept(s): y-intercept:

$$0 = \sqrt{1 - x} \qquad y = \sqrt{1 - 0}$$
$$x = 1 \qquad\qquad y = 1$$
$$x = 1$$

        The intercepts are $(1, 0)$, $(0, 1)$.

(c)

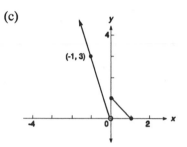

    (d)   The range is $\{y \mid 0 \le y < \infty\}$ or $[0, \infty)$.

48. $g(x) = \dfrac{1}{x + 2} - 2$

    (a)   The domain is $\{x \mid x \ne -2\}$.

    (b) x-intercept(s): y-intercept:

$$0 = \frac{1}{x + 2} - 2 \qquad y = \frac{1}{0 + 2} - 2$$
$$2 = \frac{1}{x + 2} \qquad\qquad y = \frac{-3}{2}$$
$$2x + 4 = 1$$
$$2x = -3$$
$$x = \frac{-3}{2}$$

        The intercepts are $\left[\dfrac{-3}{2}, 0\right]$, $\left[0, \dfrac{-3}{2}\right]$

    (d)   The range is $\{y \mid y \ne -2\}$

(c)

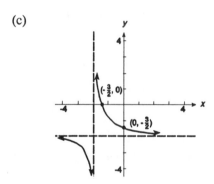

$$g(x) = \frac{1}{x + 2} - 2$$

50. $h(x) = -[\![x]\!]$

    (a)   The domain is the set of all real numbers.

    (b) x-intercept(s): y-intercept:

$$0 = -[\![x]\!] \qquad y = -[\![0]\!]$$
$$0 \le x < 1 \qquad y = 0$$
$$\{(x, 0) \mid 0 \le x < 1\}$$

(c)

$h(x) = -[\![x]\!]$

    (d)   The range is the set of integers.

52. $$f(x) = \frac{2 - x}{3 + x}$$

$$y = \frac{2 - x}{3 + x}$$

$$x = \frac{2 - y}{3 + y}$$

$$3x + xy = 2 - y$$
$$xy + y = 2 - 3x$$
$$y(x + 1) = 2 - 3x$$

$$y = \frac{2 - 3x}{x + 1}$$

$$f^{-1}(x) = \frac{2 - 3x}{x + 1}$$

Verify: $\displaystyle f(f^{-1}(x)) = f\left(\frac{2 - 3x}{x + 1}\right) = \frac{2 - \left[\dfrac{2 - 3x}{x + 1}\right]}{3 + \left[\dfrac{2 - 3x}{x + 1}\right]} = \frac{\dfrac{2x + 2 - 2 + 3x}{x + 1}}{\dfrac{3x + 3 + 2 - 3x}{x + 1}} = \frac{5x}{5} = x$

$\displaystyle f^{-1}(f(x)) = f^{-1}\left(\frac{2 - x}{3 + x}\right) = \frac{2 - 3\left[\dfrac{2 - x}{x + 3}\right]}{1 + \left[\dfrac{2 - x}{x + 3}\right]} = \frac{\dfrac{2x + 6 - 6 + 3x}{x + 3}}{\dfrac{x + 3 + 2 - x}{x + 3}} = \frac{5x}{5} = x$

Domain of f = range of f^{-1} = all real numbers except -3
Range of f = domain of f^{-1} = all real numbers except -1

54. $f(x) = \sqrt{x - 2}, \ x \ge 2, \ y \ge 0$

$y = \sqrt{x - 2}$

$x = \sqrt{y - 2}$

$x^2 = y - 2$

$y = x^2 + 2$

$f^{-1}(x) = x^2 + 2, \ x \ge 0$

Verify: $f(f^{-1}(x)) = f(x^2 + 2) = \sqrt{x^2 + 2 - 2} = x$

$f^{-1}(f(x)) = f^{-1}\left(\sqrt{x - 2}\right) = \left(\sqrt{x - 2}\right)^2 + 2 = x$

Domain of f = range of f^{-1} = $[2, \infty)$
Range of f = domain of f^{-1} = $[0, \infty)$

56. $f(x) = x^{1/3} + 1$

       $y = x^{1/3} + 1$

       $x = y^{1/3} + 1$

   $x - 1 = y^{1/3}$

       $y = (x - 1)^3$

   $f^{-1}(x) = (x - 1)^3$

Verify: $f(f^{-1}(x)) = f[(x - 1)^3] = [(x - 1)^3]^{1/3} + 1 = x$

               $f^{-1}(f(x)) = f^{-1}(x^{1/3} + 1) = (x^{1/3} + 1 - 1)^3 = x$

Domain of f = range of f^{-1} = all real numbers

Range of f = domain of f^{-1} = all real numbers

58. $f(x) = 4 - x, g(x) = 1 + x^2$

   (a) $(f \circ g)(2) = f(g(2)) = f(5) = -1$

                  $\uparrow$

              $g(2) = 5$

   (b) $(g \circ f)(-2) = g(f(-2)) = g(6) = 37$

                     $\uparrow$

               $f(-2) = 6$

   (c) $(f \circ f)(4) = f(f(4)) = f(0) = 4$

                $\uparrow$

             $f(4) = 0$

   (d) $(g \circ g)(-1) = g(g(-1)) = g(2) = 5$

                   $\uparrow$

              $g(-1) = 2$

60. $f(x) = 1 - 3x^2, g(x) = \sqrt{4 - x}$

   (a) $(f \circ g)(2) = f(g(2)) = 1 - 3\sqrt{2^2} = 1 - 3 \cdot 2 = -5$

                $\uparrow$

              $g(2) = \sqrt{2}$

   (b) $(g \circ f)(-2) = g(f(-2)) = \sqrt{4 - (-11)} = \sqrt{15}$

                   $\uparrow$

               $f(-2) = -11$

   (c) $(f \circ f)(4) = f(f(4)) = 1 - 3(-47)^2 = -6626$

                $\uparrow$

              $f(4) = -47$

   (d) $(g \circ g)(-1) = g(g(-1)) = \sqrt{4 - \sqrt{5}}$

                  $\uparrow$

              $g(-1) = \sqrt{5}$

62. $f(x) = \dfrac{2}{1 + 2x^2}, g(x) = 3x$

   (a) $(f \circ g)(2) = f(g(2)) = f(6) = \dfrac{2}{73}$

                $\uparrow$

              $g(2) = 6$

   (b) $(g \circ f)(-2) = g(f(-2)) = g\left[\dfrac{2}{9}\right] = \dfrac{2}{3}$

                  $\uparrow$

              $f(-2) = \dfrac{2}{9}$

(c) $(f \circ f)(4) = f(f(4)) = f\left[\dfrac{2}{33}\right] = \dfrac{2}{11}$

$\uparrow$

$f(4) = \dfrac{2}{33}$

(d) $(g \circ g)(-1) = g(g(-1)) = g(-3) = -9$

$\uparrow$

$g(-1) = -3$

64. $f(x) = \dfrac{2x}{x + 1}, \; g(x) = \dfrac{2x}{x - 1}$

$(f \circ g)(x) = f(g(x)) = f\left[\dfrac{2x}{x - 1}\right] = \dfrac{2\dfrac{2x}{x - 1}}{\dfrac{2x}{x - 1} + 1} = \dfrac{\dfrac{4x}{x - 1}}{\dfrac{2x + x - 1}{x - 1}} = \dfrac{4x}{x - 1} \cdot \dfrac{x - 1}{3x - 1} = \dfrac{4x}{3x - 1}$

$(g \circ f)(x) = g(f(x)) = g\left[\dfrac{2x}{x + 1}\right] = \dfrac{2\dfrac{2x}{x + 1}}{\dfrac{2x}{x + 1} - 1} = \dfrac{\dfrac{4x}{x + 1}}{\dfrac{2x - (x + 1)}{x + 1}} = \dfrac{4x}{x + 1} \cdot \dfrac{x + 1}{x - 1} = \dfrac{4x}{x - 1}$

$(f \circ f)(x) = f(f(x)) = f\left[\dfrac{2x}{x + 1}\right] = \dfrac{2\dfrac{2x}{x + 1}}{\dfrac{2x}{x + 1} + 1} = \dfrac{\dfrac{4x}{x + 1}}{\dfrac{2x + x + 1}{x + 1}} = \dfrac{4x}{x + 1} \cdot \dfrac{x + 1}{3x + 1} = \dfrac{4x}{3x + 1}$

$(g \circ g)(x) = g(g(x)) = g\left[\dfrac{2x}{x - 1}\right] = \dfrac{2\dfrac{2x}{x - 1}}{\dfrac{2x}{x - 1} - 1} = \dfrac{\dfrac{4x}{x - 1}}{\dfrac{2x - (x - 1)}{x - 1}} = \dfrac{4x}{x - 1} \cdot \dfrac{x - 1}{x + 1} = \dfrac{4x}{x + 1}$

66. $f(x) = \sqrt{3x}, \; g(x) = 1 + x + x^2$

$(f \circ g)(x) = f(g(x)) = f(1 + x + x^2) = \sqrt{3(1 + x + x^2)} = \sqrt{3 + 3x + 3x^2}$

$(g \circ f)(x) = g(f(x)) = g\left(\sqrt{3x}\right) = 1 + \sqrt{3x} + \sqrt{3x}^2 = 1 + \sqrt{3}x + 3x$

$(f \circ f)(x) = f(f(x)) = f\left(\sqrt{3x}\right) = \sqrt{3\sqrt{3x}}$

$(g \circ g)(x) = g(g(x)) = g(1 + x + x^2) + 1 + (1 + x + x^2) + (1 + x + x^2)^2$

$= 2 + x + x^2 + 1 + 2x + 3x^2 + 2x^3 + x^4$

$= 3 + 3x + 4x^2 + 2x^3 + x^4$

68. $f(x) = \sqrt{x^2 - 3}, \; g(x) = \sqrt{3 - x^2}$

$f \circ g = f(g(x)) = f\left(\sqrt{3 - x^2}\right) = \sqrt{\left(\sqrt{3 - x^2}\right)^2 - 3}$

$g \circ f = g(f(x)) = g\left(\sqrt{x^2 - 3}\right) = \sqrt{3 - \left(\sqrt{x^2 - 3}\right)^2} = \sqrt{3 - (x^2 - 3)} = \sqrt{6 - x^3}$

$f \circ f = f(f(x)) = f\left(\sqrt{x^2 - 3}\right) = \sqrt{\left(\sqrt{x^2 - 3}\right)^3 - 3}$

$g \circ g = g(g(x)) = g\left(\sqrt{3 - x^2}\right) = \sqrt{3 - \left(\sqrt{3 - x^2}\right)^2} = \sqrt{3 - (3 - x^2)} = \sqrt{x^2} = |x|$

70. (a) $y = f(-x)$

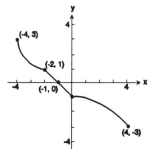

(b) $y = -f(x)$

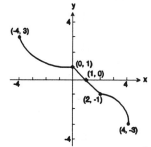

(c) $y = f(x + 2)$

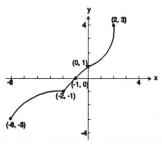

(d) $y = f(x) + 2$

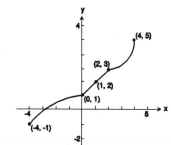

(e) $y = f(2 - x)$

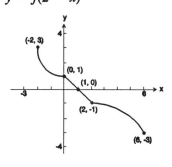

(f) f^{-1}

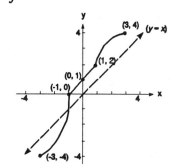

72. $v(20) = 80, \ m = 5$

$v - v_1 = m(t - t_1)$

$v - 80 = 5(t - 20)$

$v = v(t) = 5t - 20$

$v(30) = 5 \cdot 30 - 20 = 130$

The car is going 130 feet per second after 30 seconds.

2. $y = 10x + \sqrt{(4 + (5 - x)^2)}$

3.
x	y
0	$75.39
1	$72.61
2	$70.50
3	$69.60
4	$71.24
5	$78.00

Yes, going 3 miles along the highway before turning toward the house seems to cost the least.

4. All the installation possibilities seem to be less than $80, so the company would not lose money on any choice the Stevens made.

5. There appears to be a minimum at $x = 2.955$, but the cost would still be $69.60.

6. If $x = 4.5$ miles, the cost would be $73.86.

7. The cost would be 5000 times the difference between $73.86 and $69.60. This is $21,300.

POLYNOMIAL AND RATIONAL FUNCTIONS

2. F   4. H   6. C   8. G

10. $f(x) = 2x^2$
   Start with the graph of $y = x^2$. Then
   vertically stretch to obtain the graph of
   $f(x) = 2x^2$.

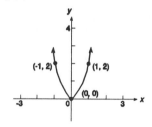

12. $f(x) = 2x^2 - 3$
   Start with the graph of $y = x^2$.
   Vertically stretch and shift down
   3 units.

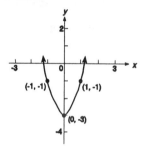

14. $f(x) = 2x^2 + 4$
   Start with the graph of $y = x^2$. Vertically
   stretch and shift up 4 units.

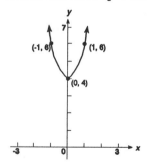

16. $f(x) = -2x^2 - 2$
   Start with the graph of $y = x^2$.
   Vertically stretch, reflect about the
   x-axis and shift down 2 units.

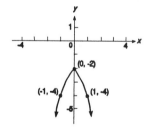

18. $f(x) = x^2 - 6x - 1$
    $f(x) = (x^2 - 6x + 9) - 1 - 9$
    $f(x) = (x - 3)^2 - 10$
    Start with the graph of $y = x^2$. Shift to the right 3 units and shift down 10 units.

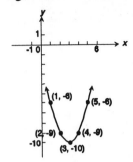

20. $f(x) = 3x^2 + 6x$
    $f(x) = 3(x^2 + 2x)$
    $f(x) = 3(x^2 + 2x + 1) - 3$
    $f(x) = 3(x + 1)^2 - 3$
    Start with the graph of $y = x^2$.

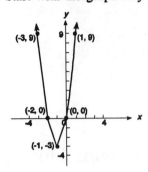

22. $f(x) = -2x^2 + 6x + 2$
    $f(x) = -2(x^2 - 3x) + 2$
    $f(x) = -2\left[x - 3x + \dfrac{9}{4}\right] + \dfrac{9}{2} + 2$
    $f(x) = -2\left[x - \dfrac{3}{2}\right]^2 + \dfrac{13}{2}$
    Start with the graph of $y = x^2$. Vertically stretch, reflect about the x-axis, shift to the right $\dfrac{3}{2}$ units and shift up $\dfrac{13}{2}$ units.

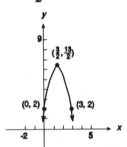

24. $f(x) = \dfrac{2}{3}x^2 + \dfrac{4}{3}x - 1$
    $f(x) = \dfrac{2}{3}(x^2 + 2x) - 1$
    $f(x) = \dfrac{2}{3}(x^2 + 2x + 1) - \dfrac{2}{3} - 1$
    $f(x) = \dfrac{2}{3}(x + 1)^2 - \dfrac{5}{3}$
    Start with the graph of $y = x^2$. Vertically compress, shift to the left 1 unit, and shift down $\dfrac{5}{3}$ units.

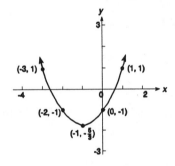

Chapter 5 Polynomial and Rational Functions

26. $f(x) = x^2 - 2x - 3$
Opens upward; vertex at $(1, -4)$; axis of symmetry $x = 1$; x-intercepts $(3, 0)$, $(-1, 0)$; y-intercept $(0, -3)$.

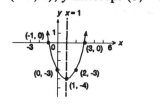

28. $f(x) = -x^2 + x + 2$
Opens downward; vertex at $\left[\dfrac{1}{2}, \dfrac{9}{4}\right]$; axis of symmetry $x = \dfrac{1}{2}$; x-intercepts $(2, 0)$, $(-1, 0)$; y-intercept $(0, 2)$.

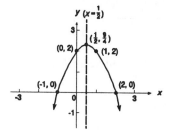

30. $f(x) = -x^2 + 4x - 4$
Opens downward; vertex at $(2, 0)$; axis of symmetry $x = 2$; x-intercepts $(2, 0)$; y-intercept $(0, -4)$.

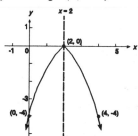

32. $f(x) = 4x^2 - 2x + 1$
Opens upward; vertex at $\left[\dfrac{1}{4}, \dfrac{3}{4}\right]$; axis of symmetry $x = \dfrac{1}{4}$; no x-intercept; y-intercept $(0, 1)$.

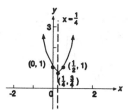

34. $f(x) = -3x^2 + 3x - 2$
Opens downward; vertex at $\left[\dfrac{1}{2}, \dfrac{-5}{4}\right]$; axis of symmetry $x = \dfrac{1}{2}$; no x-intercepts; y-intercept $(0, -2)$.

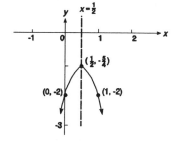

36. $f(x) = 2x^2 + 5x + 3$
Opens upward; vertex at $\left[\dfrac{-5}{4}, \dfrac{-1}{8}\right]$; axis of symmetry $x = \dfrac{-5}{4}$; x-intercepts $\left[\dfrac{-3}{2}, 0\right]$, $(-1, 0)$; y-intercept $(0, 3)$.

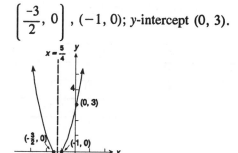

38. $f(x) = 3x^2 - 8x + 2$

Opens upward; vertex at $\left(\dfrac{4}{3}, \dfrac{-10}{3}\right)$;

axis of symmetry $x = \dfrac{4}{3}$;

x-intercepts $\left(\dfrac{4 + \sqrt{10}}{3}, 0\right)$,

$\left(\dfrac{4 - \sqrt{10}}{3}, 0\right)$; y-intercept $(0, 2)$.

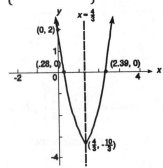

40. $f(x) = 4x^2 - 8x + 3$

Since $a = 4 > 0$, $f\left(-\dfrac{b}{2a}\right) = f(1) = -1$ is the minimum value of f.

42. $f(x) = -2x^2 + 8x + 3$

Since $a = -2 < 0$, $f\left(-\dfrac{b}{2a}\right) = f(2) = 11$ is the maximum value of f.

44. $f(x) = 4x^2 - 4x$

Since $a = 4 > 0$, $f\left(-\dfrac{b}{2a}\right) = f\left(\dfrac{1}{2}\right) = -1$ is the minimum value of f.

46. $f(x) = x^2 + cx + 1$
$a = 1$, $b = c$, $c = 1$

Each parabola opens up and has its vertex at $x = \dfrac{-b}{2a} = \dfrac{-c}{2}$;

$y = \left(\dfrac{-c}{2}\right)^2 + c\left(\dfrac{-c}{2}\right) + 1 = \dfrac{c^2}{4} - \dfrac{c^2}{2} + 1 = \dfrac{-c^2}{4} + 1$.

Each has axis of symmetry $x = \dfrac{-c}{2}$. The y-intercept is $f(0) = 1$.

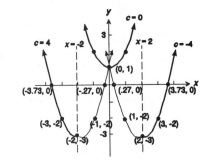

c	Vertex	Axis of Symmetry	x-intercept
-4	$(2, -3)$	$x = 2$	$(.27, 0)$ $(3.73, 0)$
0	$(0, 1)$	$x = 0$	None
4	$(-2, -3)$	$x = -2$	$(-3.73, 0)$ $(-.27, 0)$

48. Each parabola opens up and passes through $(0, 1)$. The graphs are compressed or stretched depending on the coefficient of x^2.

50. We are given $R = \dfrac{-1}{2}p^2 + 1900p$, which represents a parabola that opens downward. Thus, R will be a maximum at the vertex. Now, $h = \dfrac{-1900}{-1} = 1900$, and $k = f(1900) = \dfrac{-1}{2}(1900)^2 + 1900(1900) = 5{,}415{,}000$. The maximum revenue is $\$5{,}415{,}000$. It occurs when $p = \$1{,}900$.

52. The perimeter P of a rectangle is $P = 2\ell + 2w$. Thus, we have $2\ell = P - 2w$ or $\ell = \dfrac{P}{2} - w$.

The area $A = \ell w = \left[\dfrac{P}{2} - w\right]w = -w^2 + \dfrac{P}{2}w$. The maximum area occurs when $w = \dfrac{-\dfrac{P}{2}}{-2} = \dfrac{P}{4}$

and $\ell = \dfrac{P}{2} - w = \dfrac{P}{2} - \dfrac{P}{4} = \dfrac{P}{4}$. (Since $\ell = w$, the rectangle of fixed perimeter P that results in the largest area is a square.)

54. Let x be the width of the rectangular plot. Then, $2000 - 2x$ is the length. We want to maximize the area
$$A = x(2000 - 2x)$$
$$A = -2x^2 + 2000x$$
Since $a < 0$, A is a maximum when
$$x = \dfrac{-b}{2a} = \dfrac{-2000}{-4} = 500$$
The largest area is $A = 750[2000 - 2(500)]$
$$= 750(2000 - 1000)$$
$$= 750(1000)$$
$$= 750,000 \text{ m}^2$$

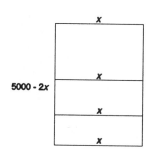

56. Let x be the width of the rectangle.

Then the length is $\dfrac{1}{2}(10,000 - 4x) = 5,000 - 2x$.

The area is $A = x(5000 - 2x)$
$$A = -2x^2 + 5000x$$
Since $a < 0$, A is a maximum when $x = \dfrac{-b}{2a} = \dfrac{-5000}{-4} = 1250$
The largest area is $A = 1250[5000 - 2(1250)]$
$$= 1250(2500)$$
$$= 3,125,000 \text{ m}^2$$

58. (a) The maximum height occurs at $x = \dfrac{-b}{2a} = \dfrac{-1}{2\left[\dfrac{-32}{100^2}\right]} = 156.25$. Therefore, the maximum

height is $f(156.25) = 78.125$ feet.

(b) The projectile strikes the ground when the height is zero.

Therefore, we solve $0 = \dfrac{-32x^2}{100^2} + x$.

Using the quadratic formula, we have

$$x = \dfrac{-1 \pm \sqrt{1 - 4\left[\dfrac{-32}{100^2}\right](0)}}{2\left[\dfrac{-32}{100^2}\right]} = \dfrac{-1 \pm 1}{\dfrac{-64}{100^2}}$$

$x = 312.5$ feet or
$x = 0$

Therefore, the projectile lands 312.5 feet from where it was fired.

(c)

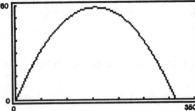

The projectile has traveled 62.5 and 250.00 feet.

60. Let d be the distance between the aircraft.
$$d^2 = (15 - 400t)^2 + (150t)^2$$
$$= 182,500t^2 - 12,000t + 225$$
Since $a > 0$, d^2, and hence d, is a minimum when
$$t = \frac{-b}{2a} = \frac{12,000}{2(182,500)} \approx 0.033.$$
Thus, $d^2 = 182,500(0.033)^2 - 12,000(0.033) + 225$
$$d^2 = 27.7425$$
$$d \approx 5.3 \text{ miles}$$
The two aircraft came about 5.3 miles apart.

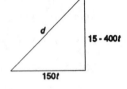

62. $y = -ax^2 + k,\ a > 0$
When $x = 0$, $y = k = 25$
When $x = 60$, $y = 0 = -a(3600) + 25$
$$a = \frac{25}{3600}$$
Thus, $y = \frac{-25}{3600}x^2 + 25$

$x = 10$: $y = \frac{-25}{3600}(100) + 25 \approx 24.3$ ft.

$x = 20$: $y = \frac{-25}{3600}(400) + 25 \approx 22.2$ ft.

$x = 40$: $y = \frac{-25}{3600}(1600) + 25 \approx 13.9$ ft.

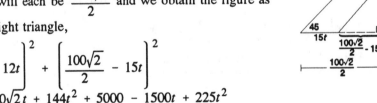

64. We want to minimize d. We complete the triangle to form 2 right triangles. Since the pleasure boat is NE of the port, the larger triangle is a 45-45-90 triangle. Thus, the legs will each be $\frac{100\sqrt{2}}{2}$ and we obtain the figure as labeled. In the smaller right triangle,

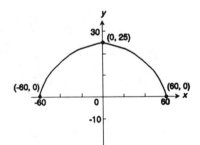

$$d^2 = \left[\frac{100\sqrt{2}}{2} - 12t\right]^2 + \left[\frac{100\sqrt{2}}{2} - 15t\right]^2$$
$$= 5000 - 1200\sqrt{2}\,t + 144t^2 + 5000 - 1500t + 225t^2$$
$$= 369t^2 - 2700\sqrt{2}\,t + 10000$$

Thus, d^2 and thus d is minimized when $t = \frac{2700\sqrt{2}}{738} = 5.17$ hours

When $t = 5.17$, $d^2 = 369(5.17)^2 - 2700\sqrt{2}\,(5.17) + 10000 \approx 121.95$
$$d \approx 11.04 \text{ nautical miles}$$

Chapter 5 Polynomial and Rational Functions

66. Since the diameter $(2r)$ equals the width of the rectangle, then $w = 2r$. The perimeter is $1500 = 2\ell + 2\pi r$ or $\ell = 750 - \pi r$. We want to maximize the area of the rectangle. $A = \ell w = 2r(750 - \pi r) = -2\pi r^2 + 1500r$. The area of the rectangle is maximum when

$$r = \frac{-1500}{-4\pi} \approx 119.37$$

Thus, $w = 2r \approx 238.7$ and $\ell = 750 - \pi r = 375$. The dimension of the rectangle should be 375 m. by 238.7 m.

68. $y = \dfrac{-32}{V_0^2}x^2 + x$

The maximum height occurs at the vertex: $x = \dfrac{-b}{2a} = \dfrac{-1}{\dfrac{-2(32)}{V_0^2}} = \dfrac{V_0^2}{64}$

(a) Thus, the maximum height in terms of V_0 is:

$$y = \frac{-32}{V_0^2}\left[\frac{V_0^2}{64}\right]^2 + \frac{V_0^2}{64} = \frac{-V_0^2}{128} + \frac{V_0^2}{64} = \frac{-V_0^2 + 2V_0^2}{128} = \frac{V_0^2}{128} \text{ feet}$$

(b) If V_0 is doubled, the maximum height is increased by a factor of 4. $\left[\dfrac{(2V_0)^2}{128} = \dfrac{4V_0^2}{128}\right]$.

(c) When $V_0 = 64$ and $y = 0$,

$$0 = \frac{-32}{64^2}x^2 + x$$
$$0 = -32x^2 + 64^2x = -32x(x - 128)$$
$$x = 0 \text{ or } x = 128$$

The projectile will land 128 feet from the starting point.

70. Let x be the number of cars not rented when the price is increased by $2x$ dollars. The income I is $(10 + 2x)(24 - x) = -2x^2 + 38x + 240$. Since $a < 0$, the maximum income occurs when

$x = \dfrac{-b}{2a} = \dfrac{-38}{2(-2)} = 9.5$. Since 0.5 cars cannot be rented, a price of $10 + 2(9) = \$28$ or $10 + 2(10) = \$30$ should be charged to maximize income.

72. Since the vertex is at $x = 1$, then $\dfrac{-b}{2a} = 1$ or $-b = 2a$. Also, since $f(x) = ax^2 + bx + c$ passes through $(0, 1)$ and $(-1, -8)$, we have $1 = c$ and $-8 = a - b + c = a - b + 1$. Thus, $-9 = a - b = a + 2a = 3a$. Thus, $a = 3$ and $-b = 2a = 2(-3) = -6$ or $b = 6$. The function is $f(x) = -3x^2 + 6x + 1$.

74. The rectangle we want to maximize has length x and height $10 - x$. Thus, the area is $A = x(10 - x) = -x^2 + 10x$. The maximum area occurs when

$$x = \frac{-10}{-2} = 5$$

and the maximum area is
$$A = 5(10 - 5) = 25$$

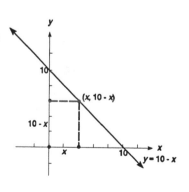

76. If x is an even integer, so is x^2. The product ax^2 is even as is the product bx. Since c is odd, the sum $ax^2 + bx + c$ will be odd.

   If x is an odd integer, so is x^2. The product ax^2 is odd as is the product bx. Since c is odd, the sum $ax^2 + bx + c$ will be odd.

78. Let x = the year in which the school year began (i.e., $x = 84$ for the 1984-85 school year). Therefore, the vertex of the parabola occurs at (84.5, 39). Two additional points on the parabola are (80, 41.5) and (88, 40.1). Using the standard form of an equation for a parabola, $y = ax^2 + bx + c$, we have
$$39 = a(84.5)^2 + b(84.5) + c$$
$$41.5 = a(80)^2 + b(80) + c$$
$$40.1 = a(88)^2 + b(88) + c$$
Solving this system of equations, we obtain:
$$a = 0.1087; \ b = -18.4417, \ c = 820.9603$$
Therefore, $y = 0.1087x^2 - 18.4417x + 820.9603$

Prediction of enrollment for 1991-92:
$$y = 0.1087(91)^2 - 18.4417(91) + 820.9603$$
$$y \approx 42.9 \text{ million students}$$

5.2   Polynomial Functions

2. $f(x) = 5x^2 + 4x^4$ f is a polynomial function of degree four.

4. $h(x) = 3 - \dfrac{1}{2}x$ h is a polynomial function of degree one.

6. $f(x) = x(x - 1)$ f is a polynomial function of degree two.

8. $h(x) = \sqrt{x}\left(\sqrt{x} - 1\right)$ h is not a polynomial function. The variable x is raised to the $\dfrac{1}{2}$ power.

10. $F(x) = \dfrac{x^2 - 5}{x^3}$ F is not a polynomial function. It is the ratio of two polynomials and the polynomial in the denominator is of positive degree.

12. $f(x) = x^4 + 2$

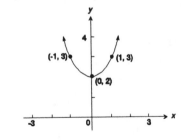

14. $f(x) = -x^4$

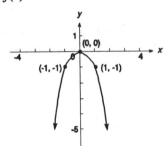

16. $f(x) = 3 - (x + 2)^4$

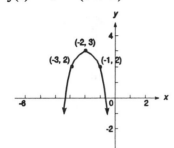

18. $f(x) = 1 - 2(x + 1)^4$

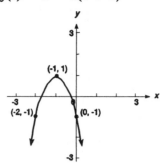

20. $f(x) = 4(x + 4)(x + 3)^3$ -4 is a zero of multiplicity one.
-3 is a zero of multiplicity three.
The graph crosses the x-axis at -4 and at -3.

22. $f(x) = 2(x - 3)(x + 4)^3$ 3 is a zero of multiplicity one.
-4 is a zero of multiplicity three.
The graph crosses the x-axis at 3 and at -4.

24. $f(x) = \left(x - \dfrac{1}{3}\right)^2 (x - 1)^3$ $\dfrac{1}{3}$ is a zero of multiplicity two.
1 is a zero of multiplicity three.

The graph touches the x-axis at $\dfrac{1}{3}$ and crosses it at 1.

26. $f(x) = \left(x + \sqrt{3}\right)^2 (x - 2)^4$ $-\sqrt{3}$ is a zero of multiplicity two.
2 is a zero of multiplicity four.

The graph touches the x-axis at $-\sqrt{3}$ and at 2.

28. $f(x) = -2(x^2 + 3)^3$ No zeros.
The graph neither crosses nor touches the x-axis.

30. (a) x-intercept: 2; y-intercept: -8 (f)
(b)   Crosses at 2
(c) $y = x^3$
(d) 2
(e)

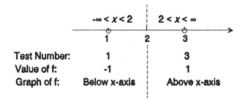

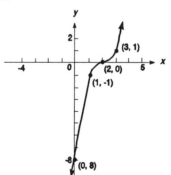

32. (a) x-intercepts: $-2, 0$; y-intercept: 0
(b) Touches at -2; crosses at 0
(c) $y = x^3$
(d) 2
(e)

	$-\infty < x < -2$	$-2 < x < 0$	$0 < x < \infty$
Test Number:	-3	-1	1
Values of f:	-3	-1	9
Graph of f:	Below x-axis	Below x-axis	Above x-axis

(f)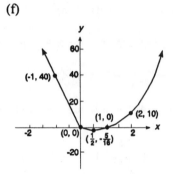

34. (a) x-intercepts: $0, 1$; y-intercept: 0
(b) Crosses at 0 and 1
(c) $y = 5x^4$
(d) 3
(e)

	$-\infty < x < 0$	$0 < x < 1$	$1 < x < \infty$
Test Number:	-1	$\frac{1}{2}$	2
Value of f:	40	$-\frac{5}{16}$	0
Graph of f:	Above x-axis	Below x-axis	Above x-axis

(f)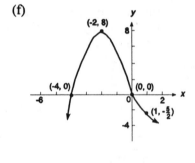

36. (a) x-intercepts: $-4, 0$; y-intercept: 0
(b) Crosses at -4 and 0
(c) $y = -\dfrac{1}{2}x^4$
(d) 3
(e)

	$-\infty < x < -4$	$-4 < x < 0$	$0 < x < \infty$
Test Number:	-5	-2	1
Value of f:	$-\frac{125}{2}$	8	$-\frac{5}{2}$
Graph of f:	Below x-axis	Above x-axis	Below x-axis

(f)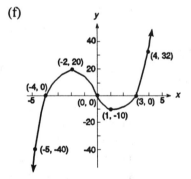

38. (a) x-intercepts: $-4, 0, 3$; y-intercept: 0
(b) Crosses at $-4, 0$, and 3
(c) $y = x^3$
(d) 2
(e)

	$-\infty < x < -4$	$-4 < x < 0$	$0 < x < 3$	$3 < x < \infty$
Test Number:	-5	-2	1	4
Values of f:	-40	20	-10	32
Graph of f:	Below x-axis	Above x-axis	Below x-axis	Above x-axis

(f)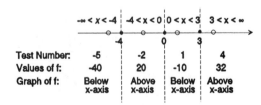

40. $f(x) = x - x^3 = -x(x^2 - 1) = -x(x - 1)(x + 1)$

    (a) x-intercepts: $-1, 0, 1$; y-intercept: 0

    (b)   Crosses at $-1, 0$, and 1

    (c) $y = -x^3$

    (d)   2

    (e)

(f)

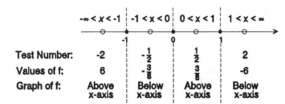

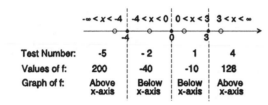

	$-\infty < x < -1$	$-1 < x < 0$	$0 < x < 1$	$1 < x < \infty$
Test Number:	-2	$-\frac{1}{2}$	$\frac{1}{2}$	2
Values of f:	6	$-\frac{3}{8}$	$\frac{3}{8}$	-6
Graph of f:	Above x-axis	Below x-axis	Above x-axis	Below x-axis

42.   (a) x-intercepts: $-4, 0, 3$; y-intercept: 0

    (b)   Crosses at -4, and 3; touches at 0

    (c) $y = x^4$

    (d)   3

    (e)

(f)

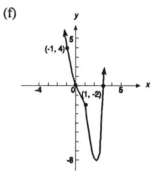

	$-\infty < x < -4$	$-4 < x < 0$	$0 < x < 3$	$3 < x < \infty$
Test Number:	-5	-2	1	4
Values of f:	200	-40	-10	128
Graph of f:	Above x-axis	Below x-axis	Below x-axis	Above x-axis

44.   (a) x-intercepts: $0, 3$; y-intercept: 0

    (b)   Crosses at 0 and 3

    (c) $y = x^4$

    (d)   3

    (e)

(f)

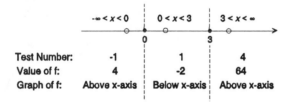

	$-\infty < x < 0$	$0 < x < 3$	$3 < x < \infty$
Test Number:	-1	1	4
Value of f:	4	-2	64
Graph of f:	Above x-axis	Below x-axis	Above x-axis

46.   (a) x-intercepts: $0, 1$, and 3; y-intercept: 0

    (b)   Touches at 0; crosses at 1 and 3

    (c) $y = x^4$

    (d)   3

    (e)

(f)

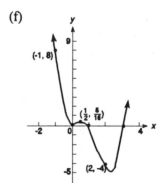

	$-\infty < x < 0$	$0 < x < 1$	$1 < x < 3$	$3 < x < \infty$
Test Number:	-1	$\frac{1}{2}$	2	4
Value of f:	8	$\frac{5}{16}$	-4	48
Graph of f:	Above x-axis	Above x-axis	Below x-axis	Above x-axis

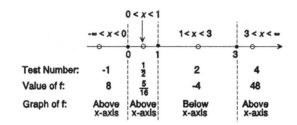

48. (a) *x*-intercepts: $-4, -2, 0, 2$; *y*-intercept: 0
    (b) Crosses at $-4, -2, 0$, and 2
    (c) $y = x^4$
    (d) 3
    (e)

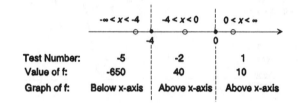

(f)

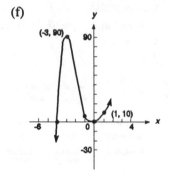

50. (a) *x*-intercepts: $-4, 0$; *y*-intercept: 0
    (b) Crosses at -4; touches at 0
    (c) $y = x^5$
    (d) 4
    (e)

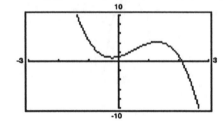

(f)

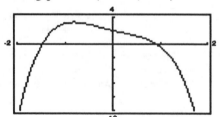

52. c, e, f

54. d, e, f

56. *x*-intercept: 2.00
    turning points: $(-0.22, 0.79); (1.27, 4.17)$

58. *x*-intercept: $-1.47, 0.91$
    turning points: $(-0.80, 3.21)$

60. *x*-intercept: -0.71

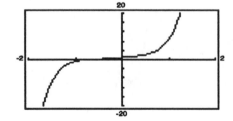

66. a, b, c, d

2. $R(x) = \dfrac{5x^2}{3 + x}$

    The domain of $R(x)$ consists of all real numbers except -3.

4. $G(x) = \dfrac{6}{(x + 3)(4 - x)}$

    The domain of $G(x)$ consists of all real numbers except -3 and 4.

6. $Q(x) = \dfrac{-x(1 - x)}{3x^2 + 5x - 2} = \dfrac{-x(1 - x)}{(3x - 1)(x + 2)}$

    The domain of $Q(x)$ consists of all real numbers except $\dfrac{1}{3}$ and -2.

8. $R(x) = \dfrac{x}{x^4 - 1} = \dfrac{x}{(x + 1)(x - 1)(x^2 + 1)}$

    The domain of $R(x)$ consists of all real numbers except -1 and 1.

10. $G(x) = \dfrac{x - 3}{x^4 + 1}$

    The domain of $G(x)$ consists of all real numbers.

12. (a) Domain: $\{x \mid x \neq -1\}$;
         Range: $\{y \mid y > 0\}$
    (b) $(0, 2)$
    (c) $y = 0$
    (d) $x = -1$
    (e) None

14. (a) Domain: $\{x \mid x \neq 0\}$;
         Range: $-\infty < y \leq -2, 2 \leq y < \infty$
    (b) None
    (c) None
    (d) $x = 0$
    (e) $y = -x$

16. (a) Domain: $\{x \mid x \neq -1, x \neq 1\}$;
         Range: All real numbers
    (b) $(0, 0)$
    (c) $y = 0$
    (d) $x = -1, x = 1$
    (e) None

18. $R(x) = \dfrac{3}{x}$

    Start with the graph of $y = \dfrac{1}{x}$. Vertically stretch.

    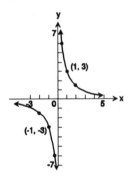

20. $G(x) = \dfrac{2}{(x + 2)^2}$

    Start with the graph of $y = \dfrac{1}{x^2}$.

    Vertically stretch and shift to the left 2 units.

    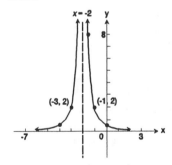

22. $R(x) = \dfrac{1}{x-1} + 1$

Start with the graph of $y = \dfrac{1}{x}$. Shift to the right one unit and shift up one unit.

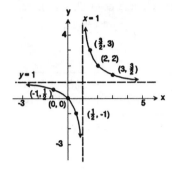

24. $Q(x) = 1 + \dfrac{1}{x}$

Start with the graph of $y = \dfrac{1}{x}$. Shift up one unit.

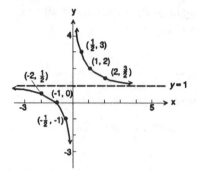

26. $R(x) = \dfrac{x-4}{x} = 1 - \dfrac{4}{x}$

Start with the graph of $y = \dfrac{1}{x}$. Reflect about the x-axis, vertically stretch and shift up one unit.

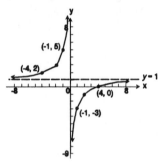

28. $R(x) = \dfrac{3x+5}{x-6}$

Vertical asymptote: $x = 6$
Horizontal asymptote: $y = 3$

30. $G(x) = \dfrac{-x^2 + 1}{x+5}$

Vertical asymptote: $x = -5$
Since

$$
\begin{array}{r}
-x + 5 \\
x + 5 \overline{\smash{)}-x^2 + 1} \\
\underline{-x^2 - 5x } \\
5x + 1 \\
\underline{5x + 25} \\
-24
\end{array}
$$

Oblique asymptote: $y = -x + 5$

32. $P(x) = \dfrac{4x^5}{x^3 - 1}$

Vertical asymptote: $x = 1$
The degree of the numerator exceeds the degree of the denominator by two, so the graph has no horizontal and no oblique asymptote.

34. $F(x) = \dfrac{-2x^2 + 1}{2x^3 + 4x^2}$

Vertical asymptote:
$$2x^3 + 4x^2 = 0$$
$$2x^2(x + 2) = 0$$
$$x = 0 \text{ and } x = -2$$

Horizontal asymptote: $y = 0$

36. $R(x) = \dfrac{6x^2 + x + 12}{3x^2 - 5x - 2}$

Vertical asymptote:
$$3x^2 - 5x - 2 = (3x + 1)(x - 2) = 0$$
$$x = \dfrac{-1}{3} \text{ and } x = 2$$

Horizontal asymptote: $y = 2$

38. $R(x) = \dfrac{x}{(x - 1)(x + 2)}$

  1. The x-intercept is 0. The y-intercept is $R(0) = 0$.

  2. Because $R(-x) = \dfrac{-x}{(-x - 1)(-x + 2)}$, we conclude that R is neither even nor odd.

  3. The vertical asymptotes are $x = 1$ and $x = -2$.

  4. The line $y = 0$ is the horizontal asymptote; not intersected.

  5. The zero of the numerator, 0, and the zeros of the denominator, 1 and -2, divide the x-axis into: $x < -2, -2 < x < 0, 0 < x\ 1, x > 1$.

Interval	Test Number	Value of R	Graph of R
$x < -2$	-3	$R(-3) = -0.75$	Below x-axis
$-2 < x < 0$	-1	$R(-1) = 0.5$	Above x-axis
$0 < x < 1$	$\dfrac{1}{2}$	$R\left[\dfrac{1}{2}\right] = -0.4$	Below x-axis
$x > 1$	2	$R(2) = 0.5$	Above x-axis

  6.

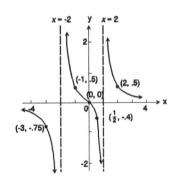

40. $R(x) = \dfrac{2x + 4}{(x - 1)} = \dfrac{2(x + 2)}{x - 1}$

  1. The x-intercept is -2. The y-intercept is $R(0) = -4$.

  2. Because $R(-x) = \dfrac{-2x + 4}{-x - 1}$, we conclude that R is neither even nor odd.

  3. The vertical asymptotes are $x = 1$.

  4. The horizontal asymptote is $y = 2$; not intersected.

  5. The zero of the numerator, -2, and the zero of the denominator, 1, divide the x-axis into: $x < -2, -2 < x < 1, x > 1$.

Interval	Test Number	Value of R	Graph of R
$x < -2$	-3	$R(-3) = 0.5$	Above x-axis
$-2 < x < 1$	0	$R(0) = -4$	Below x-axis
$x > 1$	2	$R(2) = 8$	Above x-axis

6.

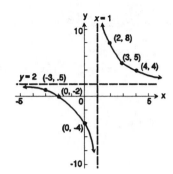

42. $R(x) = \dfrac{6}{x^2 - x - 6} = \dfrac{6}{(x - 3)(x + 2)}$

1. No x-intercept. The y-intercept is $R(0) = -1$.

2. Because $F(-x) = \dfrac{6}{x^2 + x - 6}$, we conclude that R is neither even nor odd.

3. The vertical asymptotes are $x = -2$ and $x = 3$.

4. The line $y = 0$ is the horizontal asymptote; not intersected.

5. The zero of the denominator, -2 and 3, divide the x-axis into: $x < -2$, $-2 < x < 3$, $x > 3$.

Interval	Test Number	Value of R	Graph of R
$x < -2$	-3	$R(-3) = 1$	Above x-axis
$-2 < x < 3$	0	$R(0) = -1$	Below x-axis
$x > 3$	4	$R(4) = 1$	Above x-axis

6.

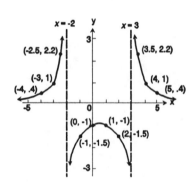

44. $Q(x) = \dfrac{x^4 - 1}{x^2 - 4} = \dfrac{(x - 1)(x + 1)(x^2 + 1)}{(x + 2)(x - 2)}$

1. The x-intercepts are 1 and -1. The y-intercept is $Q(0) = \dfrac{1}{4}$.

2. Because $Q(-x) = \dfrac{x^4 - 1}{x^2 - 4}$, we conclude Q is an even function. Thus, there is symmetry about the y-axis.

3. The vertical asymptotes are $x = -2$ and $x = 2$.

4. There is no horizontal and no oblique asymptote.

5. The zeros of the numerator, 1 and -1, and the zeros of the denominator, 2 and -2, divide the x-axis into: $x < -2$, $-2 < x < -1$, $-1 < x 1$, $1 < x < 2$, $x > 2$.

Interval	Test Number	Value of Q	Graph of Q
$x < -2$	-3	$Q(-3) = 16$	Above x-axis
$-2 < x < -1$	-1.5	$Q(-1.5) \approx -2.3$	Below x-axis
$-1 < x < 1$	0	$Q(0) = \dfrac{1}{4}$	Above x-axis
$1 < x < 2$	1.5	$Q(1.5) \approx -2.3$	Below x-axis
$x > 2$	3	$Q(3) = 16$	Above x-axis

6.

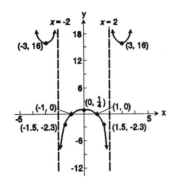

46. $G(x) = \dfrac{x^3 + 1}{x^2 + 2x} = \dfrac{(x + 1)(x^2 - x + 1)}{x(x + 2)}$

    1.   The x-intercept is -1. There is no y-intercept.

    2.   Because $G(-x) = \dfrac{-x^3 + 1}{x^2 - 2x}$, we conclude that G is neither even nor odd.

    3.   The vertical asymptotes are $x = 0$ and $x = -2$.

    4.   The line $y = x - 2$ is an oblique asymptote; intersected at $\left[\dfrac{-1}{4}, \dfrac{-9}{4} \right]$.

    5.   The zero of the numerator, -1, and the zeros of the denominators, 0 and -2, divide the x-axis into: $x < -2, -2 < x < -1, -1 < x < 0, x > 0$.

Interval	Test Number	Value of G	Graph of G
$x < -2$	-3	$G(-3) \approx -8.7$	Below x-axis
$-2 < x < -1$	-1.5	$G(-1.5) \approx 3.2$	Above x-axis
$-1 < x < 0$	-0.5	$G(-0.5) \approx -1.2$	Below x-axis
$x > 0$	1	$G(1) \approx 0.67$	Above x-axis

    6.

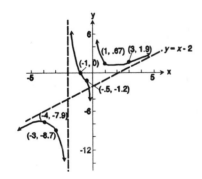

48. $R(x) = \dfrac{x^2 + x - 12}{x^2 - 4} = \dfrac{(x + 4)(x - 3)}{(x + 2)(x - 2)}$

1. The x-intercepts are -4 and 3. The y-intercept is $R(0) = 3$.

2. Because $R(-x) = \dfrac{x^2 - x - 12}{x^2 - 4}$, we conclude R is an neither even nor odd.

3. The vertical asymptotes are $x = -2$ and $x = 2$.

4. The horizontal asymptote is $y = 1$; not intersected.

5. The zeros of the numerator, -4 and 3, and the zeros of the denominator, -2 and 2, divide the x-axis into: $x < -4$, $-4 < x < -2$, $-2 < x < 2$, $2 < x < 3$, $x > 3$.

Interval	Test Number	Value of R	Graph of R
$x < -4$	-5	$R(-5) \approx 0.38$	Above x-axis
$-4 < x < -2$	-3	$R(-3) = -1.2$	Below x-axis
$-2 < x < 2$	0	$R(0) = 3$	Above x-axis
$2 < x < 3$	2.5	$R(2.5) \approx -1.4$	Below x-axis
$x > 3$	4	$R(4) \approx 0.67$	Above x-axis

6.

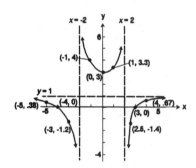

50. $G(x) = \dfrac{3x}{x^2 - 1} = \dfrac{3x}{(x - 1)(x + 1)}$

1. The x-intercept is 0. The y-intercept is $G(0) = 0$.

2. Because $G(-x) = \dfrac{-3x}{x^2 - 1}$, we conclude that G is an odd function.

3. The vertical asymptotes are $x = 1$ and $x = -1$.

4. The horizontal asymptote is $y = 0$; intersected at $(0, 0)$.

5. The zero of the numerator, 0, and the zeros of the denominator, 1 and -1, divide the x-axis into: $x < -1$, $-1 < x < 0$, $0 < x < 1$, $x > 1$.

Interval	Test Number	Value of G	Graph of G
$x < -1$	-2	$G(-2) = -2$	Below x-axis
$-1 < x < 0$	-0.5	$G(-0.5) = 2$	Above x-axis
$0 < x < 1$	0.5	$G(0.5) = -2$	Below x-axis
$x < 1$	2	$G(2) = 2$	Above x-axis

6.

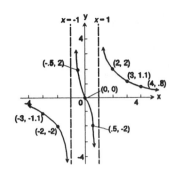

52. $R(x) = \dfrac{-4}{(x + 1)(x^2 - 9)} = \dfrac{-4}{(x + 1)(x - 3)(x + 3)}$

1. No x-intercept. The y-intercept is $R(0) = \dfrac{4}{9}$.

2. Because $R(-x) = \dfrac{-4}{(-x + 1)(x^2 - 9)}$, we conclude R is an neither even nor odd.

3. The vertical asymptotes are $x = -1$ and $x = 3$, and $x = -3$.

4. The horizontal asymptote is $y = 0$; not intersected.

5. The zeros of the denominator, -1, 3 and -3, divide the x-axis into: $x < -3$, $-3 < x < -1$, $-1 < x < 3$, $x > 3$.

Interval	Test Number	Value of R	Graph of R
$x < -3$	-4	$R(-4) \approx 0.2$	Above x-axis
$-3 < x < -1$	-2	$R(-2) = -0.8$	Below x-axis
$-1 < x < 3$	0	$R(0) = \dfrac{4}{9}$	Above x-axis
$x > 3$	4	$R(4) \approx -0.1$	Below x-axis

6.

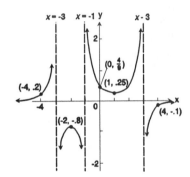

54. $H(x) = \dfrac{x^2 + 4}{x^4 - 1} = \dfrac{x^2 + 4}{(x - 1)(x + 1)(x^2 + 1)} = \dfrac{x^2 + 4}{(x - 1)(x + 1)(x^2 + 1)}$

1. No x-intercept. The y-intercept is $H(0) = -4$.

2. Because $H(-x) = \dfrac{x^2 + 4}{x^4 - 1}$, we conclude H is an even function.

    Thus, there is symmetry about the y-axis.

3. The vertical asymptotes are $x = 1$ and $x = -1$.

4. The horizontal asymptote is $y = 0$; not intersected.

5. The zeros of the denominator, 1 and -1, divide the x-axis into: $x < -1$, $-1 < x < 1$, $x > 1$.

Interval	Test Number	Value of H	Graph of H
$x < -1$	-2	$H(-2) \approx 0.53$	Above x-axis
$-1 < x < 1$	0	$H(0) = -4$	Below x-axis
$x > 1$	2	$H(2) \approx 0.53$	Above x-axis

6.

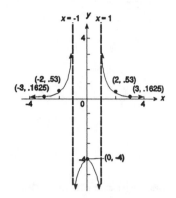

56. $F(x) = \dfrac{x^2 + 3x + 2}{x - 1} = \dfrac{(x + 2)(x + 1)}{(x - 1)}$

    1.   The x-intercepts are -2 and -1. The y-intercept is $F(0) = -2$.

    2.   Because $F(-x) = \dfrac{x^2 - 3x + 2}{-x - 1}$, we conclude F is an neither even nor odd.

    3.   The vertical asymptote is $x = 1$.

    4.   The line $y = x + 4$ is an oblique asymptote; not intersected.

    5.   The zeros of the numerator, -2, and -1, and the zero of the denominator, 1, divide the x-axis into: $x < -2$, $-2 < x < -1$, $-1 < x < 1$, $x > 1$.

Interval	Test Number	Value of F	Graph of F
$x < -2$	-3	$F(-3) = -0.5$	Below x-axis
$-2 < x < -1$	-1.5	$F(-1.5) = 0.1$	Above x-axis
$-1 < x < 1$	0	$F(0) = -2$	Below x-axis
$x > 1$	2	$F(2) = 12$	Above x-axis

6.

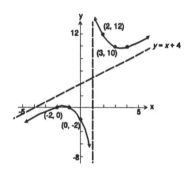

58. $R(x) = \dfrac{x^2 - x - 12}{x + 5} = \dfrac{(x - 4)(x + 3)}{(x + 5)}$

    1.   The x-intercepts are 4 and -3. The y-intercept is $R(0) = \dfrac{-12}{5}$.

    2.   Because $R(-x) = \dfrac{x^2 + x - 12}{-x + 5}$, we conclude R is neither even nor odd.

3. The vertical asymptote is $x = -5$.
4. The line $y = x - 6$ is an oblique asymptote; not intersected.
5. The zeros of the numerator and the zero of the denominator, divide the x-axis into: $x < -5$, $-5 < x < -3$, $-3 < x < 4$, $x > 4$.

Interval	Test Number	Value of R	Graph of R
$x < -5$	-6	-30	Below x-axis
$-5 < x < -3$	-4	8	Above x-axis
$-3 < x < 4$	1	-2	Below x-axis
$x > 4$	5	$4/5$	Above x-axis

6.

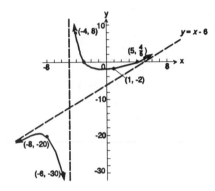

60. $G(x) = \dfrac{x^2 - x - 12}{x + 1} = \dfrac{(x - 4)(x + 3)}{(x + 1)}$

1. The x-intercepts are 4 and -3. The y-intercept is $G(0) = -12$.

2. Because $G(-x) = \dfrac{x^2 + x - 12}{-x + 1}$, we conclude G is neither even nor odd.

3. The vertical asymptote is $x = -1$.
4. The line $y = x - 2$ is an oblique asymptote; not intersected.
5. The zeros of the numerator and the zero of the denominator divide the x-axis into: $x < -3$, $-3 < x < -1$, $-1 < x < 4$, $x > 4$.

Interval	Test Number	Value of G	Graph of G
$x < -3$	-4	$-8/3$	Below x-axis
$-3 < x < -1$	-2	6	Above x-axis
$-1 < x < 4$	2	$-10/3$	Below x-axis
$x > 4$	5	$4/3$	Above x-axis

6.

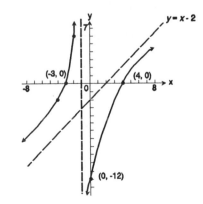

62. $R(x) = \dfrac{(x - 1)(x + 2)(x - 3)}{x(x - 4)^2}$

   1.   The x-intercepts are 1, -2 and 3. There is no y-intercept.

   2.   Because $R(-x) = \dfrac{(-x - 1)(-x + 2)(-x - 3)}{-x(-x - 4)^2}$, we conclude R is neither even nor odd.

   3.   The vertical asymptotes are $x = 0$ and $x = 4$.

   4.   Since the degree of the numerator equals the degree of the denominator, the horizontal

         asymptote is $y = 1$; intersected at $\left(\dfrac{7 + \sqrt{33}}{4},\ 1\right)$ and $\left(\dfrac{7 - \sqrt{33}}{4},\ 1\right)$

   5.   The zeros of the numerator and the zeros of the denominator, divide the x-axis into: $x < -2$, $-2 < x < 0, 0 < x < 1, 1 < x < 3, 3 < x < 4, x > 4$.

Interval	Test Number	Value of R	Graph of R
$x < -2$	-3	20/147	Above x-axis
$-2 < x < 0$	-1	$-6/25$	Below x-axis
$0 < x < 1$	1/2	15/49	Above x-axis
$1 < x < 3$	2	$-1/2$	Below x-axis
$3 < x < 4$	7/2	55/7	Above x-axis
$x > 4$	5	84/5	Above x-axis

   6.

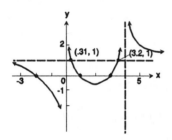

64. If $y = 2$ is a horizontal asymptote, then the result of long division is a constant. This is true when the degree of the numerator equals the degree of the denominator.

66. Each graph passes through $(0, 0)$ and has the vertical asymptote $x = 1$.

68. Minimum value: 8.48 at $x = 2.12$

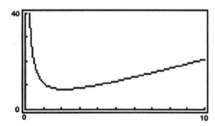

70. Minimum value: 10.30 at $x = 1.31$

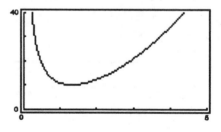

72. Minimum value: 5.11 at $x = 1.91$

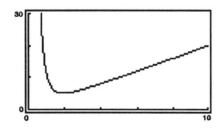

74. b, e

76. No

Chapter 5 Polynomial and Rational Functions

2. $f(-1) = -1 + 2 + 3 + 1 = 5$

4. $f(-2) = 32 + 8 + 2 + 1 = 43$

6. $f(2) = 22$

8. $f(-1) = -16$

10. $f(-2.1) = 0.241$

12.
$$
\begin{array}{r|rrrr}
-1 & 1 & 2 & -3 & 1 \\
   &   & -1 & -10 & 4 \\
\hline
   & 1 & 1 & -4 & 5
\end{array}
$$
$q(x) = x^2 + x - 4$, $R = 5$

14.
$$
\begin{array}{r|rrrr}
-2 & -4 & 2 & -1 & 1 \\
   &   & 8 & -20 & 42 \\
\hline
   & -4 & 10 & -21 & 43
\end{array}
$$
Thus, $q(x) = -4x^2 + 10x - 21$, $R = 43$

16.
$$
\begin{array}{r|rrrrr}
2 & 1 & 0 & 1 & 0 & 2 \\
  &   & 2 & 4 & 10 & 20 \\
\hline
  & 1 & 2 & 5 & 10 & 22
\end{array}
$$
Thus, $q(x) = x^3 + 2x^2 + 5x + 10$, $R = 22$

18.
$$
\begin{array}{r|rrrrrr}
-1 & 1 & 0 & 5 & 0 & 0 & -10 \\
   &   & -1 & 1 & -6 & 6 & -6 \\
\hline
   & 1 & -1 & 6 & -6 & 6 & -16
\end{array}
$$
$q(x) = x^4 - x^3 + 6x^2 - 6x + 6$, $R = -16$

20.
$$
\begin{array}{r|rrr}
-2.1 & 0.1 & 0 & -0.2 \\
     &   & -0.21 & 0.441 \\
\hline
     & 0.1 & -0.21 & 0.241
\end{array}
$$
$q(x) = 0.1x - 0.21$, $R = 0.241$

22.
$$
\begin{array}{r|rrrrr}
-1 & 1 & 0 & 0 & 0 & 1 \\
   &   & -1 & 1 & -1 & 1 & -1 \\
\hline
   & 1 & -1 & 1 & -1 & 1 & 0
\end{array}
$$
$q(x) = x^4 - x^3 + x^2 - x + 1$, $R = 0$

24. We divide by $x - (-3)$:
$$
\begin{array}{r|rrrr}
-3 & -4 & 5 & 0 & 8 \\
   &   & 12 & -51 & 153 \\
\hline
   & -4 & 17 & -51 & 161
\end{array}
$$
The remainder $= 161 \neq 0$; therefore, $x - (-3)$ is not a factor of $f(x)$.

26. We divide by $x - 2$:
$$
\begin{array}{r|rrrrr}
2 & 4 & 0 & -15 & 0 & -4 \\
  &   & 8 & 16 & 2 & 4 \\
\hline
  & 4 & 8 & 1 & 2 & 0
\end{array}
$$
The remainder $= 0$; therefore $x - 2$ is a factor of $f(x)$.

28. We divide by $x - (-3)$:
$$
\begin{array}{r|rrrrrrr}
-3 & 2 & 0 & -18 & 0 & 1 & 0 & -9 \\
   &   & -6 & 0 & 0 & 1 & -3 & 0 \\
\hline
   & 2 & -6 & 0 & 0 & 1 & -3 & 0
\end{array}
$$
The remainder $= 0$; therefore, $x - (-3)$ is a factor of $f(x)$.

30.
$$
\begin{array}{r|rrrrrrr}
-4 & 1 & 0 & -16 & 0 & 1 & 0 & -16 \\
   &   & -4 & 16 & 0 & 0 & -4 & 16 \\
\hline
   & 1 & -4 & 0 & 0 & 1 & -4 & 0
\end{array}
$$
The remainder $= 0$; therefore, $x - (-4)$ is a factor of $f(x)$.

32.
$$
\begin{array}{r|rrrrr}
-\frac{1}{3} & 3 & 1 & 0 & -3 & 1 \\
             &   & -1 & 0 & 0 & 1 \\
\hline
             & 3 & 0 & 0 & -3 & 2
\end{array}
$$
The remainder $= 2$; therefore,
$$x - \left[\frac{-1}{3}\right] \text{ is not a factor of } f(x).$$

34. To find $f(-2)$, we divide by $x - (-2)$
$= x + 2$

$$-2 \overline{)\begin{array}{cccc} -2 & 3 & 0 & 5 \\ & 4 & -14 & 28 \end{array}}$$
$$\begin{array}{cccc} -2 & 7 & -14 & 33 \end{array}$$

By the Remainder Theorem, $f(-2) = 33$.

36. To find $f(-1)$, we divide by $x - (-1)$
$= x + 1$

$$-1 \overline{)\begin{array}{ccccc} -3 & 3 & -2 & 0 & 5 \\ & 3 & -6 & 8 & -8 \end{array}}$$
$$\begin{array}{ccccc} -3 & 6 & -8 & 8 & -3 \end{array}$$

By the Remainder Theorem,
$f(-1) = -3$.

38. To find $f(-1)$, we divide by $x - (-1) = x + 1$

$$-1 \overline{)\begin{array}{ccccccccccccccccc} 10 & 0 & 0 & 4 & 0 & 0 & 0 & 0 & 0 & -2 & 0 & 0 & 1 & 0 & 0 \\ & -10 & 10 & -10 & 6 & -6 & 6 & -6 & 6 & -6 & 6 & -4 & 4 & -4 & 3 & -3 \end{array}}$$
$$\begin{array}{ccccccccccccccccc} 10 & -10 & 10 & -6 & 6 & -6 & 6 & -6 & 6 & -6 & 4 & -4 & 4 & -3 & 3 & -3 \end{array}$$

By the Remainder Theorem, $f(-1) = -3$

40. $f(x) = -4x^3 + 5x^2 + 6$
$= (-4x^2 + 5x)x + 6$
$= [(-4x + 5)x]x + 6$

42. $f(x) = 4x^4 - 15x^2 - 4$
$= (4x^3 - 15x)x - 4$
$= [(4x^2 - 15)x]x - 4$
$= \{[(4x)x - 15]x\}x - 4$

44. $f(x) = 2x^6 - 18x^4 + x^2 - 9 = (2x^5 - 18x^3 + x)x - 9$
$= [(2x^4 - 18x^2 + 1)x]x - 9 = \{[(2x^3 - 18x)x + 1]x\}x - 9$
$= [\{[(2x \cdot x \cdot x - 18)x]x + 1\}x]x - 9$

46. $f(x) = x^6 - 16x^4 + x^2 - 16 = (x^5 - 16x^3 + x)x - 16$
$= [(x^4 - 16x^2 + 1)x]x - 16 = \{[(x^3 - 16x)x + 1]x\}x - 16$
$= \{[(x \cdot x - 16)x]x + 1\}x \cdot x - 16$

48. $f(x) = 3x^4 + x^3 - 3x + 1 = (3x^3 + x^2 - 3)x + 1$
$= [(3x^2 + x)x - 3]x + 1 = \{[(3x + 1)x]x - 3\}x + 1$

50. $f(x) = -4x^3 + 5x^2 + 6$
$f(1.2) = 6.288$

52. $f(x) = 4x^4 - 15x^2 - 4$
$f(1.2) = -17.3056$

54. $f(x) = 2x^6 - 18x^4 + x^2 - 9$
$f(1.2) = -38.912832$

56. $f(x) = x^6 - 16x^4 + x^2 - 16$
$f(1.2) = -44.751616$

58. $f(x) = 3x^4 + x^3 - 3x + 1$
$f(1.2) = 5.3488$

60. By the Factor Theorem, $x + 2$ is a factor of $f(x)$ if $f(-2) = 0$.
Thus, $f(-2) = (-2)^4 - k(-2)^3 + k(-2)^2 + 1 = 0$
$$16 + 8k + 4k + 1 = 0$$
$$12k = -17$$
$$k = \frac{-17}{12}$$

62. By the Remainder Theorem, since $f(-1) = 1$, the remainder when $f(x) = -3x^{17} + x^9 - x^5 + 2x$ is divided by $x + 1$ is 1.

64. If $f(x) = x^n - c^n$, then $f(-c) = (-c)^n - (-c)^n = 0$, if $n \geq 1$ is an odd integer.
Thus, by the Factor Theorem, since $f(-c) = 0$, then $x - c$ is a factor of $f(x)$.

66. (a) $f(x) = ax^3 + bx^2 + cx + d$ has 6 multiplications and 3 additions. Since it takes $6 \cdot 33,333 + 3 \cdot 500 = 201,498$ nanoseconds for each value of x, it will take $5000 \cdot 201,498 = 1007490000$ nanoseconds for 5000 values of x.

(b) The nested form, $f(x) = [(ax + b)x + c]x + d$, has 3 multiplications and 3 additions. Since it takes $3 \cdot 33,333 + 3 \cdot 500 = 101,499$ nanoseconds for each value of x, it will take $101,499 \cdot 5000 = 507495000$ nanoseconds for 5000 values of x.

5.5   The Zeros of a Polynomial Function

2. $f(x) = 5x^4 + 2x^2 - 6x - 5$ There is one positive zero.
$f(-x) = 5x^4 + 2x^2 + 6x - 5$ There is one negative zero.
Maximum number of zeros: 4

4. $f(x) = -3x^5 + 4x^4 + 2$ There is one positive zero.
$f(-x) = 3x^5 + 4x^4 + 2$ There are no negative zeros.
Maximum number of zeros: 5

6. $f(x) = -x^3 - x^2 + x + 1$ There is one positive zero.
$f(-x) = x^3 - x^2 - x + 1$ There are two or zero negative zeros.
Maximum number of zeros: 3

8. $f(x) = x^4 + 5x^3 - 2$ There is one positive zero.
$f(-x) = x^4 - 5x^3 - 2$ There is one negative zero.
Maximum number of zeros: 4

10. $f(x) = x^5 - x^4 + x^3 - x^2 + x - 1$ There are five, three, or one positive zeros.
$f(-x) = -x^5 - x^4 - x^3 - x^2 - x - 1$ There are no negative zeros.
Maximum number of zeros: 5

12. $f(x) = x^6 + 1$ There are no positive zeros.
$f(-x) = x^6 + 1$ There are no negative zeros.
Maximum number of zeros: 6

14. $f(x) = x^5 - x^4 + 2x^2 + 3$

$\frac{p}{q}$: $\pm 1, \pm 3$ are the potential rational zeros.

16. $f(x) = 2x^5 - x^4 - x^2 + 1$

$\frac{p}{q}$: $\pm 1, \pm \frac{1}{2}$ are the potential rational zeros.

18. $f(x) = 6x^4 - x^2 + 2$

$\frac{p}{q}$: $\pm 1, \pm 2, \pm \frac{1}{3}, \pm \frac{2}{3}, \pm \frac{1}{2}, \pm \frac{1}{6}$ are the potential rational zeros.

20. $f(x) = -4x^3 + x^2 + x + 2$

$\frac{p}{q}$: $\pm 1, \pm 2, \pm \frac{1}{2}, \pm \frac{1}{4}$ are the potential rational zeros.

22. $f(x) = 3x^5 - x^2 + 2x + 3$

$\frac{p}{q}$: $\pm 1, \pm 3, \pm \frac{1}{3}$, are the potential rational zeros.

24. $f(x) = -6x^3 - x^2 + x + 3$

$\frac{p}{q}$: $\pm 1, \pm 3, \pm \frac{1}{2}, \pm \frac{3}{2}, \pm \frac{1}{3}, \pm \frac{1}{6}$ are the potential rational zeros.

26. $f(x) = x^3 + 8x^2 + 11x - 20$ There is one positive zero.
    $f(-x) = -x^3 + 8x^2 - 11x - 20$ There are two or zero negative zeros.

$\dfrac{p}{q}$: $\pm 1, \pm 2, \pm 4, \pm 5, \pm 10, \pm 20$ are the potential rational zeros.

$$
\begin{array}{r|rrr}
1 & 1 & 8 & 11 & -20 \\
  &   & 1 & 9 & 20 \\
\hline
  & 1 & 9 & 20 & 0 \\
\end{array}
$$

Thus, $x - 1$ is a factor and 1 is a zero.

$$f(x) = (x - 1)(x^2 + 9x + 20) = (x - 1)(x + 5)(x + 4)$$

The real zeros of f are -4, -5, and 1.

28. $f(x) = 2x^3 + x^2 + 2x + 1$ There are no positive zeros.
    $f(-x) = -2x^3 + x^2 - 2x + 1$ There are three or one negative zeros.

$\dfrac{p}{q}$: $\pm 1, \pm \dfrac{1}{2}$, are the potential rational zeros.

We only try the potential negative zeros.

$$
\begin{array}{r|rrrr}
-1 & 2 & 1 & 2 & 1 \\
   &   & -2 & 1 & -3 \\
\hline
   & 2 & -1 & 3 & -2 \\
\end{array}
\qquad
\begin{array}{r|rrrr}
-\frac{1}{2} & 2 & 1 & 2 & 1 \\
   &   & -1 & 0 & -1 \\
\hline
   & 2 & 0 & 2 & 0 \\
\end{array}
$$

Thus, $x + \dfrac{1}{2}$ is a factor and $-\dfrac{1}{2}$ is a zero.

$$f(x) = \left(x + \frac{1}{2}\right)(2x^2 + 2) = 2\left(x + \frac{1}{2}\right)(x^2 + 1)$$

The only real zero of f is $-\dfrac{1}{2}$.

30. $f(x) = x^4 - 3x^2 - 4$ There is one positive zero.
    $f(-x) = x^4 - 3x^2 - 4$ There is one negative zero.

$\dfrac{p}{q}$: $\pm 1, \pm 2, \pm 4$ are the potential rational zeros.

$f(x) = x^4 - 3x^2 - 4$ (a disguised quadrant)

    $= (x^2 - 4)(x^2 + 1) = (x - 2)(x + 2)(x^2 + 1)$

The real zeros of f are 2 and -2.

32. $f(x) = 4x^4 + 15x^2 - 4$ There is one positive zero.
    $f(-x) = 4x^4 + 15x^2 - 4$ There is one negative zero.

$\dfrac{p}{q}$: $\pm 1, \pm 2, \pm 4, \pm \dfrac{1}{2}, \pm \dfrac{1}{4}$ are the potential rational zeros.

$f(x) = 4x^4 + 15x^2 - 4 = (4x^2 - 1)(x^2 + 4) = (2x - 1)(2x + 1)(x^2 + 4)$

The real zeros of f are $\dfrac{1}{2}$ and $-\dfrac{1}{2}$.

34. $f(x) = x^4 - x^3 - 6x^2 + 4x + 8$ There are two or zero positive zeros.
    $f(-x) = x^4 + x^3 - 6x^2 - 4x + 8$ There are two or zero negative zeros.

$\dfrac{p}{q}$: $\pm 1, \pm 2, \pm 4, \pm 8$ are the potential rational zeros.

$$
\begin{array}{r|rrrrr}
1 & 1 & -1 & -6 & 4 & 8 \\
  &   & 1 & 0 & -6 & -2 \\
\hline
  & 1 & 0 & -6 & -2 & 6 \\
\end{array}
\qquad
\begin{array}{r|rrrrr}
-1 & 1 & -1 & -6 & 4 & 8 \\
   &   & -1 & 2 & 4 & -8 \\
\hline
   & 1 & -2 & -4 & 8 & 0 \\
\end{array}
$$

Thus, $x + 1$ is a factor and -1 is a zero.
$$f(x) = (x + 1)(x^3 - 2x^2 - 4x + 8)$$
$$q_1(x) = x^3 - 2x^2 - 4x + 8$$
There are two or zero positive zeros.
$$q_1(-x) = x^3 - 2x^2 + 4x + 8$$
There is one negative zero.

```
-1)1  -2  -4   8          2)1  -2  -4   8
     -1   3   1               2   0  -8
   ────────────             ─────────────
    1  -3  -1   9            1   0  -4   0
```

Thus, $x - 2$ is a factor and 2 is a zero.
$$f(x) = (x + 1)(x - 2)(x^2 - 4) = (x + 1)(x - 2)(x - 2)(x + 2) = (x + 1)(x + 2)(x - 2)^2$$
The real zeros of f are -1, 2, and -2.

36. $f(x) = 4x^5 + 12x^4 - x - 3$ There is one positive zero.
$f(-x) = -4x^5 + 12x^4 + x - 3$ There are two or zero negative zeros.

$\dfrac{p}{q}$: $\pm 1,\ \pm 3,\ \pm\dfrac{1}{2},\ \pm\dfrac{3}{2},\ \pm\dfrac{1}{4},\ \pm\dfrac{3}{4}$ are the potential rational zeros.

```
1)4  12   0   0  -1  -3         -1)4  12   0   0  -1  -3
     4  16  16  16  15               -4  -8   8  -8   9
   ────────────────────            ────────────────────
   4  16  16  16  15  12            4   8  -8   8  -9   6
```

```
3)4  12    0    0   -1    -3      -3)4  12   0   0  -1  -3
    12   72  216  648  1941            -12   0   0   0   3
   ──────────────────────────        ────────────────────
   4  24  72  216  647  1938          4    0   0   0  -1   0
```

Thus, $x + 3$ is a factor and -3 is a zero.
$$f(x) = (x + 3)(4x^4 - 1)$$
$$q_1(x) = 4x^4 - 1$$
There is one positive zero.
$$q_1(-x) = 4x^4 - x$$
There is one negative zero.
$$q_1(x) = 4x^4 - 1 \qquad \text{(a disguised quadratic)}$$
$$= (2x^2 - 1)(2x^2 + 1) = \left(\sqrt{2}x + 1\right)\left(\sqrt{2}x - 1\right)(2x^2 + 1)$$
Thus, $f(x) = (x + 3)\left(\sqrt{2}x + 1\right)\left(\sqrt{2}x - 1\right)(2x^2 + 1)$
The real zeros of f are -3, $-\dfrac{1}{\sqrt{2}}$ and $\dfrac{1}{\sqrt{2}}$.

38. $2x^3 + 3x^2 + 2x + 3 = 0$ The solutions of this equation are the zeros of the polynomial function $f(x)$.

$f(x) = 2x^3 + 3x^2 + 2x + 3$ There are zero positive zeros.
$f(-x) = -2x^3 + 3x^2 - 2x + 3$ There are three or one negative zeros.

$\dfrac{p}{q}$: $\pm 1,\ \pm 3,\ \pm\dfrac{1}{2},\ \pm\dfrac{3}{2}$ are the potential rational zeros.

```
-1)2   3   2   3          -3)2   3   2    3
     -2  -1  -1               -6   9  -33
   ───────────────          ────────────────
   2   1   1   2            2  -3  11  -30
```

$$-\tfrac{1}{2}\overline{\big|\;2\quad 3\quad 2\quad 3}$$

$$\begin{array}{ccccc} & -1 & -1 & -\tfrac{1}{2} \\ \hline 2 & 2 & 1 & \tfrac{5}{2} \end{array}$$

$$-\tfrac{3}{2}\overline{\big|\;2\quad 3\quad 2\quad 3}$$

$$\begin{array}{ccccc} & -3 & 0 & -3 \\ \hline 2 & 0 & 2 & 0 \end{array}$$

Thus, $x + \dfrac{3}{2}$ is a factor and $-\dfrac{3}{2}$ is a zero.

$$\left(x + \frac{3}{2}\right)(2x^2 + 2) = 2\left(x + \frac{3}{2}\right)(x^2 + 1) = 0$$

The only real zero is $-\dfrac{3}{2}$.

40. $\quad 2x^3 - 3x^2 - 3x - 5 = 0 \qquad$ The solutions of this equation are the zeros of the polynomial function $f(x)$.

$$f(x) = 2x^3 - 3x^2 - 3x - 5$$

There is one positive zero.

$$f(-x) = -2x^3 - 3x^2 + 3x - 5$$

There are two or zero negative zeros.

$\dfrac{p}{q}: \quad \pm 1, \pm 5, \pm\dfrac{1}{2}, \pm\dfrac{5}{2}$ are the potential rational zeros.

$$\begin{array}{r|rrrr} 2 & 2 & -3 & -3 & -5 \\ & & 2 & -1 & -4 \\ \hline & 2 & -1 & -4 & -9 \end{array} \qquad \begin{array}{r|rrrr} -1 & 2 & -3 & -3 & -5 \\ & & 2 & -1 & -4 \\ \hline & 2 & -1 & -4 & -9 \end{array} \qquad \begin{array}{r|rrrr} 5 & 2 & -3 & -3 & -5 \\ & & 10 & 35 & 160 \\ \hline & 2 & 7 & 32 & 155 \end{array}$$

$$\begin{array}{r|rrrr} -5 & 2 & -3 & -3 & -5 \\ & & -10 & 65 & -310 \\ \hline & 2 & -13 & 62 & -315 \end{array} \qquad \tfrac{1}{2}\overline{\big|\;2\;\;-3\;\;-3\;\;-5}$$

$$\begin{array}{ccccc} & 1 & -1 & -2 \\ \hline 2 & -2 & -4 & -7 \end{array}$$

$$-\tfrac{1}{2}\overline{\big|\;2\;\;-3\;\;-3\;\;-5}$$

$$\begin{array}{ccccc} & -1 & 2 & \tfrac{1}{2} \\ \hline 2 & -4 & -1 & -\tfrac{9}{2} \end{array}$$

$$\tfrac{5}{2}\overline{\big|\;2\;\;-3\;\;-3\;\;-5}$$

$$\begin{array}{ccccc} & 5 & 5 & 5 \\ \hline 2 & 2 & 2 & 0 \end{array}$$

Thus, $x - \dfrac{5}{2}$ is a factor and $\dfrac{5}{2}$ is a zero.

$$\left(x - \frac{5}{2}\right)(2x^2 + 2x + 2) = 2\left(x - \frac{5}{2}\right)(x^2 + x + 1) = 0$$

Since $x^2 + x + 1$ has no real solution, the only real zero is $\dfrac{5}{2}$.

42. $\quad 2x^3 - 11x^2 + 10x + 8 = 0 \qquad$ The solutions of this equation are the zeros of the polynomial function $f(x)$.

$f(x) = 2x^3 - 11x^2 + 10x + 8 \qquad$ There are two or zero positive zeros.

$f(-x) = -2x^3 - 11x^2 - 10x + 8 \qquad$ There is one negative zero.

$\dfrac{p}{q}: \quad \pm 1, \pm 2, \pm 4, \pm 8, \pm\dfrac{1}{2}$ are the potential rational zeros.

$$\begin{array}{r|rrrr} 1 & 2 & -11 & 10 & 8 \\ & & 2 & -9 & 1 \\ \hline & 2 & -9 & 1 & 9 \end{array} \qquad \begin{array}{r|rrrr} -1 & 2 & -11 & 10 & 8 \\ & & -2 & 13 & -23 \\ \hline & 2 & -13 & 23 & -15 \end{array} \qquad \begin{array}{r|rrrr} 2 & 2 & -11 & 10 & 8 \\ & & 4 & -14 & -8 \\ \hline & 2 & -7 & -4 & 0 \end{array}$$

Thus $x - 2$ is a factor and 2 is a zero.

$$f(x) = (x - 2)(2x^2 - 7x - 4) = (x - 2)(2x + 1)(x - 4)$$

The real zeros are 2, $-\dfrac{1}{2}$, and 4.

44. $x^4 - 2x^3 + 10x^2 - 18x + 9 = 0$ The solutions of this equation are the zeros of the polynomial function $f(x)$.

$$f(x) = x^4 - 2x^3 + 10x^2 - 18x + 9$$

There are four, two or zero positive zeros.

$$f(-x) = x^4 + 2x^3 + 10x^2 - 18x + 9$$

There are no negative zeros.

$\dfrac{p}{q}$: $\pm1, \pm3, \pm9$ are the potential rational zeros.

$$
\begin{array}{r|rrrrr}
1 & 1 & -2 & 10 & -18 & 9 \\
  &   & 1 & -1 & 9 & -9 \\
\hline
  & 1 & -1 & 9 & -9 & 0 \\
\end{array}
$$

Thus, $x - 1$ is a factor and 1 is a zero.

$$f(x) = (x - 1)(x^3 - x^2 + 9x - 9)$$
$$q_1(x) = x^3 - x^2 - 9x - 9$$

There are three or one positive zeros.

$$q_1(-x) = -x^3 - x^2 - 9x - 9$$

There are no negative zeros.

$$
\begin{array}{r|rrrr}
1 & 1 & -1 & 9 & -9 \\
  &   & 1 & 0 & 9 \\
\hline
  & 1 & 0 & 9 & 0 \\
\end{array}
$$

Thus, $x - 1$ is a factor and 1 is a zero.

$$(x - 1)(x - 1)(x^2 + 9) = 0$$

The only real zero is 1.

46. $x^3 + \dfrac{3}{2}x^2 + 3x - 2 = 0$ The solutions of this equation are the zeros of the polynomial function $f(x)$.

$f(x) = x^3 + \dfrac{3}{2}x^2 + 3x - 2$ There is one positive zero.

$f(-x) = -x^3 + \dfrac{3}{2}x^2 - 3x - 2$ There are two or zero negative zeros.

$f(x) = \dfrac{1}{2}(2x^3 + 3x^2 + 6x - 4)$

$\dfrac{p}{q}$: $\pm1, \pm2, \pm4, \pm\dfrac{1}{2}$ are the potential rational zeros.

$$
\begin{array}{r|rrrr}
1 & 2 & 3 & 6 & -4 \\
  &   & 2 & 5 & 11 \\
\hline
  & 2 & 5 & 11 & 7 \\
\end{array}
\qquad
\begin{array}{r|rrrr}
-1 & 2 & 3 & 6 & -4 \\
   &   & -2 & -1 & -5 \\
\hline
   & 2 & 1 & 5 & -9 \\
\end{array}
\qquad
\begin{array}{r|rrrr}
2 & 2 & 3 & 6 & -4 \\
  &   & 4 & 14 & 40 \\
\hline
  & 2 & 7 & 20 & 36 \\
\end{array}
$$

$$
\begin{array}{r|rrrr}
-2 & 2 & 3 & 6 & -4 \\
   &   & -4 & 2 & -16 \\
\hline
   & 2 & -1 & 8 & -20 \\
\end{array}
\qquad
\begin{array}{r|rrrr}
4 & 2 & 3 & 6 & -4 \\
  &   & 8 & 44 & 200 \\
\hline
  & 2 & 11 & 50 & 196 \\
\end{array}
\qquad
\begin{array}{r|rrrr}
-4 & 2 & 3 & 6 & -4 \\
   &   & -8 & 20 & -104 \\
\hline
   & 2 & -5 & 26 & -108 \\
\end{array}
$$

$$
\begin{array}{r|rrrr}
-\frac{1}{2} & 2 & 3 & 6 & -4 \\
   &   & -1 & -1 & -\frac{5}{2} \\
\hline
   & 2 & 2 & 5 & -\frac{13}{2} \\
\end{array}
\qquad
\begin{array}{r|rrrr}
\frac{1}{2} & 2 & 3 & 6 & -4 \\
   &   & 1 & 2 & 4 \\
\hline
   & 2 & 4 & 8 & 0 \\
\end{array}
$$

Thus, $x - \dfrac{1}{2}$ is a factor and $\dfrac{1}{2}$ is a zero.

$$\frac{1}{2}\left[x - \frac{1}{2}\right](2x^2 + 4x + 8) = \left[x - \frac{1}{2}\right](x^2 + 2x + 4) = 0$$

Since $x^2 + 2x + 4$ has no real solutions, the only zero is $\frac{1}{2}$.

48. $f(x) = 2x^3 + x^2 + 2x + 1 = 2\left[x + \frac{1}{2}\right](x^2 + 1)$

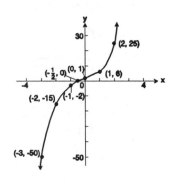

The x-intercept is $-\frac{1}{2}$; the y-intercept is $f(0) = 1$

x	Test Number	Value of f	Graph of f
$x < -\dfrac{1}{2}$	-1	$f(-1) = -2$	Below x-axis
$x > -\dfrac{1}{2}$	0	$f(0) = 1$	Above x-axis

50. $f(x) = x^4 - 3x^2 - 4 = (x - 2)(x + 2)(x^2 + 1)$
The x-intercepts are 2 and -2; the y-intercept is $f(0) = -4$.

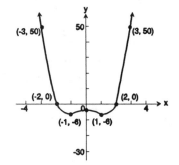

x	Test Number	Value of f	Graph of f
$x < -2$	-3	$f(-3) = 50$	Above x-axis
$-2 < x < 2$	0	$f(0) = -4$	Below x-axis
$x > 2$	3	$f(3) = 50$	Above x-axis

Since $f(x) = f(-x)$, the function is even.

52. $f(x) = 4x^4 + 15x^2 - 4 = (2x - 1)(2x + 1)(x^2 + 4)$

The x-intercepts are $\frac{1}{2}$ and $-\frac{1}{2}$; the y-intercept is $f(0) = -4$.

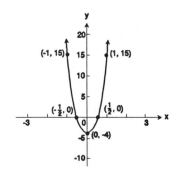

x	Test Number	Value of f	Graph of f
$x < -\dfrac{1}{2}$	-1	$f(-1) = 15$	Above x-axis
$-\dfrac{1}{2} < x < \dfrac{1}{2}$	0	$f(0) = -4$	Below x-axis
$x > \dfrac{1}{2}$	1	$f(1) = 15$	Above x-axis

Since $f(x) = f(-x)$, the function is symmetric with respect to the y-axis.

54. $f(x) = x^4 - x^3 - 6x^2 + 4x + 8 = (x + 1)(x - 2)(x + 2)$
The x-intercepts are -1, 2, and -2; the y-intercept is $f(0) = 8$.

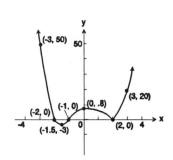

x	Test Number	Value of f	Graph of f
$x < -2$	-3	$f(-3) = 50$	Above x-axis
$-2 < x < -1$	-1.5	$f(-1.5) \approx -3$	Below x-axis
$-1 < x < 2$	0	$f(0) = 8$	Above x-axis
$x > 2$	3	$f(3) = 20$	Above x-axis

56. $f(x) = 4x^5 + 12x^4 - x - 3 = (x + 3)\left(\sqrt{2}x + 1\right)\left(\sqrt{2}x - 1\right)(2x^2 + 1)$

The x-intercepts are -3, $-\dfrac{1}{\sqrt{2}}$, and $\dfrac{1}{\sqrt{2}}$; the y-intercept is $f(0) = -3$.

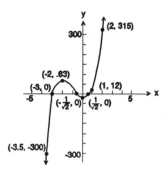

x	Test Number	Value of f	Graph of f
$x < -3$	-3.5	$f(-3.5) \approx -300$	Below x-axis
$-3 < x < -\dfrac{1}{\sqrt{2}}$	-2	$f(-2) = 63$	Above x-axis
$\dfrac{-1}{\sqrt{2}} < x < \dfrac{1}{\sqrt{2}}$	0	$f(0) = -3$	Below x-axis
$x > \dfrac{1}{\sqrt{2}}$	1	$f(1) = 12$	Above x-axis

58. $x^4 - 5x^2 - 36 = 0$ The solutions of this equation are the zeros of the polynomial function $f(x)$.
$f(x) = x^4 - 5x^2 - 36$ There is one positive zero.
$f(-x) = x^4 - 5x^2 - 36$ There is one negative zero.

$\dfrac{p}{q}$: $\pm 1, \pm 2, \pm 3, \pm 4, \pm 6, \pm 9, \pm 12, \pm 18, \pm 36$ are the potential rational zeros.

```
1)1  0  -5   0  -36    -1)1   0  -5  0  -36    2)1  0  -5   0  -36    3)1  0  -5   0  -36
     1   1  -4   -4           -1   1   4  -4          2   4  -2   -4          3   9  12   36
   -----------------        ------------------       ------------------       ------------------
   1   1  -4  -4  -40        1  -1  -4   4  -40        1  2  -1  -2  -40        1  3   4  12    0
```

Thus, $x - 3$ is a factor and 3 is a zero.

$q_1(x) = x^3 + 3x^2 + 4x + 12$

```
3)1  3   4  12      -3)1   3   4   12
     3  18  66            -3   0  -12
   ------------         --------------
   1  6  22  88         1   0   4    0
```

Thus, $x + 3$ is a factor and -3 is a zero.

$q_2(x) = x^2 + 4$

Thus, we have $(x - 3)(x + 3)(x^2 + 4) = 0$

For $x^2 + 4$, $x^2 = -4$ or $x = \pm\sqrt{-4} = \pm 2i$.

In the complex number system, the zeros are 3, -3, $2i$, and $-2i$.

60. $x^4 - x^3 + x - 1 = 0$ The solutions of this equation are the zeros of the polynomial function $f(x)$.
$f(x) = x^4 - x^3 + x - 1$ There are three or one zero(s).
$f(-x) = x^4 + x^3 - x - 1$ There is one negative zero.

$\dfrac{p}{q}$: ± 1, are the potential rational zeros.

```
1)1  -1   0   1  -1
      1   0   0   1
   ------------------
   1   0   0   1   0
```

Thus, $x - 1$ is a factor and 1 is a zero.

$q_1(x) = x^3 + 1 = (x + 1)(x^2 - x + 1)$

Thus, we have $(x - 1)(x + 1)(x^2 - x + 1) = 0$

For $x^2 - x + 1 = 0$, $x = \dfrac{1 \pm \sqrt{1 - 4}}{2} = \dfrac{1}{2} \pm \dfrac{\sqrt{3}\,i}{2}$

In the complex number system, the zeros are 1, -1, $\dfrac{1}{2} + \dfrac{\sqrt{3}}{2}i$ and $\dfrac{1}{2} - \dfrac{\sqrt{3}}{2}i$.

62. $x^4 - 3x^3 - 5x^2 + 27x - 36 = 0$ The solutions of this equation are the zeros of the polynomial function $f(x)$.

$f(x) = x^4 - 3x^3 - 5x^2 + 27x - 36$ There are three or one positive zero(s).
$f(-x) = x^4 + 3x^3 - 5x^2 - 27x - 36$ There is one negative zero.

$\dfrac{p}{q}$: $\pm 1, \pm 2, \pm 3, \pm 4, \pm 6, \pm 9, \pm 12, \pm 16, \pm 36$ are the potential rational zeros.

$1\rfloor$	1	-3	-5	27	-36
		1	-2	-7	20
	1	-2	-7	20	-16

$-1\rfloor$	1	-3	-5	27	-36
		-1	4	1	-28
	1	-4	-1	28	-64

$2\rfloor$	1	-3	-5	27	-36
		2	-2	-14	26
	1	-1	-7	13	-10

$-2\rfloor$	1	-3	-5	27	-36
		-2	10	-10	-34
	1	-5	5	17	-70

$3\rfloor$	1	-3	-5	27	-36
		3	0	-15	36
	1	0	-5	12	0

Thus, $x - 3$ is a factor and 3 is a zero.
$q_1(x) = x^3 - 5x + 12$

$3\rfloor$	1	0	-5	12
		3	9	12
	1	3	4	24

$-3\rfloor$	1	0	-5	12
			-3	9	-12
	1	-3	4	0

Thus, $x + 3$ is a factor and -3 is a zero.
$q_2(x) = x^2 - 3x + 4$
Thus, we have $(x - 3)(x + 3)(x^2 - 3x + 4) = 0$

For $x^2 - 3x + 4$, $x = \dfrac{3 \pm \sqrt{9 - 16}}{2} = \dfrac{3}{2} \pm \dfrac{\sqrt{7}}{2}i$

In the complex number system, the zeros are 3, -3, $\dfrac{3}{2} + \dfrac{\sqrt{7}}{2}i$, $\dfrac{3}{2} - \dfrac{\sqrt{7}}{2}i$.

64. $x^5 + x^4 + x^3 + x^2 - 2x - 2 = 0$ The solutions of this equation are the zeros of the polynomial function $f(x)$.

$f(x) = x^5 + x^4 + x^3 + x^2 - 2x - 2$ There is one positive zero.
$f(-x) = -x^5 + x^4 - x^3 + x^2 + 2x - 2$ There are four, two, or zero negative zeros.

$\dfrac{p}{q}$: $\pm 1, \pm 2$ are the potential rational zeros.

$1\rfloor$	1	1	1	1	-2	-2
		1	2	3	4	2
	1	2	3	4	2	0

Thus, $x - 1$ is a factor and 1 is a zero.
$q_1(x) = x^4 + 2x^3 + 3x^2 + 4x + 2$

$1\rfloor$	1	2	3	4	2
		1	3	6	10
	1	3	6	10	12

$-1\rfloor$	1	2	3	4	2
		-1	-1	-2	-2
	1	1	2	2	0

Thus, $x + 1$ is a factor and -1 is a zero.
$q_2(x) = x^3 + x^2 + 2x + 2$

$1\rfloor$	1	1	2	2
		-1	0	-2
	1	0	2	0

Thus, $x + 1$ is a factor and -1 is a zero (of multiplicity two.)
$q_3(x) = x^2 + 2$

Thus, we have $(x - 1)(x + 1)(x + 1)(x^2 + 2) = 0$

For $x^2 + 2 = 0$, $x = \pm\sqrt{-2} = \pm\sqrt{2}\,i$

In the complex number system, the zeros are 1, -1, $\sqrt{2}\,i$, and $-\sqrt{2}\,i$.

66. If -2 is a solution of $x^3 + 5x^2 + 5x - 2 = 0$, then $x + 2$ is a factor of $f(x) = x^3 + 5x^2 + 5x - 2$. Thus, using synthetic division, we find

$$
\begin{array}{r|rrrr}
-2 & 1 & 5 & 5 & -2 \\
   &   & -2 & -6 & 2 \\
\hline
   & 1 & 3 & -1 & 0
\end{array}
$$

The depressed equation is $x^2 + 3x - 1 = 0$

Its solutions are: $\dfrac{-3 \pm \sqrt{9 + 4}}{2} = \dfrac{-3 \pm \sqrt{13}}{2}$

The sum of these solutions is $\dfrac{-3 + \sqrt{13}}{2} + \dfrac{-3 - \sqrt{13}}{2} = -3$

68. No (use the Rational Zeros Theorem).   70. No (use the Rational Zeros Theorem).

72. Let x be the length of the edge of a cube and V be its volume. Then $x^3 = V$ and
$$(x + 6)(x + 12)(x - 4) = 2(x^3)$$
$$x^3 + 14x^2 - 288 = 2x^3$$
$$x^3 - 14x^2 + 288 = 0$$

$\dfrac{p}{q}$: $\pm 1, \pm 2, \pm 4, \pm 6, \pm 8, \pm 12, \pm 16, \pm 18, \pm 24, \pm 36, \pm 48, \pm 72, \pm 144, \pm 288$ are the

potential rational zeros.
Since $x > 0$,

$$
\begin{array}{r|rrrr}
1 & 1 & -14 & 0 & 288 \\
  &   & 1 & -13 & -13 \\
\hline
  & 1 & -13 & -13 & 275
\end{array}
\qquad
\begin{array}{r|rrrr}
2 & 1 & -14 & 0 & 288 \\
  &   & 2 & -24 & -48 \\
\hline
  & 1 & -12 & -24 & 240
\end{array}
$$

$$
\begin{array}{r|rrrr}
4 & 1 & -14 & 0 & 288 \\
  &   & 4 & -40 & -160 \\
\hline
  & 1 & -10 & -40 & 128
\end{array}
\qquad
\begin{array}{r|rrrr}
6 & 1 & -14 & 0 & 288 \\
  &   & 6 & -48 & -288 \\
\hline
  & 1 & -8 & -48 & 0
\end{array}
$$

Thus, $x - 6$ is a factor and 6 is a zero. The edge of the cube is 6 centimeters.

74. Let $\dfrac{p}{q}$, where p and q have no common factors except 1 and -1, be a solution of the polynomial
$$f(x) = a_n x^n + a_{n-1}x^{n-1} + \ldots + a_1 x + a_0$$
whose coefficients are all integers. Then

$$f\left[\frac{p}{q}\right] = a_n\left[\frac{p}{q}\right]^n + a_{n-1}\left[\frac{p}{q}\right]^{n-1} + \ldots + a_1\left[\frac{p}{q}\right] + a_0 = 0$$
$$= \frac{1}{q}(a_n p^n + a_{n-1}p^{n-1}q + \ldots + a_1 pq^{n-1} + a_0 q^n) = 0$$

Because p is a factor of the first n terms of this equation, p must also be a factor of $a_0 q^n$. Since p is not a factor of q (p and q have no common factors except 1 and -1), p must be a factor of a_0. Similarly, q must be a factor of a_n.

76. We start with $x^3 + px + q = 0$ and let $x = H + K$, to obtain:
$$(H + K)^3 + p(H + K) + q = 0$$
$$H^3 + 3H^2K + 3HK^2 + K^3 + pH + pK + q = 0$$
$$H^3 + K^3 + 3H^2K + 3HK^2 + pH + pK = -q \qquad (1)$$

Section 5.5 The Zeros of a Polynomial Function   177

Now we let $3HK = -p$. If we multiply both sides of $3HK = -p$ by H, we obtain $3H^2K = -pH$. If we multiply both sides by K, we obtain $3HK^2 = -pK$. This allows us to rewrite equation (1):

$$H^3 + K^3 - pH - pK + pH + pK = -q$$

or

$$H^3 + K^3 = -q, \text{ as desired.}$$

Notice we now need to find H or K such that

$$H^3 + K^3 = -q,$$

and

$$3HK = -p$$

If we can find H and K, then we will have our x, since $x = H + K$.

78. From Problem 74 we have $H^3 + K^3 = -q$, and from Problem 75,

$$H = \sqrt[3]{\frac{-q}{2} + \sqrt{\frac{q^2}{4} + \frac{p^3}{27}}}$$

Since $H^3 + K^3 = -q$, we have

$$\frac{-q}{2} + \sqrt{\frac{q^2}{4} + \frac{p^3}{27}} + K^3 = -q$$

$$K^3 = -q + \frac{q}{2} - \sqrt{\frac{q^2}{4} + \frac{p^3}{27}}$$

$$K^3 = \frac{-q}{2} - \sqrt{\frac{q^2}{4} + \frac{p^3}{27}}$$

or $\quad K = \sqrt[3]{\dfrac{-q}{2} + \sqrt{\dfrac{q^2}{4} + \dfrac{p^3}{27}}}$

80. From Problem 77, a solution of the equation $x^3 + px + q = 0$ is given by

$$x = \sqrt[3]{\frac{-q}{2} + \sqrt{\frac{q^2}{4} + \frac{p^3}{27}}} + \sqrt[3]{\frac{-q}{2} - \sqrt{\frac{q^2}{4} + \frac{p^3}{27}}}$$

We are asked to solve $x^3 - 6x - 9 = 0$

Here $p = -6$, $q = -9$

Then $\dfrac{q^2}{4} + \dfrac{p^3}{27} = \dfrac{81}{4} - \dfrac{216}{27} = \dfrac{81}{4} - 8 = \dfrac{49}{4}$

and we have $x = \sqrt[3]{\dfrac{-(-9)}{2} + \sqrt{\dfrac{49}{4}}} + \sqrt[3]{\dfrac{-(-9)}{2} - \sqrt{\dfrac{49}{4}}} = \sqrt[3]{\dfrac{9}{2} + \dfrac{7}{2}} + \sqrt[3]{\dfrac{9}{2} - \dfrac{7}{2}} = \sqrt[3]{8} + \sqrt[3]{1}$

or $x = 2 + 1 = 3$

Now we have one solution, $x = 3$. Hence $x - 3$ is a factor, and we can perform synthetic division to find the other factor:

```
-3)1   0  -6  -9
      -3  -9  -9
    1  3   3   0
```

Quotient $= x^2 + 3x + 3$

The solutions to the quadratic $x^2 + 3x + 3 = 0$ are $x = \dfrac{-3 \pm \sqrt{9 - 4(3)}}{2} = \dfrac{-3 \pm \sqrt{-3}}{2}$, which involves complex numbers (refer to Sections 1.9 and 2.4). The three solutions are:

$$x = 3, \quad x = \frac{-3 + \sqrt{3}\,i}{2}, \quad x = \frac{-3 - \sqrt{3}\,i}{2}$$

82. Solve $f(x) = x^3 + 3x - 14 = 0$ There is one positive zero.

$f(-x) = -x^3 - 3x - 14$ There are no negative zeros.

$\dfrac{p}{q}$: $\pm 1, \pm 2, \pm 7, \pm 14$ are the potential rational zeros.

```
1|1  0  3  -14      -1|1   0  3  -14      2|1  0  3  -14
  |   1  3   4        |   -1  1   -4       |   2  4   14
   1  1  4  -10          1  -1  4  -18       2  2  7    0
```

Thus, $x - 2$ is a factor and 2 is a zero.

$$q_1(x) = 2x^2 + 2x + 7$$

Thus, $f(x) = (x - 2)(2x^2 + 2x + 7) = 0$

For $2x^2 + 2x + 7$, $x = \dfrac{-2 \pm \sqrt{4 - 56}}{4} = \dfrac{-1}{2} \pm \dfrac{\sqrt{13}}{2}\,i$

84. $f(2 + x) = f(2 - x)$

$\begin{aligned}
f(x) &= (x - r_1)(x - r_2)(x - r_3)(x - r_4)\\
&= x^4 - (r_3 + r_4)x^3 + r_3 r_4 x^2 - (r_1 + r_2)x^3 + (r_1 + r_2)(r_3 + r_4)x^2\\
&\quad - r_3 r_4 (r_1 + r_2)x + r_1 r_2 x^2 - r_1 r_2 (r_3 + r_4)x + r_1 r_2 r_3 r_4\\
&= x^4 - (r_1 + r_2 + r_3 + r_4)x^3 + [(r_1 r_2 + r_3 r_4 + (r_1 + r_2)(r_3 + r_4))]x^2\\
&\quad - [r_1 r_2 (r_3 + r_4) + r_3 r_4 (r_1 + r_2)]x + r_1 r_2 r_3 r_4
\end{aligned}$

It is clear that $f(0) = f(4)$ and $f(3) = f(1)$ for $x = 2$ and $x = 1$, respectively. Therefore,

$\begin{aligned}
f(0) = r_1 r_2 r_3 r_4 = f(4) &= 256 - 64(r_1 + r_2 + r_3 + r_4) + 16[(r_1 r_2 + r_3 r_4\\
&\quad + (r_1 + r_2)(r_3 + r_4))] - 4[r_1 r_2 (r_3 + r_4) + r_3 r_4 (r_1 + r_2)] + r_1 r_2 r_3 r_4\\
0 &= 64 - 16(r_1 + r_2 + r_3 + r_4) + 4[(r_1 r_2 + r_3 r_4 + (r_1 + r_2)(r_3 + r_4))]\\
&\quad - [r_1 r_2 (r_3 + r_4) + r_3 r_4 (r_1 + r_2)]
\end{aligned}$

$f(1) = f(3)$, so

$\begin{aligned}
&1 - (r_1 + r_2 + r_3 + r_4) + [r_1 r_2 + r_3 r_4 + (r_1 + r_2)(r_3 + r_4)] - [r_1 r_2 (r_3 + r_4)\\
&\quad + (r_3 r_4)(r_1 + r_2)] + r_1 r_2 r_3 r_4 = 81 - 27(r_1 + r_2 + r_3 + r_4) + 9(r_1 r_2 + r_3 r_4\\
&\quad + (r_1 + r_2)(r_3 + r_4)) - 3[r_1 r_2 (r_3 + r_4) + r_3 r_4 (r_1 + r_2)] + r_1 r_2 r_3 r_4\\
&26(r_1 + r_2 + r_3 + r_4) = 80 + 8(r_1 r_2 + r_3 r_4 + (r_1 + r_2)(r_3 + r_4)) - 2[r_1 r_2 (r_3 + r_4)\\
&\quad + r_3 r_4 (r_1 + r_2)]\\
&16(r_1 + r_2 + r_3 + r_4) = 64 + 4(r_1 r_2 + r_3 r_4 + (r_1 + r_2)(r_3 + r_4)) - [r_1 r_2 (r_3 + r_4)\\
&\quad + r_3 r_4 (r_1 + r_2)]\\
&26(r_1 + r_2 + r_3 + r_4) = 80 + 8(r_1 r_2 + r_3 r_4 + (r_1 + r_2)(r_3 + r_4)) - 2[r_1 r_2 (r_3 + r_4)\\
&\quad + r_3 r_4 (r_1 + r_2)]\\
&-6(r_1 + r_2 + r_3 + r_4) = -48\\
&r_1 + r_2 + r_3 + r_4 = 8
\end{aligned}$

2. $f(x) = 3x^3 - 2x^2 + x + 4$

$$
\begin{array}{r|rrrr}
1 & 3 & -2 & 1 & 4 \\
& & 3 & 1 & 2 \\
\hline
& 3 & 1 & 2 & 6
\end{array}
$$

Thus, 1 is an upper bound.

$$
\begin{array}{r|rrrr}
-1 & 3 & -2 & 1 & 4 \\
& & -3 & 5 & -6 \\
\hline
& 3 & -5 & 6 & -2
\end{array}
$$

Thus, -1 is a lower bound. The zeros of f lie between -1 and 1.

4. $f(x) = 2x^3 - x^2 - 11x - 6$

Upper Bound

$$
\begin{array}{r|rrrr}
1 & 2 & -1 & -11 & -6 \\
& & 2 & 1 & -10 \\
\hline
& 2 & 1 & -10 & -16
\end{array}
$$

$$
\begin{array}{r|rrrr}
2 & 2 & -1 & -11 & -6 \\
& & 4 & 6 & -10 \\
\hline
& 2 & 3 & -5 & -16
\end{array}
$$

$$
\begin{array}{r|rrrr}
3 & 2 & -1 & -11 & -6 \\
& & 6 & 15 & 12 \\
\hline
& 2 & 5 & 4 & 6
\end{array}
$$

Lower Bound

$$
\begin{array}{r|rrrr}
-1 & 2 & -1 & -11 & -6 \\
& & -2 & 3 & 8 \\
\hline
& 2 & -3 & -8 & 2
\end{array}
$$

$$
\begin{array}{r|rrrr}
-2 & 2 & -1 & -11 & -6 \\
& & -4 & 10 & 2 \\
\hline
& 2 & -5 & -1 & -4
\end{array}
$$

$$
\begin{array}{r|rrrr}
-3 & 2 & -1 & -11 & -6 \\
& & -6 & 21 & -30 \\
\hline
& 2 & -7 & 10 & -36
\end{array}
$$

Upper Bound: 3; Lower Bound: -3

6. $f(x) = 4x^4 - 12x^3 + 27x^2 - 54x + 81$

$$
\begin{array}{r|rrrrr}
1 & 4 & -12 & 27 & -54 & 81 \\
& & 4 & -8 & 19 & -35 \\
\hline
& 4 & -8 & 19 & -35 & 46
\end{array}
$$

$$
\begin{array}{r|rrrrr}
2 & 4 & -12 & 27 & -54 & 81 \\
& & 8 & -8 & 38 & -32 \\
\hline
& 4 & -4 & 19 & -16 & 49
\end{array}
$$

$$
\begin{array}{r|rrrrr}
3 & 4 & -12 & 27 & -54 & 81 \\
& & 12 & 0 & 81 & 81 \\
\hline
& 4 & 0 & 27 & 27 & 162
\end{array}
$$

Thus, 3 is an upper bound

$$
\begin{array}{r|rrrrr}
-1 & 4 & -12 & 27 & -97 & 178 \\
& & -4 & 16 & -43 & 97 \\
\hline
& 4 & -16 & 43 & -97 & 178
\end{array}
$$

Thus, -1 is a lower bound. The zeros of f lie between -1 and 3.

8. $f(x) = x^4 + 8x^3 - x^2 + 2;\ [-1, 0]$
$f(-1) = -6$ and $f(0) = 2$
Because $f(-1) < 0$ and $f(0) > 0$, it follows from the Intermediate Value Theorem that f has a zero between -1 and 0.

10. $f(x) = 3x^3 - 10x + 9;\ [-3, -2]$
$f(-3) = -42$ and $f(-2) = 5$
Because $f(-3) < 0$ and $f(-2) > 0$, it follows from the Intermediate Value Theorem that f has a zero between -3 and -2.

12. $f(x) = x^5 - 3x^4 - 2x^3 + 6x^2 + x + 2$; [1.7, 1.8]

$f(1.7) = 0.35627$ and $f(1.8) = -1.02112$

Because $f(1.7) > 0$ and $f(1.8) < 0$, it follows from the Intermediate Value Theorem that f has a zero between 1.7 and 1.8.

14. $f(x) = 2x^4 + x^2 - 1$

Since $f(0) = -1$ and $f(1) = 2$, it follows from the Intermediate Value Theorem that f has a zero between 0 and 1.

$$f(0.0) = -1$$
$$f(0.1) = -0.9898$$
$$f(0.2) = -0.9568$$
$$f(0.3) = -0.8938$$
$$f(0.4) = -0.7888$$
$$f(0.5) = -0.625$$
$$f(0.6) = -0.3808$$
$$f(0.7) = -0.0298$$
$$f(0.8) = 0.4592$$

There is a zero between 0.7 and 0.8.

$$f(0.70) = -0.0298$$
$$f(0.71) = 0.01233362$$

The positive zero is approximately 0.70.

16. $f(x) = 3x^3 - 2x^2 - 20$

Since $f(2) = -4$ and $f(3) = 43$, it follows from the Intermediate Value Theorem that f has a zero between 2 and 3.

$$f(2.0) = -4$$
$$f(2.1) = -1.037$$
$$f(2.2) = 2.264$$

There is a zero between 2.1 and 2.2.

$$f(2.10) = -1.037$$
$$f(2.11) = -0.722407$$
$$f(2.12) = -0.404416$$
$$f(2.13) = -0.083009$$
$$f(2.14) = 0.241832$$

The positive zero is approximately 2.13.

18. $f(x) = x^4 + 8x^3 - x^2 + 2 = \{[(x + 8)x - 1]x\}x + 2; -1 \le r \le 0$

$$f(-1) = -6; f(0) = 2$$
$$f(0.1) = 1.9821$$
$$f(-0.2) = 1.8976$$
$$f(-0.3) = 1.7021$$
$$f(-0.4) = 1.2770$$
$$f(-0.5) = 0.8125$$
$$f(-0.6) = 0.0416$$
$$f(-0.7) = -0.9939$$
$$f(-0.61) = -0.0495$$

Thus, the zero is approximately -0.60.

20. $f(x) = 3x^3 - 10x + 9 = (3x^2 - 10)x + 9; \ -3 \le r \le -2$
     $f(-3) = -42; f(-2) = 5$
     $f(-2.1) = 2.217$
     $f(-2.2) = -0.944$
     $f(-2.11) = 1.9182$
     $f(-2.12) = 1.6156$
     $f(-2.13) = 1.3092$
     $f(-2.14) = 0.9989$
     $f(-2.15) = 0.6849$
     $f(-2.16) = 0.3669$
     $f(-2.17) = 0.0451$
     $f(-2.18) = -0.2807$

Thus, the zero is approximately -2.17.

22. $f(x) = 2x^4 + x^2 - 1$

Since $f(0) = -1$ and $f(1) = 2$, it follows from the Intermediate Value Theorem that f has a zero between 0 and 1.
     $f(0.0) = -1$
     $f(0.1) = -0.9898$
     $f(0.2) = -0.9568$
     $f(0.3) = -0.8938$
     $f(0.4) = -0.7888$
     $f(0.5) = -0.625$
     $f(0.6) = -0.3808$
     $f(0.7) = -0.0298$
     $f(0.8) = 0.4592$

There is a zero between 0.7 and 0.8.
     $f(0.70) = -0.0298$
     $f(0.71) = 0.01233362$

The positive zero is approximately 0.70.

24. $f(x) = 3x^3 - 2x^2 - 20$

Since $f(2) = -4$ and $f(3) = 43$, it follows from the Intermediate Value Theorem that f has a zero between 2 and 3.
     $f(2.0) = -4$
     $f(2.1) = -1.037$
     $f(2.2) = 2.264$

There is a zero between 2.1 and 2.2.
     $f(2.10) = -1.037$
     $f(2.11) = -0.722407$
     $f(2.12) = -0.404416$
     $f(2.13) = -0.083009$
     $f(2.14) = 0.241832$

The positive zero is approximately 2.13.

26. The zeros are $-.60$ and -8.11 correct to two decimal places.

28. The zero is -2.17 correct to two decimal places.

5.7 Complex Polynomials; Fundamental Theorem of Algebra

2. Since complex zeros appear as conjugate pairs, it follows that $3 - i$, the conjugate of $3 + i$, is the remaining zero of f.

4. $2 - i$, the conjugate of $2 + i$, is the remaining zero of f.

6. $-i$, the conjugate of i, is the remaining zero of f.

8. $2 + i$, the conjugate of $2 - i$ and i, the conjugate of $-i$, are the remaining zeros of f.

10. $-i$, the conjugate of i, $3 + 2i$, the conjugate of $3 - 2i$, and $-2 - i$, the conjugate of $-2 + i$, are the remaining zeros of f.

12. Zeros that are complex numbers must occur in conjugate pairs.

14. A missing zero is $4 + i$. If the remaining zero were a complex number, then its conjugate would also be a zero.

16. $f(z) = z^4 - 1 = (z^2 - 1)(z^2 + 1)$
$= (z - 1)(z + 1)(z^2 + 1)$
Zeros: $1, -1, i, -i$

18. $f(z) = 3z + i$
$f(1 + i) = 3(1 + i) + i = 3 + 4i$

20. $f(z) = (4 + i)z^2 + 5 - 2i$
$f(1 + i) = (4 + i)(1 + i)^2 + 5 - 2i$
$= (4 + i)(2i) + 5 - 2i$
$= 8i - 2 + 5 - 2i = 3 + 6i$

22. $f(z) = iz^3 - 2z^2 + 1$
$f(1 + i) = i(1 + i)^3 - 2(1 + i)^2 + 1$
$= i(2i - 2) = 2(2i) + 1$
$= 2i^2 + 2i - 4i + 1 = -1 - 6i$

24. $f(z) = iz^4 + (2 + i)z^2 - z, r = 1 - i$

$1 - i$	i	0	$2 + i$	-1	0
		$1 + 8$	2	$-5 - 3i$	$1 - 7i$
	i	$1 + i$	$4 + i$	$4 - 3i$	$1 - 7i$

$f(r) = 1 - 7i$

26. $f(z) = 2iz^3 + 8z^2 - 4iz + 1, r = 2 + i$

$2 + i$	$2i$	8	$-4i$	1
		$-2 + 4i$	$8 + 14i$	$6 + 28i$
	$2i$	$6 + 4i$	$8 + 10i$	$7 - 28i$

$f(r) = 7 + 28i$

28. $f(z) = z^4 + z^2 + 1, r = 1 - 2i$

$1 - 2i$	1	0	1	0	1
		$1 - 2i$	$-2 - 4i$	-10	$-10 + 20i$
	1	$1 - 3i$	$-2 - 4i$	-10	$-9 + 20i$

$f(r) = -9 + 20i$

30. $f(z) = (z - i)^2[z - (1 + 2i)] = (z^2 - 2zi + i^2)[z - (1 + 2i)]$
$= z^3 - z^2 - 2z^2i - 2z^2i + 2zi + 4zi^2 - z + 1 + 2i$
$= z^3 - (1 + 4i)z^2 + (-5 + 2i)z + 1 + 2i$

32. $f(z) = (z - i)[z - (4 - i)][z - (2 + i)]$
$= [z^2 - 4z + 4i + 1][z - (2 + i)]$
$= z^3 - (6 + i)z^2 + (9 + 8i)z - 9i + 2$

34. $f(z) = (z - 1)^3[z - (1 + i)]$
$= [z^3 - 3z^2 + 3z - 1][z - (1 + i)]$
$= z^4 - (4 + i)z^3 + (6 + 3i)z^2 - (4 + 3i)z + 1 + i$

5 Chapter Review

2. $f(x) = (x + 1)^2 - 4$
opens upward, vertex $(-1, -4)$,
axis of symmetry $x = -1$
y-intercept $(0, -3)$,
x-intercepts $(1, 0)$, $(-3, 0)$.

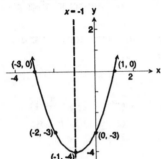

4. $f(x) = -\frac{1}{2}x^2 + 2$
opens downward, vertex $(0, 2)$,
axis of symmetry $x = 0$,
y-intercept $(0, 2)$,
x-intercepts $(2, 0)$, $(-2, 0)$.

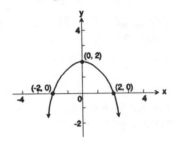

6. $f(x) = 9x^2 - 6x + 3$
opens upward, vertex $\left[\frac{1}{3}, 2\right]$
axis of symmetry $x = \frac{1}{3}$,
y-intercept $(0, 3)$,
no x-intercept.

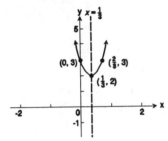

8. $f(x) = -x^2 + x + \frac{1}{2}$
opens downward, vertex $\left[\frac{1}{2}, \frac{3}{4}\right]$
axis of symmetry $x = \frac{1}{2}$,
y-intercept $\left[0, \frac{1}{2}\right]$
x-intercepts: $\left[\frac{1 - \sqrt{3}}{2}, 0\right]$, $\left[\frac{1 + \sqrt{3}}{2}, 0\right]$

10. $f(x) = -2x^2 - x + 4$
opens downward, vertex $\left[-\frac{1}{4}, \frac{33}{8}\right]$
axis of symmetry $x = -\frac{1}{4}$,
y-intercept $(0, 4)$,
x-intercepts: $\left[\frac{-1 + \sqrt{33}}{4}, 0\right]$, $\left[\frac{-1 - \sqrt{33}}{4}, 0\right]$

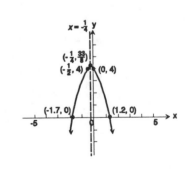

12. $f(x) = -x^3 + 3$

Start with the graph of $y = x^3$. Shift up 3 units and reflect about the x-axis.

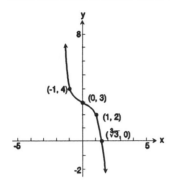

14. $f(x) = (x - 1)^4 - 2$

Start with the graph of $y = x^4$. Shift to the right one unit and shift down two units.

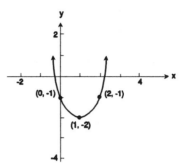

16. $f(x) = (1 - x)^3$

Start with the graph of $y = x^3$. Reflect about the y-axis and shift to the left 1 unit.

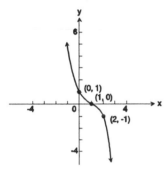

18. $f(x) = 2x^2 + 8x + 5$

Since $a = 2 > 0$, the parabola opens upward and the function has a minimum value. It occurs at

$$x = \frac{-b}{2a} = \frac{-8}{4} = -2$$

The minimum value is
$$f(-2) = 2(-2)^2 + 8(-2) + 5$$
$$= 8 - 16 + 5 = -3$$

20. $f(x) = -x^2 - 10x - 3$

Since $a = -1 < 0$, the parabola opens downward and the function has a maximum value. It occurs at

$$x = \frac{-b}{2a} = \frac{10}{-2} = -5$$

The maximum value is
$$f(-5) = -(-5)^2 - 10(-5) - 3$$
$$= -25 + 50 - 3 = 22$$

22. $f(x) = -2x^2 + 4$

Since $a = -2 < 0$, the parabola opens downward and the function has a maximum value. It occurs at

$$x = \frac{-b}{2a} = \frac{0}{-4} = 0$$

The maximum value is $f(0) = 0 + 4 = 4$

24. (a) x-intercepts: 0, 2, 4; y-intercept: 8
    (b) crosses x-axis at 0, 2, and 4
    (c) $y = x^3$
    (d) 2
    (e)

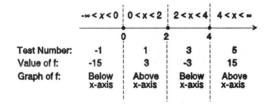

	$-\infty < x < 0$	$0 < x < 2$	$2 < x < 4$	$4 < x < \infty$
Test Number:	-1	1	3	5
Value of f:	-15	3	-3	15
Graph of f:	Below x-axis	Above x-axis	Below x-axis	Above x-axis

(f)

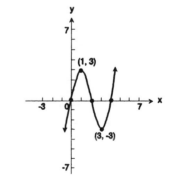

26. (a) x-intercepts: $-4, 2$; y-intercept: -32
   (b) touches at -4; crosses at 2
   (c) $y = x^3$
   (d) 2
   (e)

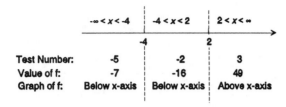

	$-\infty < x < -4$	$-4 < x < 2$	$2 < x < \infty$
Test Number:	-5	-2	3
Value of f:	-7	-16	49
Graph of f:	Below x-axis	Below x-axis	Above x-axis

(f)

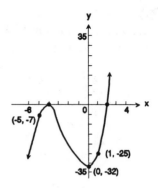

28. $f(x) = x(x^2 + 4)$
   (a) x-intercepts: 0; y-intercept: 0
   (b) crosses at 0
   (c) $y = x^3$
   (d) 2
   (e)

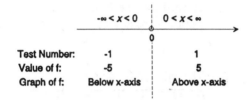

	$-\infty < x < 0$	$0 < x < \infty$
Test Number:	-1	1
Value of f:	-5	5
Graph of f:	Below x-axis	Above x-axis

(f)

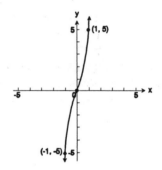

30. (a) x-intercepts: $-2, 2, 4$; y-intercept: 32
   (b) crosses at 2 and 4; touches at -2
   (c) $y = x^4$
   (d) 3
   (e)

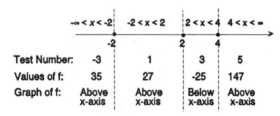

	$-\infty < x < -2$	$-2 < x < 2$	$2 < x < 4$	$4 < x < \infty$
Test Number:	-3	1	3	5
Values of f:	35	27	-25	147
Graph of f:	Above x-axis	Above x-axis	Below x-axis	Above x-axis

(f)

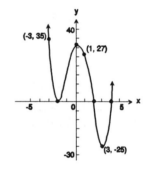

32. $R(x) = \dfrac{4 - x}{x}$

   1. The x-intercept is 4. There is no y-intercept.
   2. No symmetry is present.
   3. The graph of R has one vertical asymptote: $x = 0$.
   4. There is one horizontal asymptote: $y = -1$; not intersected.
   5. The zero of the numerator, 4, and the zero of the denominator, 0, divide the x-axis into:
      $x < 0, 0 < x < 4, x > 4$.

x	Test Number	Value of R	Graph of R
$x < 0$	-1	$R(-1) = -5$	Below x-axis
$0 < x < 4$	1	$R(1) = 3$	Above x-axis
$x > 4$	5	$R(5) = -\dfrac{1}{5}$	Below x-axis

Chapter 5 Polynomial and Rational Functions

6.

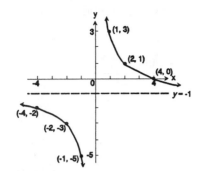

34. $H(x) = \dfrac{x}{x^2 - 1} = \dfrac{x}{(x + 1)(x - 1)}$

 1. The x-intercept is 0. The y-intercept is $H(0) = 0$.
 2. Since $H(-x) = -H(x)$, the function is odd and the graph is symmetric with respect to the origin.
 3. The graph has two vertical asymptotes: $x = -1$ and $x = 1$.
 4. The line $y = 0$ is a horizontal asymptote intersected at $(0, 0)$.
 5. The zero of the numerator, 0, and the zeros of the denominator, 1 and -1, divide the x-axis into: $x < -1$, $-1 < x < 0$, $0 < x < 1$, $x > 1$.

x	Test Number	Value of H	Graph of H
$x < -1$	-2	$H(-2) = -\dfrac{2}{3}$	Below x-axis
$-1 < x < 0$	-0.5	$H(-0.5) = \dfrac{2}{3}$	Above x-axis
$0 < x < 1$	0.5	$H(0.5) = -\dfrac{2}{3}$	Below x-axis
$x > 1$	2	$H(2) = \dfrac{2}{3}$	Above x-axis

6.

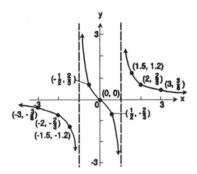

36. $R(x) = \dfrac{x^2 - 6x + 9}{x^2} = \dfrac{(x - 3)^2}{x^2}$

 1. The x-intercept is 3. There is no y-intercept.
 2. No symmetry is present.
 3. The graph has one vertical asymptote: $x = 0$.
 4. There is one horizontal asymptote: $y = 1$; intersected at $\left[\dfrac{3}{2}, 1\right]$.

5. The zero of the numerator, 3, and the zero of the denominator, 0, divide the x-axis into: $x < 0, 0 < x < 3, x > 3$.

x	Test Number	Value of R	Graph of R
$x < 0$	-1	$R(-1) = 16$	Above x-axis
$0 < x < 3$	1	$R(1) = 4$	Above x-axis
$x > 3$	4	$R(4) = \dfrac{1}{16}$	Above x-axis

6.
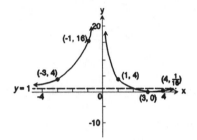

38. $F(x) = \dfrac{3x^2}{(x-1)^2}$

1. The x-intercept is 0. The y-intercept is $F(0) = 0$.
2. No symmetry is present.
3. The graph has one vertical asymptote: $x = 1$.
4. The line $y = 3x + 6$ is an oblique asymptote; intersected at $\left[\dfrac{2}{3}, 8\right]$.
5. The zero of the numerator, 0, and the zero of the denominator, 1, divide the x-axis into: $x < 0, 0 < x < 1, x > 1$.

x	Test Number	Value of F	Graph of F
$x < 0$	-1	$F(-1) = -\dfrac{3}{4}$	Below x-axis
$0 < x < 1$	0.5	$F(0.5) = 1.5$	Above x-axis
$x > 1$	2	$F(2) = 24$	Above x-axis

6.
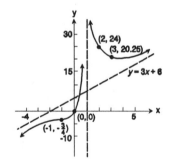

40. $R(x) = \dfrac{x^4}{x^2 - 9} = \dfrac{x^4}{(x+3)(x-3)}$

1. The x-intercept is 0. The y-intercept is $R(0) = 0$.
2. Since $R(-x) = R(x)$, the function is even and the graph is symmetric with respect to the y-axis.
3. The graph has two vertical asymptotes: $x = 3$ and $x = -3$.
4. There are no horizontal and no oblique asymptotes.

Chapter 5 Polynomial and Rational Functions

5. The zero of the numerator, 0, and the zeros of the denominator, -3 and 3, divide the x-axis into: $x < -3, \; -3 < x < 0, \; 0 < x < 3, \; x > 3$.

x	Test Number	Value of R	Graph of R
$x < -3$	-4	$R(-4) \approx 36.6$	Above x-axis
$-3 < x < 0$	-1	$R(-1) = -\dfrac{1}{8}$	Below x-axis
$0 < x < 3$	1	$R(1) = -\dfrac{1}{8}$	Below x-axis
$x > 3$	4	$R(4) \approx 36.6$	Above x-axis

6.

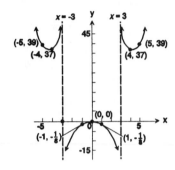

42. $f(x) = 2x^3 + 8x^2 - 5x + 5; \; g(x) = x - 2$

$$\begin{array}{r|rrrr} 2 & 2 & 8 & -5 & 5 \\ & & 4 & 24 & 38 \\ \hline & 2 & 12 & 19 & 43 \end{array}$$

$q(x) = 2x^2 + 12x + 19; \; R = 43$

44. $f(x) = x^4 - x^2 + 3x;$
$g(x) = x + 1$

$$\begin{array}{r|rrrrr} -1 & 1 & 0 & -1 & 3 & 0 \\ & & -1 & 1 & 0 & -3 \\ \hline & 1 & -1 & 0 & 3 & -3 \end{array}$$

$q(x) = x^3 - x^2 + 3; \; R = -3$

46. $f(x) = -16x^3 + 18x^2 - x + 2$
$f(-2) = 204$

48. $f(x) = -6x^5 + x^4 + 5x^3 - x + 1$
There is one positive zero.
$f(-x) = 6x^5 + x^4 - 5x^3 - x + 1$
There are two or no negative zeros.

50. $f(x) = -6x^5 + x^4 + 2x^3 - x + 1$
$\dfrac{p}{q}$: $\;\pm 1, \; \pm \dfrac{1}{2}, \; \pm \dfrac{1}{3}, \; \pm \dfrac{1}{6}$ are the potential rational zeros.

52. $f(x) = x^3 - x^2 - 10x - 8$
$f(-x) = -x^3 - x^2 + 10x - 8$
  There is one positive zero.
  There are two or zero negative zeros.

$\dfrac{p}{q}$: $\;\pm 1, \; \pm 2, \; \pm 4, \; \pm 8$ are the potential rational zeros.

$$\begin{array}{r|rrrr} 1 & 1 & -1 & -10 & -8 \\ & & 1 & -10 & -18 \\ \hline & 1 & 0 & -10 & -18 \end{array} \qquad \begin{array}{r|rrrr} -1 & 1 & -1 & -10 & -8 \\ & & -1 & 2 & 8 \\ \hline & 1 & -2 & -8 & 0 \end{array}$$

Thus, $x + 1$ is a factor and -1 is a zero.
$$f(x) = (x + 1)(x^2 - 2x - 8) = (x + 1)(x - 4)(x + 2)$$
The real zeros of f are -1, 4, and -2.

54. $f(x) = 4x^3 - 4x^2 - 7x - 2$
$f(-x) = -4x^3 - 4x^2 + 7x - 2$
  There is one positive zero.
  There are two or zero negative zeros.

$\dfrac{p}{q}$: $\;\pm 1, \; \pm 2, \; \pm \dfrac{1}{2}, \; \pm \dfrac{1}{4}$ are the potential rational zeros.

$$
\begin{array}{r|rrrr}
1 & 4 & -4 & -7 & -2 \\
  &   & 4 & 0 & -7 \\
\hline
  & 4 & 0 & -7 & -9
\end{array}
\qquad
\begin{array}{r|rrrr}
-1 & 4 & -4 & -7 & -2 \\
   &   & -4 & 8 & -1 \\
\hline
   & 4 & -8 & 1 & -3
\end{array}
\qquad
\begin{array}{r|rrrr}
-2 & 4 & -4 & -7 & -2 \\
   &   & 8 & 8 & 2 \\
\hline
   & 4 & 4 & 1 & 0
\end{array}
$$

Thus, $x - 2$ is a factor and 2 is a zero.

$$f(x) = (x - 2)(4x^2 + 4x + 1) = (x - 2)(2x + 1)(2x + 1) = (x - 2)(2x + 1)^2$$

The real zeros of f are 2 and $-\dfrac{1}{2}$

56. $f(x) = x^4 + 6x^3 + 11x^2 + 12x + 18$ There are no positive zeros.

     $f(-x) = x^4 - 6x^3 + 11x^2 - 12x + 18$ There are 4, 2 or 0 negative zeros.

$\dfrac{p}{q}$: $\pm 1, \pm 2, \pm 3, \pm 6, \pm 9, \pm 18$ are the potential rational zeros.

$$
\begin{array}{r|rrrrr}
1 & 1 & 6 & 11 & 12 & 18 \\
  &   & 1 & 7 & 18 & 30 \\
\hline
  & 1 & 7 & 18 & 30 & 48
\end{array}
\qquad
\begin{array}{r|rrrrr}
-1 & 1 & 6 & 11 & 12 & 18 \\
   &   & -1 & 5 & 6 & -6 \\
\hline
   & 1 & 5 & 6 & 6 & 12
\end{array}
\qquad
\begin{array}{r|rrrrr}
2 & 1 & 6 & 11 & 12 & 18 \\
  &   & 2 & 16 & 54 & 132 \\
\hline
  & 1 & 8 & 27 & 66 & 150
\end{array}
$$

$$
\begin{array}{r|rrrrr}
-2 & 1 & 6 & 11 & 12 & 18 \\
   &   & -2 & -8 & -6 & -12 \\
\hline
   & 1 & 4 & 3 & 6 & 6
\end{array}
\qquad
\begin{array}{r|rrrrr}
3 & 1 & 6 & 11 & 12 & 18 \\
  &   & 3 & 27 & 114 & 378 \\
\hline
  & 1 & 9 & 38 & 126 & 396
\end{array}
\qquad
\begin{array}{r|rrrrr}
-3 & 1 & 6 & 11 & 11 & 18 \\
   &   & -3 & -9 & -6 & -18 \\
\hline
   & 1 & 3 & 2 & 6 & 0
\end{array}
$$

Thus, $x + 3$ is a factor and -3 is a zero.

$$f(x) = (x + 3)(x^3 + 3x^2 + 2x + 6)$$

$$
\begin{array}{r|rrrr}
-3 & 1 & 3 & 2 & 6 \\
   &   & -3 & 0 & -6 \\
\hline
   & 1 & 0 & 2 & 0
\end{array}
$$

Thus, $x + 3$ is a factor of $x^3 + 3x^2 + 2x + 6$, and -3 is a zero.

$$f(x) = (x + 3)(x + 3)(x^2 + 2) = (x + 3)^2(x^2 + 2)$$

The only real zero of f is -3.

58. $3x^4 + 3x^3 - 17x^2 + x - 6 = 0$ The solutions of the equation are the zeros of the polynomial

                                  function $f(x)$.

     $f(x) = 3x^4 + 3x^3 - 17x^2 + x - 6$ There are three or one positive zeros.

     $f(-x) = 3x^4 - 3x^3 - 17x^2 - x - 6$ There is one negative zeros.

$\dfrac{p}{q}$: $\pm 1, \pm 2, \pm 3, \pm 6, \pm \dfrac{1}{3}, \pm \dfrac{2}{3}$ are the potential rational zeros.

$$
\begin{array}{r|rrrrr}
1 & 3 & 3 & -17 & 1 & -6 \\
  &   & 3 & 6 & -11 & -10 \\
\hline
  & 3 & 6 & -11 & -10 & -16
\end{array}
\qquad
\begin{array}{r|rrrrr}
-1 & 3 & 3 & -17 & 1 & -6 \\
   &   & -3 & 0 & 17 & -18 \\
\hline
   & 3 & 0 & -17 & 18 & -24
\end{array}
\qquad
\begin{array}{r|rrrrr}
2 & 3 & 3 & -17 & 1 & -6 \\
  &   & -6 & -18 & -2 & -6 \\
\hline
  & 3 & 9 & 1 & 3 & 0
\end{array}
$$

Thus, $x - 2$ is a factor and 2 is a zero.

$$f(x) = (x - 2)(3x^3 + 9x^2 + x + 3)$$

$$
\begin{array}{r|rrrr}
2 & 3 & 9 & 1 & 3 \\
  &   & 6 & 30 & 62 \\
\hline
  & 3 & 15 & 31 & 65
\end{array}
\qquad
\begin{array}{r|rrrr}
-2 & 3 & 9 & 1 & 3 \\
   &   & -6 & -6 & 10 \\
\hline
   & 3 & 3 & -5 & 13
\end{array}
\qquad
\begin{array}{r|rrrr}
2 & 3 & 9 & 1 & 3 \\
  &   & 9 & 54 & 165 \\
\hline
  & 3 & 18 & 55 & 168
\end{array}
\qquad
\begin{array}{r|rrrr}
-3 & 3 & 9 & 1 & 3 \\
   &   & -9 & 0 & -3 \\
\hline
   & 3 & 0 & 1 & 0
\end{array}
$$

Thus, $x + 3$ is a factor of $3x^3 + 9x^2 + x + 3$, and -3 is a zero.

$$(x - 2)(x + 3)(3x^2 + 1) = 0$$

The real zeros are 2 and -3.

60. $2x^4 + 7x^3 - 5x^2 - 28x - 12 = 0$ The solutions of this equation are the zeros of the

                                  polynomial function $f(x)$.

     $f(x) = 2x^4 + 7x^3 - 5x^2 - 28x - 12$ There is one positive zero.

     $f(-x) = 2x^4 - 7x^3 - 5x^2 + 28x - 12$ There are three or one negative zeros.

$\dfrac{p}{q}$: $\pm 1, \pm 2, \pm 3, \pm 4, \pm 6, \pm 12, \pm \dfrac{1}{2}, \pm \dfrac{3}{2}$ are the potential rational zeros.

$$
\begin{array}{r|rrrr}
1 & 2 & 7 & -5 & -28 & -12 \\
  &   & 2 & 9 & 4 & -24 \\
\hline
  & 2 & 9 & 4 & -24 & -26
\end{array}
\qquad
\begin{array}{r|rrrr}
-1 & 2 & 7 & -5 & -28 & -12 \\
   &   & -2 & -5 & 10 & 18 \\
\hline
   & 2 & 5 & -10 & -18 & 6
\end{array}
\qquad
\begin{array}{r|rrrr}
2 & 2 & 7 & -5 & -28 & -12 \\
  &   & 4 & 22 & 34 & 12 \\
\hline
  & 2 & 11 & 17 & 6 & 0
\end{array}
$$

Thus, $x - 2$ is a factor and 2 is a zero.

$$(x - 2)(2x^3 + 11x^2 + 17x + 6) = 0$$

$$
\begin{array}{r|rrr}
2 & 2 & 11 & 17 & 6 \\
  &   & 4 & 30 & 94 \\
\hline
  & 2 & 15 & 47 & 100
\end{array}
\qquad
\begin{array}{r|rrr}
-2 & 2 & 11 & 17 & 6 \\
   &   & -4 & -14 & -6 \\
\hline
   & 2 & 7 & 3 & 0
\end{array}
$$

Thus, $x + 2$ is a factor of $2x^3 + 11x^2 + 17x + 6$, and -2 is a zero.

$$(x - 2)(x + 2)(2x^2 + 7x + 3) = (x - 2)(x + 2)(2x + 1)(x + 3) = 0$$

The real zeros are 2, -2, $-\dfrac{1}{2}$ and -3.

62. $f(x) = x^3 - x^2 - 10x - 8 = (x + 1)(x - 4)(x + 2)$

The x-intercepts are -1, 4, and -2; the y-intercept is $f(0) = -8$.

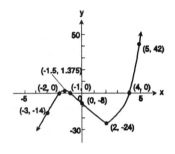

x	Test Number	Value of f	Graph of f
$x < -2$	-3	$f(-3) = -14$	Below x-axis
$-2 < x < -1$	-1.5	$f(-1.5) = 1.375$	Above x-axis
$-1 < x < 4$	0	$f(0) = -8$	Below x-axis
$x > 4$	5	$f(5) = 42$	Above x-axis

64. $f(x) = 4x^3 - 4x^2 - 7x - 2 = (x - 2)(2x + 1)(2x + 1)$

The x-intercepts are 2, and $-\dfrac{1}{2}$; the y-intercept is $f(0) = -2$.

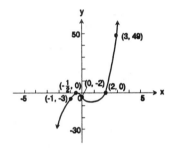

x	Test Number	Value of f	Graph of f
$x < -\dfrac{1}{2}$	-1	$f(-1) = -3$	Below x-axis
$-\dfrac{1}{2} < x < 2$	0	$f(0) = -2$	Below x-axis
$x > 2$	3	$f(3) = 49$	Above x-axis

66. $f(x) = x^4 + 6x^3 + 11x^2 + 12x + 18 = (x + 3)(x + 3)(x^2 + 2)$

The x-intercept is -3; the y-intercept is $f(0) = 18$.

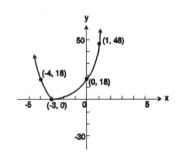

x	Test Number	Value of f	Graph of f
$x < 3$	-4	$f(-4) = 18$	Above x-axis
$x > -3$	0	$f(0) = 18$	Above x-axis

68. $f(x) = 3x^4 + 3x^3 - 17x^2 + x - 6$
     $= (x - 2)(x + 3)(3x^2 + 1)$
     The x-intercepts are 2 and -3; the y-intercept is $f(0) = -6$.

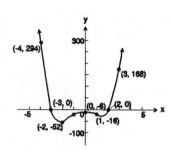

x	Test Number	Value of f	Graph of f
$x < 3$	-4	$f(-4) = 294$	Above x-axis
$-3 < x < 2$	0	$f(0) = -6$	Below x-axis
$x > 2$	3	$f(3) = 168$	Above x-axis

70. $f(x) = 2x^4 + 7x^3 - 5x^2 - 28x - 12 = (x - 2)(x + 2)(2x + 1)(x + 3)$
     The x-intercepts are 2, -2, $-\dfrac{1}{2}$, and -3; the y-intercept is $f(0) = -12$.

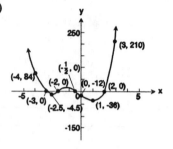

x	Test Number	Value of f	Graph of f
$x < -3$	-4	$f(-4) = 84$	Above x-axis
$-3 < x < -2$	-2.5	$f(-2.5) = -4.5$	Below x-axis
$-2 < x < -\dfrac{1}{2}$	-1	$f(-1) = 6$	Above x-axis
$-\dfrac{1}{2} < x < 2$	0	$f(0) = -12$	Below x-axis
$x > 2$	3	$f(3) = 210$	Above x-axis

72. $f(x) = 2x^3 - x^2 - 3$
     $f(1) = -2$ and $f(2) = 9$

74. $f(x) = 3x^4 + 4x^3 - 8x - 2$
     $f(1) = -3$ and $f(2) = 62$

76. $f(x) = 2x^3 + x^2 - 10x - 5$

```
1)2   1   -10   -5        2)2   1   -10   -5        3)2   1   -10   -5
      2    3    -5              4   10    0              6   21   33
  ──────────────────        ──────────────────        ──────────────────
   2   3   -7   -12          2   5    0    -5           2   7   11   28
```
Thus, 3 is an upper bound

```
-1)2   1   -10   -5        -2)2   1   -10   -5        -3)2   1   -10   -5
       -2    1    9               -4    6    8               -6   15  -15
   ──────────────────         ──────────────────         ──────────────────
    2   -1   -9    4           2   -3   -4    3            2   -5    5   -20
```
Thus, -3 is a lower bound.

78. $f(x) = 3x^3 - 7x^2 - 6x + 14$

```
1)3   -7   -6   14        2)3   -7   -6   14        3)3   -7   -6   14
      3   -4  -10             -1   -8   -2             9    6    0
  ──────────────────       ──────────────────       ──────────────────
   3   -4  -10    4          3   -1   -8   -2          3    2    0   14
```
Thus, 3 is an upper bound.

```
-1)3   -7   -6   14        -2)3   -7   -6   14
       -3   10   -4              -6   26  -40
   ──────────────────        ──────────────────
    3  -10    4   10          3  -13   20  -26
```
Thus, -2 is a lower bound.

Chapter 5 Polynomial and Rational Functions

80. $f(x) = 2x^3 - x^2 - 3$
     Since $f(1) = -2$ and $f(2) = 9$, the positive zero is between 1 and 2.
     $$f(1.0) = -2$$
     $$f(1.1) = -1.548$$
     $$f(1.2) = -0.984$$
     $$f(1.3) = -0.296$$
     $$f(1.4) = 0.528$$
     The zero is between 1.3 and 1.4
     $$f(1.30) = -0.296$$
     $$f(1.31) = -0.219918$$
     $$f(1.32) = -0.142464$$
     $$f(1.33) = -0.063626$$
     $$f(1.34) = 0.016608$$

     The zero is approximately 1.33 correct to two decimal places.

82. $f(x) = 3x^4 + 4x^3 - 8x - 2$
     Since $f(1) = -3$ and $f(2) = 62$, the positive zero is between 1 and 2.
     $$f(1.0) = -3$$
     $$f(1.1) = -1.0837$$
     $$f(1.2) = 1.5328$$
     The zero is between 1.1 and 1.2
     $$f(1.10) = -1.0837$$
     $$f(1.11) = -0.8553$$
     $$f(1.12) = -0.6198$$
     $$f(1.13) = -0.3770$$
     $$f(1.14) = -0.1269$$
     $$f(1.15) = 0.1305$$
     The zero is approximately 1.14 correct to two decimal places.

84. $3 - 4i$, the conjugate of $3 + 4i$, is the remaining zero.

86. $1 - i$, the conjugate of $1 + i$, is the remaining zero.

88. $f(z) = (z - i)^2(z - 2)^2 = (z^2 - 2iz - 1)(z^2 - 4z + 4)$
     $= z^4 - (4 + 2i)z^3 + (3 + 8i)z^2 + (4 - 4i)z - 4$

90. $f(z) = (z - 1)[z - (1 + i)][z - (1 + 2i)]$
     $= (z - 1)[z^2 - (2 + 3i)z - 1 + 3i]$
     $= z^3 - (3 + 3i)z^2 + (1 + 6i)z + 1 - 3i$

92.

$$
\begin{array}{r}
x^2 + x + 1 \\
x^2 + x - 6 \overline{\smash{\big)}\, x^4 + 2x^3 - 4x^2 - 5x - 6} \\
\underline{x^4 + x^3 - 6x^2} \\
x^3 + 2x^2 - 5x - 6 \\
\underline{x^3 + x^2 - 6x} \\
x^2 + x - 6 \\
\underline{x^2 + x - 6} \\
0
\end{array}
$$

     $q(x) = x^2 + x + 1; R = 0$

94. $x^3 - 3x^2 - 4x + 12 = 0$

$\dfrac{p}{q}$: $\pm 1, \pm 2, \pm 3, \pm 4, \pm 6, \pm 12$

```
1|1  -3  -4   12        -1|1  -3  -3   12        2|1  -3  -4   12
     1  -2   -6              -1   4    0             2  -2  -12
  ─────────────            ─────────────          ─────────────
   1  -2  -6    6           1  -4   0   12          1  -1  -6    0
```

Thus, $(x - 2)(x^2 - x - 6) = 0$

$(x - 2)(x - 3)(x + 2) = 0$

$x = 2$ or $x = -2$ or $x = 3$

96. $x^4 + 4x^3 + 2x^2 - 8x - 8 = 0$

$\dfrac{p}{q}$: $\pm 1, \pm 2, \pm 4, \pm 8$

```
1|1  4   2   -8   -8     -1|1   4   2  -8   -8     2|1   4    2   -8   -8    -2|1   4   2   -8   -8
    1   5   7   -1             -1  -2   1    7          2  12   28   40           -2  -4    4    8
 ─────────────────         ─────────────────        ──────────────────        ──────────────────
  1   5   7  -1   -9         1   3  -1  -7   11        1   6   14   20   32       1   2  -2   -4    0
```

Thus, $(x + 2)(x^3 + 2x^2 - 2x - 4) = 0$

```
-2|1   2   -2   -4
      -2    0    4
   ───────────────
    1   0   -2    0
```

Thus, $(x + 2)(x + 2)(x^2 - 2) = 0$

$(x + 2)(x + 2)(x - \sqrt{2})(x + \sqrt{2}) = 0$

$x = -2$ or $x = \sqrt{2}$ or $x = -\sqrt{2}$

98. $f(x) = 5x^3 + 4x^2 - 6x + 8 = (5x^2 + 4x - 6)x + 8$

$\quad = [(5x + 4)x - 6]x + 8$

$f(1.5) = 24.875$

100. $f(x) = x^4 - x^2 + 3x = (x^3 - x + 3)x = [(x \cdot x - 1)x + 3]x$

$f(1.5) = 7.3125$

102. $f(x) = 8x^6 - x^4 + 6x^2 + 24x + 15$

```
1|8   0   -1   0    6   24   15
     8   8    7    7   12   37
  ──────────────────────────────
   8   8    7   7   13   37   52
```

Thus, 1 is an upper bound.

```
-1|8   0   -1   0    6    24    15        -2|8    0    -1    0     6     24    15
     -8   8   -7    7  -13   -11               -16   32   -62  124  -260   472
   ───────────────────────────────           ────────────────────────────────────
    8  -8    7   -7   13    11    4            8   -16   31   -62   130   236   487
```

Thus, -2 is a lower bound.

104. Let (x_0, y_0) be a point on the line $y = x + 1$.

Then the distance, d, from (x_0, y_0) to $(4, 1)$ is $\sqrt{(4 - x_0)^2 + (1 - y_0)^2}$.

The minimum value of the function $f(x) = d^2$ is

$f(x) = (4 - x_0)^2 + (1 - y_0)^2$, where $y_0 = x_0 + 1$

$\quad = (4 - x_0)^2 + [1 - (x_0 + 1)]^2 = 16 - 8x_0 + x_0^2 + x_0^2$

$\quad = 2x_0^2 - 8x_0 + 16$

The minimum value is $x = \dfrac{-b}{2a} = \dfrac{8}{4} = 2$ and $y_0 = 3$.

Thus, $(2, 3)$ is the point on $y = x + 1$ that is closest to the point $(4, 1)$.

106. The rectangle of length ℓ and width w has perimeter $20 = 2\ell + 2w$ so that $\ell = 10 - w$. The area of the rectangle is $16 = \ell w = (10 - w)w = -w^2 + 10w$ or $-w^2 + 10w - 16 = 0$. Thus, $(-w + 2)(w - 8) = 0$. $w = 2$ ft. and $\ell = 10 - w = 10 - 2 = 8$ ft.

Mission Possible

1. The vertical asymptotes are $x = -1$ and $x = 3$. So the denominator must be a multiple of $(x + 1)(x - 3)$.

2. The x-intercepts come from the zeros of the numerator. So, the numerator must be a multiple of $(x - 1)(x + 3)$.

3. So far our guess probably looks like $\dfrac{x^2 + 2x - 3}{x^2 - 2x - 3}$ However, as viewed on the graphing utility, the middle section is decreasing instead of having a maximum at $x = 1$. This could be fixed by changing the multiplicity of the zero to two. $\dfrac{(x - 1)^2(x + 3)}{(x + 1)(x - 3)}$

4. To adjust the behavior near the left vertical asymptote, square the $(x + 1)$ factor. Now the function looks like $\dfrac{(x - 1)^2(x + 3)}{(x + 1)^2(x - 3)}$.

5. This new result will not be consistent with a horizontal asymptote of $y = 0$ unless the degree of the denominator is greater than the degree of the numerator. Therefore, the denominator needs to be of degree ≥ 3. However, we already have the two real roots of the denominator. So, we need to consider imaginary roots for the denominator that will raise the degree to 3 or more without changing the number of real roots. Imaginary roots of polynomials with real coefficients necessarily occur in pairs. So one possibility is an additional factor of $x^2 + 1$.

6. We need a y-intercept of -4. At present, it is -1. Multiplying the whole fraction by 4 will adjust the graph as desired.

$$f(x) = \dfrac{4(x - 1)^2(x + 3)}{(x + 1)^2(x - 3)(x^2 + 1)}$$

7–9. Students should be encouraged to reason together to get the answer to this dreadful rational function. After teams have worked on the problem for a while, long enough to get a foothold but not so long as to get frustrated, it might be a good idea to bring the class together to use all the IQ's to defeat the Cardassian challenge.

Chapter 6

EXPONENTIAL AND LOGARITHMIC FUNCTIONS

6.1 Exponential Functions

2. (a) 15.426   (b) 16.189   (c) 16.241   (d) 16.242

4. (a) 6.498   (b) 6.543   (c) 6.580   (d) 6.581

6. (a) 21.738   (b) 22.884   (c) 23.119   (d) 23.141

8. 0.273   10. 8.166   12. F   14. H   16. C   18. G

20. $y = -e^x$
Reflect the graph $y = e^x$ about the x-axis.

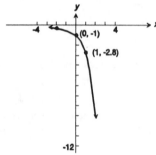

22. $y = e^x - 1$
Using the graph $y = e^x$, vertically shift downward 1.

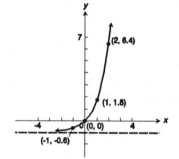

24.

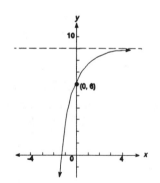

26.

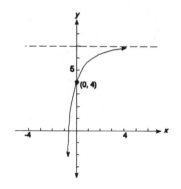

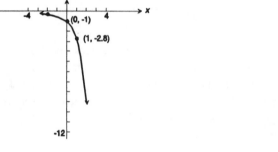

28.
$$2^x = 3$$
$$2^{-x} = 3^{-1}$$
$$2^{-x} = \frac{1}{3}$$
$$(2^{-x})^2 = \left[\frac{1}{3}\right]^2$$
$$4^{-x} = \frac{1}{9}$$

30.
$$5^{-x} = 3$$
$$(5^{-x})^{-1} = 3^{-1}$$
$$5^x = \frac{1}{3}$$
$$(5^x)^3 = \left[\frac{1}{3}\right]^3$$
$$5^{3x} = \frac{1}{27}$$

32. (a) $p = 760e^{-0.145(2)}$
≈ 568.68 mm Hg
(b) $p = 760e^{-0.145(10)}$
≈ 178.27 mm Hg

34. $A = A_0 \cdot e^{-0.35n}$
(a) $A = 100 \cdot e^{-0.35(3)}$
≈ 34.99 cm^2
(b) $A = 100 \cdot e^{-0.35(10)}$
≈ 3.02 cm^2

36. $N = 1000(1 - e^{-.15(3)}) \approx 362$ students

38. (a) $P = \$100,000 - \$60,000\left[\frac{1}{2}\right]^5$
$= \$98125$

(b) $P = \$100,000 - \$60,000\left[\frac{1}{2}\right]^{10}$
$= \$99941.41$

(c) The company can expect to make $100,000 in maximum profit from this product.

(d)
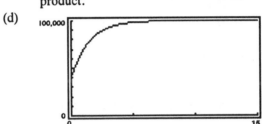

It takes about $\frac{3}{4}$ of a year before a profit of $65,000 is obtained.

40. (a) For $t = 0$ ms,
$$I_1 = \frac{120}{2000} \cdot e^{-0/2000}$$
$$I_1 = 0.06 \text{ milliamperes}$$
For $t = 1000$ ms,
$$I_1 = \frac{120}{2000} \cdot e^{-1000/2000}$$
$$I_1 \approx 0.0364 \text{ milliamperes}$$
For $t = 3000$ ms,
$$I_1 = \frac{120}{2000} \cdot e^{-3000/2000}$$
$$I_1 \approx 0.0134 \text{ milliamperes}$$

(b) $I_{1(max)} = \frac{120}{2000} = 0.06$ milliamperes

(c)
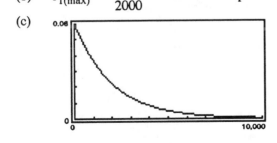

(d) For $t = 0$ ms, $I_2 = \frac{120}{1000} \cdot e^{-0/2000}$
$= 0.12$ milliamperes
For $t = 1000$ ms, $I_2 = \frac{120}{1000} \cdot e^{-1000/2000}$
≈ 0.0728 milliamperes
For $t = 3000$ ms, $I_2 = \frac{120}{1000} \cdot e^{-3000/2000}$
≈ 0.0268 milliamperes

(e) $I_{2(max)} = \frac{120}{1000} = 0.12$ milliamperes

(f)
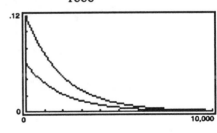

42. (a) $y = 78e^{0.025(1995-1848)}$
$= 78e^{0.025(147)}$
≈ 3077 stamps

(b) $y = 78e^{0.025(1998-1848)}$
$= 78e^{0.025(150)}$
≈ 3317 stamps

(c)

Year	Actual Number of Stamps Issued	Predicted
1848	2	78
1868	88	129
1888	218	212
1908	341	350

The predictions appear to get better with time.

44. For $n = 2$, expression $= 2 + \dfrac{1}{1 + \dfrac{1}{2}} = 2.\overline{666}$

$n = 3$, expression $= 2 + \dfrac{1}{1 + \dfrac{1}{2 + \dfrac{2}{3}}} = 2.7\overline{272}$

n	Expression
2	$2.\overline{666}$
3	$2.7\overline{2}$
4	2.71698
5	2.71845
6	2.71826

As n increases, the expression gets closer to e which is approximately 2.71828182846.

46. $f(x) = a^x$
$f(A + B) = a^{A+B} = a^A \cdot a^B = f(A) \cdot f(B)$

48. $f(x) = a^x$
$f(\alpha x) = a^{\alpha x} = (a^x)^\alpha = [f(x)]^\alpha$

50. $\cosh x = \dfrac{1}{2}\left(e^x + e^{-x}\right)$

(a) $f(x) = \cosh x$

$f(-x) = \dfrac{1}{2}\left(e^{-x} + e^{-(-x)}\right)$

$= \dfrac{1}{2}\left(e^x + e^{-x}\right)$

$= f(x)$

$= \cosh x$

(b)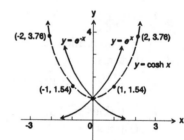

(c) $(\cosh x)^2 - (\sinh x)^2$

$= \left[\dfrac{1}{2}\left(e^x + e^{-x}\right)\right]^2 - \left[\dfrac{1}{2}\left(e^x - e^{-x}\right)\right]^2 = \dfrac{1}{4}\left[\left(e^x + e^{-x}\right)^2 - \left(e^x - e^{-x}\right)^2\right]$

$= \dfrac{1}{4}\left(e^{2x} + 2e^x e^{-x} + e^{-2x} - 4^{2x} + 2e^x e^{-x} - e^{-2x}\right) = \dfrac{1}{4}\left(4e^x e^{-x}\right)$

$= \dfrac{1}{4}\left(4e^{x-x}\right) = \dfrac{1}{4}\left(4e^0\right) = \dfrac{1}{4}(4) = 1$

52. 59 minutes

2. $2 = \log_4 16$ 4. $3 = \log_a 2.1$ 6. $3 = \log_{2.2} N$ 8. $x = \log_3 4.6$

10. $\pi = \log_x e$ 12. $2.2 = \ln M$ 14. $3^{-2} = \dfrac{1}{9}$ 16. $b^2 = 4$

18. $2^x = 6$ 20. $3^{2.1} = N$ 22. $\pi^{1/2} = x$ 24. $e^4 = x$

26. $\log_8 8 = 1$ 28. $\log_3\left(\dfrac{1}{9}\right) = -2$ 30. $\log_{1/3} 9 = -2$

32. $\log_5 \sqrt[3]{25} = \log_5 5^{2/3} = \dfrac{2}{3}$ 34. $\log_{\sqrt{3}} 9 = 4$ 36. $\ln e^3 = 3$

38. $g(x) = \ln(x^2 - 1)$
The domain of g consists of all x for which $x^2 - 1 > 0$; that is, all $x < -1$ or $x > 1$.

40. $H(x) = \log_5 x^3$
The domain of H consists of all x for which $x^3 > 0$; that is, all $x > 0$.

42. $G(x) = \log_{1/2} \dfrac{1}{x}$
The domain of G consists of all x for which $\dfrac{1}{x} > 0$; that is, all $x > 0$.

44. $g(x) = \ln(x - 5)$
The domain of g consists of all x for which $x - 5 > 0$; that is, all $x > 5$.

46. $h(x) = \log_3\left(\dfrac{x^2}{x - 1}\right)$
The domain of h consists of all x for which $\dfrac{x^2}{x - 1} > 0$; that is, all $x > 1$.

48. $\dfrac{\ln 5}{3} \approx 0.536$

50. $\dfrac{\ln \dfrac{2}{3}}{-0.1} \approx 4.055$

52. $f(x) = \log_a x$
$-4 = \log_a\left(\dfrac{1}{2}\right)$ at the point $\left(\dfrac{1}{2}, -4\right)$
$a^{-4} = \dfrac{1}{2}$
$\dfrac{1}{a^4} = \dfrac{1}{2}$
$a^4 = 2$
$a = \sqrt[4]{2}$

54.   F   56.   H   58.   C   60.   G

62.

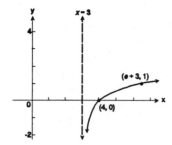

64.

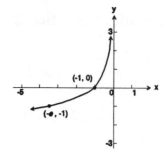

66.

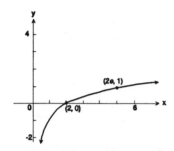

68.

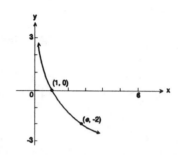

70.

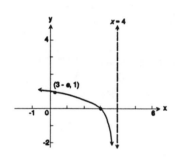

72. (a) $\text{pH} = -\log_{10}(.0000001) = 7$

(b) $-\log_{10}[H^-] = 4.2$

$$\log_{10}[H^-] = -4.2$$
$$H^- = 10^{-4.2}$$
$$H^- = 0.0000631$$

74. (a) $A = A_0 e^{-0.35n}$

Since $A = \frac{1}{2}A_0$,

we substitute $\frac{1}{2}A_0$ for A:

$$\frac{1}{2}A_0 = A_0 e^{-0.35n}$$

$$\frac{1}{2} = e^{-0.35n}$$

$$\ln\frac{1}{2} = -0.35n$$

$$n = \frac{\ln\frac{1}{2}}{-0.35}$$

$n \approx 1.98$ days, so about 2 days

(b) $A = 0.10A_0$.

Therefore, substitute $0.10A_0$ for A:

$$0.10A_0 = A_0 e^{-0.35n}$$
$$0.10 = e^{-0.35n}$$
$$\ln 0.10 = -0.35n$$

$$n = \frac{\ln 0.10}{-0.35}$$

$n \approx 6.58$ days, so about $6\frac{1}{2}$ days

76. $$N = P(1 - e^{-0.15d})$$
$$450 = 1000(1 - e^{-0.15d})$$
$$0.45 = 1 - e^{-0.15d}$$
$$-0.55 = -e^{-0.15d}$$
$$0.55 = e^{-0.15d}$$
$$\ln 0.55 = -0.15d$$
$$d \approx 3.99 \text{ days, so about 4 days}$$

78. $L(t) = A(1 - e^{-kt})$

(a)
$$20 = 200(1 - e^{-k(5)})$$
$$0.10 = 1 - e^{-5k}$$
$$-0.9 = -e^{-5k}$$
$$0.9 = e^{-5k}$$
$$\ln 0.9 = -5k$$
$$k \approx 0.0211$$

(b) $L(10) = 200(1 - e^{-.0211(10)})$
≈ 38 words

(c) $L(15) = 200(1 - e^{-.0211(15)})$
≈ 54 words

(d)
$$180 = 200(1 - e^{-.0211t})$$
$$0.9 = 1 - e^{-.0211t}$$
$$-0.1 = -e^{-.0211t}$$
$$t \approx 109.13 \text{ minutes,}$$
so about 1 hour, 50 minutes

82. For $t = 1$, $\$38000 = \$36,600e^{R \cdot (1)}$
$$1.03825 = e^R$$
$$\ln 1.03825 = R$$
$$R = 0.0375$$

Similar calculations for $t = 2, 3, 4$ and 5 years result in the following:

t	R
1	0.0375, 3.75%
2	0.0797, 7.97%
3	0.0930, 9.30%
4	0.1007, 10.07%
5	0.1167, 11.67%

6.3 Properties of Logarithms

Problems 2-12 use the fact ln 2 = a and ln 3 = b.

2. $\ln \dfrac{2}{3} = \ln 2 - \ln 3 = a - b$

4. $\ln 0.5 = \ln \dfrac{1}{2} = \ln 2^{-1} = -\ln 2 = -a$

6. $\ln \left(\dfrac{3}{e} \right) = \ln 3 - \ln e = b - 1$

8. $\ln 24 = \ln(3 \cdot 2^3) = \ln 3 + \ln 2^3$
$= \ln 3 + 3 \ln 2 = b + 3a$

10. $\ln \sqrt[4]{48} = \dfrac{1}{4} \ln(3 \cdot 2^4) = \dfrac{1}{4}[\ln 3 + 4 \ln 2] = \dfrac{1}{4}[b + 4(a)] = \dfrac{1}{4}b + a$

12. $\log_3 2 = \dfrac{\ln 2}{\ln 3} = \dfrac{a}{b}$

14. $\ln\left[x\sqrt{1 + x^2}\right] = \ln x + \ln\sqrt{1 + x^2} = \ln x + \dfrac{1}{2} \ln(1 + x^2)$

16. $\log_5 \left[\dfrac{\sqrt[3]{x^2 + 1}}{x^2 - 1} \right] = \log_5 \sqrt[3]{x^2 + 1} - \log_5(x^2 - 1) = \dfrac{1}{3} \log_5(x^2 + 1) - \log_5(x^2 - 1)$

18. $\log \dfrac{x^3\sqrt{x + 1}}{(x - 2)^2} = \log x^3 + \log\sqrt{x + 1} - \log(x - 2)^2 = 3 \log x + \dfrac{1}{2} \log(x + 1) - 2 \log(x - 2)$

20. $\ln \left[\dfrac{(x-4)^2}{x^2-1}\right]^{2/3} = \dfrac{2}{3} \ln \dfrac{(x-4)^2}{x^2-1} = \dfrac{2}{3}\left[\ln(x-4)^2 - \ln(x^2-1)\right] = \dfrac{2}{3}\left[2\ln(x-4) - \ln(x^2-1)\right]$

22. $\ln \left[\dfrac{5x^2\sqrt[3]{1-x}}{4(x+1)^2}\right] = \ln 5x^2 + \ln\sqrt[3]{1-x} - \ln\left[4(x+1)^2\right]$

$\qquad\qquad = \ln 5 + 2\ln x + \dfrac{1}{3}\ln(1-x) - \ln 4 - 2\ln(x+1)$

24. $\log_3 u^2 - \log_3 v = \log_3 \dfrac{u^2}{v}$

26. $\log_2\left[\dfrac{1}{x}\right] + \log_2\left[\dfrac{1}{x^2}\right] = \log_2\left[\dfrac{1}{x} \cdot \dfrac{1}{x^2}\right] = \log_2(x)^{-3} = -3\log_2 x$

28. $\log\left[\dfrac{x^2+2x-3}{x^2-4}\right] - \log\left[\dfrac{x^2+7x+6}{x+2}\right] = \log\dfrac{\dfrac{(x+3)(x-1)}{(x+2)(x-2)}}{\dfrac{(x+6)(x+1)}{(x+2)}}$

$\qquad = \log\left[\dfrac{(x+3)(x-1)}{(x+2)(x-2)} \cdot \dfrac{(x+2)}{(x+6)(x+1)}\right] = \log\dfrac{(x+3)(x-1)}{(x-2)(x+6)(x+1)}$

30. $21\log_3 \sqrt[3]{x} + \log_3 9x^2 - \log_5 25 = \log_3 x^7 + \log_3 9x^2 - \log_5 5^2 = \log_3(x^7 \cdot 9x^2) - 2$

$\qquad\qquad = \log_3(9x^9) - 2 = 9\log_3 9x - 2$

32. $\dfrac{1}{3}\log(x^3+1) + \dfrac{1}{2}\log(x^2+1) = \log(x^3+1)^{1/3} + \log(x^2+1)^{1/2} = \log\left[\sqrt[3]{x^3+1}\ \sqrt{x^2+1}\right]$

34. $\log_5 18 = \dfrac{\ln 18}{\ln 5} \approx 1.796$

36. $\log_{1/2} 15 \dfrac{\ln 15}{\ln \dfrac{1}{2}} \approx -3.907$

38. $\log_{\sqrt 5} 8 = \dfrac{\ln 8}{\ln \sqrt 5} \approx 2.584$

40. $\log_\pi \sqrt 2 = \dfrac{\ln \sqrt 2}{\ln \pi} \approx 0.303$

42. $\log_a(\sqrt x + \sqrt{x-1}) + \log_a(\sqrt x - \sqrt{x-1}) = \log_a\left[(\sqrt x + \sqrt{x-1})(\sqrt x - \sqrt{x-1})\right]$

$\qquad\qquad = \log_a[x - (x-1)] = \log_a 1 = 0$

44. If $f(x) = \log_a x$, then $\dfrac{f(x+h) - f(x)}{h} = \dfrac{\log_a(x+h) - \log_a x}{h} = \dfrac{\log_a\left[\dfrac{x+h}{x}\right]}{h}$

$\qquad\qquad = \dfrac{1}{h}\log_a\left[1 + \dfrac{h}{x}\right] = \log_a\left[1 + \dfrac{h}{x}\right]^{1/h}, \ h \neq 0$

46. $f(x) = \log_a x$

$f\left[\dfrac{1}{x}\right] = \log_a\left[\dfrac{1}{x}\right] = \log_a x^{-1} = -\log_a x = -f(x)$

48. $f(x) = \log_a x$

$f(x^\alpha) = \log_a x^\alpha = \alpha \log_a x = \alpha f(x)$

50. $\ln y = \ln(x + C)$
    $y = x + C$

52. $\ln y = 2 \ln x - \ln(x + 1) + \ln C$

$\ln y = \ln \dfrac{Cx^2}{(x + 1)}$

$y = \dfrac{Cx^2}{(x + 1)}$

54. $\ln y = -2x + \ln C$
    $\ln y - \ln C = -2x$

$\ln\left[\dfrac{y}{c}\right] = -2x$

$\dfrac{y}{C} = e^{-2x}$

$y = Ce^{-2x}$

56. $\ln(y + 4) = 5x + \ln C$
    $\ln(y + 4) - \ln C = 5x$

$\ln\dfrac{y + 4}{C} = 5x$

$\dfrac{y + 4}{C} = e^{5x}$

$y = Ce^{5x} - 4$

58. $2 \ln y = -\dfrac{1}{2} \ln x + \dfrac{1}{3} \ln(x^2 + 1) + \ln C$

$\ln y^2 = \ln\left[Cx^{-1/2}(x^2 + 1)^{1/3}\right]$
$y^2 = Cx^{-1/2}(x^2 + 1)^{1/3}$
$y = \sqrt{C}\, x^{-1/4}(x^2 + 1)^{1/6}, \ y > 0$

60. $\log_2 4 \cdot \log_4 6 \cdot \log_6 8 = \dfrac{\ln 4}{\ln 2} \cdot \dfrac{\ln 6}{\ln 4} \cdot \dfrac{\ln 8}{\ln 6} = 3$

62. $\log_2 2 \cdot \log_2 4 \cdots \log_2 2^n = \log_2 2 \cdot \log_2 2^2 \cdots \log_2 2^n = \log_2 2 \cdot 2 \log_2 2 \cdots n \log_2 2$
    $1 \cdot 2 \cdot \cdots n = n!$

64. $\log_a\left[\dfrac{1}{N}\right] = \log_a N^{-1} = -\log_a N, \text{ with } a \neq 1$

6.4 Logarithmic and Exponential Equations

2. $\log_3(3x - 2) = 2$
   $3x - 2 = 3^2$
   $3x = 9 + 2$
   $x = \dfrac{11}{3}$

4. $\log_5(x^2 + x + 4) = 2$
   $x^2 + x + 4 = 25$
   $x^2 + x - 21 = 0$
   $x = \dfrac{-1 + \sqrt{85}}{2} \text{ or } x = \dfrac{-1 - \sqrt{85}}{2}$

6. $-2 \log_4 x = \log_4 9$
   $\log_4 x^{-2} = \log_4 9$
   $x^{-2} = 9$
   $x = \dfrac{1}{3}$

8. $3 \log_2 x = -\log_2 27$
   $\log_2 x^3 = \log_2 27^{-1}$
   $x^3 = \dfrac{1}{27}$
   $x = \dfrac{1}{3}$

10. $2 \log_3(x + 4) - \log_3 9 = 2$

$$\log_3 \frac{(x + 4)^2}{9} = 2$$

$$\frac{(x + 4)^2}{9} = 3^2$$

$$(x + 4)^2 = 81$$

$x = 5$ or $x = -13$ (impossible)

Thus, $x = 5$

12. $\log_4 x + \log_4(x - 3) = 1$

$$\log_4[x(x - 3)] = 1$$
$$\log_4(x^2 - 3x) = 1$$
$$x^2 - 3x = 4$$
$$x^2 - 3x - 4 = 0$$
$$(x - 4)(x + 1) = 0$$

$x = 4$ or $x = -1$ (impossible)

Thus, $x = 4$

14. $\log_x \left(\frac{1}{8} \right) = 3$

$$x^3 = \frac{1}{8}$$

$$x = \frac{1}{2}$$

16. $\log_2(x + 4)^3 = 6$

$$(x + 4)^3 = 2^6$$
$$(x + 4)^3 = 64$$
$$x + 4 = 4$$
$$x = 0$$

18. $\log_{1/3}(1 - 2x)^{1/2} = -1$

$$(1 - 2x)^{1/2} = \left[\frac{1}{3} \right]^{-1}$$

$$(1 - 2x)^{1/2} = 3$$
$$1 - 2x = 9$$
$$x = -4$$

20. $5^{1-2x} = \frac{1}{5}$

$$5^{1-2x} = 5^{-1}$$
$$1 - 2x = -1$$
$$-2x = -2$$
$$x = 1$$

22. $4^{x^2} = 2^x$

$$2^{2x^2} = 2^x$$
$$2x^2 = x$$
$$x = 0$$

24. $9^{-x} = \frac{1}{3}$

$$3^{-2x} = 3^{-1}$$
$$-2x = -1$$
$$x = \frac{1}{2}$$

26. $\left[\frac{1}{2} \right]^{1-x} = 4$

$$2^{-1(1-x)} = 2^2$$
$$x - 1 = 2$$
$$x = 3$$

28. $3^{2x} + 3^x - 2 = 0$

$$(3^x)^2 + 3^x - 2 = 0$$
$$(3^x + 2)(3^x - 1) = 0$$

$3^x + 2 = 0$ or $3^x - 1 = 0$

$\quad 3^x = -2 \qquad\qquad 3^x = 1$

No solution $\qquad\qquad x = 0$

30. $4^x - 2^x = 0$

$$2^{2x} = 2^x$$
$$2x = x$$
$$x = 0$$

32. $9^{2x} = 27$

$$3^{4x} = 3^3$$
$$4x = 3$$

$$x = \frac{3}{4}$$

34. $3^x = 14$

$$x = \log_3 14 = \frac{\ln 14}{\ln 3} \approx 2.402$$

36. $2^{-x} = 1.5$

$$-x = \log_2 1.5 = \frac{\ln 1.5}{\ln 2} \approx 0.585$$

$$x = -0.585$$

38.

$$2^{x+1} = 5^{1-2x}$$
$$\ln 2^{x+1} = \ln 5^{1-2x}$$
$$(x+1)\ln 2 = (1-2x)\ln 5$$
$$x\ln 2 + \ln 2 = \ln 5 - 2x\ln 5$$
$$(\ln 2 + 2\ln 5)x = \ln 5 - \ln 2$$
$$x = \frac{\ln 5 - \ln 2}{\ln 2 + 2\ln 5}$$
$$\approx 0.234$$

40.

$$\left(\frac{4}{3}\right)^{1-x} = 5^x$$
$$\ln\left(\frac{4}{3}\right)^{1-x} = \ln 5^x$$
$$(1-x)\ln\left(\frac{4}{3}\right) = x\ln 5$$
$$\ln\left(\frac{4}{3}\right) - x\ln\left(\frac{4}{3}\right) = x\ln 5$$
$$\left[\ln 5 + \ln\left(\frac{4}{3}\right)\right]x = \ln\left(\frac{4}{3}\right)$$
$$x = \frac{\ln\left(\frac{4}{3}\right)}{\ln 5 + \ln\left(\frac{4}{3}\right)}$$
$$\approx 0.152$$

42.

$$(0.3)^{1+x} = 1.7^{2x-1}$$
$$\ln (0.3)^{1+x} = \ln 1.7^{2x-1}$$
$$(1+x)\ln(0.3) = (2x-1)\ln 1.7$$
$$\ln(0.3) + x\ln(0.3) = 2x\ln 1.7 - \ln 1.7$$
$$[\ln(0.3) - 2\ln 1.7]x = -\ln 1.7 - \ln(0.3)$$
$$x = \frac{-\ln 1.7 - \ln(0.3)}{\ln(0.3) - 2\ln 1.7} \approx -0.297$$

44.

$$e^{x+3} = \pi^x$$
$$\ln e^{x+3} = \ln \pi^x$$
$$(x+3)\ln e = x\ln \pi$$
$$x + 3 = x\ln \pi$$
$$x - x\ln \pi = -3$$
$$x(1 - \ln \pi) = -3$$
$$x = \frac{-3}{1 - \ln \pi} \approx 20.728$$

46.

$$0.3 \times 4^{0.2x} = 0.2$$
$$4^{0.2x} = \frac{0.2}{0.3} = \frac{2}{3}$$
$$\ln 4^{0.2x} = \ln \frac{2}{3}$$
$$0.2x \ln 4 = \ln \frac{2}{3}$$
$$x = \frac{\ln \frac{2}{3}}{.2 \ln 4} = -1.462$$

48.

$$500e^{0.3x} = 600$$
$$e^{0.3x} = \frac{600}{500} = \frac{6}{5}$$
$$0.3x = \ln \frac{6}{5}$$
$$x = \frac{\ln \frac{6}{5}}{0.3}$$
$$= 0.608$$

50.

$$\log_a x + \log_a(x-2) = \log_a(x+4)$$
$$\log_a x + \log_a(x-2) - \log_a(x+4) = 0$$
$$\log_a \frac{x(x-2)}{x+4} = 0$$
$$\frac{x(x-2)}{x+4} = 1$$
$$x^2 - 2x = x + 4$$
$$x^2 - 3x - 4 = 0$$
$$(x-4)(x+1) = 0$$
$$x = 4 \text{ or } x = -1 \text{ (impossible)}$$
$$\text{Thus, } x = 4$$

52. $\log_4(x^2 - 9) - \log_4(x + 3) = 3$

$$\log_4 \frac{x^2 - 9}{x + 3} = 3$$

$$\log_4 \frac{(x + 3)(x - 3)}{x + 3} = 3$$

$$\log_4(x - 3) = 3$$

$$x - 3 = 4^3$$

$$x = 67$$

54. $\log_3 3^x = -1$

$$x = -1$$

56. $\log_2(3x + 2) - \log_4 x = 3$

$$\log_2(3x + 2) - \frac{\log 2x}{\log_2 4} = 3$$

$$\log_2(3x + 2) - \frac{\log_2 x}{2} = 3$$

$$\log_2(3x + 2) - \frac{1}{2}\log_2 x = 3$$

$$\log_2 \frac{3x + 2}{x^{1/2}} = 3$$

$$2^3 = \frac{3x + 2}{\sqrt{x}}$$

$$8\sqrt{x} = 3x + 2$$

$$64x = 9x^2 + 12x + 4$$

$$0 = 9x^2 - 52x + 4$$

$$x = \frac{52 \pm \sqrt{2704 - 144}}{18}$$

$$= \frac{52 \pm \sqrt{2560}}{18}$$

58. $\log_9 x + 3 \log_3 x = 14$

$$\frac{\log_3 x}{\log_3 9} + 3 \log_3 x = 14$$

$$\frac{\log_3 x}{2} + 3 \log_3 x = 14$$

$$\left[\frac{1}{2} + 3\right] \log_3 x = 14$$

$$\frac{7}{2}\log_3 x = 14$$

$$\log_3 x^{\frac{7}{2}} = 14$$

$$x^{\frac{7}{2}} = 3^{14}$$

$$x^{\frac{1}{2}} = 3^2 = 9$$

$$x = 81$$

60. $0.44, -1.98$

62. $1.85, 4.53$

64. 1.15

66. 0.65

68. 1.47

70. 0.56

6.5 Compound Interest

2. $A = P\left(1 + \frac{r}{n}\right)^{nt} = 50\left(1 + \frac{.06}{12}\right)^{12\cdot3}$

$\approx \$59.83$

4. $A = P\left(1 + \frac{r}{n}\right)^{nt} = 300\left(1 + \frac{.12}{12}\right)^{(12)(1.5)}$

$\approx \$358.84$

6. $A = P\left(1 + \frac{r}{n}\right)^{nt} = 700\left(1 + \frac{.06}{365}\right)^{(365)(2)}$

$= \$789.24$

8. $A = Pe^{rt} = 40e^{(.07)(3)} \approx \$49.35$

10. $A = Pe^{rt} = 100e^{(.12)(3.75)} \approx \$156.83$

12. $V = A\left(1 + \frac{r}{n}\right)^{-nt} = 75\left(1 + \frac{.08}{4}\right)^{-4(3)}$

$= \$59.14$

206   Chapter 6 Exponential and Logarithmic Functions

14. $V = A\left[1 + \dfrac{r}{n}\right]^{-nt} = 800\left[1 + \dfrac{.07}{12}\right]^{-12(3.5)}$

$\approx \$626.61$

16. $V = A\left[1 + \dfrac{r}{n}\right]^{-nt}$

$= 300\left[1 + \dfrac{.03}{365}\right]^{-365(4)} \approx \$266.08$

18. $V = Ae^{-rt} = 800e^{-(.08)(2.5)} \approx \$654.98$

20. $V = Ae^{-rt} = 1000e^{-(.12)} \approx \$886.92$

22. $.07 = \left[1 + \dfrac{r}{4}\right]^4 - 1$

$1.07 = \left[1 + \dfrac{r}{4}\right]^4$

$1.0170585 = 1 + \dfrac{r}{4}$

$r \approx 6.82\%$

24. $2P = P(1 + r)^{10}$

$2 = (1 + r)^{10}$

$1.07177 = 1 + r$

$r \approx 7.18\%$

26. $r_e = \left[1 + \dfrac{.09}{4}\right]^4 - 1$

$= 1.093 - 1$

$= 9.3\%$

9% compounded quarterly is a better investment.

28. r_e (8% semiannually)

$= \left[1 + \dfrac{.08}{2}\right] - 1 = 8.16\%$

r_e (7.9% daily)

$= \left[1 + \dfrac{.079}{365}\right]^{365} - 1 = 8.22\%$

7.9% compounded daily.

30. $2P = P\left[1 + \dfrac{.10}{12}\right]^t$

$2 = (1 + .00833)^t$

$\ln 2 = t \ln 1.00833$

$t \approx 83.56$ months

$2P = Pe^{.10t}$

$2 = e^{.10t}$

$\ln 2 = .10t$

$t = 6.93$ years

$= 83.18$ months

32. $175 = 100\left[1 + \dfrac{.10}{12}\right]^t$

$1.75 = (1 + .00833)^t$

$\ln 1.75 = t \ln(1.00833)$

$t \approx 67.46$ months

$175 = 100e^{.10t}$

$1.75 = e^{.10t}$

$\ln 1.75 = .10t$

$t = 5.596$ years

$= 67.15$ months

34. $\$80000 = \$25000e^{.07t}$

$3.2 = e^{.07t}$

$\ln 3.2 = .07t$

$t \approx 16.62$ years or 16 years, 7 months

36. $A = \$200(1 + .0125)^{12(5/12)} = \$212.82$

(You get a 1 month grace period.)

38. $\$3000 = P\left[1 + \dfrac{.03}{12}\right]^{6 \cdot (1/2)}$

$\$3000 = P(1.00752)$

$P \approx \$2977.61$

40. $\$20 = \$15(1 + r)^2$

$1.3333 = (1 + r)^2$

$1.1547 = 1 + r$

$r \approx 15.47\%$

42. $\$5000 = P(1 + .08)^{10}$
$P = \$5000(1 + .08)^{-10}$
$P \approx \$2315.97$

44. CD matures on April 1.
On April 1, you have:
$A = \$1000e^{.068(1/4)}$
$A = \$1017.15$
Place \$1017.15 into passbook account for 1 month. On May 1 you have:
$$A = \$1017.15 \left[1 + \frac{.0525}{12}\right]^{12(1/12)}$$
$\approx \$1021.60$

46. $A = \$1000e^{.10(3)} = \$1349.86$
Take the \$1000 now; it will be worth more.

48. (a)
$$r_e = \left[1 + \frac{.0425}{360}\right]^{360} - 1$$
$$= 1.04341344 - 1$$
$$\approx 4.341344\%$$

(b)
$$r_e = \left[1 + \frac{.0425}{365}\right]^{365} - 1$$
$$= 1.04341347 - 1$$
$$\approx 4.341347\%$$

50. $\$40000 = P(1 + .08)^{17}$
$P = \$40000(1 + .08)^{-17}$
$P \approx \$10810.76$

52. $\$25000 = \$12485.52(1 + r)^8$
$2.00232 = (1 + r)^8$
$1.09067 = 1 + r$
$r \approx 9.07\%$

60. (a) $y = \dfrac{\ln 8000 - \ln 1000}{.10} = 20.79$ years

(b) $35 = \dfrac{\ln 30000 - \ln 2000}{r}$

$r = \dfrac{\ln 30000 - \ln 2000}{35}$

$r \approx .0774$
$r \approx 7.74\%$

(c)
$$A = Pe^{rt}$$
$$\frac{A}{P} = e^{rt}$$
$$\ln\left[\frac{A}{P}\right] = rt$$
$$\ln A - \ln P = rt$$
$$t = \frac{\ln A - \ln P}{r}$$

2.
$$N = 1000e^{0.01t}$$
$$1500 = 1000e^{0.01t}$$
$$e^{0.01t} = 1.5$$
$$0.01t = \ln 1.5$$
$$t = \frac{\ln 1.5}{0.01} \approx 40.5465$$

The population will equal 1500 in 40.5465 hours.

$$2000 = 1000e^{0.01t}$$
$$e^{0.01t} = 2$$
$$0.01t = \ln 2$$
$$t = \frac{\ln 2}{0.01} \approx 69.3147$$

The population will equal 2000 in 69.3147 hours.

4.
$$A = A_0 e^{-0.087t}$$
$$\frac{1}{2} A_0 = A_0 e^{-0.087t}$$
$$\frac{1}{2} = e^{-0.087t}$$
$$-0.087t = \ln \frac{1}{2}$$
$$t = \frac{\ln \frac{1}{2}}{-0.087t} \approx 7.9672$$

The half-life of iodine-131 is 7.9672 years.

6.
$$A = A_0 e^{-0.087t}$$
$$10 = 100e^{-0.087t}$$
$$0.1 = e^{-0.087}$$
$$-0.087t = \ln 0.1$$
$$t = \frac{\ln 0.1}{-0.087} \approx 26.4665$$

It takes 26.4665 years for 100 grams of iodine-131 to decay to 10 grams.

8.
$$A = A_0 e^{kt}$$
$$800 = 500e^{k(1)}$$
$$\frac{8}{5} = e^k$$
$$k = \ln \frac{8}{5} \approx 0.4700$$

When $t = 5$,
$$A = 500e^{0.4700(5)} = 5242.88$$

There are approximately 5243 bacteria cells after 5 hours.

$$20{,}000 = 500e^{0.4700t}$$
$$40 = e^{0.4700t}$$
$$0.4700t = \ln 40$$
$$t = \frac{\ln 40}{0.4700} \approx 7.8487$$

There are 20,000 bacteria after 7.8487 hours.

10.
$$A = A_0 e^{kt}$$
When $t = 2$,
$$800{,}000 = 900{,}000e^{2k}$$
$$0.888\ldots = e^{2k}$$
$$2k = \ln 0.888\ldots$$
$$k = \frac{\ln 0.888\ldots}{2}$$
$$\approx -0.05889$$

When $t = 4$,
$$A = 900{,}000e^{-0.05889(4)}$$
$$\approx 711{,}111.1111$$

The population will be 711,111 in 1997.

12.
$$A = A_0 e^{kt}$$
$$\frac{1}{2} A_0 = A_0 e^{k(1.3 \times 10^9)}$$
$$0.5 = e^{k(1.3 \times 10^9)}$$
$$k(1.3 \times 10^9) = \ln 0.5$$
$$k = -5.33 \times 10^{-10}$$
$$A = 10e^{(-5.33 \times 10^{-10})(100)}$$
$$= 9.99999947 \text{ grams}$$
$$A = 10e^{-5.33 \times 10^{-10})(1000)}$$
$$= 9.9999947 \text{ grams}$$

14.
$$A = A_0 e^{kt}$$
$$\frac{1}{2}A_0 = A_0 e^{k(5600)}$$
$$\frac{1}{2} = e^{5600k}$$
$$5600k = \ln \frac{1}{2}$$
$$k = \frac{1}{5600} \ln \frac{1}{2} \approx -0.000124$$
$$0.7A_0 = A_0 e^{-0.000124t}$$
$$0.7 = e^{-0.000124t}$$
$$-0.000124t = \ln 0.7$$
$$t = \frac{1}{-0.000124} \ln 0.7 \approx 2876.41 \text{ years}$$

The fossil is about 2876.41 years old.

16. $u = T + (u_0 - T)e^{kt}$

If $u_0 = 72$, $T = 38$ and $u = 60$ when $t = 2$, then
$$60 = 38 + (72 - 38)e^{2k}$$
$$22 = 34e^{2k}$$
$$e^{2k} = \frac{11}{17}$$
$$2k = \ln \frac{11}{17}$$
$$k = \frac{1}{2} \ln \frac{11}{17} \approx -0.217659$$

After 7 minutes,
$$u = 38 + (72 - 38)e^{(-0.217659)(7)}$$
$$\approx 45.41$$

The thermometer will read 45.41°F after 7 minutes.
$$39 = 38 + (72 - 38)e^{-0.216759t}$$
$$1 = 34e^{-0.216759t}$$
$$e^{-0.216759t} = \frac{1}{34}$$
$$-0.216759t = \ln \frac{1}{34}$$
$$t = \frac{1}{-0.216759} \ln \frac{1}{34}$$
$$\approx 16.3 \text{ minutes}$$

The thermometer will read 39°F after about 16.3 minutes.

18. Using the formula $T - u = (T - u_0)e^{kt}$, if $u_0 = 28$, $T = 70$, and $u = 35$ when $t = 10$, then
$$70 - 35 = (70 - 28)e^{10k}$$
$$35 = 42e^{10k}$$
$$e^{10k} = \frac{35}{42}$$
$$10k = \ln \frac{35}{42}$$
$$k = \frac{1}{10} \ln \frac{35}{42} \approx -0.018232$$

When $t = 30$,
$$70 - u = (70 - 28)e^{-0.018232(30)}$$
$$-u \approx -45.69$$
$$u \approx 45.69$$

After 30 minutes, the steak will be about 45.69°F.
$$70 - 45 = (70 - 28)e^{-0.018232t}$$
$$e^{-0.018232t} = \frac{25}{42}$$
$$-0.018232t = \ln \frac{25}{42}$$
$$t = \frac{1}{-0.018232} \ln \frac{25}{42} \approx 28.3$$

The steak will thaw after about 28.3 minutes.

20. $$A = A_0 e^{kt}$$
$$10 = 40e^{2k}$$
$$e^{2k} = \frac{1}{4}$$
$$2k = \ln \frac{1}{4}$$
$$k = \frac{1}{2} \ln \frac{1}{4}$$
$$\approx -0.693147$$
When $t = 5$,
$$A = 40e^{-0.693147(5)}$$
$$\approx 1.25$$
The voltage is 1.25 volts after 5 seconds.

22. u = temperature of a heated object at time t
T = constant temperature of surrounding medium
u_0 = initial temperature of heated object
$$T = 325°, \; u_0 = 75°$$
$$u = T + (u_0 - T)e^{kt}$$
$$u = 325 + (75 - 325)e^{kt}$$
$$u = 325 - 250e^{kt}$$
$$100 = 325 - 250e^{k(2)}$$
$$e^{2k} = \frac{225}{250} = \frac{9}{10}$$
$$2k = \ln \frac{9}{10}$$
$$k = \frac{1}{2} \ln \frac{9}{10} \approx -0.0527$$
We want to find t when $u = 175°$.
$$175° = 325 - 250e^{-0.0527t}$$
$$-150 = -250e^{-0.0527t}$$
$$e^{-0.0527t} = \frac{3}{5}$$
$$-0.0527t = \ln \frac{3}{5}$$
$$t = \frac{1}{-0.0527} \ln \frac{3}{5} \approx 9.7 \text{ hours}$$
The hotel may serve its guests at about 9:35, 9.7 hours after noon.

6.7 Logarithmic Scales

2. $L(10^{-3}) = 10 \log \dfrac{10^{-3}}{10^{-12}} = 10 \log 10^9 = 90$ decibels

4. $L(10^{-9.8}) = 10 \log \dfrac{10^{-9.8}}{10^{-12}} = 10 \log 10^{2.2} = 22$ decibels

6. $L(x) = 10 \log \dfrac{x}{10^{-12}}, \; L(50x) = 10 \log \dfrac{50x}{10^{-12}}$

$$10 \, \frac{\log \frac{50x}{10^{-12}}}{} - 10 \log \frac{x}{10^{-12}} = 10 \left(\log \frac{\frac{50x}{10^{-12}}}{\frac{x}{10^{-12}}} \right) = 10 \log 50 \approx 16.99 \text{ decibels}$$

8. $M(1210) = \log \dfrac{1210}{0.001} \approx 6.0827$
This earthquake thus measures 6.0827 on the Richter scale.

10. Let x_1 and x_2 denote the seismographic readings of the two earthquakes, then

$$\log\frac{x_1}{x_0} - \log\frac{x_2}{x_0} = 1$$

$$\log\frac{x_1}{x_2} = 1$$

$$\frac{x_1}{x_2} = 10$$

$$x_1 = 10x_2$$

x_1 is 10 times as intense as x_2

6 Chapter Review

2. $\log_3 81 = 4$

4. $e^{\ln 0.1} = 0.1$

6. $\log_2 2^{\sqrt{3}} = \sqrt{3}$

8. $-2\log_3\left[\dfrac{1}{x}\right] + \dfrac{1}{3}\log_3\sqrt{x} = \log_3\left[\dfrac{1}{x}\right]^{-2} + \log_3 x^{1/6} = \log_3(x^2 \cdot x^{1/6}) = \log_3 x^{1/3}$

10. $\log(x^2 - 9) - \log(x^2 + 7x + 12) = \log\dfrac{x^2 - 9}{x^2 + 7x + 12} = \log\dfrac{(x - 3)(x + 3)}{(x + 3)(x + 4)} = \log\dfrac{x - 3}{x + 4}$

12. $\dfrac{1}{2}\ln(x^2 + 1) - 4\ln\dfrac{1}{2} - \dfrac{1}{2}\big[\ln(x - 4) + \ln x\big]$

$$= \ln(x^2 + 1)^{1/2} - \ln\left[\dfrac{1}{2}\right]^4 - \big[\ln(x - 4) + \ln x\big]^{1/2} = \ln\dfrac{(x^2 + 1)^{1/2}}{\dfrac{1}{16}\big[(x - 4)(x)\big]^{1/2}}$$

14.
$$\ln(y - 3) = \ln 2x^2 + \ln C$$
$$\ln(y - 3) - \ln 2x^2 - \ln C = 0$$
$$\ln\dfrac{y - 3}{(2x^2)(C)} = 0$$
$$\dfrac{y - 3}{2Cx^2} = e^0 = 1$$
$$y - 3 = 2Cx^2$$
$$y = 3 + 2Cx^2$$

16.
$$\ln 2y = \ln(x + 1) + \ln(x + 2) + \ln C$$
$$\ln 2y - \ln(x + 1) - \ln(x + 2) - \ln C = 0$$
$$\ln\dfrac{2y}{C(x + 1)(x + 2)} = 0$$
$$e^0 = \ln\dfrac{2y}{C(x + 1)(x + 2)}$$
$$2y = C(x^2 + 3x + 2)$$
$$y = \dfrac{C(x^2 + 3x + 2)}{2}$$

Chapter 6 Exponential and Logarithmic Functions

18. $$\ln(y - 1) + \ln(y + 1) = -x + C$$
$$\ln[(y - 1)(y + 1)] = -x + C$$
$$y^2 - 1 = e^{-x+C}$$
$$y^2 = e^{-x+C} + 1$$
$$y = \sqrt{e^{-x+C} + 1}$$

20. $$e^{3y-C} = (x + 4)^2$$
$$\ln e^{3y-C} = \ln(x + 4)^2$$
$$3y - C = 2 \ln(x + 4)$$
$$3y = 2 \ln(x + 4) + C$$
$$y = \frac{2 \ln(x + 4) + C}{3}$$

22. $f(x) = \ln(-x)$

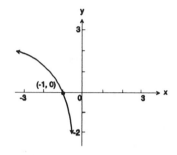

24. $f(x) = 3 + \ln x$

26. $f(x) = \dfrac{1}{2} \ln x$

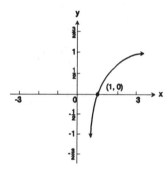

28. $f(x) = \ln |x|$

30. $f(x) = 4 - \ln(-x)$

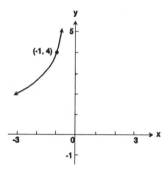

32. $$8^{6+3x} = 4$$
$$2^{3(6+3x)} = 2^2$$
$$18 + 9x = 2$$
$$9x = -16$$
$$x = \frac{-16}{9}$$

34.

$$4^{x-x^2} = \frac{1}{2}$$

$$2^{2(x-x^2)} = 2^{-1}$$

$$2x - 2x^2 = -1$$

$$2x^2 - 2x - 1 = 0$$

$$x = \frac{2 \pm \sqrt{12}}{4} = \frac{2 \pm 2\sqrt{3}}{4} = \frac{1 \pm \sqrt{3}}{2}$$

$$x = \frac{1 + \sqrt{3}}{2} \quad \text{or} \quad x = \frac{1 - \sqrt{3}}{2}$$

36.

$$\log_{\sqrt{2}} x = -6$$

$$\left(\sqrt{2}\right)^{-6} = x$$

$$x = 0.125$$

38.

$$5^{x+2} = 7^{x-2}$$

$$\ln 5^{x+2} = \ln 7^{x-2}$$

$$(x + 2)\ln 5 = (x - 2)\ln 7$$

$$(\ln 5)x + 2\ln 5 = (\ln 7)x - 2\ln 7$$

$$(\ln 5 - \ln 7)x = -2\ln 7 - 2\ln 5$$

$$x = \frac{-2\left[\ln 7 + \ln 5\right]}{\ln 5 - \ln 7}$$

$$\approx 21.1331$$

40.

$$25^{2x} = 5^{x^2-12}$$

$$5^{2(2x)} = 5^{x^2-12}$$

$$4x = x^2 - 12$$

$$x^2 - 4x - 12 = 0$$

$$(x - 6)(x + 2) = 0$$

$$x = 6 \quad \text{or} \quad x = -2$$

42.

$$2^{x+1} \cdot 8^{-x} = 4$$

$$2^{x+1} \cdot 2^{-3x} = 2^2$$

$$2^{-2x+1} = 2^2$$

$$-2x + 1 = 2$$

$$-2x = 1$$

$$x = \frac{-1}{2}$$

44.

$$2^x \cdot 5 = 10^x$$

$$\ln(2^x \cdot 5) = \ln 10^x$$

$$\ln 2^x + \ln 5 = \ln 10^x$$

$$x\ln 2 + \ln 5 = x\ln 10$$

$$x(\ln 2 - \ln 10) = -\ln 5$$

$$x = \frac{-\ln 5}{\ln 2 - \ln 10} = 1$$

46.

$$\log_{10}(7x - 12) = 2\log_{10} x$$

$$\log_{10}(7x - 12) = \log_{10} x^2$$

$$7x - 12 = x^2$$

$$x^2 - 7x + 12 = 0$$

$$(x - 3)(x - 4) = 0$$

$$x = 3 \quad \text{or} \quad x = 4$$

48.

$$e^{1-2x} = 4$$

$$\ln e^{1-2x} = \ln 4$$

$$1 - 2x = \ln 4$$

$$-2x = \ln 4 - 1$$

$$x = -\frac{1}{2}(\ln 4 - 1)$$

$$\approx -0.193$$

50.

$$2^{x^3} = 3^{x^2}$$

$$x^3 = x^2\log_2 3$$

$$x = \log_2 3$$

$$x = \frac{\ln 3}{\ln 2} \approx 1.585$$

52.

$$h(500) = (30 \cdot 5 + 8000)\log\left[\frac{760}{500}\right]$$

$$= 8150 \cdot (.1818436)$$

$$\approx 1482 \text{ m}$$

54.
$$8900 = (30 \cdot 5 + 8000)\log\left[\frac{760}{x}\right]$$
$$8900 = 8150 \cdot \log\left[\frac{760}{x}\right]$$
$$1.09202454 = \log\left[\frac{760}{x}\right]$$
$$10^{1.09202454} = \left[\frac{760}{x}\right]$$
$$x \approx 61.5 \text{ mm Hg}$$

56. (a) $L = 9 + 5.1 \log 3.5$
≈ 11.8

(b) $14 = 9 + 5.1 \log d$
$5 = 5.1 \log d$
$\log d = .980392$
$d = 10^{.980392}$
$d \approx 9.56 \text{ inches}$

58.
$$m = 55.3 - 6\ln(1000 - 5000)$$
$$= 55.3 - 51.1$$
$$= 4.2 \text{ months}$$

60.
$$A = \$10,000\left[1 + \frac{.04}{2}\right]^{18 \cdot 2}$$
$$= \$10,000(2.03989)$$
$$\approx \$20,398.90$$

62. (a)
$$\$5000 = \$620.17e^{r(20)}$$
$$8.0623 = e^{20r}$$
$$\ln 8.0623 = 20r$$
$$r = 10.436\%$$

(b) $P = \$4000, r = 10.436\%, t = 20$
$$A = Pe^{rt} = 4000e^{(.10436)(20)}$$
$$= \$32,249.23$$
The actual value is $32,249.23

64. Let x_1 and x_2 denote the seismographic readings of the Chicago earthquake and the San Francisco earthquake respectively. Then
$$\frac{x_1}{x_2} = \frac{10^3}{10^{8.9}} = 10^{-5.9} = 0.000001258$$
$$x_1 = 0.000001258x_2$$
The Chicago earthquake was 0.000001258 times as intense as the San Francisco earthquake.

66. Using Newton's Law of Cooling, $u = T + (u_0 - T)e^{kt}$, with $T = 70$ and $u_0 = 450$. When $t = 5$, $u = 400$. Thus,
$$400 = 70 + (450 - 70)e^{5k}$$
$$330 = 380e^{5k}$$
$$e^{5k} = \frac{330}{380}$$
$$5k = \ln\frac{33}{38}$$
$$k = \frac{1}{5}\ln\frac{33}{38} \approx -0.0282$$
Thus,
$$150 = 70 + 380e^{-0.0282t}$$
$$80 = 380e^{-0.0282t}$$
$$e^{-0.0282t} = \frac{80}{380}$$
$$-0.0282t = \ln\frac{8}{38}$$
$$t = \frac{-1}{0.0282}\ln\frac{8}{38} \approx 55.25 \text{ minutes}$$

Mission Possible

1. $170°F = 76.67°C$; $140°F = 60°C$; $120°F = 48.89°C$

2. Say 2 minutes. Then $T = 57.39°C$. Once brewed, if the coffee container is surrounded by a circulating liquid at $57.39°C$, the coffee will cool to $140°F$ in 2 minutes.

3. Suppose you wish to hold the coffee above $120°F$ for 40 minutes. Then T must be $\geq 48.68°C$. Continuing to circulate the original liquid at $57.39°C$ would still work as would any temperature down to $48.68°$.

   An inner container heats the water to $170°F$ and brews the coffee. As soon as the brewing is complete, the outer container is filled with liquid heated to $57.39°C$. This liquid circulating around the inner container drops the temperature of the coffee to $140°$ in 2 minutes. One could probably just continue the circulation of the $57.39°C$ liquid and the coffee would stay above $120°F$.

TRIGONOMETRIC FUNCTIONS

2.

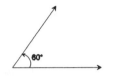

4.

6.

8.

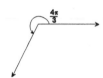

10.

12.

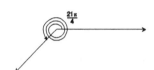

14. $120° = 120 \cdot \dfrac{\pi}{180}$ radians $= \dfrac{2\pi}{3}$ radians

16. $330° = 330 \cdot \dfrac{\pi}{180}$ radians $= \dfrac{11\pi}{6}$ radians

18. $-30° = -30 \cdot \dfrac{\pi}{180}$ radians $= \dfrac{-\pi}{6}$ radians

20. $270° = 270 \cdot \dfrac{\pi}{180}$ radians $= \dfrac{3\pi}{2}$ radians

22. $-225° = -225 \cdot \dfrac{\pi}{180}$ radians

$= \dfrac{-5\pi}{4}$ radians

24. $\dfrac{5\pi}{6}$ radians $= \dfrac{5\pi}{6} \cdot \dfrac{180}{\pi}$ degrees $= 150°$

26. $\dfrac{-2\pi}{3}$ radians $= \dfrac{-2\pi}{6} \cdot \dfrac{180}{\pi}$ degrees

$= -120°$

28. 4π radians $= 4\pi \cdot \dfrac{180}{\pi}$ degrees $= 720°$

30. $\dfrac{5\pi}{12}$ radians $= \dfrac{5\pi}{12} \cdot \dfrac{180}{\pi}$ degrees $= 75°$

32. $\dfrac{5\pi}{4}$ radians $= \dfrac{5\pi}{4} \cdot \dfrac{180}{\pi}$ degrees $= 225°$

34. $s = r\theta$
$s = (6 \text{ feet})(2 \text{ radians})$
$s = 12 \text{ feet}$

36. $r = \dfrac{s}{\theta}$

$r = \dfrac{6}{\frac{1}{4}} = 6 \cdot 4 = 24 \text{ centimeters}$

38. $\theta = \dfrac{s}{r}$

$\theta = \dfrac{8}{6} = \dfrac{4}{3}$ radian

40. $s = r\theta$

$s = 3 \cdot \dfrac{2\pi}{3}$

$s = 2\pi \approx 6.28$ meters

42. $73° = 1.27$ radians

44. $-51° = -0.89$ radians

46. $200° = 3.49$ radians

48. $350° = 6.11$ radians

50. π radians $= 180.0°$

52. 0.75 radians $= 42.97°$

54. 3 radians $= 171.89°$

56. $\sqrt{2}$ radians $= 81.03°$

58. $61° \, 42' \, 21'' = 61° + 42 \cdot \dfrac{1°}{60} + 21 \cdot \left[\dfrac{1}{60} \cdot \dfrac{1}{60}\right]° = 61° + 0.7° + 0.0058° = 61.71°$

60. $73° \, 40' \, 40'' = 73° + 40 \cdot \dfrac{1°}{60} + 40 \cdot \left[\dfrac{1}{60} \cdot \dfrac{1}{60}\right]° = 73° + 0.6667° + 0.0111° = 73.68°$

62. $98° \, 22' \, 45'' = 98° + 22 \cdot \dfrac{1°}{60} + 45 \cdot \left[\dfrac{1}{60} \cdot \dfrac{1}{60}\right]° = 98° + 0.3667° + 0.0125° = 98.38°$

64. $0.24° = (0.24)(60') = 14.4$

$0.4' = (0.4)(60'') = 24''$

Thus, $61.24° = 61° \, 14' \, 24''$

66. $0.411° = (0.411)(60') = 24.66'$

$0.66' = (0.66)(60'') = 39.6''$

Thus, $29.411° = 29° \, 24' \, 40''$

68. $0.01° = (0.01)(60') = 0.6'$

$0.6' = (0.6)(60'') = 36''$

Thus, $44.01° = 44° \, 0' \, 36''$

70. $s = r\theta = 40 \cdot 20° \cdot \dfrac{\pi}{180} = \dfrac{40\pi}{9}$ inches

72. $V = \dfrac{s}{t} = \dfrac{5 \, m}{20 \, \text{sec}} = 0.25 \, \dfrac{m}{\text{sec}}$; $\quad \omega = \dfrac{V}{r} = \dfrac{0.25}{2} = 0.125 \, \dfrac{\text{radians}}{\text{sec}}$

74. $\omega = \dfrac{V}{r}$, so $V = \omega r$

We have $\omega = 3$ rev/sec and $r = 15''$.

Thus, $V = \dfrac{3 \text{ rev}}{\text{sec}} \cdot 15 \text{ inches} \cdot \dfrac{2\pi \text{ radians}}{1 \text{ rev}} = 282.74$ inches/second

$\dfrac{282.74 \text{ inches}}{\text{second}} \cdot \dfrac{3600 \text{ seconds}}{1 \text{ hour}} \cdot \dfrac{1 \text{ foot}}{12 \text{ inches}} \cdot \dfrac{1 \text{ mile}}{5280 \text{ feet}} = 16.06$ mph

76. $V = r\omega = \dfrac{18 \cdot \dfrac{2\pi}{3}}{1} = 12\pi \approx 37.7$ inches per second

78. $V = r\omega = 9.29 \cdot 10^7 \text{ miles} \cdot \dfrac{1 \text{ rev}}{365 \text{ days}} \cdot \dfrac{1 \text{ day}}{24 \text{ hrs}} \cdot \dfrac{2\pi \text{ radians}}{\text{rev}} = 66{,}633$ mph

80. r_1 rotates ω_1 rev/min; r_2 rotates ω_2 rev/min

$$V = r_1\omega_1 = r_2\omega_2$$

Therefore, $\dfrac{r_1}{r_2} = \dfrac{\omega_2}{\omega_1}$

82. $\omega = \dfrac{480 \text{ revolutions}}{\text{minute}} = \dfrac{480 \text{ rev}}{1 \text{ min}} \cdot \dfrac{2\pi \text{ radians}}{\text{rev}} = 960\dfrac{\pi \text{ radians}}{\text{minute}}$

$V = r\omega$ where $r = \dfrac{26}{2} = 13$ inches

$V = 13\left[960\dfrac{\pi \text{ radians}}{\text{minute}}\right] = \dfrac{12,480\pi \text{ inches}}{\text{minute}} \cdot \dfrac{1 \text{ ft}}{12 \text{ in}} \cdot \dfrac{1 \text{ mile}}{5280 \text{ ft}} \cdot \dfrac{60 \text{ min}}{1 \text{ hr}} \approx 37.13$ mph

$V = r\omega$

80 mph $= 13$ inches $\cdot \omega$

$\omega = \dfrac{80 \text{ miles}}{1 \text{ hour}} \cdot \dfrac{1}{13 \text{ inches}} \cdot \dfrac{12 \text{ inches}}{1 \text{ foot}} \cdot \dfrac{5280 \text{ feet}}{1 \text{ mile}} \cdot \dfrac{1 \text{ hour}}{60 \text{ min}} \cdot \dfrac{1 \text{ rev}}{2\pi \text{ rad}} = 1034$ rpm

84. $\dfrac{t}{90} = \dfrac{24}{24\pi(3960)}$

$t = .0868$ hours $= 5.21$ minutes

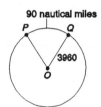

7.2 Right Triangle Trigonometry

2.
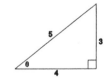

$$a^2 + b^2 = c^2$$
$$9 + 16 = c^2$$
$$25 = c^2$$
$$5 = c$$

$\sin\theta = \dfrac{3}{5}$ $\csc\theta = \dfrac{5}{3}$

$\cos\theta = \dfrac{4}{5}$ $\sec\theta = \dfrac{5}{4}$

$\tan\theta = \dfrac{3}{4}$ $\cot\theta = \dfrac{4}{3}$

4.

$$a^2 + b^2 = c^2$$
$$9 + 9 = c^2$$
$$18 = c^2$$
$$3\sqrt{2} = c$$

$\sin\theta = \dfrac{3}{3\sqrt{2}} = \dfrac{\sqrt{2}}{2}$ $\csc\theta = \sqrt{2}$

$\cos\theta = \dfrac{3}{3\sqrt{2}} = \dfrac{\sqrt{2}}{2}$ $\sec\theta = \sqrt{2}$

$\tan\theta = \dfrac{3}{3} = 1$ $\cot\theta = 1$

6.

$$a^2 + b^2 = c^2$$
$$9 + b^2 = 16$$
$$b^2 = 7$$
$$b = \sqrt{7}$$

$$\sin \theta = \frac{3}{4} \qquad \csc \theta = \frac{4}{3}$$

$$\cos \theta = \frac{\sqrt{7}}{4} \qquad \sec \theta = \frac{4}{\sqrt{7}} = \frac{4\sqrt{7}}{7}$$

$$\tan \theta = \frac{3}{\sqrt{7}} = \frac{3\sqrt{7}}{7} \qquad \cot \theta = \frac{\sqrt{7}}{3}$$

8.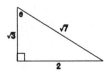

$$a^2 + b^2 = c^2$$
$$4 + 3 = c^2$$
$$7 = c^2$$
$$\sqrt{7} = c$$

$$\sin \theta = \frac{2}{\sqrt{7}} = \frac{2\sqrt{7}}{7} \qquad \csc \theta = \frac{\sqrt{7}}{2}$$

$$\cos \theta = \frac{\sqrt{3}}{\sqrt{7}} = \frac{\sqrt{21}}{7} \qquad \sec \theta = \frac{\sqrt{7}}{\sqrt{3}} = \frac{\sqrt{21}}{3}$$

$$\tan \theta = \frac{2}{\sqrt{3}} = \frac{2\sqrt{3}}{3} \qquad \cot \theta = \frac{\sqrt{3}}{2}$$

10.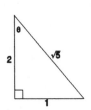

$$a^2 + b^2 = c^2$$
$$a^2 + 4 = 5$$
$$a^2 = 1$$
$$a = 1$$

$$\sin \theta = \frac{1}{\sqrt{5}} = \frac{\sqrt{5}}{5} \qquad \csc \theta = \sqrt{5}$$

$$\cos \theta = \frac{2}{\sqrt{5}} = \frac{2\sqrt{5}}{5} \qquad \sec \theta = \frac{\sqrt{5}}{2}$$

$$\tan \theta = \frac{1}{2} \qquad \cot \theta = 2$$

12. $\sin \theta = \frac{\sqrt{3}}{2}$, $\cos \theta = \frac{1}{2}$

$$\tan \theta = \frac{\sin \theta}{\cos \theta} = \frac{\frac{\sqrt{3}}{2}}{\frac{1}{2}} = \sqrt{3} \qquad \csc \theta = \frac{1}{\sin \theta} = \frac{1}{\frac{\sqrt{3}}{2}} = \frac{2\sqrt{3}}{3}$$

$$\sec \theta = \frac{1}{\cos \theta} = \frac{1}{\frac{1}{2}} = 2 \qquad \cot \theta = \frac{1}{\tan \theta} = \frac{1}{\sqrt{3}} = \frac{\sqrt{3}}{3}$$

14. $\sin \theta = \frac{1}{3}$, $\cos \theta = \frac{2\sqrt{2}}{3}$

$$\tan \theta = \frac{\sin \theta}{\cos \theta} = \frac{\frac{1}{3}}{\frac{2\sqrt{2}}{3}} = \frac{1}{2\sqrt{2}} = \frac{\sqrt{2}}{4} \qquad \csc \theta = \frac{1}{\sin \theta} = \frac{1}{\frac{1}{3}} = 3$$

$$\sec \theta = \frac{1}{\cos \theta} = \frac{1}{\frac{2\sqrt{2}}{3}} = \frac{3}{2\sqrt{2}} = \frac{3\sqrt{2}}{4} \qquad \cot \theta = \frac{1}{\tan \theta} = \frac{1}{\frac{1}{2\sqrt{2}}} = 2\sqrt{2}$$

16. $\cos\theta = \dfrac{\text{adjacent}}{\text{hypotenuse}} = \dfrac{\sqrt{2}}{2}$

$(\text{opposite})^2 = (\text{hypotenuse})^2 - (\text{adjacent})^2$

$$b^2 = 4 - 2$$
$$b^2 = 2$$
$$b = \sqrt{2}$$

$\sec\theta = \dfrac{\text{hypotenuse}}{\text{adjacent}} = \dfrac{2}{\sqrt{2}} = \sqrt{2}$

$\sin\theta = \dfrac{\text{opposite}}{\text{adjacent}} = \dfrac{\sqrt{2}}{2}$ $\qquad$ $\csc\theta = \dfrac{\text{hypotenuse}}{\text{opposite}} = \dfrac{2}{\sqrt{2}} = \sqrt{2}$

$\tan\theta = \dfrac{\text{opposite}}{\text{adjacent}} = \dfrac{\sqrt{2}}{\sqrt{2}} = 1$ $\qquad$ $\cot\theta = \dfrac{\text{adjacent}}{\text{opposite}} = \dfrac{\sqrt{2}}{\sqrt{2}} = 1$

18. $\sin\theta = \dfrac{\sqrt{3}}{4} = \dfrac{b}{c}$

$$c^2 = a^2 + b^2$$
$$16 = a^2 + 3$$
$$\sqrt{13} = a$$

$\cos\theta = \dfrac{\sqrt{13}}{4}$ $\qquad$ $\sec\theta = \dfrac{4\sqrt{13}}{13}$

$\tan\theta = \dfrac{\sqrt{3}}{\sqrt{13}} = \dfrac{\sqrt{39}}{13}$ $\qquad$ $\cot\theta = \dfrac{\sqrt{13}}{\sqrt{3}} = \dfrac{\sqrt{39}}{3}$ $\qquad$ $\csc\theta = \dfrac{4}{\sqrt{3}} = \dfrac{4\sqrt{3}}{3}$

20. $\cot\theta = \dfrac{1}{2} = \dfrac{a}{b}$

$$c^2 = a^2 + b^2$$
$$c^2 = 1 + 4$$
$$c = \sqrt{5}$$

$\sin\theta = \dfrac{2}{\sqrt{5}} = \dfrac{2\sqrt{5}}{5}$ $\qquad$ $\csc\theta = \dfrac{\sqrt{5}}{2}$

$\cos\theta = \dfrac{1}{\sqrt{5}} = \dfrac{\sqrt{5}}{5}$ $\qquad$ $\sec\theta = \sqrt{5}$ $\qquad$ $\tan\theta = \dfrac{2}{1} = 2$

22. $\csc\theta = 5 = \dfrac{c}{b}$

$$c^2 = a^2 + b^2$$
$$25 = a^2 + 1$$
$$24 = a^2$$
$$2\sqrt{6} = a$$

$\cos\theta = \dfrac{2\sqrt{6}}{5}$ $\qquad$ $\sec\theta = \dfrac{5}{2\sqrt{6}} = \dfrac{5\sqrt{6}}{12}$

$\tan\theta = \dfrac{1}{2\sqrt{6}} = \dfrac{\sqrt{6}}{12}$ $\qquad$ $\cot\theta = \dfrac{2\sqrt{6}}{1} = 2\sqrt{6}$ $\qquad$ $\sin\theta = \dfrac{1}{5}$

24. $\sec \theta = \dfrac{5}{3} = \dfrac{c}{a}$

$a^2 + b^2 = c^2$

$9^2 + b^2 = 25$

$b^2 = 16$

$b = 4$

$\sin \theta = \dfrac{4}{5}$ $\qquad\qquad$ $\csc \theta = \dfrac{5}{4}$

$\tan \theta = \dfrac{4}{3}$ $\qquad\qquad$ $\cot \theta = \dfrac{3}{4}$ $\qquad\qquad\qquad$ $\cos \theta = \dfrac{3}{5}$

26. $\sec^2 28° - \tan^2 28° = 1.$ $\qquad$ Since $\tan^2 \theta + 1 = \sec^2 \theta$ or $\sec^2 \theta - \tan^2 \theta = 1$

28. $\tan 10° \cot 10° = \tan 10° \left[\dfrac{1}{\tan 10°}\right] = \dfrac{\tan 10°}{\tan 10°} = 1$

30. $\cot 25° - \dfrac{\cos 25°}{\sin 25°} = \cot 25° - \cot 25° = 0$

32. $\tan 12° - \cot 78° = 0$

Since $\tan 12° = \cot(90° - 12°)$ or $\tan 12° = \cot 78°$ by the cofunction formula.

34. $\dfrac{\cos 40°}{\sin 50°} = 1$ $\qquad$ Since $\cos 40° = \sin(90° - 40°)$ or $\cos 40° = \sin 50°$.

36. $1 + \tan^2 5° - \csc^2 85°$

$\qquad$ By the fundamental identity: $\qquad \tan^2 5° + 1 = \sec^2 5°$

$\qquad\qquad\qquad\qquad\qquad\qquad\qquad\quad \sec^2 5° - \csc^2 85° = 0$

Since $\sec^2 5° = \csc^2(90° - 5°)$ by the cofunction formula.

38. $\cot 40° - \dfrac{\sin 50°}{\sin 40°} = \cot 40° - \dfrac{\cos 40°}{\sin 40°} = \cot 40° - \cot 40° = 0$

40. $\sec 35° \csc 55° - \tan 35° \cot 55° = \sec 35° \sec 35° - \tan 35° \tan 35° = \sec^2 35° - \tan^2 35° = 1$

42. $\sin 60° = \dfrac{\sqrt{3}}{2}$

(a) $\cos 30° = \dfrac{\sqrt{3}}{2}$ (using cofunctions)

(b) $\cos^2 60° = 1 - \sin^2 60° = 1 - \left[\dfrac{\sqrt{3}}{2}\right]^2 = 1 - \dfrac{3}{4} = \dfrac{1}{4}$

(c) $\sec \dfrac{\pi}{6} = \dfrac{1}{\cos\dfrac{\pi}{6}} = \dfrac{1}{\cos 30°} = \dfrac{1}{\dfrac{\sqrt{3}}{2}} = \dfrac{2\sqrt{3}}{3}$ $\qquad$ (d) $\csc \dfrac{\pi}{3} = \sec \dfrac{\pi}{6} = \dfrac{2\sqrt{3}}{3}$

44. $\sec \theta = 3$

(a) $\cos \theta = \dfrac{1}{3}$ $\qquad\qquad\qquad\qquad\qquad$ (c) $\csc(90° - \theta) = \sec \theta = 3$

(b) $\tan^2 \theta = \sec^2 \theta - 1 = (3)^2 - 1 = 8$ $\qquad$ (d) $\sin^2 \theta = 1 - \cos^2 \theta = 1 - \left[\dfrac{1}{3}\right]^2 = \dfrac{8}{9}$

46. $\cot \theta = 2$

    (a) $\tan \theta = \dfrac{1}{2}$
                                      (c) $\tan\left[\dfrac{\pi}{2} - \theta\right] = \cot \theta = 2$

    (b) $\csc^2 \theta = \cot^2 \theta + 1 = 2^2 + 1 = 5$
          (d) $\sec^2 \theta = \tan^2 \theta + 1 = \left[\dfrac{1}{2}\right]^2 + 1 = \dfrac{5}{4}$

48.   (a) $\sin 21° = \sqrt{1 - \cos^2 21°} = \sqrt{1 - (0.93)^2} \approx \sqrt{0.1351} = 0.37$

       (b) $\tan 21° = \dfrac{\sin 21°}{\cos 21°} \approx \dfrac{0.37}{0.93} \approx 0.40$

       (c) $\cot 21° = \dfrac{1}{\tan 21°} \approx \dfrac{1}{0.40} \approx 2.5$

       (d) $\sec 21° = \dfrac{1}{\cos 21°} \approx \dfrac{1}{0.93} \approx 1.08$

       (e) $\csc 21° = \dfrac{1}{\sin 21°} \approx \dfrac{1}{0.37} \approx 2.70$

       (f) $\sin 69° = \cos 21° \approx 0.93$

       (g) $\cos 69° = \sin 21° \approx 0.37$

       (h) $\tan 69° = \dfrac{\sin 69°}{\cos 69°} \approx \dfrac{0.93}{0.37} \approx 2.51$

50. $\tan \theta + \tan\left[\dfrac{\pi}{2} - \theta\right] = \tan \theta + \cot \theta = \tan \theta + \dfrac{1}{\tan \theta} = 4 + \dfrac{1}{4} = \dfrac{17}{4}$

52. $\tan \theta = \cot(\theta + 45°)$
       Since $\tan \theta = \cot(90° - \theta)$, then: $\theta + 45° = 90° - \theta$
$$2\theta = 45°$$
$$\theta = 22.5°$$

54.   From geometry, we know m $\angle ABC = \theta$. In addition, we know the length of the side opposite $\angle ABC$ is 4. Thus, $\sin \theta = \dfrac{4}{\overline{AB}}$ so $\overline{AB} = \dfrac{4}{\sin \theta}$. We also know from geometry $\theta = $ m $\angle DAE$ and the length of the side adjacent to $\angle DAE$ is 3. Thus, $\cos \theta = \dfrac{3}{\overline{AE}}$. The length of the ladder is $\overline{AB} + \overline{AE} = \dfrac{4}{\sin \theta} + \dfrac{3}{\cos \theta}$.

56.

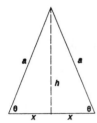

$A = \dfrac{1}{2}(2x)h = xh$

Since $\sin \theta = \dfrac{h}{a}$ and $\cos \theta = \dfrac{x}{a}$, then

$A = a^2 \sin \theta \cos \theta = a^2 \cdot \dfrac{h}{a} \cdot \dfrac{x}{a} = xh$

58. $\sin \theta = \dfrac{a}{x + a} = \dfrac{b}{x + 2a + b}$

$xb + ab = xa + 2a^2 + ba$

$(b - a)x = 2a^2$

$\quad x = \dfrac{2a^2}{b - a}$

$\sin \theta = \dfrac{a}{x + a} = \dfrac{a}{\dfrac{2a^2}{b - a} + a} = \dfrac{a}{\dfrac{2a^2 + ba - a^2}{b - a}} = \dfrac{a(b - a)}{a^2 + ba} = \dfrac{b - a}{b + a}$

$\cos \theta = \sqrt{1 - \sin^2 \theta} = \sqrt{1 - \left[\dfrac{b - a}{b + a}\right]^2} = \sqrt{1 - \dfrac{b^2 - 2ab + a^2}{b^2 + 2ab + a^2}}$

$\quad = \sqrt{\dfrac{b^2 + 2ab + a^2 - (b^2 - 2ab + a^2)}{(a + b)^2}} = \sqrt{\dfrac{4ab}{(a + b)^2}} = \dfrac{2\sqrt{ab}}{a + b} = \dfrac{\sqrt{ab}}{\dfrac{a + b}{2}}$

60. **(a)** Area of triangle $OBC = \dfrac{1}{2} \, |\, OB\,| \, |\, AC \,| = \dfrac{1}{2} \, |\, AC \,| = \dfrac{1}{2} \dfrac{|AC|}{|OC|} = \dfrac{1}{2} \sin \theta$

(b) Area of triangle $OBD = \dfrac{1}{2} \, |\, OB \,| \cdot |\, BD \,| = \dfrac{1}{2} \, |\, BD \,| = \dfrac{1}{2} \dfrac{|BD|}{|OD|} \cdot \dfrac{|OD|}{1}$

$\quad\quad = \dfrac{1}{2} \dfrac{|BD|}{|OD|} \cdot \dfrac{|OD|}{|OC|} = \dfrac{1}{2} \sin \theta \, \dfrac{1}{\cos \theta}$

(c) Area $\triangle OBC <$ Area $\overset{\frown}{OBC} <$ Area $\triangle OBD$

$\quad \dfrac{1}{2} \sin \theta < \dfrac{1}{2} \theta < \dfrac{1}{2} \sin \theta \cdot \dfrac{1}{\cos \theta}$

$\quad 1 < \dfrac{\theta}{\sin \theta} < \dfrac{1}{\cos \theta}$

62. If θ is an acute angle of a right triangle, then the ratio between the hypotenuse to the side adjacent the angle will always be greater than one due to the fact that the hypotenuse is always the longest side of the triangle.

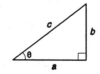

$\quad \sec \theta = \dfrac{c}{a} > 1$

7.3 Computing the Values of Trigonometric Functions of Given Angles

2. $\sin 30° = \dfrac{1}{2}$ $\qquad\qquad$ $\csc 30° = 2$

$\quad \cos 30° = \dfrac{\sqrt{3}}{2}$ $\qquad\qquad$ $\sec 30° = \dfrac{2\sqrt{3}}{3}$

$\quad \tan 30° = \dfrac{\sqrt{3}}{3}$ $\qquad\qquad$ $\cot 30° = \sqrt{3}$

4. $2 \sin 45° + 4 \sin 30° = 2\left[\dfrac{\sqrt{2}}{2}\right] + 4\left[\dfrac{1}{2}\right] = \sqrt{2} + 2$

6. $\sin 30° \cdot \tan 60° = \left[\dfrac{1}{2}\right]\left(\sqrt{3}\right) = \dfrac{\sqrt{3}}{2}$

8. $\tan \dfrac{\pi}{4} + \cot \dfrac{\pi}{4} = 1 + 1 = 2$

10. $4 + \tan^2 \dfrac{\pi}{3} = 4 + \left(\sqrt{3}\right)^2 = 4 + 3 = 7$

12. $\sec^2 18° - \tan^2 18° = 1$

14. $1 + \tan^2 5° - \csc^2 85°$
$\quad = \sec^2 5° - \csc^2 85°$
$\quad = \sec^2 5° - \sec^2 5° = 0$

16. $\cos \theta = \cos 60° = \dfrac{1}{2}$

18. $\cos \dfrac{\theta}{2} = \cos \dfrac{60°}{2} = \cos 30° = \dfrac{\sqrt{3}}{2}$

20. $(\cos \theta)^2 = (\cos 60°)^2 = \left[\dfrac{1}{2}\right]^2 = \dfrac{1}{4}$

22. $2 \cos \theta = 2 \cos 60° = 2\left[\dfrac{1}{2}\right] = 1$

24. $\dfrac{\cos \theta}{2} = \dfrac{\cos 60°}{2} = \dfrac{\frac{1}{2}}{2} = \dfrac{1}{4}$

26. $\cos 14° = 0.97$

28. $\sin 15° = 0.26$

30. $\csc 55° = 1.22$

32. $\tan 80° = 5.67$

34. $\cos \dfrac{\pi}{8} = 0.92$

36. $\sin \dfrac{3\pi}{10} = 0.81$

38. $\csc \dfrac{5\pi}{13} = 1.07$

40. $\sin \dfrac{\pi}{18} = 0.17$

42. $\tan 1 = 1.56$

44. $\tan 1° = 0.02$

46. $\cos 35.2° = 0.82$

48. $\tan 0.1 = 0.10$

50. $R = \dfrac{2\left[150\ \frac{m}{\sec}\right]^2 \sin 30° \cos 30°}{9.8\ \frac{m}{\sec^2}} = \dfrac{45{,}000\frac{m^2}{\sec^2}\sin 30° \cos 30°}{9.8\frac{m}{\sec^2}} = \dfrac{45{,}000(0.4330)}{9.8}$

$\quad = 1988.3236\ m$

$H = \dfrac{(150)^2\frac{m^2}{\sec^2}\sin^2 30°}{2\left[9.8\frac{m}{\sec^2}\right]} = \dfrac{22{,}500(0.25)}{19.6} = 286.9898\ m$

52. $R = \dfrac{2\left[\dfrac{200 \text{ ft.}}{\sec}\right]^2 \sin 50° \cos 50°}{\dfrac{32.2 \text{ ft}}{\sec^2}} = \dfrac{80{,}000(0.4924)\text{ft.}}{32.2} = 1223.3637 \text{ ft.}$

$H = \dfrac{\left[\dfrac{200 \text{ ft.}}{\sec}\right]^2 \sin^2 50°}{2\left[\dfrac{32.2 \text{ ft.}}{\sec^2}\right]} = \dfrac{(40{,}000)(0.5868241)\text{ ft.}}{64.4} = 364.4870 \text{ ft.}$

54. When $\theta = 30°$, $x = \cos 30° + \sqrt{16 + 0.5\,(2\cos^2 30° - 1)}$

$= \dfrac{\sqrt{3}}{2} + \sqrt{16 + 0.5\left[2\left[\dfrac{\sqrt{3}}{2}\right]^2 - 1\right]} = \dfrac{\sqrt{3}}{2} + \sqrt{16 + 0.5\left[\dfrac{1}{2}\right]} \approx 4.897$

When $\theta = 45°$, $x = \cos 45° + \sqrt{16 + 0.5\,(2\cos^2 45° - 1)}$

$= \dfrac{\sqrt{2}}{2} + \sqrt{16 + 0.5\left[2\left[\dfrac{\sqrt{2}}{2}\right]^2 - 1\right]} = \dfrac{\sqrt{2}}{2} + \sqrt{16 + 0.5(0)} = \dfrac{\sqrt{2}}{2} + 4$

≈ 4.707

56. (a) $V = \dfrac{1}{3}\pi r^2 h$

We can see that $\tan\theta = \dfrac{h}{r}$ where r is the radius of the cone. In addition, $\triangle ABO$ is similar to $\triangle BCD$. Thus, $m\angle BOA = \theta$. So, $\cos\theta = \dfrac{R}{h - R}$.

We want to substitute for h and r in $V = \dfrac{1}{3}\pi r^2 h$.

Solving $\cos\theta = \dfrac{R}{h - R}$ for h, we get $h = R + R\sec\theta = R(1 + \sec\theta)$.

Solving $\dfrac{h}{r} = \tan\theta$ for r, we get $r = \dfrac{h}{\tan\theta}$.

Thus, $V = \dfrac{1}{3}\pi r^2 h = \dfrac{1}{3}\pi\dfrac{h^3}{\tan^2\theta} = \dfrac{1}{3}\pi R^3\dfrac{(1 + \sec\theta)^3}{\tan^2\theta}$

(b) For $\theta = 30°$, $V = \dfrac{\pi(2)^3(1 + \sec 30°)^3}{3\tan^2 30°} \approx 251.4 \text{ cm}^3$

For $\theta = 45°$, $V = \dfrac{\pi(2)^3(1 + \sec 45°)^3}{3\tan^2 45°} \approx 117.9 \text{ cm}^3$

For $\theta = 60°$, $V = \dfrac{\pi(2)^3(1 + \sec 60°)^3}{3\tan^2 60°} \approx 75.4 \text{ cm}^3$

(c)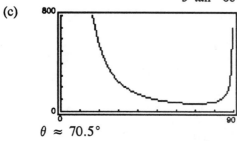

$\theta \approx 70.5°$

58.

θ	0.5	0.4	0.2	0.1	0.01	0.001	0.0001	0.00001
$\cos\theta - 1$	$-.1224$	$-.0789$	$-.0199$	$-.0050$	$-.00005$	.0000	.0000	.0000
$\dfrac{\cos\theta - 1}{\theta}$	$-.2448$	$-.1973$	$-.0997$	$-.0500$	$-.0050$	$-.0005$	$-.00005$	.0000

$\dfrac{\cos\theta - 1}{\theta}$ approaches 0 as $\theta \to 0$.

60. $\cot 1° \cdot \cot 2° \cdot \cot 3° \cdot \cdots \cdot \cot 89°$
$= \cot 1° \cdot \cot 2° \cdot \cot 3° \cdot \cdots \cdot \tan(90° - 46°) \cdot \cdots \cdot \tan(90° - 89°)$
$= (\cot 1° \cdot \tan 1°) \cdot (\cot 2° \cdot \tan 2°) \cdot \cdots \cdot (\cot 44° \cdot \tan 44°) \cdot \cot 45°$
$= 1 \cdot 1 \cdot \cdots \cdot 1 \cdot 1 = 1$

62. $\sin 1° \cdot \sin 2° \cdot \cdots \cdot \sin 45° \cdot \sec 46° \cdot \cdots \cdot \sec 89°$
$= \sin 1° \cdot \sin 2° \cdot \cdots \sin 45° \cdot \csc(90° - 46°) \cdot \cdots \cdot \csc(90° - 89°)$
$= (\sin 1° \cdot \csc 1°) \cdot (\sin 2° \cdot \csc 2°) \cdot \cdots \cdot (\sin 44° \cdot \csc 44°) \cdot \sin 45°$
$= 1 \cdot 1 \cdot \cdots \cdot 1 \cdot \dfrac{\sqrt{2}}{2} = \dfrac{\sqrt{2}}{2}$

7.4 Applications

2.
$\sin 10° = \dfrac{4}{c}$ $\qquad$ $\tan 10° = \dfrac{4}{a}$ $\qquad$ $\alpha = 90° - \beta$
$c \sin 10° = 4$ $\qquad$ $a \tan 10° = 4$ $\qquad\quad$ $= 90° - 10°$
$\qquad c = \dfrac{4}{\sin 10°}$ $\qquad\quad a = \dfrac{4}{\tan 10°}$ $\qquad \alpha = 80°$
$\qquad c \approx 23.0$ $\qquad\qquad a \approx 22.7$

4.
$\cos 50° = \dfrac{7}{c}$ $\qquad$ $\tan 50° = \dfrac{b}{7}$ $\qquad$ $\alpha = 90° - \beta$
$c \cos 50° = 7$ $\qquad$ $7 \tan 50° = b$ $\qquad\quad = 90° - 50°$
$\qquad c = \dfrac{7}{\cos 50°}$ $\qquad\quad 8.3 \approx b$ $\qquad \alpha = 40°$
$\qquad c \approx 10.89$

6.
$\cos 20° = \dfrac{6}{c}$ $\qquad$ $\tan 20° = \dfrac{a}{6}$ $\qquad$ $\beta = 90° - \alpha$
$c \cos 20° = 6$ $\qquad$ $6 \tan 20° = a$ $\qquad\quad = 90° - 20°$
$\qquad c = \dfrac{6}{\cos 20°}$ $\qquad\quad 2.2 \approx a$ $\qquad \beta = 70°$
$\qquad c \approx 6.4$

8.
$\sin 40° = \dfrac{6}{c}$ $\qquad$ $\tan 40° = \dfrac{6}{b}$ $\qquad$ $\beta = 90° - \alpha$
$c \sin 40° = 6$ $\qquad$ $b \tan 40° = 6$ $\qquad\quad = 90° - 40°$
$\qquad c = \dfrac{6}{\sin 40°}$ $\qquad\quad b = \dfrac{6}{\tan 40°}$ $\qquad \beta = 50°$
$\qquad c \approx 9.3$ $\qquad\qquad b \approx 7.2$

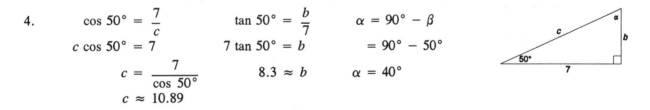

10.
$$\cos 40° = \frac{b}{10} \qquad \sin 40° = \frac{a}{10} \qquad \beta = 90° - \alpha$$
$$10 \cos 40° = b \qquad 10 \sin 40° = a \qquad = 90° - 40°$$
$$7.7 \approx b \qquad\qquad 6.4 \approx a \qquad \beta = 50°$$

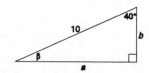

12.
$$c^2 = a^2 + b^2 = 2^2 + 8^2 = 68$$
$$c = \sqrt{68} \approx 8.25$$
$$\sin \alpha = \frac{a}{c} = \frac{2}{8.25}$$
$$\alpha = 14.0°$$
$$\beta = 90° - \alpha = 90° - 14° = 76.0°$$

14.
$$a^2 = c^2 - b^2 = 6^2 - 4^2 = 20$$
$$a = \sqrt{20} \approx 4.47$$
$$\sin \alpha = \frac{a}{c} = \frac{4.47}{6}$$
$$\alpha = 48.2°$$
$$\beta = 90° - \alpha = 90° - 48.2° = 41.8°$$

16.
$$\sin 40° = \frac{a}{2} \qquad \cos 40° = \frac{b}{2}$$
$$2 \sin 40° = a \qquad 2 \cos 40° = b$$
$$1.3 \approx a \qquad\qquad 1.5 \approx b$$

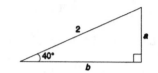

18.

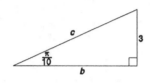

$$\sin \frac{\pi}{10} = \frac{3}{c}$$
$$c \sin \frac{\pi}{10} = 3$$
$$c = \frac{3}{\sin \frac{\pi}{10}}$$
$$c \approx \frac{3}{0.30902}$$
$$c \approx 9.71 \text{ meters}$$

or

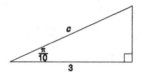

$$\cos \frac{\pi}{10} = \frac{3}{c}$$
$$c \cos \frac{\pi}{10} = 3$$
$$c = \frac{3}{\cos \frac{\pi}{10}}$$
$$c \approx \frac{3}{0.9511}$$
$$c \approx 3.15 \text{ meters}$$

20. $c = 3$. Suppose $a = 1$. Then, $\sin \alpha = \dfrac{a}{c} = \dfrac{1}{3}$, so $\alpha = 19.5°$

$\beta = 90° - \alpha \approx 90° - 19.5° \approx 70.5°$

22.

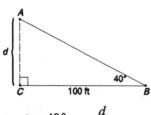

$$\tan 40° = \frac{d}{100}$$
$$100 \tan 40° = d$$
$$83.9 \text{ ft.} \approx d$$

24.

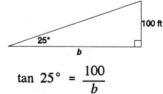

$$\tan 25° = \frac{100}{b}$$
$$b \tan 25° = 100$$
$$b = \frac{100}{\tan 25°}$$
$$b \approx 214.5 \text{ ft.}$$

Chapter 7 Trigonometric Functions

26.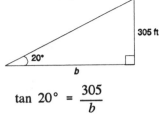

$$\tan 20° = \frac{305}{b}$$

$$b \tan 20° = 305$$

$$b = \frac{305}{\tan 20°}$$

$$b \approx 837.98 \text{ ft.}$$

28.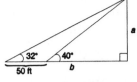

$$\cot 32° = \frac{b + 50}{a}, \ \cot 40° = \frac{b}{a}$$

$$b = a \cot 40°$$

$$\cot 32° = \frac{a \cot 40° + 50}{a}$$

$$a \cot 32° = a \cot 40° + 50$$

$$a(\cot 32° - \cot 40°) = 50$$

$$a = \frac{50}{\cot 32° - \cot 40°} \approx 122.4 \text{ ft.}$$

30. If α is the angle of elevation, then $\tan \alpha = \frac{10}{35}$, so $\alpha = 15.9°$.

32.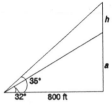

$$\tan 32° = \frac{a}{800}$$

$$\tan 35° = \frac{a + h}{800}$$

$$800 \tan 32° = a$$

$$499.9 \approx a$$

$$800 \tan 35° = a + h$$

$$560.2 \approx a + h$$

$$h = 560.2 - 499.9$$

$$h \approx 60.3 \text{ ft.}$$

34.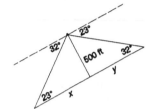

$$\tan 23° = \frac{500}{x}$$

$$x \tan 23° = 500$$

$$x = \frac{500}{\tan 23°} \approx 1178 \text{ ft.}$$

$$\tan 32° = \frac{500}{y}$$

$$y \tan 32° = 500$$

$$y = \frac{500}{\tan 32°} \approx 800 \text{ ft.}$$

$$\text{Distance} = x + y$$
$$= 1178 + 800 = 1978 \text{ ft.}$$

36.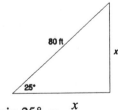

$$\sin 25° = \frac{x}{80}$$

$$x = 80 \sin 25° \approx 34 \text{ ft.}$$

38.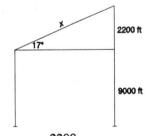

$$\sin 17° = \frac{2200}{x}$$

$$x \sin 17° = 2200$$

$$x = \frac{2200}{\sin 17°} \approx 7525 \text{ ft. or } 1.4 \text{ mi.}$$

40. Ship: First we must find the angle formed between the radius of the earth and the lighthouse, call it α. (See Figure.)

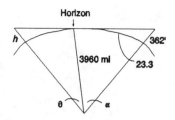

$$\cos \alpha = \frac{3960}{3960 + \dfrac{362}{5280}} = 0.99998269$$

$$\alpha = 0.00588439$$

If the distance measured is statute miles, we find the arc length from the base of the lighthouse to the horizon.

$$s = \theta r$$
$$s = 0.00588439(3960) = 23.3022 \text{ miles}$$

Thus, the distance from the horizon to the ship is $40 - 23.3022 = 16.6978$ miles. Thus,

$$s = \theta r$$
$$16.6978 = \theta(3960)$$
$$\theta = 0.00421662$$

So, $\cos 0.00421662 = \dfrac{3960}{3960 + h}$, where h is the height of the ship.

$$0.99999111 = \frac{3960}{3960 + h}$$
$$.99999111h = 0.0352044316$$
$$h = 0.0352044316 \text{ miles}$$
$$h \approx 186 \text{ feet}$$

If s is in nautical miles, then

$$23.3022 = \theta \cdot 3960$$
$$\theta = 0.0058838384 \text{ radians} = 0.337119° = 20.23'$$

So, there are 20.23 nautical miles from the lighthouse to the horizon. Thus, from the horizon to the ship is $40 - 20.23 = 19.77'$. So, $\theta = 19.77'$.

$$\cos 19.77' = \frac{3960}{3960 + h}$$
$$h \approx 345.76 \text{ feet}$$

Plane: The distance from the horizon to the plane is $120 - 23.3 = 96.7$ if the distance is measured in statute miles. So,

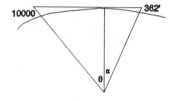

$$s = \theta r$$
$$96.7 = \theta \cdot 3960$$
$$\theta = 0.0244192$$

So, $\cos 0.0244192 = \dfrac{3960}{3960 + h}$, where h is the height of the plane.

$$h = 6235.48 \text{ feet}$$

A plane 6235.48 feet above sea level could see the lighthouse 120 miles away. It seems the brochure understates the distance if the distance is measured in statute miles.

If the distance is measured in nautical miles, then

$$120 - 20.23 = \theta \cdot 3960$$
$$\theta = 0.0251944 \text{ radians} = 1.443535°$$
$$\cos 1.443535° = \frac{3960}{3960 + h} \quad h \text{ is the height of the plane}$$
$$h \approx 1.25716 \text{ miles} \approx 6637.79 \text{ feet}$$

Again, the distance is understated.

42. (a) $\angle NOP = 100°$ and $\angle OQR = 100° + 90° = 190°$
Thus, the bearing of the ship is $190°$.

(b) After two hours, $\overline{PR} = 2(15) = 30$. Let $\theta = \angle POR$. Then,

$$\tan \theta = \frac{30}{15} = 2$$
$$\theta \approx 63.4°$$

Thus, the bearing of the ship is $100° + 63.4° = 163.4°$.

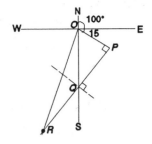

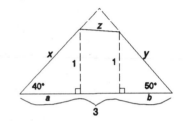

44. The height of the beam above the wall is $46 - 20 = 26$ feet. So,

$$\tan \theta = \frac{26}{10} = 2.6$$
$$\theta \approx 69°$$

46. Length of highway $= x + y + z$

$$\sin 40° = \frac{1}{x}$$
$$x = \frac{1}{\sin 40°} \approx 1.56 \text{ miles}$$
$$\sin 50° = \frac{1}{y}$$
$$y = \frac{1}{\sin 50°} \approx 1.31 \text{ miles}$$

To find z, we must find a and b.

$$\tan 40° = \frac{1}{a}, \text{ so } a = \frac{1}{\tan 40°} \approx 1.19 \text{ miles}$$
$$\tan 50° = \frac{1}{b}, \text{ so } b = \frac{1}{\tan 50°} \approx 0.84 \text{ miles}$$

Therefore, $z = 3 - a - b = 3 - 1.19 - 0.84 = 0.97$ miles.
So, the length of the highway is $1.56 + 1.31 + 0.97 = 3.84$ miles.

7.5 Trigonometric Functions of General Angles

2. $(5, -12)$, $a = 5$, $b = -12$, $r = \sqrt{25 + 144} = \sqrt{169} = 13$. Thus,

$$\sin \theta = \frac{-12}{13} \qquad \csc \theta = \frac{-13}{12}$$
$$\cos \theta = \frac{5}{13} \qquad \sec \theta = \frac{13}{5}$$
$$\tan \theta = \frac{-12}{5} \qquad \cot \theta = \frac{-5}{12}$$

4. $(-1, -2)$, $a = -1$, $b = -2$, $r = \sqrt{1 + 4} = \sqrt{5}$. Thus,

$$\sin \theta = \frac{-2}{\sqrt{5}} = \frac{-2\sqrt{5}}{5} \qquad \csc \theta = \frac{-\sqrt{5}}{2}$$

$$\cos \theta = \frac{-1}{\sqrt{5}} = \frac{-\sqrt{5}}{5} \qquad \sec \theta = -\sqrt{5}$$

$$\tan \theta = 2 \qquad \cot \theta = \frac{1}{2}$$

6. $(2, -2)$, $a = 2$, $b = -2$, $r = \sqrt{4 + 4} = \sqrt{8} = 2\sqrt{2}$. Thus,

$$\sin \theta = \frac{-2}{2\sqrt{2}} = \frac{-\sqrt{2}}{2} \qquad \csc \theta = -\sqrt{2}$$

$$\cos \theta = \frac{2}{2\sqrt{2}} = \frac{\sqrt{2}}{2} \qquad \sec \theta = \sqrt{2}$$

$$\tan \theta = -1 \qquad \cot \theta = -1$$

8. $(2, 2)$, $a = 2$, $b = 2$, $r = \sqrt{4 + 4} = 2\sqrt{2}$. Thus,

$$\sin \theta = \frac{2}{2\sqrt{2}} = \frac{\sqrt{2}}{2} \qquad \csc \theta = \sqrt{2}$$

$$\cos \theta = \frac{2}{2\sqrt{2}} = \frac{\sqrt{2}}{2} \qquad \sec \theta = \sqrt{2}$$

$$\tan \theta = \frac{2}{2} = 1 \qquad \cot \theta = 1$$

10. We know that $\sin \theta < 0$ for points P in quadrants III and IV, and $\cos \theta > 0$ for points P in quadrants I and IV. Both conditions are satisfied only if P lies in quadrant IV.

12. Since $\cos \theta > 0$ for points P in quadrants I and IV, and $\tan \theta > 0$ for points P in quadrants I and III, P lies in quadrant I.

14. Since $\sin \theta < 0$ for points P in quadrants III and IV, and $\cot \theta > 0$ for points P in quadrants I and III, P lies in quadrant III.

16. Since $\csc \theta > 0$ for points P in quadrants I and II, and $\cot \theta < 0$ for points P in quadrants II and IV, P lies in quadrant II.

18. $60°$

20. $360° - 300° = 60°$

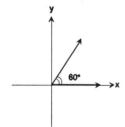

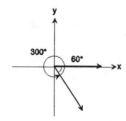

22. $360° - 330° = 30°$

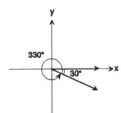

24. $\pi - \dfrac{5\pi}{6} = \dfrac{\pi}{6}$

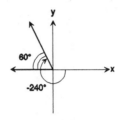

26. $2\pi - \dfrac{7\pi}{4} = \dfrac{\pi}{4}$

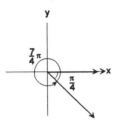

28. $-180° - (-240°) = 60°$

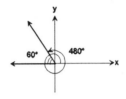

30. $-\pi - \left[\dfrac{-7\pi}{6} \right] = \dfrac{\pi}{6}$

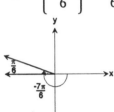

32. $490° - 360° = 130°$
    $180° - 130° = 50°$

34. $\cos 420° = \cos(360° + 60°) = \cos 60° = \dfrac{1}{2}$

36. $\sin 390° = \sin(360° + 30°) = \sin 30° = \dfrac{1}{2}$

38. $\sec 540° = \sec(360° + 180°) = \sec 180° = -1$

40. $\sec 420° = \sec(360° + 60°) = \sec 60° = 2$

42. $\sin \dfrac{9\pi}{4} = \sin\left[\dfrac{\pi}{4} + \dfrac{8\pi}{4} \right] = \sin\left[\dfrac{\pi}{4} + 2\pi \right] = \sin \dfrac{\pi}{4} = \dfrac{\sqrt{2}}{2}$

44. $\csc \dfrac{9\pi}{2} = \csc\left[\dfrac{\pi}{2} + \dfrac{8\pi}{2} \right] = \csc\left[\dfrac{\pi}{2} + 2\pi \cdot 2 \right] = \csc \dfrac{\pi}{2} = 1$

46. $\cot \dfrac{17\pi}{4} = \cot\left[\dfrac{\pi}{4} + \dfrac{16\pi}{4} \right] = \cot\left[\dfrac{\pi}{4} + 2\pi \cdot 2 \right] = \cot \dfrac{\pi}{4} = 1$

48. $\sec \dfrac{25\pi}{6} = \sec\left[\dfrac{\pi}{6} + \dfrac{24\pi}{6} \right] = \sec\left[\dfrac{\pi}{6} + 2\pi \cdot 2 \right] = \sec \dfrac{\pi}{6} = \dfrac{2\sqrt{3}}{3}$

50. $\cos 210° = -\cos 30° = \dfrac{-\sqrt{3}}{2}$

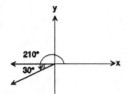

52. $\sin 120° = \sin 60° = \dfrac{\sqrt{3}}{2}$

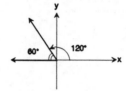

54. $\csc 300° = -\csc 60° = \dfrac{-2}{\sqrt{3}} = \dfrac{-2\sqrt{3}}{3}$

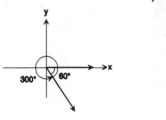

56. $\tan 225° = \tan 45° = 1$

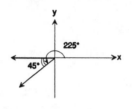

58. $\cos \dfrac{2\pi}{3} = -\cos \dfrac{\pi}{3} = \dfrac{-1}{2}$

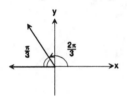

60. $\csc \dfrac{7\pi}{4} = -\csc \dfrac{\pi}{4} = -\sqrt{2}$

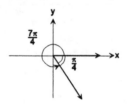

62. $\tan(-120°) = \tan 60° = \sqrt{3}$

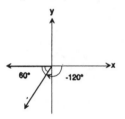

64. $\cot\left[\dfrac{-\pi}{6}\right] = -\cot \dfrac{\pi}{6} = -\sqrt{3}$

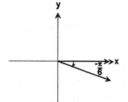

66. $\sec \dfrac{11\pi}{4} = -\sec \dfrac{\pi}{4} = -\sqrt{2}$

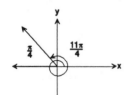

68. $\sec(-225°) = -\sec 45° = -\sqrt{2}$

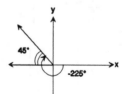

Chapter 7 Trigonometric Functions

70. $\cos \theta = \dfrac{3}{5}$, $270° < \theta < 360°$; $\sin \theta = \pm\sqrt{1 - \cos^2 \theta}$

Since $270° < \theta < 360°$, $\sin \theta < 0$

$$\sin \theta = -\sqrt{1 - \left(\dfrac{3}{5}\right)^2} = -\sqrt{1 - \dfrac{9}{25}} = -\sqrt{\dfrac{16}{15}} = \dfrac{-4}{5}$$

$$\tan \theta = \dfrac{\sin \theta}{\cos \theta} = \dfrac{\dfrac{-4}{5}}{\dfrac{3}{5}} = \dfrac{-4}{3} \qquad\qquad \csc \theta = \dfrac{1}{\sin \theta} = \dfrac{1}{\dfrac{-4}{5}} = \dfrac{-5}{4}$$

$$\sec \theta = \dfrac{1}{\cos \theta} = \dfrac{1}{\dfrac{3}{5}} = \dfrac{5}{3} \qquad\qquad \cot \theta = \dfrac{1}{\tan \theta} = \dfrac{1}{\dfrac{-4}{3}} = \dfrac{-3}{4}$$

72. $\sin \theta = \dfrac{-5}{13}$, $\pi < \theta < \dfrac{3\pi}{2}$; $\cos \theta = \pm\sqrt{1 - \sin^2 \theta}$

Since $\pi < \theta < \dfrac{3\pi}{2}$, $\cos \theta < 0$

$$\cos \theta = -\sqrt{1 - \left(\dfrac{-5}{13}\right)^2} = -\sqrt{1 - \dfrac{25}{169}} = \sqrt{\dfrac{144}{169}} = \dfrac{-12}{13}$$

$$\tan \theta = \dfrac{\sin \theta}{\cos \theta} = \dfrac{\dfrac{-5}{13}}{\dfrac{-12}{13}} = \dfrac{5}{12} \qquad\qquad \csc \theta = \dfrac{1}{\sin \theta} = \dfrac{1}{\dfrac{-5}{13}} = \dfrac{-13}{5}$$

$$\sec \theta = \dfrac{1}{\cos \theta} = \dfrac{1}{\dfrac{-12}{13}} = \dfrac{-13}{12} \qquad\qquad \cot \theta = \dfrac{1}{\tan \theta} = \dfrac{1}{\dfrac{5}{12}} = \dfrac{12}{5}$$

74. $\cos \theta = \dfrac{4}{5}$, $\sin \theta < 0$. First, we solve for $\sin \theta$:

$$\sin^2 \theta + \cos^2 \theta = 1$$
$$\sin^2 \theta = 1 - \cos^2 \theta$$

$$\sin \theta = -\sqrt{1 - \cos^2 \theta} = -\sqrt{1 - \left(\dfrac{4}{5}\right)^2} = -\sqrt{1 - \dfrac{16}{25}} = \dfrac{-3}{5}$$

$$\tan \theta = \dfrac{\sin \theta}{\cos \theta} = \dfrac{\dfrac{-3}{5}}{\dfrac{4}{5}} = \dfrac{-3}{4} \qquad\qquad \cot \theta = \dfrac{1}{\tan \theta} = \dfrac{1}{\dfrac{-3}{4}} = \dfrac{-4}{3}$$

$$\csc \theta = \dfrac{1}{\sin \theta} = \dfrac{1}{\dfrac{-3}{5}} = \dfrac{-5}{3} \qquad\qquad \sec \theta = \dfrac{1}{\cos \theta} = \dfrac{1}{\dfrac{4}{5}} = \dfrac{5}{4}$$

76. $\sin\theta = \dfrac{-2}{3}$, $\sec\theta > 0$

Because $\sec\theta = \dfrac{1}{\cos\theta} > 0$, $\cos\theta > 0$.

To solve for $\cos\theta$, we use $\sin^2\theta + \cos^2\theta = 1$

$$\cos^2\theta = 1 - \sin^2\theta$$

$$\cos\theta = \sqrt{1 - \sin^2\theta} = \sqrt{1 - \left[\dfrac{-2}{3}\right]^2} = \sqrt{1 - \dfrac{4}{9}} = \dfrac{\sqrt{5}}{3}$$

$$\tan\theta = \dfrac{\sin\theta}{\cos\theta} = \dfrac{\dfrac{-2}{3}}{\dfrac{\sqrt{5}}{3}} = \dfrac{-2\sqrt{5}}{5} \qquad \cot\theta = \dfrac{1}{\tan\theta} = \dfrac{1}{\dfrac{-2\sqrt{5}}{5}} = \dfrac{-\sqrt{5}}{2}$$

$$\csc\theta = \dfrac{1}{\sin\theta} = \dfrac{1}{\dfrac{-2}{3}} = \dfrac{-3}{2} \qquad \sec\theta = \dfrac{1}{\cos\theta} = \dfrac{1}{\dfrac{\sqrt{5}}{3}} = \dfrac{3\sqrt{5}}{5}$$

78. $\cos\theta = \dfrac{-1}{4}$, $\tan\theta > 0$

Because $\tan\theta = \dfrac{\sin\theta}{\cos\theta} > 0$ and $\cos\theta < 0$, it follows that $\sin\theta < 0$.

We solve for $\sin\theta$:

$$\sin\theta = -\sqrt{1 - \cos^2\theta} = -\sqrt{1 - \left[\dfrac{-1}{4}\right]^2} = -\sqrt{1 - \dfrac{1}{16}} = \dfrac{-\sqrt{15}}{4}$$

$$\csc\theta = \dfrac{1}{\sin\theta} = \dfrac{1}{\dfrac{-\sqrt{15}}{4}} = \dfrac{-4\sqrt{15}}{15} \qquad \sec\theta = \dfrac{1}{\cos\theta} = \dfrac{1}{\dfrac{-1}{4}} = -4$$

$$\tan\theta = \dfrac{\sin\theta}{\cos\theta} = \dfrac{\dfrac{-\sqrt{15}}{4}}{\dfrac{-1}{4}} = \sqrt{15} \qquad \cot\theta = \dfrac{1}{\tan\theta} = \dfrac{1}{\sqrt{15}} = \dfrac{\sqrt{15}}{15}$$

80. $\csc\theta = 3$, $\cot\theta < 0$

$$\cot^2\theta + 1 = \csc^2\theta$$
$$\cot^2\theta = \csc^2\theta - 1$$

$$\cot\theta = -\sqrt{\csc^2\theta - 1} = -\sqrt{3^2 - 1} = -\sqrt{8} = -2\sqrt{2}$$

$$\tan\theta = \dfrac{1}{\cot\theta} = \dfrac{1}{-2\sqrt{2}} = \dfrac{-\sqrt{2}}{4} \qquad \sin\theta = \dfrac{1}{\csc\theta} = \dfrac{1}{3}$$

Because $\cot\theta = \dfrac{\cos\theta}{\sin\theta} < 0$ and $\sin\theta > 0$, it follows that $\cos\theta < 0$. Therefore, we use the minus sign:

$$\cos\theta = -\sqrt{1 - \sin^2\theta} = -\sqrt{1 - \left[\dfrac{1}{3}\right]^2} = -\sqrt{1 - \dfrac{1}{9}} = -\sqrt{\dfrac{8}{9}} = \dfrac{-2\sqrt{2}}{3}$$

$$\sec = \dfrac{1}{\cos\theta} = \dfrac{1}{\dfrac{-2\sqrt{2}}{3}} = \dfrac{-3\sqrt{2}}{4}$$

82. $\cot \theta = \dfrac{4}{3}$, $\cos \theta < 0$

$$\csc^2 \theta = \cot^2 \theta + 1$$

$$\csc \theta = \pm\sqrt{\cot^2 \theta + 1}$$

Because $\cot \theta = \dfrac{\cos \theta}{\sin \theta} = \dfrac{4}{3} > 0$ and $\cos \theta < 0$, it follows that $\sin \theta < 0$.

Because $\sin \theta = \dfrac{1}{\csc \theta} < 0$, it follows that $\csc \theta < 0$. Therefore, we use the minus sign:

$$\csc \theta = -\sqrt{\cot^2 \theta + 1} = -\sqrt{\left[\dfrac{4}{3}\right]^2 + 1} = \sqrt{\dfrac{16}{9} + 1} = \dfrac{-5}{3}$$

$$\sin \theta = \dfrac{1}{\csc \theta} = \dfrac{1}{\dfrac{-5}{3}} = \dfrac{-3}{5} \qquad \tan \theta = \dfrac{1}{\cot \theta} = \dfrac{1}{\dfrac{4}{3}} = \dfrac{3}{4}$$

$$\cos \theta = -\sqrt{1 - \sin^2 \theta} = -\sqrt{1 - \left[\dfrac{-3}{5}\right]^2} = -\sqrt{1 - \dfrac{9}{25}} = \dfrac{-4}{5}$$

$$\sec \theta = \dfrac{1}{\cos \theta} = \dfrac{1}{\dfrac{-4}{5}} = \dfrac{-5}{4}$$

84. $\sec \theta = -2$, $\tan \theta > 0$

$$\tan^2 \theta + 1 = \sec^2 \theta$$

$$\tan \theta = \sqrt{\sec^2 \theta - 1} = \sqrt{(-2)^2 - 1} = \sqrt{3}$$

$$\cos \theta = \dfrac{1}{\sec \theta} = \dfrac{1}{-2} = \dfrac{-1}{2}$$

$$\sin \theta = \pm\sqrt{1 - \cos^2 \theta}$$

Because $\tan \theta = \dfrac{\sin \theta}{\cos \theta} > 0$ and $\cos \theta < 0$, it follows that $\sin \theta < 0$. Therefore, we use the minus sign:

$$\sin \theta = -\sqrt{1 - \left[\dfrac{-1}{2}\right]^2} = -\sqrt{1 - \dfrac{1}{4}} = \dfrac{-\sqrt{3}}{2}$$

$$\csc \theta = \dfrac{1}{\sin \theta} = \dfrac{1}{\dfrac{-\sqrt{3}}{2}} = \dfrac{-2\sqrt{3}}{3} \qquad \cot \theta = \dfrac{1}{\tan \theta} = \dfrac{1}{\sqrt{3}} = \dfrac{\sqrt{3}}{3}$$

86. $\tan 60° + \tan 150° = \sqrt{3} - \dfrac{1}{\sqrt{3}} = \sqrt{3} - \dfrac{\sqrt{3}}{3} = \dfrac{2}{3}\sqrt{3}$

88. $\cos(\theta + \pi) = -0.4$

90. $\cot(\theta + \pi) = -2$

92. $\sec \theta = \dfrac{3}{2}$

94. $\cos 1° + \cos 2° + \cos 3° + \cdots + \cos 358° + \cos 359°$
$$= (\cos 1° + \cos 179°) + (\cos 2° + \cos 178°) + \cdots + (\cos 89° + \cos 91°) + \cos 90°$$
$$+ \cos 180° + \cos 270° + (\cos 181° + \cos 359°) + (\cos 182° + \cos 358°)$$
$$+ \cdots + (\cos 269° + \cos 275°) = -1$$

Since $\cos 1° + \cos 179° = \cos 1° - \cos 1° = 0$ and $\cos 180° = -1$

2. $P = \left[\dfrac{\sqrt{2}}{2}, \dfrac{\sqrt{2}}{2} \right] = (a, b)$

$\sin t = b = \dfrac{\sqrt{2}}{2}$ $\csc t = \dfrac{1}{b} = \dfrac{1}{\dfrac{\sqrt{2}}{2}} = \dfrac{2\sqrt{2}}{2} = \sqrt{2}$

$\cos t = a = \dfrac{\sqrt{2}}{2}$ $\sec t = \dfrac{1}{a} = \dfrac{1}{\dfrac{\sqrt{2}}{2}} = \sqrt{2}$

$\tan t = \dfrac{b}{a} = \dfrac{\dfrac{\sqrt{2}}{2}}{\dfrac{\sqrt{2}}{2}} = 1$ $\cot t = \dfrac{a}{b} = \dfrac{\dfrac{\sqrt{2}}{2}}{\dfrac{\sqrt{2}}{2}} = 1$

4. $P = \left[\dfrac{\sqrt{2}}{2}, \dfrac{-\sqrt{2}}{2} \right] = (a, b)$

$\sin t = b = \dfrac{-\sqrt{2}}{2}$ $\csc t = \dfrac{1}{b} = \dfrac{1}{\dfrac{-\sqrt{2}}{2}} = -\sqrt{2}$

$\cos t = a = \dfrac{\sqrt{2}}{2}$ $\sec t = \dfrac{1}{a} = \dfrac{1}{\dfrac{\sqrt{2}}{2}} = \sqrt{2}$

$\tan t = \dfrac{b}{a} = \dfrac{\dfrac{\sqrt{2}}{2}}{\dfrac{-\sqrt{2}}{2}} = -1$ $\cot t = \dfrac{a}{b} = \dfrac{\dfrac{-\sqrt{2}}{2}}{\dfrac{\sqrt{2}}{2}} = -1$

6. $P = \left[\dfrac{-12}{13}, \dfrac{5}{13} \right] = (a, b)$

$\sin t = b = \dfrac{5}{13}$ $\csc t = \dfrac{1}{b} = \dfrac{13}{5}$

$\cos t = a = \dfrac{-12}{13}$ $\sec t = \dfrac{1}{a} = \dfrac{-13}{12}$

$\tan t = \dfrac{b}{a} = \dfrac{\dfrac{5}{13}}{\dfrac{-12}{13}} = \dfrac{-5}{12}$ $\cot t = \dfrac{a}{b} = \dfrac{-12}{5}$

8. $P = \left(\dfrac{-2\sqrt{5}}{5}, \dfrac{-\sqrt{5}}{5} \right) = (a, b)$

$\sin t = b = \dfrac{-\sqrt{5}}{5}$ $\qquad\qquad$ $\csc t = \dfrac{1}{b} = -\sqrt{5}$

$\cos t = a = \dfrac{-2\sqrt{5}}{5}$ $\qquad\qquad$ $\sec t = \dfrac{1}{a} = \dfrac{-5}{2\sqrt{5}} = \dfrac{-\sqrt{5}}{2}$

$\tan t = \dfrac{b}{a} = \dfrac{\frac{-\sqrt{5}}{5}}{\frac{-2\sqrt{5}}{5}} = \dfrac{1}{2}$ $\qquad$ $\cot t = \dfrac{a}{b} = \dfrac{\frac{-2\sqrt{5}}{5}}{\frac{-\sqrt{5}}{5}} = 2$

10. $\cos \theta = \dfrac{1}{3}, \dfrac{3\pi}{2} < \theta < 2\pi$

$a^2 + b^2 = r^2$ $\qquad\qquad$ In quadrant IV, $\cos \theta$ and $\sec \theta$ are positive. The other trig functions
$1^2 + b^2 = 3^2$ $\qquad\qquad$ are negative.
$\quad b^2 = 8$

$\qquad b = 2\sqrt{2}$

$\sin \theta = \dfrac{-2\sqrt{2}}{3}$ $\qquad\qquad$ $\csc \theta = \dfrac{-3}{2\sqrt{2}} = \dfrac{-3\sqrt{2}}{4}$

$\tan \theta = -2\sqrt{2}$ $\qquad\qquad$ $\cot \theta = \dfrac{-1}{2\sqrt{2}} = \dfrac{-\sqrt{2}}{4}$ $\qquad\qquad$ $\sec \theta = 3$

12. $\sec \theta = 3 = \dfrac{r}{a}$ $\qquad\qquad$ $0 < \theta < \dfrac{\pi}{2}$

$a^2 + b^2 = r^2$ $\qquad\qquad$ In quadrant I, all six trig functions are positive.
$1^2 + b^2 = 3^2$

$\qquad b = 2\sqrt{2}$

$\sin \theta = \dfrac{2\sqrt{2}}{3}$ $\qquad\qquad$ $\csc \theta = \dfrac{3}{2\sqrt{2}} = \dfrac{3\sqrt{2}}{4}$ $\qquad\qquad$ $\cos \theta = \dfrac{1}{3}$

$\tan \theta = 2\sqrt{2}$ $\qquad\qquad$ $\cot \theta = \dfrac{1}{2\sqrt{2}} = \dfrac{\sqrt{2}}{4}$

14. $\cos \theta = \dfrac{-3}{4}, \sin \theta < 0$

The only quadrant where $\cos \theta$ and $\sin \theta$ are negative is in quadrant III. Only $\tan \theta$ and $\cot \theta$ are positive.

$a^2 + b^2 = r^2$
$3^2 + b^2 = 4^2$
$\qquad b^2 = 7$

$\qquad b = \sqrt{7}$

$\sin \theta = \dfrac{-\sqrt{7}}{4}$ $\qquad\qquad$ $\csc \theta = \dfrac{-4}{\sqrt{7}} = \dfrac{-4\sqrt{7}}{7}$

$\tan \theta = \dfrac{\sqrt{7}}{3}$ $\qquad\qquad$ $\cot \theta = \dfrac{3}{\sqrt{7}} = \dfrac{3\sqrt{7}}{7}$ $\qquad\qquad$ $\sec \theta = \dfrac{-4}{3}$

16. $\csc \theta = -4$, $\tan \theta > 0$
The only quadrant where csc is negative and $\tan \theta$ is positive is in quadrant III.
$$a^2 + b^2 = r^2$$
$$a^2 + 1^2 = 4^2$$
$$a^2 = 15$$
$$a = \sqrt{15}$$

$\cos \theta = \dfrac{-\sqrt{15}}{4}$ $\sec \theta = \dfrac{-4}{\sqrt{15}} = \dfrac{-4\sqrt{15}}{15}$

$\tan \theta = \dfrac{1}{\sqrt{15}} = \dfrac{\sqrt{15}}{15}$ $\cot \theta = \dfrac{\sqrt{15}}{1} = \sqrt{15}$ $\sin \theta = \dfrac{-1}{4}$

18. The domain of the cosine function is all real numbers.

20. $f(\theta) = \cot \theta$ is not defined for integral multiples of π (180°).

22. $f(\theta) = \csc \theta$ is not defined for integral multiples of π (180°).

24. The range of the cosine function are values between -1 and 1 and including -1 and 1. We write $-1 \le \cos \theta \le 1$ or $[-1, 1]$.

26. The range of the cotangent function is all real numbers.

28. The range of the cosecant function includes values greater than or equal to one and values less than or equal to negative one. We write $(-\infty, -1] \cup [1, \infty)$.

30. The cosine function is even. Its graph is symmetric with respect to the y-axis.

32. The cotangent function is odd. Its graph is symmetric with respect to the origin.

34. The cosecant function is odd. Its graph is symmetric with respect to the origin.

36. Let $P = (x, y)$ be the point on the unit circle that corresponds to the angle θ.

Consider that equation $\cot \theta = \dfrac{x}{y} = a$. Then $x = ay$. But $x^2 + y^2 = 1$ so that $a^2y^2 + y^2 = 1$.

Thus, $x = \pm \dfrac{a}{\sqrt{1 + a^2}}$, $y = \pm \dfrac{1}{\sqrt{1 + a^2}}$; that is, for any real number a, there is a point $P = (x, y)$ on the unit circle for which $\cot \theta = a$. In other words, $-\infty < \cot \theta < \infty$, and the range of the cotangent function is the set of all real numbers.

38. Suppose there is a number p, $0 < p < 2\pi$, for which $\cos(\theta + p) = \cos \theta$ for all θ. If $\theta = \dfrac{\pi}{2}$, then

$\cos\left(\dfrac{\pi}{2} + p\right) = \cos \dfrac{\pi}{2} = 0$, so that $p = \pi$. If $\theta = 0$, then $\cos(0 + p) = \cos 0$. But $p = \pi$.
Thus, $\cos \pi = -1 = \cos 0 = 1$. This is impossible. The smallest positive number p for which $\cos(\theta + p) = \cos \theta$ for all θ is therefore 2π.

40. $\csc \theta = \dfrac{1}{\sin \theta}$; since $\sin \theta$ has period 2π, so does $\csc \theta$.

42. If $P = (a, b)$ is the point on the unit circle corresponding to the angle θ, then $Q = (-a, -b)$ is the point on the unit circle corresponding to $\theta + \pi$.

Thus, $\cot(\theta + \pi) = \dfrac{-a}{-b} = \dfrac{a}{b} = \cot\theta$; that is, the period of the cotangent function is π.

7 Chapter Review

2. $210° \cdot \dfrac{\pi}{180} = \dfrac{7\pi}{6}$ radians

4. $15° \cdot \dfrac{\pi}{180} = \dfrac{5\pi}{60} = \dfrac{\pi}{12}$ radians

6. $\dfrac{2\pi}{3} \cdot \dfrac{180}{\pi} = 120°$

8. $\dfrac{-3\pi}{2} \cdot \dfrac{180}{\pi} = -270°$

10. $\cos\dfrac{\pi}{3} + \sin\dfrac{\pi}{2} = \dfrac{1}{2} + 1 = \dfrac{3}{2}$

12. $4\cos 60° + 3\tan\dfrac{\pi}{3} = 4\left(\dfrac{1}{2}\right) + 3\left(\sqrt{3}\right) = 2 + 3\sqrt{3}$

14. $3\sin\dfrac{2\pi}{3} - 4\cos\dfrac{5\pi}{2} = 3\left(\dfrac{\sqrt{3}}{2}\right) - 4(0) = \dfrac{3\sqrt{3}}{2}$

16. $4\csc\dfrac{3\pi}{4} - \cot\left(\dfrac{-\pi}{4}\right) = 4\left(\sqrt{2}\right) - (-1) = 4\sqrt{2} + 1$

18. $\cos\dfrac{\pi}{2} - \csc\left(-\dfrac{\pi}{2}\right) = 0 - (-1) = 1$

20. $\sin 270° + \cos(-180°) = -1 + -1 = -2$

22. $\dfrac{1}{\cos^2 40°} - \dfrac{1}{\cot^2 40°} = \sec^2 40° - \tan^2 40° = 1$ (Since $\tan^2\theta + 1 = \sec^2\theta$)

24. $\tan 10° \cot 10° = \tan 10° \cdot \dfrac{1}{\tan 10°} = 1$

26. $\dfrac{\tan 20°}{\cot 70°} = \dfrac{\tan 20°}{\tan(90° - 70°)} = \dfrac{\tan 20°}{\tan 20°} = 1$

28. $\tan(-20°)\cot 20° = -\tan 20° \cdot \dfrac{1}{\tan 20°} = -1$

30. $\cot 200° \cot(-70°) = -\cot 20° \cot 70° = -\tan(90° - 20°)\cot 70° = -\tan 70° \cot 70°$

$= -\tan 70° \cdot \dfrac{1}{\tan 70°} = -1$

32. $\cos \theta = \dfrac{-3}{5}$, $\sin \theta < 0$. First, we solve for $\sin \theta$:

$$\sin^2 \theta = 1 - \cos^2 \theta$$

$$\sin \theta = -\sqrt{1 - \cos^2 \theta} = -\sqrt{1 - \left[\dfrac{-3}{5}\right]^2} = -\sqrt{1 - \dfrac{9}{25}} = \dfrac{-4}{5}$$

$$\tan \theta = \dfrac{\sin \theta}{\cos \theta} = \dfrac{\dfrac{-4}{5}}{\dfrac{-3}{5}} = \dfrac{4}{3} \qquad \csc \theta = \dfrac{1}{\sin \theta} = \dfrac{1}{\dfrac{-4}{5}} = \dfrac{-5}{4}$$

$$\sec \theta = \dfrac{1}{\cos \theta} = \dfrac{1}{\dfrac{-3}{5}} = \dfrac{-5}{3} \qquad \cot \theta = \dfrac{1}{\tan \theta} = \dfrac{1}{\dfrac{4}{3}} = \dfrac{3}{4}$$

34. $\cot \theta = \dfrac{12}{5}$, $\cos \theta < 0$

Because $\cot \theta = \dfrac{\cos \theta}{\sin \theta} > 0$ and $\cos \theta < 0$, it follows that $\sin \theta < 0$.

Since $\sin \theta = \dfrac{1}{\csc \theta}$, it follows that $\csc \theta < 0$.

$$\csc^2 \theta = \cot^2 \theta + 1$$

$$\csc \theta = -\sqrt{\cot^2 \theta + 1} = -\sqrt{\left[\dfrac{12}{5}\right]^2 + 1} = -\sqrt{\dfrac{144}{25} + 1} = -\sqrt{\dfrac{169}{25}} = \dfrac{-13}{5}$$

$$\sin \theta = \dfrac{1}{\csc \theta} = \dfrac{1}{\dfrac{-13}{5}} = \dfrac{-5}{13}$$

$$\cos \theta = -\sqrt{1 - \sin^2 \theta} = -\sqrt{1 - \left[\dfrac{-5}{13}\right]^2} = -\sqrt{1 - \dfrac{25}{169}} = -\sqrt{\dfrac{144}{169}} = \dfrac{-12}{13}$$

$$\sec \theta = \dfrac{1}{\cos \theta} = \dfrac{1}{\dfrac{-12}{13}} = \dfrac{-13}{12} \qquad \tan \theta = \dfrac{1}{\cot \theta} = \dfrac{1}{\dfrac{12}{5}} = \dfrac{5}{12}$$

36. $\csc \theta = \dfrac{-5}{3}$, $\cot \theta < 0$

$$\cot^2 \theta + 1 = \csc^2 \theta$$
$$\cot^2 \theta = \csc^2 \theta - 1$$

$$\cot \theta = \pm\sqrt{\csc^2 \theta - 1}$$

Because $\cot \theta < 0$, we use the minus sign:

$$\cot \theta = -\sqrt{\csc^2 \theta - 1} = -\sqrt{\left[\dfrac{-5}{3}\right]^2 - 1} = -\sqrt{\dfrac{25}{9} - 1} = -\sqrt{\dfrac{16}{9}} = \dfrac{-4}{3}$$

$$\tan \theta = \dfrac{1}{\cot \theta} = \dfrac{1}{\dfrac{-4}{3}} = \dfrac{-3}{4}$$

$$\sin \theta = \dfrac{1}{\csc \theta} = \dfrac{1}{\dfrac{-5}{3}} = \dfrac{-3}{5} \qquad \cos \theta = \pm\sqrt{1 - \sin^2 \theta}$$

Since $\cot \theta = \dfrac{\cos \theta}{\sin \theta} < 0$ and $\sin \theta < 0$, it follows that $\cos \theta > 0$.

$$\cos \theta = \sqrt{1 - \left[\dfrac{-3}{5}\right]^2} = \sqrt{1 - \dfrac{9}{25}} = \dfrac{4}{5} \qquad \sec \theta = \dfrac{1}{\cos \theta} = \dfrac{1}{\dfrac{4}{5}} = \dfrac{5}{4}$$

38. $\cos \theta = \dfrac{-3}{5}$, θ in quadrant III

   $\sin \theta = \dfrac{-4}{5}$ $\csc \theta = \dfrac{-5}{4}$

   $\cos \theta = \dfrac{-3}{5}$ $\sec \theta = \dfrac{-5}{3}$

   $\tan \theta = \dfrac{4}{3}$ $\cot \theta = \dfrac{3}{4}$

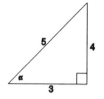

40. $\cos \theta = \dfrac{12}{13}$, $\dfrac{3\pi}{2} < \theta < 2\pi$

   $\sin \theta = \dfrac{-5}{13}$ $\csc \theta = \dfrac{-13}{5}$

   $\cos \theta = \dfrac{12}{13}$ $\sec \theta = \dfrac{13}{12}$

   $\tan \theta = \dfrac{-5}{12}$ $\cot \theta = \dfrac{-12}{5}$

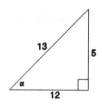

42. $\tan \theta = \dfrac{-2}{3}$, $90° < \theta < 180°$

   $\sin \theta = \dfrac{2\sqrt{13}}{13}$ $\csc \theta = \dfrac{\sqrt{13}}{2}$

   $\cos \theta = \dfrac{-3\sqrt{13}}{13}$ $\sec \theta = \dfrac{-\sqrt{13}}{3}$

   $\tan \theta = \dfrac{-2}{3}$ $\cot \theta = \dfrac{-3}{2}$

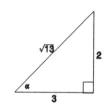

44. $\csc \theta = -4$, $\pi < \theta < \dfrac{3\pi}{2}$

   $\sin \theta = \dfrac{-1}{4}$ $\csc \theta = -4$

   $\cos \theta = \dfrac{-\sqrt{15}}{4}$ $\sec \theta = \dfrac{-4\sqrt{15}}{15}$

   $\tan \theta = \dfrac{\sqrt{15}}{15}$ $\cot \theta = \sqrt{15}$

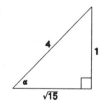

46. $\tan\theta = -2$, $\dfrac{3\pi}{2} < \theta < 2\pi$

$$\sin\theta = \frac{-2}{\sqrt{5}} = \frac{-2\sqrt{5}}{5} \qquad \csc\theta = \frac{-\sqrt{5}}{2}$$

$$\cos\theta = \frac{1}{\sqrt{5}} = \frac{\sqrt{5}}{5} \qquad \sec\theta = \sqrt{5}$$

$$\tan\theta = -2 \qquad \cot\theta = \frac{-1}{2}$$

48. $\beta = 90° - 35° = 55°$

$\sin 35° = \dfrac{5}{c}$ so $c = \dfrac{5}{\sin 35°} \approx 8.72$

$b^2 = c^2 - a^2 = (8.72)^2 - 5^2 \approx 50.99$

$b \approx 7.14$

50. $c^2 = a^2 + b^2 = 3^2 + 1^2 = 10$

$c \approx 3.16$

$\tan\alpha = \dfrac{3}{1}$, so $\alpha = 71.6°$

$\beta = 90° - \alpha = 18.4°$

52. $s = r\theta$

In 30 minutes $\theta = \pi$, $r = 8$ inches.

Hence, $s = 8\pi$ inches.

In 20 minutes $\theta = \dfrac{2\pi}{3}$.

Hence, $s = 8\left[\dfrac{2\pi}{3}\right] = \dfrac{16\pi}{3}$ inches.

54. $V = r\omega$

150 mph $= \dfrac{1}{4}$ mile ω

$\omega = 600\,\dfrac{\text{rad}}{\text{hour}} = 600\,\dfrac{\text{rad}}{\text{hour}} \cdot \dfrac{1\text{ rev}}{2\pi\text{ rad}}$

$= 95.5$ rev/hour

56.

$\tan 40° = \dfrac{b_1}{200}$

$200\tan 40° = b_1$

167.82 feet $\approx b_1$

$\tan 10° = \dfrac{b_2}{200}$

$200\tan 10° = b_2$

35.27 feet $\approx b_2$

$b \approx 167.82 - 35.27 = 132.55$ ft.

$\text{speed} \approx \dfrac{132.55\text{ ft.}}{\text{min}} \cdot \dfrac{1\text{ mi}}{5280\text{ ft.}} \cdot \dfrac{60\text{ minutes}}{1\text{ hr.}} \approx \dfrac{1.5\text{ miles}}{\text{hour}}$

58.

$\tan 25° = \dfrac{h}{80}$

$80\tan 25° = h$

37.3 ft. $\approx h$

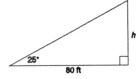

60. $\sin\theta = \dfrac{900}{4100}$

$\theta \approx 12.7°$

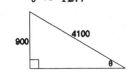

1. This is due to the fact that the earth is round and higher heights are necessary to see further distances.

2. Circumference $= 2\pi r = 2\pi(3960 \text{ miles}) = 7920\pi$ miles ≈ 24881.4 miles
    Distances from point to point on the surface of the earth are measured in arc length, which is a fraction of the circumference. For example, $\frac{1}{4}$ around the earth is $\frac{1}{4}(24881.4 \text{ miles})$.

3. Let $\alpha =$ the angle formed at the center of the earth. Then,

    $$\tan \alpha = \frac{190}{3960}$$
    $$\alpha \approx 0.047943$$

    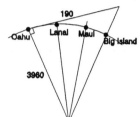

4. $$\cos 0.047943 = \frac{3960}{h}$$
    $$h \approx 3964.555 \text{ miles}$$
    So, to see Mauna Kea, it must be 4.555 miles or 24052.8 feet tall. Thus, you cannot see Mauna Kea from Oahu.

5. Lanaihale: $$\tan \alpha = \frac{65}{3960}$$
    $$\alpha = 0.0164$$
    $$\cos 0.0164 = \frac{3960}{h}$$
    $$h \approx 3960.533424$$
    So, Lanaihale must be 0.533424 miles or 2816.5 feet high to see it from Oahu. Thus, Lanaihale is visible from Oahu.

    Haleakala: $$\tan \alpha = \frac{110}{3960}$$
    $$\alpha = 0.02777$$
    $$\cos 0.02777 = \frac{3960}{h}$$
    $$h \approx 3961.5275$$
    So, Haleakala must be 1.5275 miles or 8065.1 feet high to see it from Oahu. Thus, Haleakala is visible from Oahu.

6. $$\tan \alpha = \frac{40}{3960}$$
    $$\alpha \approx 0.0101 \text{ radians}$$
    $$\cos 0.0101 = \frac{3960}{h}$$
    $$h \approx 3960.202015 \text{ miles}$$
    So, Kamakou must be 0.202015 miles or 1066.6 feet high to see it from Oahu. Thus, Kamakou is visible from Oahu.

7. The three volcanic peaks are Haleakala, Kamakou, and Lanaihale. Consult a map.

GRAPHS OF TRIGONOMETRIC FUNCTIONS

2. 1

4. The graph of $y = \cos x$ is decreasing for $0 \le x \le \pi$.

6. -1

8. For numbers $\dfrac{\pi}{2}$, $\dfrac{3\pi}{2}$, $\cos x = 0$

10. For $x = -2\pi, 0, 2\pi$, $\cos x = 1$
    For $x = -\pi, \pi$, $\cos x = -1$

12.
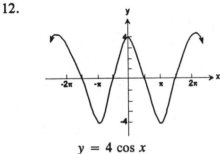
$$y = 4 \cos x$$

14.
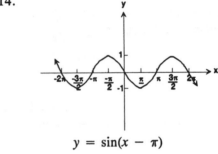
$$y = \sin(x - \pi)$$

16.
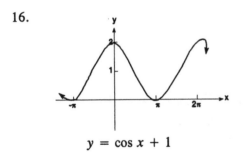
$$y = \cos x + 1$$

18.
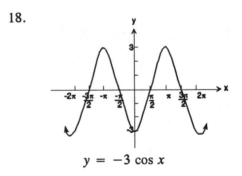
$$y = -3 \cos x$$

20.
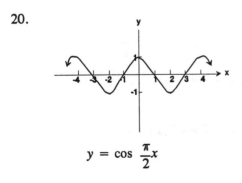
$$y = \cos \frac{\pi}{2}x$$

22.
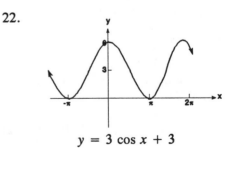
$$y = 3 \cos x + 3$$

24.

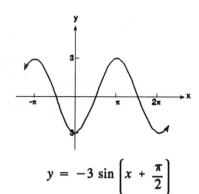

$$y = -3 \sin\left[x + \frac{\pi}{2}\right]$$

26.

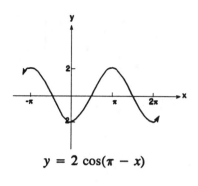

$$y = 2 \cos(\pi - x)$$

28. The graph of $y = A \sin x$, $A > 0$, lies between $-A$ and A.

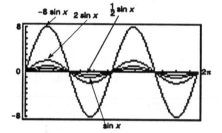

30. The graph of $y = \sin(x - \phi)$, $\phi > 0$, is the graph of $y = \sin x$ shifted horizontally to the right ϕ units.

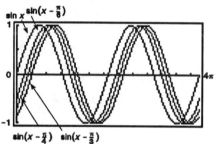

8.2 Sinusoidal Graphs

2.
$$y = 3 \cos x$$
Amplitude = 3
Period = 2π

4.
$$y = -\sin \frac{1}{2}x$$
Amplitude = 1
Period = 4π

6.
$$y = -3 \cos 3x$$
Amplitude = 3
Period = $\dfrac{2\pi}{3}$

8.
$$y = \frac{4}{3} \sin \frac{2}{3}x$$
Amplitude = $\dfrac{4}{3}$
Period = 3π

10. $\qquad y = \dfrac{9}{5} \cos\left[\dfrac{-3\pi}{2}x\right]$

Amplitude $= \dfrac{9}{5}$

Period $= \dfrac{4}{3}$

12. $\qquad y = 2 \cos \dfrac{\pi}{2}x$

Amplitude $= 2$

Period $= \dfrac{2\pi}{\dfrac{\pi}{2}} = 4$

(E)

14. $\qquad y = 3 \cos 2x$

Amplitude $= 3$

Period $= \dfrac{2\pi}{2} = \pi$

(I)

16. $\qquad y = 2 \sin \dfrac{1}{2}x$

Amplitude $= 2$

Period $= \dfrac{2\pi}{\dfrac{1}{2}} = 4\pi$

(B)

18. $\qquad y = -2 \cos \dfrac{\pi}{2}x$

Amplitude $= 2$

Period $= \dfrac{2\pi}{\dfrac{\pi}{2}} = 4$

(G)

20. $\qquad y = -2 \sin \dfrac{1}{2}x$

Amplitude $= 2$

Period $= \dfrac{2\pi}{\dfrac{1}{2}} = 4\pi$

(D)

22. $\qquad$ Amplitude $= 4$

Period $= \dfrac{2\pi}{6} = \dfrac{\pi}{3}$

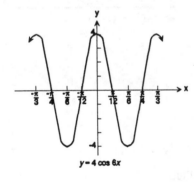

$y = 4 \cos 6x$

24. $\qquad$ Amplitude $= 2$

Period $= \dfrac{2\pi}{\pi} = 2$

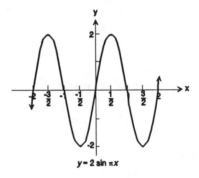

$y = 2 \sin \pi x$

Chapter 8 Graphs of Trigonometric Functions

26. Amplitude $= 5(A = -5 < 0)$

Period $= \dfrac{2\pi}{2\pi} = 1$

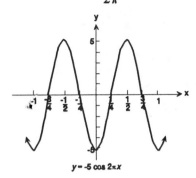

$y = -5\cos 2\pi x$

28. Amplitude $= 2(A = -2 < 0)$

Period $= \dfrac{2\pi}{\dfrac{1}{2}} = 4\pi$

$y = -2\cos \frac{1}{2}x$

30.

$$y = \frac{4}{3}\cos\left[\frac{-1}{3}x\right]$$

$$= \frac{4}{3}\cos\frac{1}{3}x$$

Amplitude $= \dfrac{4}{3}$

Period $= \dfrac{2\pi}{\dfrac{1}{3}} = 6\pi$

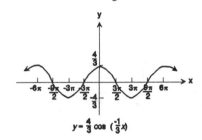

$y = \frac{4}{3}\cos\left(\frac{-1}{3}x\right)$

32. sin function

Amplitude $= 4$

Period $= 8\pi = \dfrac{2\pi}{\omega}$

$\omega = \dfrac{2\pi}{8\pi} = \dfrac{1}{4}$

$y = A\sin \omega x$

$y = 4\sin \dfrac{x}{4}$

34. reflection of the sin function about the x-axis

Amplitude $= 2$

Period $= 4 = \dfrac{2\pi}{\omega}$

$\omega = \dfrac{2\pi}{4} = \dfrac{\pi}{2}$

$y = -A\sin \omega x$

$y = -2\sin \dfrac{\pi}{2}x$

36. reflection of the cos function about the x-axis

Amplitude $= \dfrac{5}{2}$

Period $= 2 = \dfrac{2\pi}{\omega}$

$\omega = \dfrac{2\pi}{2} = \pi$

$y = -A\cos \omega x$

$y = -\dfrac{5}{2}\cos \pi x$

38. reflection of the cos function about the *x*-axis

Amplitude $= \pi$

$$\text{Period} = 2\pi = \frac{2\pi}{\omega}$$

$$\omega = \frac{2\pi}{2\pi} = 1$$

$y = -A \cos \omega x$

$y = -\pi \cos x$

40. reflection of the sin function about the *x*-axis

Amplitude $= \dfrac{1}{2}$

$$\text{Period} = \frac{4\pi}{3} = \frac{2\pi}{\omega}$$

$$\omega = 2\pi \left[\frac{3}{4\pi} \right] = \frac{3}{2}$$

$y = -A \sin \omega x$

$y = -\dfrac{1}{2} \sin \dfrac{3}{2}x$

42. reflection of the cos function about the *x*-axis

Amplitude $= 2$

$$\text{Period} = 2 = \frac{2\pi}{\omega}$$

$$\omega = \frac{2\pi}{2} = \pi$$

$y = -A \cos \omega x$

$y = -2 \cos \pi x$

44. sin function

Amplitude $= 4$

$$\text{Period} = \pi = \frac{2\pi}{\omega}$$

$$\omega = 2$$

$y = A \sin \omega x$

$y = 4 \sin 2x$

46. Amplitude $= 3$

Period $= \dfrac{2\pi}{3}$

Phase shift: $3x - \pi = 0$

$$x = \frac{\pi}{3}$$

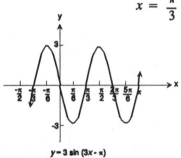

$y = 3 \sin (3x - \pi)$

48. Amplitude $= 3$

Period $= \pi$

Phase shift: $3x + \pi = 0$

$$x = \frac{-\pi}{2}$$

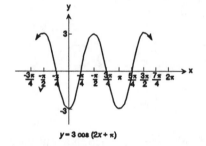

$y = 3 \cos (2x + \pi)$

50. Amplitude $= 2$

Period $= \pi$

Phase shift: $2x - \dfrac{\pi}{2} = 0$

$$x = \frac{\pi}{4}$$

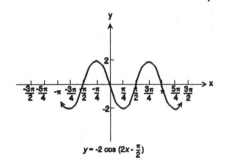

$y = -2 \cos \left(2x - \dfrac{\pi}{2}\right)$

52. $y = 2 \cos(2\pi x + 4)$

$y = A \cos(\omega x - \phi)$

$A = 2, \ \omega = 2\pi, \ \phi = -4$

Amplitude: 2

Period: $\dfrac{2\pi}{\omega} = \dfrac{2\pi}{2\pi} = 1$

Phase Shift: $\dfrac{\phi}{\omega} = \dfrac{-4}{2\pi} = \dfrac{-2}{\pi}$

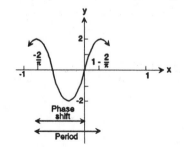

54. $y = 2 \cos(2\pi x - 4)$
$y = A \cos(\omega x - \phi)$
$A = 2$, $\omega = 2\pi$, $\phi = 4$
Amplitude: 2

Period: $\dfrac{2\pi}{\omega} = \dfrac{2\pi}{2\pi} = 1$

Phase Shift: $\dfrac{\phi}{\omega} = \dfrac{4}{2\pi} = \dfrac{2}{\pi}$

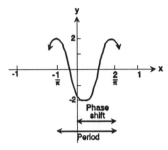

56. $y = 3 \cos\left[-2x + \dfrac{\pi}{2}\right] = 3 \cos\left[2x - \dfrac{\pi}{2}\right]$
$y = A \cos(\omega x - \phi)$

$A = 3$, $\omega = 2$, $\phi = \dfrac{\pi}{2}$

Amplitude: 3

Period: $\dfrac{2\pi}{\omega} = \dfrac{2\pi}{2} = \pi$

Phase Shift: $\dfrac{\phi}{\omega} = \dfrac{\dfrac{\pi}{2}}{2} = \dfrac{\pi}{4}$

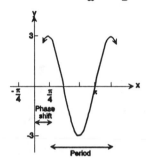

58. $I = 120 \sin 30\,\pi t,\ t \geq 0$

Period $= \dfrac{2\pi}{\omega} = \dfrac{2\pi}{30\pi} = \dfrac{1}{15}$

Amplitude $= 120$

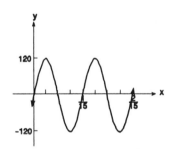

60. $I = 220 \sin\left[60\pi t - \dfrac{\pi}{6}\right],\ t \geq 0$

Period $= \dfrac{2\pi}{60\pi} = \dfrac{1}{30}$

Amplitude $= 220$

Phase Shift $= \dfrac{1}{360}$

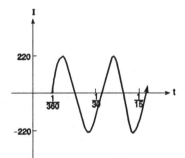

(Compare by ratio the period $2\pi : \dfrac{1}{30}$

and then the shift $\dfrac{\pi}{6}$ compares to $\dfrac{1}{360}$)

$\left(\dfrac{\dfrac{2\pi}{1}}{\dfrac{1}{30}} \times \dfrac{\dfrac{\pi}{6}}{x}\right)$

$\left[\begin{array}{l} 2\pi x = \dfrac{\pi}{180} \\[2mm] x = \dfrac{1}{360} \end{array}\right]$

62. $V = 120 \sin 120\pi t$

    (a)   Amplitude $= 120$

         Period $= \dfrac{2\pi}{120\pi} = \dfrac{1}{60}$

    (b)

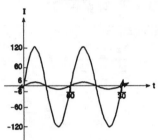

    (c) $V = IR$

$$\frac{120 \sin 120\pi t}{20} = \frac{20}{20}I$$

$$I = 6 \sin 120\,\pi t$$

    (d)   Amplitude $= 6$

         Period $= \dfrac{1}{60}$

    (e)   See graph in (b).

64. $P = 100 \sin \omega t, \; \omega = \dfrac{2\pi}{t}$

    (a)   Physical potential: $\omega = \dfrac{2\pi}{23}$

         Emotional potential: $\omega = \dfrac{2\pi}{28} = \dfrac{\pi}{14}$

         Intellectual potential: $\omega = \dfrac{2\pi}{33}$

    (b)

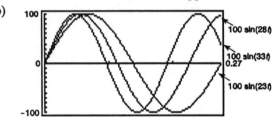

    (c) $(23)(28)(33) = 21{,}252$ days

    (d)

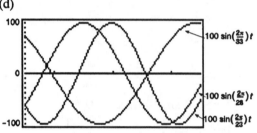

8.3   Applications

2.

x	$-\pi$	$\dfrac{-\pi}{2}$	0	$\dfrac{\pi}{2}$	π	$\dfrac{3\pi}{2}$	2π
$f_1(x) = x$	$-\pi$	$\dfrac{-\pi}{2}$	0	$\dfrac{\pi}{2}$	π	$\dfrac{3\pi}{2}$	2π
$f_2(x) = \cos 2x$	1	-1	1	-1	1	-1	1
$f(x) = x + \cos 2x$	$-\pi + 1$	$\dfrac{-\pi}{2} - 1$	1	$\dfrac{\pi}{2} - 1$	$\pi + 1$	$\dfrac{3\pi}{2} - 1$	$2\pi + 1$
Point on graph of f	$(-\pi, -\pi + 1)$	$\left(\dfrac{-\pi}{2}, \dfrac{-\pi}{2} - 1\right)$	$(0, 1)$	$\left(\dfrac{\pi}{2}, \dfrac{\pi}{2} - 1\right)$	$(\pi, \pi + 1)$	$\left(\dfrac{3\pi}{2}, \dfrac{3\pi}{2} - 1\right)$	$(2\pi, 2\pi + 1)$

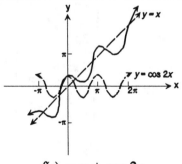

$$f(x) = x + \cos 2x$$

4.

x	$-\pi$	$\dfrac{-\pi}{2}$	0	$\dfrac{\pi}{2}$	π	$\dfrac{3\pi}{2}$	2π
$f_1(x) = x$	$-\pi$	$\dfrac{-\pi}{2}$	0	$\dfrac{\pi}{2}$	π	$\dfrac{3\pi}{2}$	2π
$f_2(x) = -\cos x$	1	0	-1	0	1	0	-1
$f(x) = x - \cos x$	$-\pi + 1$	$\dfrac{-\pi}{2}$	-1	$\dfrac{\pi}{2}$	$\pi + 1$	$\dfrac{3\pi}{2}$	$2\pi - 1$
Point on graph of f	$(-\pi, -\pi + 1)$	$\left[\dfrac{-\pi}{2}, \dfrac{-\pi}{2}\right]$	$(0, -1)$	$\left[\dfrac{\pi}{2}, \dfrac{\pi}{2}\right]$	$(\pi, \pi + 1)$	$\left[\dfrac{3\pi}{2}, \dfrac{3\pi}{2}\right]$	$(2\pi, 2\pi - 1)$

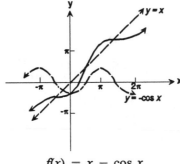

$$f(x) = x - \cos x$$

6.

x	$-\pi$	$\dfrac{-\pi}{2}$	0	$\dfrac{\pi}{2}$	π	$\dfrac{3\pi}{2}$	2π
$f_1(x) = \sin 2x$	0	0	0	0	0	0	0
$f_2(x) = \cos x$	-1	0	1	0	-1	0	1
$f(x) = \sin 2x + \cos x$	-1	0	1	0	-1	0	1
Point on graph of f	$(-\pi, -1)$	$\left[\dfrac{-\pi}{2}, 0\right]$	$(0, 1)$	$\left[\dfrac{\pi}{2}, 0\right]$	$(\pi, -1)$	$\left[\dfrac{3\pi}{2}, 0\right]$	$(2\pi, 1)$

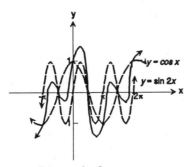

$$f(x) = \sin 2x + \cos x$$

8.

x	$-\pi$	$\dfrac{-\pi}{2}$	0	$\dfrac{\pi}{2}$	π	$\dfrac{3\pi}{2}$	2π
$g_1(x) = \cos 2x$	1	−1	1	−1	1	−1	1
$g_2(x) = \cos x$	−1	0	1	0	−1	0	1
$g(x) = \cos 2x + \cos x$	0	−1	2	−1	0	−1	2
Point on graph of g	$(-\pi, 0)$	$\left[\dfrac{-\pi}{2}, -1\right]$	$(0, 2)$	$\left[\dfrac{\pi}{2}, -1\right]$	$(\pi, 0)$	$\left[\dfrac{3\pi}{2}, -1\right]$	$(2\pi, 2)$

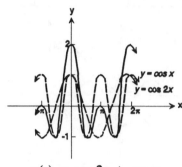

$$g(x) = \cos 2x + \cos x$$

10.

x	0	$\dfrac{\pi}{4}$	$\dfrac{\pi}{2}$	π	$\dfrac{3\pi}{2}$	2π
$h_1(x) = \sqrt{x}$	0	≈0.89	≈1.25	≈1.77	≈2.17	≈2.51
$h_2(x) = \cos x$	1	$\dfrac{\sqrt{2}}{2}$	0	−1	0	1
$h(x) = \sqrt{x} + \cos x$	1	≈1.6	≈1.25	≈.77	≈2.17	≈3.51
Point on graph of h	$(0, 1)$	$\left[\dfrac{\pi}{4}, 1.6\right]$	$\left[\dfrac{\pi}{2}, 1.25\right]$	$(\pi, .77)$	$\left[\dfrac{3\pi}{2}, 2.17\right]$	$(2\pi, 3.51)$

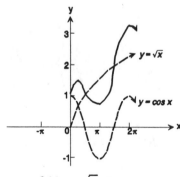

$$h(x) = \sqrt{x} + \cos x$$

12.

x	$-\pi$	$\dfrac{-\pi}{2}$	0	$\dfrac{\pi}{2}$	π	$\dfrac{3\pi}{2}$	2π
$F_1(x) = 2\cos 2x$	2	-2	2	-2	2	-2	2
$F_2(x) = -\sin x$	0	1	0	-1	0	1	0
$F(x) = 2\cos 2x - \sin x$	2	-1	2	-3	2	-1	2
Point on graph of F	$(-\pi, 2)$	$\left[\dfrac{-\pi}{2}, -1\right]$	$(0, 2)$	$\left[\dfrac{\pi}{2}, -3\right]$	$(\pi, 2)$	$\left[\dfrac{3\pi}{2}, -1\right]$	$(2\pi, 2)$

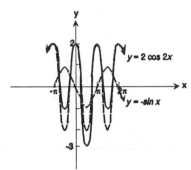

$$F(x) = 2\cos 2x - \sin x$$

14.

x	-2	-1	0	1	2	3	4
$f_1(x) = 2\cos\dfrac{\pi}{2}x$	-2	0	2	0	-2	0	2
$f_2(x) = \sin\dfrac{\pi}{2}x$	0	-1	0	1	0	-1	0
$f(x) = 2\cos\dfrac{\pi}{2}x + \sin\dfrac{\pi}{2}x$	-2	-1	2	1	-2	-1	2
Point on graph of f	$(-2, -2)$	$(-1, -1)$	$(0, 2)$	$(1, 1)$	$(2, -2)$	$(3, -1)$	$(4, 2)$

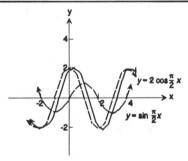

$$f(x) = 2\cos\dfrac{\pi}{2}x + \sin\dfrac{\pi}{2}x$$

16.

x	$\dfrac{-\pi}{2}$	0	$\dfrac{\pi}{2}$	π
$f_1(x) = \dfrac{x^2}{\pi^2}$	$\dfrac{1}{4}$	0	$\dfrac{1}{4}$	1
$f_2(x) = -\cos 2x$	1	-1	1	-1
$f(x) = \dfrac{x^2}{\pi^2} - \cos 2x$	$\dfrac{5}{4}$	-1	$\dfrac{5}{4}$	0
Point on graph of f	$\left[\dfrac{-\pi}{2}, \dfrac{5}{4}\right]$	$(0,\,-1)$	$\left[\dfrac{\pi}{2}, \dfrac{5}{4}\right]$	$(\pi,\,0)$

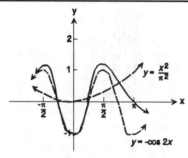

$$f(x) = \frac{x^2}{\pi^2} - \cos 2x$$

18.

x	-1	$\dfrac{-1}{2}$	0	$\dfrac{1}{2}$	1	$\dfrac{3}{2}$	2
$f_1(x) = |x|$	1	$\dfrac{1}{2}$	0	$\dfrac{1}{2}$	1	$\dfrac{3}{2}$	2
$f_2(x) = \cos \pi x$	-1	0	1	0	-1	0	1
$f(x) = |x| + \cos \pi x$	0	$\dfrac{1}{2}$	1	$\dfrac{1}{2}$	0	$\dfrac{3}{2}$	3
Point on graph of f	$(-1,\,0)$	$\left[\dfrac{-1}{2}, \dfrac{1}{2}\right]$	$(0,\,1)$	$\left[\dfrac{1}{2}, \dfrac{1}{2}\right]$	$(1,\,0)$	$\left[\dfrac{3}{2}, \dfrac{3}{2}\right]$	$(2,\,3)$

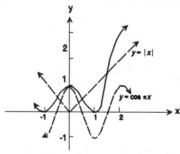

$$f(x) = |x| + \cos \pi x$$

20.

x	$\dfrac{-\pi}{3}$	$\dfrac{-\pi}{6}$	0	$\dfrac{\pi}{6}$	$\dfrac{\pi}{3}$
$f_1(x) = 2\sin 3x$	0	-2	0	2	0
$f_2(x) = 3\cos 2x$	$\dfrac{-3}{2}$	$\dfrac{3}{2}$	3	$\dfrac{3}{2}$	$\dfrac{-3}{2}$
$f(x) = 2\sin 3x + 3\cos 2x$	$\dfrac{-3}{2}$	$\dfrac{-1}{2}$	3	$\dfrac{7}{2}$	$\dfrac{-3}{2}$
Point on graph of f	$\left[\dfrac{-\pi}{3}, \dfrac{-3}{2}\right]$	$\left[\dfrac{-\pi}{6}, \dfrac{-1}{2}\right]$	$(0, 3)$	$\left[\dfrac{\pi}{6}, \dfrac{7}{2}\right]$	$\left[\dfrac{\pi}{3}, \dfrac{-3}{2}\right]$

x	$\dfrac{\pi}{2}$	$\dfrac{2\pi}{3}$	$\dfrac{5\pi}{6}$	π
$f_1(x) = 2\sin 3x$	-2	0	2	0
$f_2(x) = 3\cos 2x$	-3	$\dfrac{-3}{2}$	$\dfrac{3}{2}$	3
$f(x) = 2\sin 3x + 3\cos 2x$	-5	$\dfrac{-3}{2}$	$\dfrac{7}{2}$	3
Point on graph of f	$\left[\dfrac{\pi}{2}, -5\right]$	$\left[\dfrac{2\pi}{3}, \dfrac{-3}{2}\right]$	$\left[\dfrac{5\pi}{6}, \dfrac{7}{2}\right]$	$(\pi, 3)$

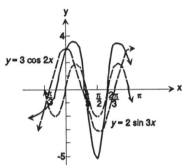

$$f(x) = 2\sin 3x + 3\cos 2x$$

22.

x	$\dfrac{-\pi}{2}$	$\dfrac{-\pi}{4}$	0	$\dfrac{\pi}{4}$	$\dfrac{\pi}{2}$	$\dfrac{3\pi}{4}$	π
$f_1(x) = x$	$\dfrac{-\pi}{2}$	$\dfrac{-\pi}{4}$	0	$\dfrac{\pi}{4}$	$\dfrac{\pi}{2}$	$\dfrac{3\pi}{4}$	π
$f_2(x) = \sin 2x$	0	-1	0	1	0	-1	0
$f(x) = x\sin 2x$	0	$\dfrac{\pi}{4}$	0	$\dfrac{\pi}{4}$	0	$\dfrac{-3\pi}{4}$	0
Point on graph of f	$\left[\dfrac{-\pi}{2}, 0\right]$	$\left[\dfrac{-\pi}{4}, \dfrac{\pi}{4}\right]$	$(0, 0)$	$\left[\dfrac{\pi}{4}, \dfrac{\pi}{4}\right]$	$\left[\dfrac{\pi}{2}, 0\right]$	$\left[\dfrac{3\pi}{4}, \dfrac{-3\pi}{4}\right]$	$(\pi, 0)$

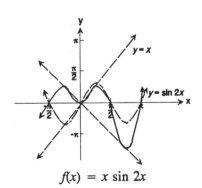

$$f(x) = x\sin 2x$$

24.

x	$-\pi$	$\dfrac{-\pi}{2}$	0	$\dfrac{\pi}{2}$	π	$\dfrac{3\pi}{2}$	2π
$f_1(x) = x^2$	π^2	$\dfrac{\pi^2}{4}$	0	$\dfrac{\pi^2}{4}$	π^2	$\dfrac{9\pi^2}{4}$	$4\pi^2$
$f_2(x) = \cos x$	-1	0	1	0	-1	0	1
$f(x) = x^2 + \cos x$	$-\pi^2$	0	0	0	$-\pi^2$	0	$4\pi^2$
Point on graph of f	$(-\pi, -\pi^2)$	$\left[\dfrac{-\pi}{2}, 0\right]$	$(0, 0)$	$\left[\dfrac{\pi}{2}, 0\right]$	$(\pi, -\pi^2)$	$\left[\dfrac{3\pi}{2}, 0\right]$	$(2\pi, 4\pi^2)$

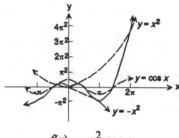

$$f(x) = x^2 \cos x$$

26.

x	$-\pi$	$\dfrac{-\pi}{2}$	0	$\dfrac{\pi}{2}$	π	$\dfrac{3\pi}{2}$	2π
$f_1(x) = \lvert x \rvert$	π	$\dfrac{\pi}{2}$	0	$\dfrac{\pi}{2}$	π	$\dfrac{3\pi}{2}$	2π
$f_2(x) = \sin x$	0	-1	0	1	0	-1	0
$f(x) = \lvert x \rvert \sin x$	0	$\dfrac{-\pi}{2}$	0	$\dfrac{\pi}{2}$	0	$\dfrac{-3\pi}{2}$	0
Point on graph of f	$(\pi, 0)$	$\left[\dfrac{-\pi}{2}, \dfrac{-\pi}{2}\right]$	$(0, 0)$	$\left[\dfrac{\pi}{2}, \dfrac{\pi}{2}\right]$	$(\pi, 0)$	$\left[\dfrac{3\pi}{2}, \dfrac{-3\pi}{2}\right]$	$(2\pi, 0)$

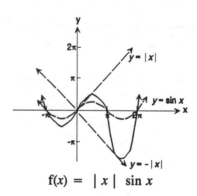

$$f(x) = \lvert x \rvert \sin x$$

28.

x	0	$\dfrac{\pi}{4}$	$\dfrac{\pi}{2}$	$\dfrac{3\pi}{4}$	π
$f_1(x) = e^{-x}$	1	0.46	0.21	0.09	0.04
$f_2(x) = \sin 2x$	0	1	0	-1	0
$f(x) = e^{-x}\sin 2x$	0	0.46	0	-0.09	0
Point on graph of f	$(0, 0)$	$\left[\dfrac{\pi}{4},\ 0.46\right]$	$\left[\dfrac{\pi}{2},\ 0\right]$	$\left[\dfrac{3\pi}{4},\ -0.09\right]$	$(\pi, 0.04)$

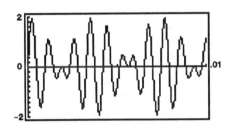

$$f(x) = e^{-x}\sin 2x,\ x \ge 0$$

30.

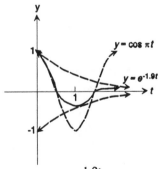

32. (a)

t	0	$\dfrac{1}{2}$	1	$\dfrac{3}{2}$	2
$V_1(t) = e^{-1.9t}$	1	$\approx .007$	$\approx .00005$	≈ 0	≈ 0
$V_2(t) = \cos \pi t$	1	0	-1	0	1
$V(t) = e^{-1.9t}\cos \pi t$	1	0	$\approx -.00005$	0	0
Point on graph of V	$(0, 1)$	$\left[\dfrac{1}{2},\ 0\right]$	$(1, -.00005)$	$\left[\dfrac{3}{2},\ 0\right]$	$(2, 0)$

$$V(t) = e^{-1.9t}\cos \pi t$$

(b) The graph V touches the graph of $y = e^{-1.9t}$ when $t = 0, 2, 4, \ldots$.
The graph V touches the graph of $y = -e^{-1.9t}$ when $t = 1, 3, 5, \ldots$.

(c)

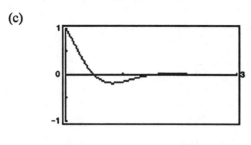

$-0.1 < V < 0.1$ for $t > 1.15$.

In Problems 34-48, we use the formulas $d = a\cos \omega t$ or $d = a \sin \omega t$, at time t, with amplitude $|a|$, period $2\pi/\omega$, and frequency $\omega/2\pi$.

34. $d = 10 \cos \dfrac{2\pi}{3}t$

36. $d = 4 \cos 4t$

38. $d = 10 \sin \dfrac{2\pi}{3}t$

40. $d = 4 \sin 4t$

42.   (a)   Simple harmonic
     (b)   4 m
     (c) π sec
     (d) $\dfrac{1}{\pi}$ oscillation/sec

44.   (a)   Simple harmonic
     (b)   5 m
     (c)   4 sec
     (d) $\dfrac{1}{4}$ oscillation/sec

46.   (a)   Simple harmonic
     (b)   2 m
     (c) π sec
     (d) $\dfrac{1}{\pi}$ oscillation/sec

48.   (a)   Simple harmonic
     (b)   3 m
     (c)   2 sec
     (d) $\dfrac{1}{2}$ oscillation/sec

50. The graph will lie between the bounding curves $y = \pm x$, $y = \pm x^2$, and $y = \pm x^3$, respectively, touching them at odd multiples of $\dfrac{\pi}{2}$. The x-intercepts of each graph are the multiples of π.

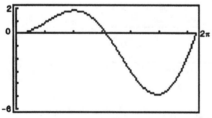

$y = x \sin x$

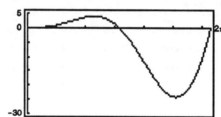

$y = x^2 \sin x$

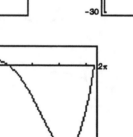

$y = x^3 \sin x$

Chapter 8   Graphs of Trigonometric Functions

2. No y-intercept.

4. No y-intercept.

6. For $x = \dfrac{-3\pi}{2}$, $\dfrac{\pi}{2}$, $\csc x = 1$

   For $x = \dfrac{-\pi}{2}$, $\dfrac{3\pi}{2}$, $\csc x = -1$

8. For $x = -2\pi$, $-\pi$, 0, π, 2π, $y = \csc x$ has vertical asymptotes.

10. For $x = -2\pi$, $-\pi$, 0, π, 2π, $y = \cot x$ has vertical asymptotes.

12. C   14. A

16.

18.

20.

22.

24.

26.

28.

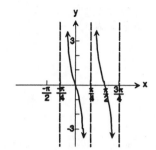

30.

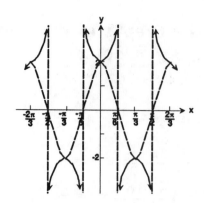

32. It would appear that $\tan x = -\cot\left[x + \dfrac{\pi}{2}\right]$ because if the cotangent function was reflected through the origin it would have the same shape as the tangent function. Also, if the cotangent function was shifted to the left $\dfrac{\pi}{2}$ units, its asymptotes would match up with the asymptotes of the tangent function. The graphs are identical.

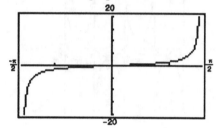

8.5 The Inverse Trigonometric Functions

2.
$$\cos \theta = 1$$
$$\theta = 0$$
$$\cos^{-1} 1 = 0$$

4.
$$\cos \theta = -1$$
$$\theta = \pi$$
$$\cos^{-1}(-1) = \pi$$

6.
$$\tan \theta = -1$$
$$\theta = \dfrac{-\pi}{4}$$
$$\tan^{-1}(-1) = \dfrac{-\pi}{4}$$

8.
$$\tan \theta = \dfrac{\sqrt{3}}{3}$$
$$\theta = \dfrac{\pi}{6}$$
$$\tan^{-1} \dfrac{\sqrt{3}}{3} = \dfrac{\pi}{6}$$

10.
$$\sin \theta = \dfrac{-\sqrt{3}}{2}$$
$$\theta = \dfrac{-\pi}{3}$$
$$\sin^{-1}\left[-\dfrac{\sqrt{3}}{2}\right] = \dfrac{-\pi}{3}$$

12.
$$\sin \theta = -\dfrac{\sqrt{2}}{2}$$
$$\theta = \dfrac{-\pi}{4}$$
$$\sin^{-1}\left[-\dfrac{\sqrt{2}}{2}\right] = \dfrac{-\pi}{4}$$

14. $\cos^{-1} 0.6 = 0.9272952 \approx 0.93$

16. $\tan^{-1} 0.2 = 0.1973956 \approx 0.20$

18. $\sin^{-1} \dfrac{1}{8} = 0.1253278 \approx 0.13$

20. $\tan^{-1}(-3) = -1.2490458 \approx -1.25$

22. $\cos^{-1}(-0.44) = 2.026395 \approx 2.03$

24. $\sin^{-1} \dfrac{\sqrt{3}}{5} = 0.3537416 \approx 0.35$

26. $\sin\left[\cos^{-1} \dfrac{1}{2}\right] = \sin \dfrac{\pi}{3} = \dfrac{\sqrt{3}}{2}$

28. $\tan\left[\sin^{-1}\left(-\dfrac{1}{2}\right)\right] = \tan \dfrac{-\pi}{6} = \dfrac{-\sqrt{3}}{3}$

30. $\cot\left[\sin^{-1}\left(-\dfrac{1}{2}\right)\right] = \cot\dfrac{-\pi}{6} = -\sqrt{3}$

32. $\sec\left(\tan^{-1}\sqrt{3}\right) = \sec \dfrac{\pi}{3} = 2$

34. $\cos\left[\sin^{-1}\left(-\dfrac{\sqrt{3}}{2}\right)\right] = \cos\dfrac{-\pi}{3} = \dfrac{1}{2}$

36. $\csc\left[\cos^{-1}\left(-\dfrac{\sqrt{3}}{2}\right)\right] = \csc\dfrac{5\pi}{6} = 2$

38. $\cos \theta = \dfrac{1}{3}$, where $0 \le \theta \le \pi$

$\cos \theta > 0$, so $0 \le \theta \le \dfrac{\pi}{2}$

$\tan \theta = \tan\left[\cos^{-1} \dfrac{1}{3}\right] = 2\sqrt{2}$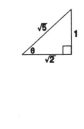

40. $\sin \theta = \dfrac{\sqrt{2}}{3}$, where $\dfrac{-\pi}{2} \le \theta \le \dfrac{\pi}{2}$

$\sin \theta > 0$, so $0 \le \theta \le \dfrac{\pi}{2}$

$\cos \theta = \cos\left[\sin^{-1} \dfrac{\sqrt{2}}{3}\right]$

$= \dfrac{\sqrt{7}}{3}$

42. $\tan \theta = -2$, where $\dfrac{-\pi}{2} \le \theta \le \dfrac{\pi}{2}$

$\csc \theta < 0$

$\csc \theta = -\sqrt{1 + \cot^2 \theta} = -\sqrt{1 + \dfrac{1}{\tan^2 \theta}} = -\sqrt{1 + \left(-\dfrac{1}{2}\right)^2} = -\sqrt{\dfrac{5}{4}} = \dfrac{-\sqrt{5}}{2}$

Hence, $\csc\left[\tan^{-1}(-2)\right] = \dfrac{-\sqrt{5}}{2}$

44. $\cos \theta = \dfrac{-\sqrt{3}}{3}$ where $0 \le \theta \le \pi$

$\cos \theta < 0$ so $\dfrac{\pi}{2} \le \theta \le \pi$

$\cot \theta < 0$

$\cot \theta = -\cot\left[\cos^{-1}\left[\dfrac{-\sqrt{3}}{3}\right]\right]$

$= \dfrac{-\sqrt{3}}{\sqrt{6}} = \dfrac{-\sqrt{2}}{2}$

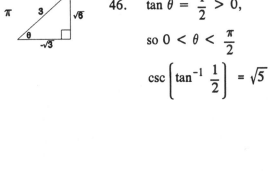

46. $\tan \theta = \dfrac{1}{2} > 0$,

so $0 < \theta < \dfrac{\pi}{2}$

$\csc\left[\tan^{-1} \dfrac{1}{2}\right] = \sqrt{5}$

48. $\cos^{-1}(\tan 0.4) = 1.1342710 \approx 1.13$

50. $\tan^{-1}(\cos 0.2) = 0.7753315 \approx 0.78$

52. $\tan^{-1}(\cos 1) = 0.4953673 \approx 0.50$

54. $\cos^{-1}\left[\sin \dfrac{\pi}{8}\right] = 1.1780972 \approx 1.18$

56. $\tan^{-1}\left[\cos\dfrac{\pi}{8}\right] = 0.6802150 \approx 0.68$

58. Let $\theta = \sin^{-1}\nu$. Then $\sin\theta = \nu,\ -\dfrac{\pi}{2} \le \theta \le \dfrac{\pi}{2}$

$$\tan\theta = \dfrac{\nu}{\sqrt{1-\nu^2}}$$

so $\tan(\sin^{-1}\nu) = \dfrac{\nu}{\sqrt{1-\nu^2}}$

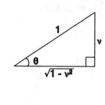

60. Let $\theta = \cos^{-1}\nu$. Then $\cos\theta = \nu,\ 0 < \theta < \pi$.

$$\sin(\cos^{-1}\nu) = \sin\theta = \sqrt{1-\cos^2\theta} = \sqrt{1-\nu^2}$$

62. Let $\theta = \tan^{-1}\nu$. Then $\tan\theta = \nu,\ \dfrac{-\pi}{2} < \theta < \dfrac{\pi}{2}$

$$\cos(\tan^{-1}\nu) = \cos\theta = \dfrac{1}{\sqrt{1+\nu^2}}$$

64. Let $\alpha = \tan^{-1}\nu$ and $\beta = \cot^{-1}\nu$.
Then $\tan\alpha = \nu = \cot\beta$, so $\alpha,\ \beta$ are

complementary. Thus, $\alpha + \beta = \dfrac{\pi}{2}$.

66. Let $\theta = \cot^{-1}e^\nu$
Then, $\cot\theta = e^\nu$
$\tan\theta = e^{-\nu}$
$\theta = \tan^{-1}e^{-\nu}$

68. $\csc^{-1}5$

Let $\theta = \csc^{-1}5$. Then $\csc\theta = 5,\ \dfrac{-\pi}{2} \le \theta \le \dfrac{\pi}{2},\ \theta \ne 0$

Thus, $\sin\theta = \dfrac{1}{5}$ and $\csc^{-1}5 = \theta = \sin^{-1}\dfrac{1}{5} \approx 0.20$

70. $\sec^{-1}(-3)$

Let $\theta = \sec^{-1}(-3)$. Then $\sec\theta = -3,\ 0 \le \theta \le \pi,\ \theta \ne \dfrac{\pi}{2}$.

Thus, $\cos\theta = \dfrac{-1}{3}$ and $\sec^{-1}(-3) = \theta = \cos^{-1}\left[\dfrac{-1}{3}\right] \approx 1.91$.

72. $\cot^{-1}\left[\dfrac{-1}{2}\right]$

Let $\theta = \cot^{-1}\left[\dfrac{-1}{2}\right]$. Then $\cot\theta = \dfrac{-1}{2},\ 0 < \theta < \pi$.

Thus, $\cos\theta = \dfrac{-1}{\sqrt{5}}$ and $\cot^{-1}\left[\dfrac{-1}{2}\right] = \theta = \cos^{-1}\left[\dfrac{-1}{\sqrt{5}}\right] \approx 2.03$.

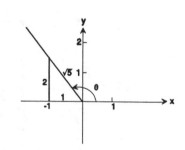

74. $\cot^{-1}(-8.1)$

Let $\theta = \cot^{-1}(-8.1)$. Then $\cot \theta = \dfrac{-81}{10}$, $0 < \theta < \pi$.

Thus, $\cos \theta = \dfrac{-81}{81.61}$ and $\cot^{-1}\left[\dfrac{-81}{10}\right] = \theta - \cos^{-1}\left[\dfrac{-81}{81.61}\right]$
≈ 3.02

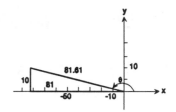

76. Let $\theta = \sec^{-1}\left[-\dfrac{4}{3}\right]$. Then $\sec \theta = \dfrac{-4}{3}$ so $\cos \theta = -\dfrac{3}{4}$ and $\theta = \cos^{-1}\left[-\dfrac{3}{4}\right] \approx 2.42$.

78. Let $\theta = \cot^{-1}\left(-\sqrt{10}\right)$. Then $\cot \theta = -\sqrt{10}$, $0 < \theta < \pi$. Thus, $\cos \theta = \dfrac{-\sqrt{10}}{\sqrt{11}}$, $\dfrac{\pi}{2} < \theta < \pi$,

and $\theta = \cos^{-1}\left[\dfrac{-\sqrt{10}}{\sqrt{11}}\right] \approx 2.84$.

80. $\dfrac{6.5}{2.5} = \dfrac{x}{24-x}$

$2.5x = 156 - 6.5x$
$9x = 156$
$x = 17.3$

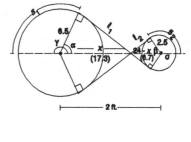

$\cos \alpha = \dfrac{6.5}{17.3} \approx 0.3757$ $\qquad$ $\cos \beta = \dfrac{2.5}{6.7}$

$\alpha = 1.18$ radians $\qquad\qquad$ $\beta = 1.18$
$\gamma = \pi - 1.18 = 1.9559$ radians $\qquad \sigma = 1.9559$
$s_1 = 6.5\gamma = 12.71$ in $\qquad\qquad$ $s_2 = 2.5\sigma \approx 4.90$
$6.5 \tan \alpha = \ell_1 = 15.78$ in $\qquad\quad$ $2.5 \tan \beta = \ell_2 = 6.22$

Length of belt $= 2(s_1 + \ell_1) + 2(s_2 + \ell_2) = 2(12.71 + 15.78) + 2(6.22 + 4.90)$
$\qquad\qquad = 2(28.48) + 2(11.12) = 56.98 + 22.24 = 79.22$ inches

82. $-1 \le x \le 1$ $\qquad\qquad$ 84. $0 \le x \le \pi$ $\qquad\qquad$ 86.

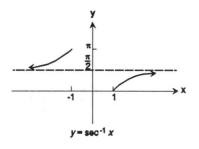

$y = \sec^{-1} x$

88. 1530 ft. $\dfrac{1 \text{ mile}}{5280 \text{ ft}} = .29$ miles

$\cos \theta = \dfrac{3960}{3960.29}$

$\theta \approx 0.6934 = 0.0121$ rad
$\qquad s = r\theta = 3960(0.0121) = 47.9047$ miles

$\dfrac{2\pi(3960)}{24} = \dfrac{47.9047}{t}$

$\qquad t = .046$ hours $= 2.77$ minutes

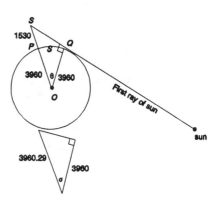

2.
$$y = \sin 2x$$
Amplitude = 1
$$\text{Period} = \frac{2\pi}{2} = \pi$$

4.
$$y = -2 \cos 3\pi x$$
Amplitude = 2
$$\text{Period} = \frac{2\pi}{3\pi} = \frac{2}{3}$$

6.
Amplitude = 2
$$\text{Period} = \frac{2\pi}{\frac{1}{3}} = 6\pi$$

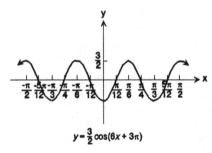

$$y = 2 \cos \tfrac{1}{3}x$$

8.
Amplitude = 6 $(A = -6 < 0)$
$$\text{Period} = \frac{2\pi}{2\pi} = 1$$
Phase Shift: $2\pi x - 2 = 0$
$$x = \frac{1}{\pi}$$

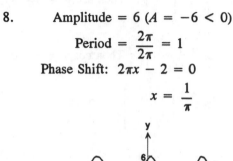

$$y = -6 \sin(2\pi x - 2)$$

10.
Amplitude = $\frac{3}{2}$
$$\text{Period} = \frac{2\pi}{6} = \frac{\pi}{3}$$
Phase Shift: $6x + 3\pi = 0$
$$x = \frac{-\pi}{2}$$

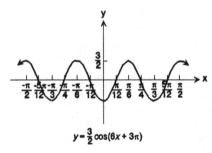

$$y = \frac{3}{2} \cos(6x + 3\pi)$$

12.
Amplitude = 7 $(A = -7 < 0)$
$$\text{Period} = \frac{2\pi}{\frac{\pi}{3}} = 6$$
Phase Shift: $\frac{\pi}{3}x + \frac{4}{3} = 0$
$$x = \frac{-4}{\pi}$$

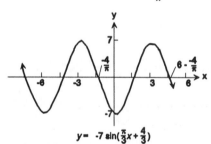

$$y = -7 \sin\left(\frac{\pi}{3}x + \frac{4}{3}\right)$$

14.
Amplitude = 4
$$\text{Period} = 8\pi = \frac{2\pi}{\omega}$$
$$\omega = \frac{1}{4}$$
$$y = 4 \sin \frac{x}{4}$$

16.
Amplitude = 7
$$\text{Period} = 8 = \frac{2\pi}{\omega}$$
$$\omega = \frac{\pi}{4}$$
$$y = -7 \sin \frac{\pi}{4}x$$

18. Amplitude $= 1\ (A < 0)$
Period $= \pi$

Phase Shift $x - \dfrac{\pi}{2} = 0$

$$x = \dfrac{\pi}{2}$$

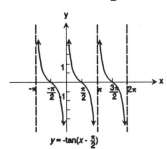

$y = -\tan\left(x - \dfrac{\pi}{2}\right)$

20. Amplitude $= 4$

Period $= \dfrac{\pi}{2}$

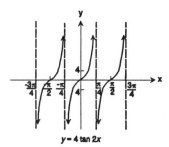

$y = 4 \tan 2x$

22.

x	$-\pi$	$\dfrac{-\pi}{2}$	0	$\dfrac{\pi}{2}$	π	$\dfrac{3\pi}{2}$	2π
$f_1(x) = 2x$	-2π	$-\pi$	0	π	2π	3π	4π
$f_2(x) = \cos 2x$	1	-1	1	-1	1	-1	1
$f(x) = 2x + \cos 2x$	$-2\pi + 1$	$-\pi - 1$	1	$\pi - 1$	$2\pi + 1$	$3\pi - 1$	$4\pi + 1$
Point on graph of f	$(-\pi, -2\pi + 1)$	$\left[\dfrac{-\pi}{2}, -\pi - 1\right]$	$(0, 1)$	$\left[\dfrac{\pi}{2}, \pi - 1\right]$	$(\pi, 2\pi + 1)$	$\left[\dfrac{3\pi}{2}, 3\pi - 1\right]$	$(2\pi, 4\pi + 1)$

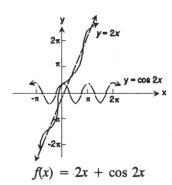

$f(x) = 2x + \cos 2x$

24.

x	0	$\dfrac{1}{2}$	1	$\dfrac{3}{2}$	2	$\dfrac{5}{2}$	3	$\dfrac{7}{2}$	4
$f_1(x) = \sin \dfrac{\pi}{2} x$	0	$\dfrac{\sqrt{2}}{2}$	1	$\dfrac{\sqrt{2}}{2}$	0	$\dfrac{-\sqrt{2}}{2}$	-1	$\dfrac{-\sqrt{2}}{2}$	0
$f_2(x) = \cos \pi x$	1	0	-1	0	1	0	-1	0	1
$f(x) = \sin \dfrac{\pi}{2} x + \cos \pi x$	1	$\dfrac{\sqrt{2}}{2}$	0	$\dfrac{\sqrt{2}}{2}$	1	$\dfrac{-\sqrt{2}}{2}$	-2	$\dfrac{-\sqrt{2}}{2}$	1
Point on graph of f	$(0, 1)$	$\left[\dfrac{1}{2}, \dfrac{\sqrt{2}}{2}\right]$	$(1, 0)$	$\left[\dfrac{3}{2}, \dfrac{\sqrt{2}}{2}\right]$	$(2, 1)$	$\left[\dfrac{5}{2}, \dfrac{-\sqrt{2}}{2}\right]$	$(3, -2)$	$\left[\dfrac{7}{2}, \dfrac{-\sqrt{2}}{2}\right]$	$(4, 1)$

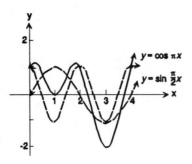

$$f(x) = \sin \frac{\pi}{2} x + \cos \pi x$$

26.

x	0	$\frac{1}{4}$	$\frac{1}{2}$	$\frac{3}{4}$	1	$\frac{5}{4}$	$\frac{3}{2}$	$\frac{7}{4}$	2
$f_1(x) = 3 \cos 2\pi x$	3	0	-3	0	3	0	-3	0	3
$f_2(x) = 4 \sin 2\pi x$	0	4	0	-4	0	4	0	-4	0
$f(x) = 3 \cos 2\pi x + 4 \sin 2\pi x$	3	4	-3	-4	3	4	-3	-4	3
Point on graph of f	$(0, 3)$	$\left[\frac{1}{4}, 4\right]$	$\left[\frac{1}{2}, -3\right]$	$\left[\frac{3}{4}, -4\right]$	$(1, 3)$	$\left[\frac{5}{4}, 4\right]$	$\left[\frac{3}{2}, -3\right]$	$\left[\frac{7}{4}, -4\right]$	$(2, 3)$

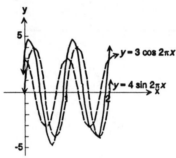

$$f(x) = 3 \cos 2\pi x + 4 \sin 2\pi x$$

28.

x	0	$\frac{1}{4}$	$\frac{1}{2}$	$\frac{3}{4}$	1	$\frac{5}{4}$	$\frac{3}{2}$	$\frac{7}{4}$	2
$f_1(x) = x$	0	$\frac{1}{4}$	$\frac{1}{2}$	$\frac{3}{4}$	1	$\frac{5}{4}$	$\frac{3}{2}$	$\frac{7}{4}$	2
$f_2(x) = \sin \pi x$	0	$\frac{\sqrt{2}}{2}$	1	$\frac{\sqrt{2}}{2}$	0	$\frac{-\sqrt{2}}{2}$	-1	$\frac{-\sqrt{2}}{2}$	0
$f(x) = x \sin \pi x$	0	$\frac{\sqrt{2}}{8}$	$\frac{1}{2}$	$\frac{3\sqrt{2}}{8}$	0	$\frac{-5\sqrt{2}}{8}$	$\frac{-3}{2}$	$\frac{-7\sqrt{2}}{8}$	0
Point on graph of f	$(0, 0)$	$\left[\frac{1}{4}, \frac{\sqrt{2}}{8}\right]$	$\left[\frac{1}{2}, \frac{1}{2}\right]$	$\left[\frac{3}{4}, \frac{3\sqrt{2}}{8}\right]$	$(1, 0)$	$\left[\frac{5}{4}, \frac{-5\sqrt{2}}{8}\right]$	$\left[\frac{3}{2}, \frac{-3}{2}\right]$	$\left[\frac{7}{4}, \frac{-7\sqrt{2}}{8}\right]$	$(2, 0)$

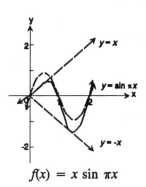

$$f(x) = x \sin \pi x$$

30.

x	0	$\frac{1}{4}$	$\frac{1}{2}$	$\frac{3}{4}$	1	$\frac{5}{4}$	$\frac{3}{2}$	$\frac{7}{4}$	2
$f_1(x) = e^{-x}$	1	.78	.61	.47	.37	.29	.22	.17	.14
$f_2(x) = \cos \pi x$	1	$\frac{\sqrt{2}}{2}$	0	$\frac{-\sqrt{2}}{2}$	-1	$\frac{-\sqrt{2}}{2}$	0	$\frac{\sqrt{2}}{2}$	1
$f(x) = e^{-x} \cos \pi x$	1	.55	0	$-.33$	$-.37$	$-.21$	0	.12	.14
Point on graph of f	(0, 1)	$\left[\frac{1}{4}, .55\right]$	$\left[\frac{1}{2}, 0\right]$	$\left[\frac{3}{4}, -.33\right]$	$(1, -.37)$	$\left[\frac{5}{4}, -.21\right]$	$\left[\frac{3}{2}, 0\right]$	$\left[\frac{7}{4}, .12\right]$	(2, .14)

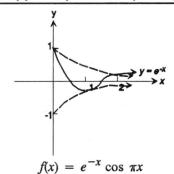

$$f(x) = e^{-x} \cos \pi x$$

32.

x	0	$\frac{1}{2}$	1	$\frac{3}{2}$	2
$f_1(x) = e^x$	1	1.6	2.7	4.6	7.4
$f_2(x) = \cos \pi x$	1	0	-1	0	1
$f(x) = e^x \cos \pi x$	1	0	-2.7	0	7.4
Point on graph of f	(0, 1)	$\left[\frac{1}{2}, 0\right]$	$(1, -2.7)$	$\left[\frac{3}{2}, 0\right]$	(2, 7.4)

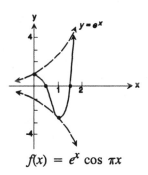

$$f(x) = e^x \cos \pi x$$

34.
$$\cos \theta = 0$$
$$\theta = \frac{\pi}{2}$$
$$\cos^{-1} 0 = \frac{\pi}{2}$$

36.
$$\sin \theta = -\frac{1}{2}$$
$$\theta = \frac{-\pi}{6}$$
$$\sin^{-1}\left(-\frac{1}{2}\right) = \frac{-\pi}{6}$$

38.
$$\tan \theta = -\sqrt{3}$$
$$\theta = \frac{-\pi}{3}$$
$$\tan^{-1}\left(-\sqrt{3}\right) = \frac{-\pi}{3}$$

40. $\cos(\sin^{-1} 0) = \cos(0) = 1$

42. $\tan\left[\cos^{-1}\left(-\frac{1}{2}\right)\right] = \tan\left(\frac{2\pi}{3}\right) = -\sqrt{3}$

44. $\csc\left[\sin^{-1}\frac{\sqrt{3}}{2}\right] = \csc\frac{\pi}{3} = \frac{2\sqrt{3}}{3}$

46. $\sin \theta = \frac{3}{5}$

$\cos \theta = \cos\left[\sin^{-1}\frac{3}{5}\right] = \frac{4}{5}$

48. $\cos \theta = -\frac{3}{5} \Rightarrow \sec \theta = \frac{-5}{3}$

$\tan \theta = -\sqrt{\sec^2 \theta - 1}$

$\tan \theta = -\sqrt{\left(\frac{-5}{3}\right)^2 - 1} = -\sqrt{\frac{25}{9} - \frac{9}{9}} = \frac{4}{3}$

$\tan\left[\cos^{-1}\left(-\frac{3}{5}\right)\right] = -\frac{4}{3}$

50.
(a) Simple harmonic
(b) 2 ft
(c) $\frac{\pi}{2}$ sec
(d) $\frac{2}{\pi}$ oscillation/sec

52.
(a) Simple harmonic
(b) 3 ft
(c) 4 sec
(d) $\frac{1}{4}$ oscillation/sec

54. $I = 220 \sin(30 \pi t + \frac{\pi}{6}),\ t \geq 0$

(a) Period $= \frac{2\pi}{30\pi} = \frac{1}{15}$

(b) Amplitude $= 220$

(c)   Phase shift compares to $\dfrac{-\pi}{6}$,

so $\dfrac{2\pi}{\frac{1}{15}} = \dfrac{-\dfrac{\pi}{6}}{x}$

$x = \dfrac{-\pi}{90} \cdot \dfrac{1}{2\pi} = -\dfrac{1}{180}$

The phase shift is $-\dfrac{1}{180}$

(d)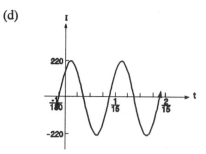

Mission Possible

1.   High tide occurs every 12.5 hours. So, the next high tide will be at 3:30 p.m. Monday. Following this pattern, high tide will be at 6:00 a.m. on Thursday.

2.   The captain of a fishing boat might want to leave harbor at high tide in order to get out of a narrow entrance to the harbor when the water level is at its highest to avoid scraping bottom. Even if the harbor is not connected to the bay or ocean by means of a narrow inlet, there may be high sand bars to get over.

3.   There will be 5 high tides between those times (not including the one at 3 a.m. Monday).

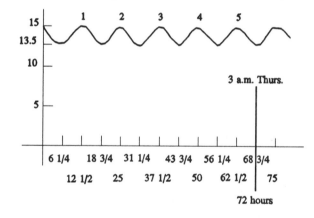

4.   Amplitude $= \dfrac{15 - 13.5}{2} = 0.75$. Since the period is 12.5, we have $k = \dfrac{2\pi}{12.5}$. In addition, the graph is vertically shifted upward 14.25. Since our graph starts at a high point, it would be easier to use cosine. Thus, the equation of the graph would be:

$$y = 0.75 \cos\left[\dfrac{2\pi}{12.5}x\right] + 14.25$$

5.   Tides are affected by the phases of the moon and the configuration of other planets. For example, at the time of a syzygy around Christmas 1985, when plants were aligned, there were floods on Cape Cod. Weather patterns also play a big role. Winds coming from a certain direction, big storms, and hurricanes can cause much higher high tides.

6.   Slack is needed to allow for the change in height of the water. A tight bowline at high tide might break when forced to take the weight of the boat as the water level declines. On the other hand, a tight bowline at low tide might cause the rising tide to swamp the boat.

ANALYTIC TRIGONOMETRY

9.1 Trigonometric Identities

2. $\csc \theta \tan \theta = \dfrac{1}{\sin \theta} \cdot \dfrac{\sin \theta}{\cos \theta} = \dfrac{1}{\cos \theta} = \sec \theta$

4. $1 + \cot^2(-\theta) = 1 + \cot^2 \theta = \csc^2 \theta$

6. $\sin \theta(\cot \theta + \tan \theta) = \sin \theta \left[\dfrac{\cos \theta}{\sin \theta} + \dfrac{\sin \theta}{\cos \theta} \right] = \sin \theta \left[\dfrac{\cos^2 \theta + \sin^2 \theta}{\sin \theta \cos \theta} \right] = \dfrac{1}{\cos \theta} = \sec \theta$

8. $\sin \theta \csc \theta - \cos^2 \theta = \sin \theta \left[\dfrac{1}{\sin \theta} \right] - \cos^2 \theta = 1 - \cos^2 \theta = \sin^2 \theta$

10. $(\csc \theta - 1)(\csc \theta + 1) = \csc^2 \theta - 1 = \cot^2 \theta$

12. $(\csc \theta + \cot \theta)(\csc \theta - \cot \theta) = \csc^2 \theta - \cot^2 \theta = 1$

14. $(1 - \sin^2 \theta)(1 + \tan^2 \theta) = (\cos^2 \theta)(\sec^2 \theta) = \cos^2 \theta \left[\dfrac{1}{\cos^2 \theta} \right] = 1$

16. $\tan^2 \theta \cos^2 \theta + \cot^2 \theta \sin^2 \theta = \dfrac{\sin^2 \theta}{\cos^2 \theta} \cdot \cos^2 \theta + \dfrac{\cos^2 \theta}{\sin^2 \theta} \cdot \sin^2 \theta = \sin^2 \theta + \cos^2 \theta = 1$

18. $\csc^4 \theta - \csc^2 \theta = \csc^2 \theta(\csc^2 \theta - 1) = \csc^2 \theta(\cot^2 \theta) = (1 + \cot^2 \theta)(\cot^2 \theta)$
$\qquad = \cot^2 \theta + \cot^4 \theta = \cot^4 \theta + \cot^2 \theta$

20. $\csc \theta - \cot \theta = \dfrac{1}{\sin \theta} - \dfrac{\cos \theta}{\sin \theta} = \dfrac{1 - \cos \theta}{\sin \theta} \left[\dfrac{\sin \theta}{\sin \theta} \right] = \dfrac{\sin \theta(1 - \cos \theta)}{1 - \cos^2 \theta}$
$\qquad = \dfrac{\sin \theta(1 - \cos \theta)}{(1 - \cos \theta)(1 + \cos \theta)} = \dfrac{\sin \theta}{1 + \cos \theta}$

22. $9 \sec^2 \theta - 5 \tan^2 \theta = 4 \sec^2 \theta + 5 \sec^2 \theta - 5 \tan^2 \theta = 4 \sec^2 \theta + 5(\sec^2 \theta - \tan^2 \theta)$
$\qquad = 4 \sec^2 \theta + 5 = 5 + 4 \sec^2 \theta$

24. $1 - \dfrac{\sin^2 \theta}{1 - \cos \theta} = 1 - \dfrac{(1 - \cos^2 \theta)}{1 - \cos \theta} = 1 - \dfrac{(1 - \cos \theta)(1 + \cos \theta)}{1 - \cos \theta} = 1 - (1 + \cos \theta) = -\cos \theta$

26. $\dfrac{\csc\theta - 1}{\csc\theta + 1} = \dfrac{\dfrac{1}{\sin\theta} - 1}{\dfrac{1}{\sin\theta} + 1} = \dfrac{\dfrac{1 - \sin\theta}{\sin\theta}}{\dfrac{1 + \sin\theta}{\sin\theta}} = \dfrac{1 - \sin\theta}{1 + \sin\theta}$

28. $\dfrac{\csc\theta - 1}{\cot\theta} = \dfrac{\csc\theta - 1}{\cot\theta}\left(\dfrac{\csc\theta + 1}{\csc\theta + 1}\right) = \dfrac{\csc^2\theta - 1}{\cot\theta(\csc\theta + 1)} = \dfrac{\cot^2\theta}{\cot\theta(\csc\theta + 1)} = \dfrac{\cot\theta}{\csc\theta + 1}$

30. $\dfrac{\cos\theta + 1}{\cos\theta - 1} = \dfrac{\dfrac{1}{\sec\theta} + 1}{\dfrac{1}{\sec\theta} - 1} = \dfrac{\dfrac{1 + \sec\theta}{\sec\theta}}{\dfrac{1 - \sec\theta}{\sec\theta}} = \dfrac{1 + \sec\theta}{1 - \sec\theta}$

32. $\dfrac{\cos\theta}{1 + \sin\theta} + \dfrac{1 + \sin\theta}{\cos\theta} = \dfrac{\cos^2\theta + (1 + \sin\theta)^2}{(1 + \sin\theta)\cos\theta} = \dfrac{\cos^2\theta + 1 + 2\sin\theta + \sin^2\theta}{(1 + \sin\theta)\cos\theta}$

$= \dfrac{2 + 2\sin\theta}{(1 + \sin\theta)\cos\theta} = \dfrac{2(1 + \sin\theta)}{(1 + \sin\theta)\cos\theta} = 2\sec\theta$

34. $1 - \dfrac{\sin^2\theta}{1 + \cos\theta} = \dfrac{1 + \cos\theta - \sin^2\theta}{1 + \cos\theta} = \dfrac{\cos^2\theta + \cos\theta}{1 + \cos\theta} = \dfrac{\cos\theta(\cos\theta + 1)}{1 + \cos\theta} = \cos\theta$

36. $\dfrac{1 - \cos\theta}{1 + \cos\theta} = \dfrac{1 - \cos\theta}{1 + \cos\theta}\left(\dfrac{1 - \cos\theta}{1 - \cos\theta}\right) = \dfrac{(1 - \cos\theta)^2}{1 - \cos^2\theta} = \dfrac{1 - 2\cos\theta + \cos^2\theta}{\sin^2\theta}$

$= \csc^2\theta - 2\csc\theta\cot\theta + \cot^2\theta = (\csc\theta - \cot\theta)^2$

38. $\dfrac{\cot\theta}{1 - \tan\theta} + \dfrac{\tan\theta}{1 - \cot\theta} = \dfrac{\dfrac{\cos\theta}{\sin\theta}}{1 - \dfrac{\sin\theta}{\cos\theta}} + \dfrac{\dfrac{\sin\theta}{\cos\theta}}{1 - \dfrac{\cos\theta}{\sin\theta}} = \dfrac{\dfrac{\cos\theta}{\sin\theta}}{\dfrac{\cos\theta - \sin\theta}{\cos\theta}} + \dfrac{\dfrac{\sin\theta}{\cos\theta}}{\dfrac{\sin\theta - \cos\theta}{\sin\theta}}$

$= \dfrac{\cos^2\theta}{\sin\theta(\cos\theta - \sin\theta)} + \dfrac{\sin^2\theta}{\cos\theta(\sin\theta - \cos\theta)}$

$= \dfrac{-\cos^2\theta(\cos\theta) + \sin^2\theta(\sin\theta)}{\sin\theta\cos\theta(\sin\theta - \cos\theta)}$

$= \dfrac{\sin^3\theta - \cos^3\theta}{\sin\theta\cos\theta(\sin\theta - \cos\theta)}$

$= \dfrac{(\sin\theta - \cos\theta)(\sin^2\theta + \sin\theta\cos\theta + \cos^2\theta)}{\sin\theta\cos\theta(\sin\theta - \cos\theta)}$

$= \dfrac{\sin^2\theta}{\sin\theta\cos\theta} + \dfrac{\sin\theta\cos\theta}{\sin\theta\cos\theta} + \dfrac{\cos^2\theta}{\sin\theta\cos\theta}$

$= \tan\theta + 1 + \cot\theta$

40. $\dfrac{\sin\theta\cos\theta}{\cos^2\theta - \sin^2\theta} = \dfrac{\sin\theta\cos\theta}{\cos^2\theta - \sin^2\theta} \cdot \dfrac{\dfrac{1}{\cos^2\theta}}{\dfrac{1}{\cos^2\theta}} = \dfrac{\tan\theta}{1 - \tan^2\theta}$

42.

$$\dfrac{\sin\theta - \cos\theta + 1}{\sin\theta + \cos\theta - 1} = \dfrac{(\sin\theta - \cos\theta + 1)}{(\sin\theta + \cos\theta - 1)} \cdot \dfrac{(\sin\theta + \cos\theta) + 1}{(\sin\theta + \cos\theta) + 1}$$

$$= \dfrac{(\sin^2\theta - \cos^2\theta) + (\sin\theta + \cos\theta) + (\sin\theta - \cos\theta) + 1}{(\sin\theta + \cos\theta)^2 - 1}$$

$$= \dfrac{\sin^2\theta - \cos^2\theta + 2\sin\theta + 1}{\sin^2\theta + 2\sin\theta\cos\theta + \cos^2\theta - 1} = \dfrac{\sin^2\theta - (1 - \sin^2\theta) + 2\sin\theta + 1}{2\sin\theta\cos\theta}$$

$$= \dfrac{2\sin^2\theta + 2\sin\theta}{2\sin\theta\cos\theta} = \dfrac{2\sin\theta(\sin\theta + 1)}{2\sin\theta\cos\theta} = \dfrac{\sin\theta + 1}{\cos\theta}$$

44.

$$\dfrac{\sec\theta - \cos\theta}{\sec\theta + \cos\theta} = \dfrac{\dfrac{1}{\cos\theta} - \cos\theta}{\dfrac{1}{\cos\theta} + \cos\theta} = \dfrac{\dfrac{1 - \cos^2\theta}{\cos\theta}}{\dfrac{1 + \cos^2\theta}{\cos\theta}} = \dfrac{\sin^2\theta}{1 + \cos^2\theta}$$

46.

$$\dfrac{\tan\theta - \cot\theta}{\tan\theta + \cot\theta} = \dfrac{\dfrac{\sin\theta}{\cos\theta} - \dfrac{\cos\theta}{\sin\theta}}{\dfrac{\sin\theta}{\cos\theta} + \dfrac{\cos\theta}{\sin\theta}} = \dfrac{\dfrac{\sin^2\theta - \cos^2\theta}{\cos\theta\sin\theta}}{\dfrac{\sin^2\theta + \cos^2\theta}{\cos\theta\sin\theta}} = \dfrac{\sin^2\theta - \cos^2\theta}{1}$$

$$= (1 - \cos^2\theta) - \cos^2\theta = 1 - 2\cos^2\theta$$

48.

$$\dfrac{\sec\theta}{1 + \sec\theta} = \dfrac{\dfrac{1}{\cos\theta}}{1 + \dfrac{1}{\cos\theta}} = \dfrac{\dfrac{1}{\cos\theta}}{\dfrac{\cos\theta + 1}{\cos\theta}} = \dfrac{1}{\cos\theta + 1}\left[\dfrac{1 - \cos\theta}{1 - \cos\theta}\right] = \dfrac{1 - \cos\theta}{1 - \cos^2\theta} = \dfrac{1 - \cos\theta}{\sin^2\theta}$$

50.

$$\dfrac{1 - \cot^2\theta}{1 + \cot^2\theta} = \dfrac{1 - \cot^2\theta}{\csc^2\theta} = \dfrac{1 - \dfrac{\cos^2\theta}{\sin^2\theta}}{\dfrac{1}{\sin^2\theta}} = \dfrac{\dfrac{\sin^2\theta - \cos^2\theta}{\sin^2\theta}}{\dfrac{1}{\sin^2\theta}} = \sin^2\theta - \cos^2\theta$$

$$= (1 - \cos^2\theta) - \cos^2\theta = 1 - 2\cos^2\theta$$

52.

$$\dfrac{\sin^2\theta - \tan\theta}{\cos^2\theta - \cot\theta} = \dfrac{\sin^2\theta - \dfrac{\sin\theta}{\cos\theta}}{\cos^2\theta - \dfrac{\cos\theta}{\sin\theta}} = \dfrac{\dfrac{\sin^2\theta\cos\theta - \sin\theta}{\cos\theta}}{\dfrac{\cos^2\theta\sin\theta - \cos\theta}{\sin\theta}} = \dfrac{\sin\theta(\sin\theta\cos\theta - 1)\sin\theta}{\cos\theta(\cos\theta\sin\theta - 1)\cos\theta}$$

$$= \dfrac{\sin^2\theta}{\cos^2\theta} = \tan^2\theta$$

54.

$$\tan\theta + \cot\theta = \dfrac{\sin\theta}{\cos\theta} + \dfrac{\cos\theta}{\sin\theta} = \dfrac{\sin^2\theta + \cos^2\theta}{\cos\theta\sin\theta} = \dfrac{1}{\cos\theta\sin\theta} = \sec\theta\csc\theta$$

56.

$$\dfrac{1 + \sin\theta}{1 - \sin\theta} - \dfrac{1 - \sin\theta}{1 + \sin\theta} = \dfrac{(1 + \sin\theta)^2 - (1 - \sin\theta)^2}{(1 - \sin\theta)(1 + \sin\theta)}$$

$$= \dfrac{1 + 2\sin\theta + \sin^2\theta - (1 - 2\sin\theta + \sin^2\theta)}{(1 - \sin^2\theta)}$$

$$= \dfrac{4\sin\theta}{\cos^2\theta} = 4\tan\theta\sec\theta$$

58. $\dfrac{1 - \sin\theta}{1 + \sin\theta} = \left[\dfrac{1 - \sin\theta}{1 + \sin\theta}\right]\left[\dfrac{1 - \sin\theta}{1 - \sin\theta}\right] = \dfrac{1 - 2\sin\theta + \sin^2\theta}{1 - \sin^2\theta} = \dfrac{1 - 2\sin\theta + \sin^2\theta}{\cos^2\theta}$

$= \sec^2\theta - 2\tan\theta\sec\theta + \tan^2\theta = (\sec\theta - \tan\theta)^2$

60. $\dfrac{\sec^2\theta - \tan^2\theta + \tan\theta}{\sec\theta} = \dfrac{\dfrac{1}{\cos^2\theta} - \dfrac{\sin^2\theta}{\cos^2\theta} + \dfrac{\sin\theta}{\cos\theta}}{\dfrac{1}{\cos\theta}} = \dfrac{\dfrac{1 - \sin^2\theta + \sin\theta\cos\theta}{\cos^2\theta}}{\dfrac{1}{\cos\theta}}$

$= \dfrac{\cos\theta(\cos\theta + \sin\theta)}{\cos\theta} = \sin\theta + \cos\theta$

62. $\dfrac{\sin\theta + \cos\theta}{\sin\theta} - \dfrac{\cos\theta - \sin\theta}{\cos\theta} = \dfrac{\sin\theta\cos\theta + \cos^2\theta - \sin\theta\cos\theta + \sin^2\theta}{\sin\theta\cos\theta}$

$= \dfrac{1}{\sin\theta\cos\theta} = \sec\theta\csc\theta$

64. $\dfrac{\sin^3\theta + \cos^3\theta}{1 - 2\cos^2\theta} = \dfrac{(\sin\theta + \cos\theta)(\sin^2\theta - \sin\theta\cos\theta + \cos^2\theta)}{1 - \cos^2\theta - \cos^2\theta}$

$= \dfrac{(\sin\theta + \cos\theta)(1 - \sin\theta\cos\theta)}{\sin^2\theta - \cos^2\theta} = \dfrac{(\sin\theta + \cos\theta)(1 - \sin\theta\cos\theta)}{(\sin\theta - \cos\theta)(\sin\theta + \cos\theta)}$

$= \dfrac{1 - \sin\theta\cos\theta}{\sin\theta - \cos\theta}\left[\dfrac{\dfrac{1}{\cos\theta}}{\dfrac{1}{\cos\theta}}\right] = \dfrac{\sec\theta - \sin\theta}{\tan\theta - 1}$

66. $\dfrac{\cos\theta + \sin\theta - \sin^3\theta}{\sin\theta} = \dfrac{\cos\theta}{\sin\theta} + \dfrac{\sin\theta}{\sin\theta} - \dfrac{\sin^3\theta}{\sin\theta} = \cot\theta + 1 - \sin^2\theta = \cot\theta + \cos^2\theta$

68. $\dfrac{1 - 2\cos^2\theta}{\sin\theta\cos\theta} = \dfrac{(1 - \cos^2\theta - \cos^2\theta)}{\sin\theta\cos\theta} = \dfrac{\sin^2\theta - \cos^2\theta}{\sin\theta\cos\theta} = \tan\theta - \cot\theta$

70. $\dfrac{1 + \cos\theta + \sin\theta}{1 + \cos\theta - \sin\theta} = \dfrac{(1 + \cos\theta) + \sin\theta}{(1 + \cos\theta) - \sin\theta} \cdot \dfrac{(1 + \cos\theta) + \sin\theta}{(1 + \cos\theta) + \sin\theta}$

$= \dfrac{(1 + \cos\theta)^2 + 2(1 + \cos\theta)(\sin\theta) + \sin^2\theta}{(1 + \cos\theta)^2 - \sin^2\theta}$

$= \dfrac{1 + 2\cos\theta + \cos^2\theta + 2(\sin\theta + \sin\theta\cos\theta) + \sin^2\theta}{1 + 2\cos\theta + \cos^2\theta - (1 - \cos^2\theta)}$

$= \dfrac{1 + 2\cos\theta + \cos^2\theta + 2\sin\theta + 2\sin\theta\cos\theta + (1 - \cos^2\theta)}{2\cos\theta + 2\cos^2\theta}$

$= \dfrac{2 + 2\cos\theta + 2\sin\theta + 2\sin\theta\cos\theta}{2\cos\theta + 2\cos^2\theta} = \dfrac{2(1 + \cos\theta) + 2\sin\theta(1 + \cos\theta)}{2\cos\theta(1 + \cos\theta)}$

$= \dfrac{2(1 + \sin\theta)(1 + \cos\theta)}{2\cos\theta(1 + \cos\theta)} = \dfrac{1 + \sin\theta}{\cos\theta} = \sec\theta + \tan\theta$

72. $(2a \sin \theta \cos \theta)^2 + a^2(\cos^2 \theta - \sin^2 \theta)^2$

$$= 4a^2 \sin^2 \theta \cos^2 \theta + a^2(\cos^4 \theta - 2 \sin^2 \theta \cos^2 \theta + \sin^4 \theta)$$
$$= 4a^2 \sin^2 \theta \cos^2 \theta + a^2 \cos^4 \theta - 2a^2 \sin^2 \theta \cos^2 \theta + a^2 \sin^4 \theta$$
$$= 2a^2 \sin^2 \theta \cos^2 \theta + a^2(\cos^4 \theta + \sin^4 \theta)$$
$$= a^2(\cos^4 \theta + 2 \sin^2 \theta \cos^2 \theta + \sin^4 \theta)$$
$$= a^2(\cos^2 \theta + \sin^2 \theta)^2$$
$$= a^2$$

74. $(\tan \alpha + \tan \beta)(1 - \cot \alpha \cot \beta) + (\cot \alpha + \cot \beta)(1 - \tan \alpha \tan \beta)$

$$= \tan \alpha - \cot \beta + \tan \beta - \cot \alpha + \cot \alpha - \tan \beta + \cot \beta - \tan \alpha = 0$$

76. $(\sin \alpha - \cos \beta)^2 + (\cos \beta + \sin \alpha)(\cos \beta - \sin \alpha)$

$$= \sin^2 \alpha - 2 \sin \alpha \cos \beta + \cos^2 \beta + \cos^2 \beta - \sin^2 \alpha$$
$$= -2 \sin \alpha \cos \beta + 2 \cos^2 \beta = -2 \cos \beta(\sin \alpha - \cos \beta)$$

78. $\ln | \sin \theta | \; - \ln | \cos \theta | \; = \ln \left| \dfrac{\sin \theta}{\cos \theta} \right| = \ln \; | \tan \theta |$

80. $\ln | \sec \theta + \tan \theta | \; + \ln | \sec \theta - \tan \theta |$

$$= \ln \left[|\sec \theta + \tan \theta| \, |\sec \theta - \tan \theta| \right] = \ln | \sec^2 \theta - \tan^2 \theta | \; = \ln 1 = 0$$

9.2 Sum and Difference Formulas

2. $\sin \dfrac{\pi}{12} = \sin \left(\dfrac{4\pi}{12} - \dfrac{3\pi}{12} \right) = \sin \dfrac{\pi}{3} \cos \dfrac{\pi}{4} - \cos \dfrac{\pi}{3} \sin \dfrac{\pi}{4} = \dfrac{\sqrt{3}}{2} \cdot \dfrac{\sqrt{2}}{2} - \dfrac{1}{2} \cdot \dfrac{\sqrt{2}}{2} = \dfrac{\sqrt{2}}{4}\left(\sqrt{3} - 1\right)$

4. $\tan \dfrac{7\pi}{12} = \tan \left(\dfrac{3\pi}{12} + \dfrac{4\pi}{12} \right) = \dfrac{\tan \dfrac{\pi}{4} + \tan \dfrac{\pi}{3}}{1 - \tan \dfrac{\pi}{4} \tan \dfrac{\pi}{3}} = \dfrac{1 + \sqrt{3}}{1 - (1)\left(\sqrt{3}\right)} = \dfrac{1 + \sqrt{3}}{1 - \sqrt{3}}$

6. $\sin 105° = \sin(45° + 60°) = \sin 45° \cos 60° + \cos 45° \sin 60°$

$$= \dfrac{\sqrt{2}}{2} \cdot \dfrac{1}{2} + \dfrac{\sqrt{2}}{2} \cdot \dfrac{\sqrt{3}}{2} = \dfrac{\sqrt{2}}{4}\left(1 + \sqrt{3}\right)$$

8. $\tan 195° = \tan(240° - 45°) = \dfrac{\tan 240° - \tan 45°}{1 + \tan 240° \tan 45°} = \dfrac{\sqrt{3} - 1}{1 + \sqrt{3}}$

10. $\tan \dfrac{19\pi}{12} = \tan \left(\dfrac{4\pi}{12} + \dfrac{15\pi}{12} \right) = \dfrac{\tan \dfrac{\pi}{3} + \tan \dfrac{5\pi}{4}}{1 - \tan \dfrac{\pi}{3} \tan \dfrac{5\pi}{4}} = \dfrac{\sqrt{3} + 1}{1 - \sqrt{3}}$

12. $\cot\left(\dfrac{-5\pi}{12}\right) = -\cot\left(\dfrac{5\pi}{12}\right) = -\cot\left(\dfrac{3\pi}{12} + \dfrac{2\pi}{12}\right) = \dfrac{-1}{\tan\left(\dfrac{\pi}{4} + \dfrac{\pi}{6}\right)} = -\left[\dfrac{1 - \tan\dfrac{\pi}{4}\tan\dfrac{\pi}{6}}{\tan\dfrac{\pi}{4} + \tan\dfrac{\pi}{6}}\right]$

$= -\left[\dfrac{1 - \dfrac{1}{\sqrt{3}}}{1 + \dfrac{1}{\sqrt{3}}}\right] = \left[\dfrac{\dfrac{1}{\sqrt{3}} - 1}{1 + \dfrac{1}{\sqrt{3}}}\right] = \dfrac{1 - \sqrt{3}}{\sqrt{3} + 1}$

14. $\sin 20° \cos 80° - \cos 20° \sin 80° = \sin(20° - 80°) = -\sin 60° = \dfrac{-\sqrt{3}}{2}$

16. $\cos 40° \cos 10° + \sin 40° \sin 10° = \cos(40° - 10°) = \cos 30° = \dfrac{\sqrt{3}}{2}$

18. $\dfrac{\tan 40° - \tan 10°}{1 + \tan 40° \tan 10°} = \tan(40° - 10°) = \tan 30° = \dfrac{1}{\sqrt{3}} = \dfrac{\sqrt{3}}{3}$

20. $\cos\dfrac{5\pi}{12}\cos\dfrac{7\pi}{12} - \sin\dfrac{5\pi}{12}\sin\dfrac{7\pi}{12} = \cos\left(\dfrac{5\pi}{12} + \dfrac{7\pi}{12}\right) = \cos\pi = -1$

22. $\sin\dfrac{\pi}{18}\cos\dfrac{5\pi}{18} + \cos\dfrac{\pi}{18}\sin\dfrac{5\pi}{18} = \sin\left(\dfrac{\pi}{18} + \dfrac{5\pi}{18}\right) = \sin\dfrac{\pi}{3} = \dfrac{\sqrt{3}}{2}$

24. $\cos\alpha = \dfrac{1}{\sqrt{5}}, 0 < \alpha < \dfrac{\pi}{2}; \sin\beta = \dfrac{-4}{5}, \dfrac{-\pi}{2} < \beta < 0$

$\cos\alpha = \dfrac{1}{\sqrt{5}}$ $\sin\beta = \dfrac{-4}{5}$

$\quad 1^2 + y^2 = \sqrt{5}^2, y > 0$ $x^2 + (-4)^2 = 5^2, x > 0$

$\quad y^2 = 5 - 1 = 4, y > 0$ $x^2 = 25 - 16 = 9, x > 0$

$\quad y = 2$ $x = 3$

$\sin\alpha = \dfrac{2}{\sqrt{5}}, \tan\alpha = 2$ $\cos\beta = \dfrac{3}{5}, \tan\beta = \dfrac{-4}{3}$

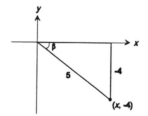

(a) $\sin(\alpha + \beta) = \sin\alpha\cos\beta + \cos\alpha\sin\beta = \dfrac{2}{\sqrt{5}} \cdot \dfrac{3}{5} + \dfrac{1}{\sqrt{5}} \cdot \dfrac{-4}{5} = \dfrac{2}{5\sqrt{5}} = \dfrac{2\sqrt{5}}{25}$

(b) $\cos(\alpha + \beta) = \cos\alpha\cos\beta - \sin\alpha\sin\beta = \dfrac{1}{\sqrt{5}} \cdot \dfrac{3}{5} - \dfrac{2}{\sqrt{5}} \cdot \dfrac{-4}{5} = \dfrac{11}{5\sqrt{5}} = \dfrac{11\sqrt{5}}{25}$

(c) $\sin(\alpha - \beta) = \sin\alpha\cos\beta - \cos\alpha\sin\beta = \dfrac{2}{\sqrt{5}} \cdot \dfrac{3}{5} - \dfrac{1}{\sqrt{5}} \cdot \dfrac{-4}{5} = \dfrac{2}{\sqrt{5}} = \dfrac{2\sqrt{5}}{5}$

(d) $\quad \tan(\alpha - \beta) = \dfrac{\tan \alpha - \tan \beta}{1 + \tan \alpha \tan \beta} = \dfrac{2 - \left[-\dfrac{4}{3}\right]}{1 + (2)\left[-\dfrac{4}{3}\right]} = \dfrac{\dfrac{10}{3}}{\dfrac{-5}{3}} = -2$

26. $\quad \tan \alpha = \dfrac{5}{12},\ \pi < \alpha < \dfrac{3\pi}{2};\ \sin \beta = \dfrac{-1}{2},\ \pi < \beta < \dfrac{3\pi}{2}$

$\tan \alpha = \dfrac{5}{12},\ \sin \alpha = \dfrac{-5}{13},\ \cos \alpha = \dfrac{-12}{13} \qquad \sin \beta = \dfrac{-1}{2},\ \cos \beta = \dfrac{-\sqrt{3}}{2},\ \tan \beta = \dfrac{1}{\sqrt{3}}$

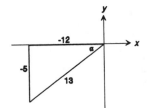

(a) $\quad \sin(\alpha + \beta) = \sin \alpha \cos \beta + \cos \alpha \sin \beta = \dfrac{-5}{13} \cdot \dfrac{-\sqrt{3}}{2} + \dfrac{-12}{13} \cdot \dfrac{-1}{2} = \dfrac{5\sqrt{3} + 12}{26}$

(b) $\quad \cos(\alpha + \beta) = \cos \alpha \cos \beta - \sin \alpha \sin \beta = \dfrac{-12}{13} \cdot \dfrac{-\sqrt{3}}{2} - \dfrac{-5}{13} \cdot \dfrac{-1}{2} = \dfrac{12\sqrt{3} - 5}{26}$

(c) $\quad \sin(\alpha - \beta) = \sin \alpha \cos \beta - \cos \alpha \sin \beta = \dfrac{-5}{13} \cdot \dfrac{-\sqrt{3}}{2} - \dfrac{-12}{13} \cdot \dfrac{-1}{2} = \dfrac{5\sqrt{3} - 12}{26}$

(d) $\quad \tan(\alpha - \beta) = \dfrac{\tan \alpha - \tan \beta}{1 + \tan \alpha \tan \beta} = \dfrac{\dfrac{5}{12} - \dfrac{1}{\sqrt{3}}}{1 + \dfrac{5}{12} \cdot \dfrac{1}{\sqrt{3}}} = \dfrac{5\sqrt{3} - 12}{12\sqrt{3} + 5}$

28. $\quad \cos \alpha = \dfrac{1}{2},\ \dfrac{-\pi}{2} < \alpha < 0\ \sin \beta = \dfrac{1}{3},\ 0 < \beta < \dfrac{\pi}{2}$

$\cos \alpha = \dfrac{1}{2},\ \sin \alpha = \dfrac{-\sqrt{3}}{2},\ \tan \alpha = -\sqrt{3} \qquad \sin \beta = \dfrac{1}{3},\ \cos \beta = \dfrac{2\sqrt{2}}{3},\ \tan \beta = \dfrac{1}{2\sqrt{2}} = \dfrac{\sqrt{2}}{4}$

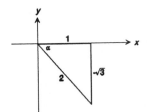 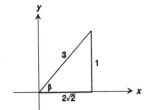

(a) $\quad \sin(\alpha + \beta) = \sin \alpha \cos \beta + \cos \alpha \sin \beta = \dfrac{-\sqrt{3}}{2} \cdot \dfrac{2\sqrt{2}}{3} + \dfrac{1}{2} \cdot \dfrac{1}{3} = \dfrac{-2\sqrt{6} + 1}{6}$

(b) $\quad \cos(\alpha + \beta) = \cos \alpha \cos \beta - \sin \alpha \sin \beta = \dfrac{1}{2} \cdot \dfrac{2\sqrt{2}}{3} - \dfrac{-\sqrt{3}}{2} \cdot \dfrac{1}{3} = \dfrac{2\sqrt{2} + \sqrt{3}}{6}$

(c) $\quad \sin(\alpha - \beta) = \sin \alpha \cos \beta - \cos \alpha \sin \beta = \dfrac{-\sqrt{3}}{2} \cdot \dfrac{2\sqrt{2}}{3} - \dfrac{1}{2} \cdot \dfrac{1}{3} = \dfrac{-2\sqrt{6} - 1}{6}$

(d) $\quad \tan(\alpha - \beta) = \dfrac{\tan \alpha - \tan \beta}{1 + \tan \alpha \tan \beta} = \dfrac{-\sqrt{3} - \dfrac{1}{2\sqrt{2}}}{1 + \left(-\sqrt{3}\right)\left[\dfrac{1}{2\sqrt{2}}\right]} = \dfrac{-2\sqrt{6} - 1}{2\sqrt{2} - \sqrt{3}}$

30. $\quad \cos\left[\dfrac{\pi}{2} + \theta\right] = \cos\dfrac{\pi}{2}\cos\theta - \sin\dfrac{\pi}{2}\sin\theta = 0 \cdot \cos\theta - 1 \cdot \sin\theta = -\sin\theta$

32. $\quad \cos(\pi - \theta) = \cos\pi\cos\theta + \sin\pi\sin\theta = -1 \cdot \cos\theta + 0 \cdot \sin\theta = -\cos\theta$

34. $\quad \cos(\pi + \theta) = \cos\pi\cos\theta - \sin\pi\sin\theta = -1 \cdot \cos\theta - 0 \cdot \sin\theta = -\cos\theta$

36. $\quad \tan(2\pi - \theta) = \dfrac{\tan 2\pi - \tan\theta}{1 + \tan 2\pi \tan\theta} = \dfrac{0 - \tan\theta}{1} = -\tan\theta$

38. $\quad \cos\left[\dfrac{3\pi}{2} + \theta\right] = \cos\dfrac{3\pi}{2}\cos\theta - \sin\dfrac{3\pi}{2}\sin\theta = 0 \cdot \cos\theta - (-1)\sin\theta = \sin\theta$

40. $\quad \cos(\alpha + \beta) + \cos(\alpha - \beta) = \cos\alpha\cos\beta - \sin\alpha\sin\beta + \cos\alpha\cos\beta + \sin\alpha\sin\beta$
$$= 2\cos\alpha\cos\beta$$

42. $\quad \dfrac{\sin(\alpha + \beta)}{\cos\alpha\cos\beta} = \dfrac{\sin\alpha\cos\beta + \cos\alpha\sin\beta}{\cos\alpha\cos\beta} = \tan\alpha + \tan\beta$

44. $\quad \dfrac{\cos(\alpha - \beta)}{\sin\alpha\cos\beta} = \dfrac{\cos\alpha\cos\beta + \sin\alpha\sin\beta}{\sin\alpha\cos\beta} = \cot\alpha + \tan\beta$

46. $\quad \dfrac{\cos(\alpha + \beta)}{\cos(\alpha - \beta)} = \dfrac{\cos\alpha\cos\beta - \sin\alpha\sin\beta}{\cos\alpha\cos\beta + \sin\alpha\sin\beta} = \left[\dfrac{\dfrac{1}{\cos\alpha\cos\beta}}{\dfrac{1}{\cos\alpha\cos\beta}}\right] \cdot \dfrac{\cos\alpha\cos\beta - \sin\alpha\sin\beta}{\cos\alpha\cos\beta + \sin\alpha\sin\beta}$

$\qquad = \dfrac{1 - \tan\alpha\tan\beta}{1 + \tan\alpha\tan\beta}$

48. $\quad \cot(\alpha - \beta) = \dfrac{1}{\tan(\alpha - \beta)} = \dfrac{1}{\dfrac{\tan\alpha - \tan\beta}{1 + \tan\alpha\tan\beta}} = \dfrac{1 + \tan\alpha\tan\beta}{\tan\alpha - \tan\beta} \cdot \left[\dfrac{\dfrac{1}{\tan\alpha\tan\beta}}{\dfrac{1}{\tan\alpha\tan\beta}}\right]$

$\qquad = \dfrac{\cot\alpha\cot\beta + 1}{\cot\beta - \cot\alpha}$

50. $\quad \sec(\alpha - \beta) = \dfrac{1}{\cos(\alpha - \beta)} = \dfrac{1}{\cos\alpha\cos\beta + \sin\alpha\sin\beta}$

$\qquad = \dfrac{1}{\cos\alpha\cos\beta + \sin\alpha\sin\beta} \cdot \left[\dfrac{\dfrac{1}{\cos\alpha\cos\beta}}{\dfrac{1}{\cos\alpha\cos\beta}}\right] = \dfrac{\sec\alpha\sec\beta}{1 + \tan\alpha\tan\beta}$

52. $\cos(\alpha - \beta)\cos(\alpha + \beta) = (\cos\alpha\cos\beta + \sin\alpha\sin\beta)(\cos\alpha\cos\beta - \sin\alpha\sin\beta)$
$$= \cos^2\alpha\cos^2\beta - \sin^2\alpha\sin^2\beta = \cos^2\alpha(1 - \sin^2\beta) - [(1 - \cos^2\alpha)\sin^2\beta]$$
$$= \cos^2\alpha - \cos^2\alpha\sin^2\beta - \sin^2\beta + \cos^2\alpha\sin^2\beta = \cos^2\alpha - \sin^2\beta$$

54. $\cos(\theta + k\pi) = \cos\theta\cos k\pi - \sin\theta\sin k\pi = \cos\theta(-1)^k - (\sin\theta)(0) = (-1)^k \cdot \cos\theta$,
k any integer

56. $\sin\left[\sin^{-1}\dfrac{\sqrt{3}}{2} + \cos^{-1}1\right] = \sin\left[\dfrac{\pi}{3} + 0\right] = \sin\dfrac{\pi}{3} = \dfrac{\sqrt{3}}{2}$

58. Let $\alpha = \sin^{-1}\left[\dfrac{-4}{5}\right]$ and $\beta = \tan^{-1}\dfrac{3}{4}$. Then

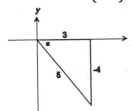 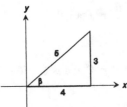

Then, $\sin\left[\sin^{-1}\left[\dfrac{-4}{5}\right] - \tan^{-1}\dfrac{3}{4}\right] = \sin(\alpha - \beta) = \sin\alpha\cos\beta - \cos\alpha\sin\beta$
$$= \left[\dfrac{-4}{5}\right]\left[\dfrac{4}{5}\right] - \left[\dfrac{3}{5}\right]\left[\dfrac{3}{5}\right] = \dfrac{-16}{25} - \dfrac{9}{25} = \dfrac{-25}{25} = -1$$

60. Let $\alpha = \tan^{-1}\dfrac{5}{12}$ and $\beta = \sin^{-1}\left[\dfrac{-3}{5}\right]$. Then

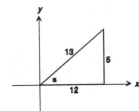 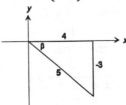

Thus, $\sin\left[\tan^{-1}\dfrac{5}{12} - \sin^{-1}\left[\dfrac{-3}{5}\right]\right] = \sin(\alpha - \beta) = \sin\alpha\cos\beta - \cos\alpha\sin\beta$
$$= \left[\dfrac{5}{13}\right]\left[\dfrac{4}{5}\right] - \left[\dfrac{12}{13}\right]\left[\dfrac{-3}{5}\right] = \dfrac{56}{65}$$

62. Let $\alpha = \tan^{-1}\dfrac{4}{3}$ and $\beta = \cot^{-1}\dfrac{5}{12}$

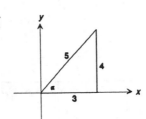 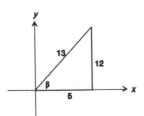

Thus, $\sec\left[\tan^{-1}\dfrac{4}{3} + \cot^{-1}\dfrac{5}{12}\right] = \sec(\alpha + \beta) = \dfrac{1}{\cos(\alpha + \beta)} = \dfrac{1}{\cos\alpha\cos\beta - \sin\alpha\sin\beta}$
$$= \dfrac{1}{\dfrac{3}{5} \cdot \dfrac{5}{13} - \dfrac{4}{5} \cdot \dfrac{12}{13}} = \dfrac{1}{\dfrac{-33}{65}} = \dfrac{-65}{33}$$

Chapter 9 Analytic Trigonometry

64. Let $\alpha = \csc^{-1} \dfrac{5}{3}$. Then

$$\cos\left[\frac{\pi}{4} - \csc^{-1}\frac{5}{3}\right] = \cos\left[\frac{\pi}{4} - \alpha\right] = \cos\frac{\pi}{4}\cos\alpha + \sin\frac{\pi}{4}\sin\alpha$$

$$= \left[\frac{1}{\sqrt{2}}\right]\left[\frac{4}{5}\right] + \left[\frac{1}{\sqrt{2}}\right]\left[\frac{3}{5}\right] = \frac{7}{5\sqrt{2}} = \frac{7\sqrt{2}}{10}$$

66. Let $\alpha = \sin^{-1} u$ and $\beta = \cos^{-1} v$. Then, $\sin\alpha = u$, $\dfrac{-\pi}{2} \le \alpha \le \dfrac{\pi}{2}$, and $\cos\beta = v$, $0 \le \beta \le \pi$.

Since $\cos\alpha \ge 0$ and $\sin\beta \ge 0$, we have $\cos\alpha = \sqrt{1 - \sin^2\alpha} = \sqrt{1 - u^2}$ and

$\sin\beta = \sqrt{1 - \cos^2\beta} = \sqrt{1 - v^2}$

Thus, $\sin(\sin^{-1}u - \cos^{-1}v) = \sin(\alpha - \beta) = \sin\alpha\cos\beta - \cos\alpha\sin\beta$

$$= uv - \sqrt{1 - u^2} \cdot \sqrt{1 - v^2}$$

68. Let $\alpha = \tan^{-1} u$ and $\beta = \tan^{-1} v$. Then, $\tan\alpha = u$, $\dfrac{-\pi}{2} < \alpha < \dfrac{\pi}{2}$, and $\tan\beta = v$,

$\dfrac{-\pi}{2} < \beta < \dfrac{\pi}{2}$. Now, $\sin\alpha = \dfrac{u}{\sqrt{1 + u^2}}$, $\cos\alpha = \dfrac{1}{\sqrt{1 + u^2}}$, $\sin\beta = \dfrac{v}{\sqrt{1 + v^2}}$,

$\cos\beta = \dfrac{1}{\sqrt{1 + v^2}}$.

Thus, $\cos(\tan^{-1}u + \tan^{-1}v) = \cos(\alpha + \beta) = \cos\alpha\cos\beta - \sin\alpha\sin\beta$

$$= \frac{1}{\sqrt{1 + u^2}\sqrt{1 + v^2}} - \frac{uv}{\sqrt{1 + u^2}\sqrt{1 + v^2}}$$

$$= \frac{1 - uv}{\sqrt{1 + u^2}\sqrt{1 + v^2}}$$

70. Let $\alpha = \tan^{-1} u$ and $\beta = \cos^{-1} v$. Then, $\tan\alpha = u$, $\dfrac{-\pi}{2} < \alpha < \dfrac{\pi}{2}$, and $\cos\beta = v$, $0 \le \beta \le$

π. Then, $\sin\alpha = \dfrac{u}{\sqrt{1 + u^2}}$, $\cos\alpha = \dfrac{1}{\sqrt{1 + u^2}}$, and $\sin\beta = \sqrt{1 - v^2}$.

Thus, $\sec(\tan^{-1}u + \tan^{-1}v) = \sec(\alpha + \beta) = \dfrac{1}{\cos(\alpha + \beta)} = \dfrac{1}{\cos\alpha\cos\beta - \sin\alpha\sin\beta}$

$$= \frac{1}{\dfrac{v}{\sqrt{1 + u^2}} - \dfrac{u\sqrt{1 - v^2}}{\sqrt{1 + u^2}}} = \frac{\sqrt{1 + u^2}}{v - u\sqrt{1 - v^2}}$$

72. $\dfrac{\cos(x + h) - \cos x}{h} = \dfrac{\cos x\cos h - \sin x\sin h - \cos x}{h} = \dfrac{-\sin x\sin h - \cos x(1 - \cos h)}{h}$

$$= -\sin x \cdot \frac{\sin h}{h} - \cos x \cdot \frac{1 - \cos h}{h}$$

74. $\cos(\sin^{-1} u + \cos^{-1} u) = \cos(\sin^{-1} u)\cos(\cos^{-1} u) - \sin(\sin^{-1} u)\sin(\cos^{-1} u)$

$$= \sqrt{1 - u^2} \cdot u - u \cdot \sqrt{1 - u^2} = 0$$

76. If $\tan \alpha = x + 1$ and $\tan \beta = x - 1$, then

$$2 \cot(\alpha - \beta) = 2 \cdot \frac{1}{\tan(\alpha - \beta)} = 2 \cdot \frac{1}{\dfrac{\tan \alpha - \tan \beta}{1 + \tan \alpha \tan \beta}} = 2 \cdot \frac{1 + \tan \alpha \tan \beta}{\tan \alpha - \tan \beta}$$

$$= 2 \cdot \frac{1 + (x + 1)(x - 1)}{(x + 1) - (x - 1)} = 2 \cdot \frac{1 + x^2 - 1}{2} = x^2$$

78. $\sin(\alpha - \theta)\sin(\beta - \theta)\sin(\gamma - \theta)$

$\quad = (\sin \alpha \cos \theta - \cos \alpha \sin \theta)(\sin \beta \cos \theta - \cos \beta \sin \theta)(\sin \gamma \cos \theta - \cos \gamma \sin \theta)$

$\quad = \sin^3 \theta[\sin \alpha \cot \theta - \cos \alpha][\sin \beta \cot \theta - \cos \beta][\sin \gamma \cot \theta - \cos \gamma]$

$\quad = \sin^3 \theta[\sin \alpha(\cot \beta + \cot \gamma)][\sin \beta(\cot \alpha + \cot \gamma)][\sin \gamma(\cot \alpha + \cot \beta)]$

$\quad = \sin^3 \theta \sin \alpha \sin \beta \sin \gamma \left[\dfrac{\cos \beta}{\sin \beta} + \dfrac{\cos \gamma}{\sin \gamma}\right]\left[\dfrac{\cos \alpha}{\sin \alpha} + \dfrac{\cos \gamma}{\sin \gamma}\right]\left[\dfrac{\cos \alpha}{\sin \alpha} + \dfrac{\cos \beta}{\sin \beta}\right]$

$\quad = \sin^3 \theta \sin \alpha \sin \beta \sin \gamma \dfrac{\sin(\gamma + \beta)}{\sin \beta \sin \gamma} \dfrac{\sin(\alpha + \gamma)}{\sin \alpha \sin \gamma} \dfrac{\sin(\alpha + \beta)}{\sin \alpha \sin \beta}$

$\quad = \sin^3 \theta \dfrac{\sin \alpha \sin \beta \sin(\alpha + \beta)}{\sin \alpha \sin \beta \sin(\alpha + \beta)} = \sin^3 \theta$

9.3 Double-Angle and Half-Angle Formulas

2. $\cos \theta = \dfrac{3}{5}, \ 0 < \theta < \dfrac{\pi}{2}$. Thus, $0 < \dfrac{\theta}{2} < \dfrac{\pi}{4}$, or $\dfrac{\theta}{2}$ lies in quadrant I.

Therefore, $\sin \theta = \dfrac{4}{5}$

(a) $\sin 2\theta = 2 \sin \theta \cos \theta = 2 \left[\dfrac{4}{5}\right]\left[\dfrac{3}{5}\right] = \dfrac{24}{25}$

(b) $\cos 2\theta = 2 \cos^2 \theta - 1 = 2 \left[\dfrac{3}{5}\right]^2 - 1 = \dfrac{-7}{25}$

(c) $\sin \dfrac{1}{2}\theta = \sqrt{\dfrac{1 - \cos \theta}{2}} = \sqrt{\dfrac{1 - \dfrac{3}{5}}{2}} = \sqrt{\dfrac{1}{5}} = \dfrac{\sqrt{5}}{5}$

(d) $\cos \dfrac{1}{2}\theta = \sqrt{\dfrac{1 + \cos \theta}{2}} = \sqrt{\dfrac{1 + \dfrac{3}{5}}{2}} = \sqrt{\dfrac{4}{5}} = \dfrac{2\sqrt{5}}{5}$

4. $\tan \theta = \dfrac{1}{2}, \ \pi < \theta < \dfrac{3\pi}{2}$. Thus, $\dfrac{\pi}{2} < \dfrac{\theta}{2} < \dfrac{3\pi}{4}$, or $\dfrac{\pi}{2}$ lies in quadrant II.

Therefore, $\sin \theta = \dfrac{-1}{\sqrt{5}} = \dfrac{-\sqrt{5}}{5}$, $\cos \theta = \dfrac{-2}{\sqrt{5}} = \dfrac{-2\sqrt{5}}{5}$

(a) $\sin 2\theta = 2 \sin \theta \cos \theta = 2 \left[\dfrac{-1}{\sqrt{5}}\right]\left[\dfrac{-2}{\sqrt{5}}\right] = \dfrac{4}{5}$

(b) $\cos 2\theta = 1 - 2 \sin^2 \theta = 1 - 2\left[\dfrac{-1}{\sqrt{5}}\right]^2 = \dfrac{3}{5}$

(c) $\sin \dfrac{1}{2}\theta = \sqrt{\dfrac{1 - \cos \theta}{2}} = \sqrt{\dfrac{1 + \dfrac{2\sqrt{5}}{5}}{2}} = \sqrt{\dfrac{5 + 2\sqrt{5}}{10}}$

(d) $\cos \dfrac{1}{2}\theta = -\sqrt{\dfrac{1 + \cos \theta}{2}} = -\sqrt{\dfrac{1 - \dfrac{2\sqrt{5}}{5}}{2}} = -\sqrt{\dfrac{5 - 2\sqrt{5}}{10}}$

6. $\sin \theta = \dfrac{-1}{\sqrt{3}}$, $\dfrac{3\pi}{2} < \theta < 2\pi$. Thus, $\dfrac{3\pi}{4} < \dfrac{\theta}{2} < \pi$, or $\dfrac{\theta}{2}$ lies in quadrant II.

Therefore, $\cos \theta = \sqrt{\dfrac{2}{3}} = \dfrac{\sqrt{6}}{3}$

(a) $\sin 2\theta = 2 \sin \theta \cos \theta = 2\left[\dfrac{-1}{\sqrt{3}}\right]\left[\dfrac{\sqrt{2}}{\sqrt{3}}\right] = \dfrac{-2\sqrt{2}}{3}$

(b) $\cos 2\theta = 1 - 2 \sin^2 \theta = 1 - 2\left[\dfrac{-1}{\sqrt{3}}\right]^2 = \dfrac{1}{3}$

(c) $\sin \dfrac{1}{2}\theta = \sqrt{\dfrac{1 - \cos \theta}{2}} = \sqrt{\dfrac{1 - \dfrac{\sqrt{6}}{3}}{2}} = \sqrt{\dfrac{3 - \sqrt{6}}{6}}$

(d) $\cos \dfrac{1}{2}\theta = -\sqrt{\dfrac{1 + \cos \theta}{2}} = -\sqrt{\dfrac{1 + \dfrac{\sqrt{6}}{3}}{2}} = -\sqrt{\dfrac{3 + \sqrt{6}}{6}}$

8. $\csc \theta = -\sqrt{5}$, $\cos \theta < 0$. Thus, θ lies in quadrant III, $\pi < \theta < \dfrac{3\pi}{2}$ and $\dfrac{\pi}{2} < \dfrac{\theta}{2} < \dfrac{3\pi}{4}$, or $\dfrac{\theta}{2}$ lies in Quadrant II.

Therefore, $\sin \theta = \dfrac{-1}{\sqrt{5}}$, $\cos \theta = \dfrac{-2}{\sqrt{5}}$

(a) $\sin 2\theta = 2 \sin \theta \cos \theta = 2\left[\dfrac{-1}{\sqrt{5}}\right]\left[\dfrac{-2}{\sqrt{5}}\right] = \dfrac{4}{5}$

(b) $\cos 2\theta = 1 - 2 \sin^2 \theta = 1 - 2\left[\dfrac{-1}{\sqrt{5}}\right]^2 = \dfrac{3}{5}$

(c) $\sin \dfrac{1}{2}\theta = \sqrt{\dfrac{1 - \cos \theta}{2}} = \sqrt{\dfrac{1 + \dfrac{2\sqrt{5}}{5}}{2}} = \sqrt{\dfrac{5 + 2\sqrt{5}}{10}}$

(d) $\cos \dfrac{1}{2}\theta = -\sqrt{\dfrac{1 + \cos \theta}{2}} = -\sqrt{\dfrac{1 - \dfrac{2\sqrt{5}}{5}}{2}} = -\sqrt{\dfrac{5 - 2\sqrt{5}}{10}}$

10. $\sec \theta = 2$, $\csc \theta < 0$. Thus, θ lies in quadrant IV, or $\dfrac{3\pi}{2} < \theta < 2\pi$ so that

$\dfrac{3\pi}{4} < \dfrac{\theta}{2} < \pi$, or $\dfrac{\theta}{2}$ lies in Quadrant II.

Therefore, $\cos \theta = \dfrac{1}{2}$, $\sin \theta = \dfrac{-\sqrt{3}}{2}$

    (a) $\sin 2\theta = 2 \sin \theta \cos \theta = 2\left[\dfrac{-\sqrt{3}}{2}\right]\left[\dfrac{1}{2}\right] = \dfrac{-\sqrt{3}}{2}$

    (b) $\cos 2\theta = \cos^2 \theta - \sin^2 \theta = \dfrac{1}{4} - \dfrac{3}{4} = \dfrac{-1}{2}$

    (c) $\sin \dfrac{1}{2}\theta = \sqrt{\dfrac{1 - \cos \theta}{2}} = \sqrt{\dfrac{1 - \dfrac{1}{2}}{2}} = \dfrac{1}{2}$

    (d) $\cos \dfrac{1}{2}\theta = -\sqrt{\dfrac{1 + \cos \theta}{2}} = -\sqrt{\dfrac{1 + \dfrac{1}{2}}{2}} = \dfrac{-\sqrt{3}}{2}$

12. $\cot \theta = 3$, $\cos \theta < 0$. Thus, θ lies in quadrant III, or $\pi < \theta < \dfrac{3\pi}{2}$ so that $\dfrac{\pi}{2} < \theta < \dfrac{3\pi}{4}$, or $\dfrac{\theta}{2}$ lies in quadrant II.

Therefore, $\cos \theta = \dfrac{-3}{\sqrt{10}} = \dfrac{-3\sqrt{10}}{10}$, $\sin \theta = \dfrac{-\sqrt{10}}{10}$

    (a) $\sin 2\theta = 2 \sin \theta \cos \theta = 2\left[\dfrac{-1}{\sqrt{10}}\right]\left[\dfrac{-3}{\sqrt{10}}\right] = \dfrac{6}{10} = \dfrac{3}{5}$

    (b) $\cos 2\theta = \cos^2 \theta - \sin^2 \theta = \dfrac{9}{10} - \dfrac{1}{10} = \dfrac{8}{10} = \dfrac{4}{5}$

    (c) $\sin \dfrac{1}{2}\theta = \sqrt{\dfrac{1 - \cos \theta}{2}} = \sqrt{\dfrac{1 - \left[\dfrac{-3\sqrt{10}}{10}\right]}{2}} = \sqrt{\dfrac{10 + 3\sqrt{10}}{20}} = \dfrac{\sqrt{10 + 3\sqrt{10}}}{2\sqrt{5}}$

                                $= \dfrac{1}{2}\sqrt{\dfrac{10 + 3\sqrt{10}}{5}}$

    (d) $\cos \dfrac{1}{2}\theta = -\sqrt{\dfrac{1 + \cos \theta}{2}} = -\sqrt{\dfrac{1 + \left[\dfrac{-3\sqrt{10}}{10}\right]}{2}} = -\sqrt{\dfrac{10 - 3\sqrt{10}}{20}} = -\dfrac{\sqrt{10 - 3\sqrt{10}}}{2\sqrt{5}}$

                                  $= -\dfrac{1}{2}\sqrt{\dfrac{10 - 3\sqrt{10}}{5}}$

14. Because $22.5° = \dfrac{45°}{2}$, $\cos 22.5° = \cos \dfrac{45°}{2} = \sqrt{\dfrac{1 + \cos 45°}{2}} = \sqrt{\dfrac{1 + \dfrac{\sqrt{2}}{2}}{2}} = \dfrac{\sqrt{2 + \sqrt{2}}}{2}$

16. Because $\dfrac{9\pi}{8} = \dfrac{\dfrac{9\pi}{4}}{2}$, $\tan\dfrac{9\pi}{8} = \tan\dfrac{\dfrac{9\pi}{4}}{2} = \dfrac{1 - \cos\dfrac{9\pi}{4}}{\sin\dfrac{9\pi}{4}} = \dfrac{1 - \dfrac{\sqrt{2}}{2}}{\dfrac{\sqrt{2}}{2}} = \dfrac{2 - \sqrt{2}}{\sqrt{2}} = \sqrt{2} - 1$

18. $\sin 195° = \sin\dfrac{390°}{2} = -\sqrt{\dfrac{1 - \cos 390°}{2}} = -\sqrt{\dfrac{1 - \dfrac{\sqrt{3}}{2}}{2}} = \dfrac{-\sqrt{2 - \sqrt{3}}}{2}$

20. $\csc\dfrac{7\pi}{8} = \dfrac{1}{\sin\dfrac{\dfrac{7\pi}{4}}{2}} = \sqrt{\dfrac{1}{\dfrac{1 - \cos\dfrac{7\pi}{4}}{2}}} = \sqrt{\dfrac{1}{\dfrac{2 - \sqrt{2}}{4}}} = \dfrac{2}{\sqrt{2 - \sqrt{2}}}$

22. $\cos\left[\dfrac{-3\pi}{8}\right] = \cos\left[\dfrac{\dfrac{-3\pi}{4}}{2}\right] = \sqrt{\dfrac{1 + \cos\left[\dfrac{-3\pi}{4}\right]}{2}} = \sqrt{\dfrac{1 + \dfrac{-\sqrt{2}}{2}}{2}} = \dfrac{\sqrt{2 - \sqrt{2}}}{2}$

24. $\cos 3\theta = \cos(2\theta + \theta) = \cos 2\theta\cos\theta - \sin 2\theta\sin\theta = (2\cos^2\theta - 1)\cos\theta - 2\sin\theta\cos\theta\sin\theta$
    $= 2\cos^3\theta - \cos\theta - 2\cos\theta(1 - \cos^2\theta) = 2\cos^3\theta - \cos\theta - 2\cos\theta + 2\cos^3\theta$
    $= 4\cos^3\theta - 3\cos\theta$

26. $\cos 4\theta = \cos(2\theta + 2\theta) = \cos 2\theta\cos 2\theta - \sin 2\theta\sin 2\theta$
    $= (2\cos^2\theta - 1)(2\cos^2\theta - 1) - (2\sin\theta\cos\theta)(2\sin\theta\cos\theta)$
    $= 4\cos^4\theta - 4\cos^2\theta + 1 - 4\sin^2\theta\cos^2\theta = 4\cos^4\theta - 4\cos^2\theta + 1 - 4(1 - \cos^2\theta)\cos^2\theta$
    $= 4\cos^4\theta - 4\cos^2\theta + 1 - 4\cos^2\theta + 4\cos^4\theta = 8\cos^4\theta - 8\cos^2\theta + 1$

28. $\cos 5\theta = \cos(3\theta + 2\theta) = \cos 3\theta\cos 2\theta - \sin 3\theta\sin 2\theta = \cos(2\theta + \theta)\cos 2\theta - \sin(2\theta + \theta)\sin 2\theta$
    $= (\cos 2\theta\cos\theta - \sin 2\theta\sin\theta)\cos 2\theta - (\sin 2\theta\cos\theta + \cos 2\theta\sin\theta)\sin 2\theta$
    $= [(2\cos^2\theta - 1)\cos\theta - 2\sin^2\theta\cos\theta](2\cos^2\theta - 1) -$
    $\quad [2\sin\theta\cos^2\theta + (2\cos^2\theta - 1)\sin\theta]2\sin\theta\cos\theta$
    $= [2\cos^3\theta - \cos\theta - 2\cos\theta(1 - \cos^2\theta](2\cos^2\theta - 1) -$
    $\quad [2\sin\theta\cos^2\theta + 2\sin\theta\cos^2\theta - \sin\theta]2\sin\theta\cos\theta]$
    $= (4\cos^3\theta - 3\cos\theta)(2\cos^2\theta - 1) - [4\cos^2\theta - 1]\cdot 2\sin^2\theta\cos\theta$
    $= (8\cos^5\theta - 10\cos^3\theta + 3\cos\theta) - 2\cos\theta(4\cos^2\theta - 1)(1 - \cos^2\theta)$
    $= 8\cos^5\theta - 10\cos^3\theta + 3\cos\theta - 2\cos\theta(-4\cos^4\theta + 5\cos^2\theta - 1)$
    $= 16\cos^5\theta - 20\cos^3\theta + 5\cos\theta$

30. $\dfrac{\cot\theta - \tan\theta}{\cot\theta + \tan\theta} = \dfrac{\dfrac{\cos\theta}{\sin\theta} - \dfrac{\sin\theta}{\cos\theta}}{\dfrac{\cos\theta}{\sin\theta} + \dfrac{\sin\theta}{\cos\theta}} = \dfrac{\dfrac{\cos^2\theta - \sin^2\theta}{\sin\theta\cos\theta}}{\dfrac{\cos^2\theta + \sin^2\theta}{\sin\theta\cos\theta}} = \dfrac{\cos 2\theta}{1} = \cos 2\theta$

32. $\cot 2\theta = \dfrac{1}{\tan 2\theta} = \dfrac{1}{\dfrac{2\tan\theta}{1 - \tan^2\theta}} = \dfrac{1 - \tan^2\theta}{2\tan\theta} = \dfrac{1}{2}\cot\theta(1 - \tan^2\theta) = \dfrac{1}{2}(\cot\theta - \tan\theta)$

34. $\csc 2\theta = \dfrac{1}{\sin 2\theta} = \dfrac{1}{2 \sin \theta \cos \theta} = \dfrac{1}{2} \sec \theta \csc \theta$

36. $(4 \sin \theta \cos \theta)(1 - 2 \sin^2 \theta) = (2 \sin 2\theta)(\cos 2\theta) = \sin 4\theta$

38. $\sin^2 \theta \cos^2 \theta = \left[\dfrac{1 - \cos 2\theta}{2}\right]\left[\dfrac{1 + \cos 2\theta}{2}\right] = \dfrac{1 - \cos^2 2\theta}{4} = \dfrac{\sin^2 2\theta}{4} = \dfrac{\dfrac{1 - \cos 4\theta}{2}}{4}$

$\qquad = \dfrac{1}{8}(1 - \cos 4\theta)$

40. $\csc^2 \dfrac{\theta}{2} = \dfrac{1}{\sin^2\left(\dfrac{\theta}{2}\right)} = \dfrac{1}{\dfrac{1 - \cos \theta}{2}} = \dfrac{2}{1 - \cos \theta}$

42. $\tan \dfrac{\theta}{2} = \sqrt{\dfrac{1 - \cos \theta}{1 + \cos \theta}} \cdot \dfrac{\sqrt{1 - \cos \theta}}{\sqrt{1 - \cos \theta}} = \dfrac{\sqrt{(1 - \cos \theta)^2}}{\sin \theta} = \dfrac{1 - \cos \theta}{\sin \theta} = \csc \theta - \cot \theta$

44. $\dfrac{\sin^3 \theta + \cos^3 \theta}{\sin \theta + \cos \theta} = \dfrac{(\sin \theta + \cos \theta)(\sin^2 \theta - \sin \theta \cos \theta + \cos^2 \theta)}{\sin \theta + \cos \theta} = 1 - \sin \theta \cos \theta$

$\qquad = 1 - \dfrac{1}{2}(2 \sin \theta \cos \theta) = 1 - \dfrac{1}{2} \sin 2\theta$

46. $\dfrac{\cos \theta + \sin \theta}{\cos \theta - \sin \theta} - \dfrac{\cos \theta - \sin \theta}{\cos \theta + \sin \theta} = \dfrac{(\cos \theta + \sin \theta)^2 - (\cos \theta - \sin \theta)^2}{\cos^2 \theta - \sin^2 \theta}$

$\qquad = \dfrac{\cos^2 \theta + 2 \cos \theta \sin \theta + \sin^2 \theta - (\cos^2 \theta - 2 \cos \theta \sin \theta + \sin^2 \theta)}{\cos 2\theta}$

$\qquad = \dfrac{1 + \sin 2\theta - (1 - \sin 2\theta)}{\cos 2\theta} = \dfrac{2 \sin 2\theta}{\cos 2\theta} = 2 \tan 2\theta$

48. $\tan \theta + \tan(\theta + 120°) + \tan(\theta + 240°) = \tan \theta + \dfrac{\tan \theta + \tan 120°}{1 - \tan \theta \tan 120°} + \dfrac{\tan \theta + \tan 240°}{1 - \tan \theta \tan 240°}$

$\qquad = \tan \theta + \dfrac{\tan \theta - \sqrt{3}}{1 + \sqrt{3} \tan \theta} + \dfrac{\tan \theta + \sqrt{3}}{1 - \sqrt{3} \tan \theta}$

$\qquad = \dfrac{\tan \theta(1 - 3 \tan^2 \theta) + \left(\tan \theta - \sqrt{3}\right)\left(1 - \sqrt{3} \tan \theta\right) + \left(\tan \theta + \sqrt{3}\right)\left(1 + \sqrt{3} \tan \theta\right)}{1 - 3 \tan^2 \theta}$

$\qquad = \dfrac{\tan \theta(1 - 3\tan^2 \theta) + \tan \theta - \sqrt{3} - \sqrt{3} \tan^2 \theta + 3 \tan \theta + \tan \theta + \sqrt{3} + \sqrt{3} \tan^2 \theta + 3 \tan \theta}{1 - 3 \tan^2 \theta}$

$\qquad = \dfrac{-3 \tan^3 \theta + 9 \tan \theta}{1 - 3 \tan^2 \theta} = \dfrac{3(3 \tan \theta - \tan^3 \theta)}{1 - 3 \tan^2 \theta} = 3 \tan 3\theta$

50. $\sin\left[2 \sin^{-1}\left(\dfrac{\sqrt{3}}{2}\right)\right] = \sin\left(2 \cdot \dfrac{\pi}{3}\right) = \sin\dfrac{2\pi}{3} = \dfrac{\sqrt{3}}{2}$

52. $\cos\left[2 \cos^{-1} \dfrac{4}{5}\right] = 2 \cos^2\left[\cos^{-1} \dfrac{4}{5}\right] - 1 = 2\left(\dfrac{4}{5}\right)^2 - 1 = 2\left[\dfrac{16}{25}\right] - 1 = \dfrac{32 - 25}{25} = \dfrac{7}{25}$

54. $\tan\left[2\tan^{-1}\dfrac{3}{4}\right] = \dfrac{2\tan\left[\tan^{-1}\dfrac{3}{4}\right]}{1-\tan^2\left[\tan^{-1}\dfrac{3}{4}\right]} = \dfrac{2\left(\dfrac{3}{4}\right)}{1-\left(\dfrac{3}{4}\right)^2} = \dfrac{\dfrac{3}{2}}{\dfrac{16-9}{16}} = \dfrac{24}{7}$

56. $\cos\left[2\tan^{-1}\left(\dfrac{-4}{3}\right)\right] = 2\cos^2\left[\tan^{-1}\left(\dfrac{-4}{3}\right)\right] - 1 = 2\left(\dfrac{3}{5}\right)^2 - 1 = \dfrac{18-25}{25} = \dfrac{-7}{25}$

58. $\cos^2\left[\dfrac{1}{2}\sin^{-1}\dfrac{3}{5}\right] = \dfrac{1+\cos\left[\sin^{-1}\dfrac{3}{5}\right]}{2} = \dfrac{1+\dfrac{4}{5}}{2} = \dfrac{9}{10}$

60. $\csc\left[2\sin^{-1}\left(\dfrac{-3}{5}\right)\right] = \dfrac{1}{\sin\left[2\sin^{-1}\left(\dfrac{-3}{5}\right)\right]} = \dfrac{1}{2\sin\left[\sin^{-1}\left(\dfrac{-3}{5}\right)\right]\cos\left[\sin^{-1}\left(\dfrac{-3}{5}\right)\right]}$

$= \dfrac{1}{2\left(\dfrac{-3}{5}\right)\left(\dfrac{4}{5}\right)} = \dfrac{1}{\dfrac{-24}{25}} = \dfrac{-25}{24}$

62. Let $\alpha = \cos^{-1}\dfrac{5}{13}$

$\cot^2\left[\dfrac{1}{2}\cos^{-1}\dfrac{5}{13}\right] = \dfrac{1}{\tan^2\left[\dfrac{1}{2}\cos^{-1}\dfrac{5}{13}\right]} = \dfrac{1}{\tan\dfrac{1}{2}\alpha} = \dfrac{1}{\left(\dfrac{1-\cos\alpha}{\sin\alpha}\right)^2}$

$= \dfrac{1}{\left(\dfrac{1-\dfrac{5}{13}}{\dfrac{12}{13}}\right)^2} = \dfrac{1}{\left(\dfrac{2}{3}\right)^2} = \dfrac{9}{4}$

64. $\dfrac{1}{2}\cos^2 x + C = \dfrac{1}{4}\cos 2x$

$C = \dfrac{1}{4}\cos 2x - \dfrac{1}{2}\cos^2 x$

$C = \dfrac{1}{4}(\cos 2x - 2\cos^2 x)$

$C = \dfrac{1}{4}(2\cos^2 x - 1 - 2\cos^2 x)$

$C = -\dfrac{1}{4}$

66. (a) $A = 2x \cdot y$ since $2x$ is the length of the base of the rectangle. The length of the hypotenuse of the triangle formed by x and y is 1 since this is the radius. Therefore, $\sin\theta = \dfrac{y}{1}$, so $y = \sin\theta$. $\cos\theta = \dfrac{x}{1}$, so $x = \cos\theta$. Thus, $A = 2\sin\theta\cos\theta$.

(b) Since $2\sin\theta\cos\theta = \sin 2\theta$, $A = \sin 2\theta$.

(c) Since $0 \le A \le 1$, A is maximized when $\sin 2\theta = 1$. Thus, $\theta = 45°$.

(d) $x = \cos 45° = \dfrac{\sqrt{2}}{2}$. So, the base of the rectangle will be $2\left[\dfrac{\sqrt{2}}{2}\right] = \sqrt{2}$.

$\quad\quad y = \sin 45° = \dfrac{\sqrt{2}}{2}$.

68. To graph $g(x) = \cos^2 x = \dfrac{(1 + \cos 2x)}{2}$ for $0 \le x \le 2\pi$, begin with the graph of $y = \cos x$. Apply a horizontal compression to obtain the graph of $y = \cos 2x$. Shift up one unit to obtain the graph of $y = 1 + \cos 2x$. Apply a vertical compression to obtain the final graph: $y = \dfrac{1}{2}(1 + \cos 2x)$.

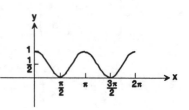

70. $\sin \dfrac{\pi}{8} = \sin \dfrac{1}{2} \cdot \dfrac{\pi}{4} = \sqrt{\dfrac{1 - \cos \dfrac{\pi}{4}}{2}} = \dfrac{\sqrt{2 - \sqrt{2}}}{2}$

$\sin \dfrac{\pi}{16} = \sin \dfrac{1}{2} \cdot \dfrac{\pi}{8} = \sqrt{\dfrac{1 - \cos \dfrac{\pi}{8}}{2}} = \sqrt{\dfrac{1 - \dfrac{\sqrt{2 - \sqrt{2}}}{2}}{2}} = \dfrac{\sqrt{2 - \sqrt{2 - \sqrt{2}}}}{2}$

$\cos \dfrac{\pi}{16} = \cos \dfrac{1}{2} \cdot \dfrac{\pi}{8} = \sqrt{\dfrac{1 + \cos \dfrac{\pi}{8}}{2}} = \sqrt{\dfrac{1 + \dfrac{\sqrt{2 - \sqrt{2}}}{2}}{2}} = \dfrac{\sqrt{2 + \sqrt{2 - \sqrt{2}}}}{2}$

72. $\tan \theta = \tan 3\left(\dfrac{\theta}{3}\right) = \dfrac{3 \tan \dfrac{\theta}{3} + \tan^3 \dfrac{\theta}{3}}{1 - 3 \tan^2 \dfrac{\theta}{3}} = a \tan \dfrac{\theta}{3}$

$3 \tan \dfrac{a}{3} - \tan^3 \dfrac{\theta}{3} = a \tan \dfrac{\theta}{3} - 3a \tan^3 \dfrac{\theta}{3}$

$3 - \tan^2 \dfrac{\theta}{3} = a - 3 a \tan^2 \dfrac{\theta}{3}$

$(3a - 1)\tan^2 \dfrac{\theta}{3} = a - 3$

$\tan^2 \dfrac{\theta}{3} = \dfrac{a - 3}{3a - 1}$

74. $\dfrac{1}{2}(\ln|1 + \cos 2\theta| - \ln 2) = \ln \left[\dfrac{|1 + \cos 2\theta|}{2}\right]^{1/2} = \ln\left|\cos \dfrac{2\theta}{\theta}\right| = \ln |\cos \theta|$

76. $y = \dfrac{1}{2} \sin 2\pi x + \dfrac{1}{4} \sin 4\pi x = \dfrac{1}{2} \sin 2\pi x + \dfrac{1}{4} \cdot 2 \sin 2\pi x \cos 2\pi x$

$= \dfrac{1}{2} \sin 2\pi x(1 + \cos 2\pi x) = \sin 2\pi x \cos^2 \pi x$

9.4 Product-To-Sum and Sum-To-Product Formulas

For Problems 2–10, we are using the formulas:

$$\sin \alpha \sin \beta = \frac{1}{2}\left[\cos(\alpha - \beta) - \cos(\alpha + \beta)\right]$$

$$\cos \alpha \cos \beta = \frac{1}{2}\left[\cos(\alpha + \beta) + \cos(\alpha - \beta)\right]$$

$$\sin \alpha \cos \beta = \frac{1}{2}\left[\sin(\alpha + \beta) + \sin(\alpha - \beta)\right]$$

2. $\cos 4\theta \cos 2\theta = \frac{1}{2}(\cos 6\theta + \cos 2\theta)$

4. $\sin 3\theta \sin 5\theta = \frac{1}{2}(\cos 2\theta - \cos 8\theta)$

6. $\sin 4\theta \cos 6\theta = \frac{1}{2}(\sin 10\theta - \sin 2\theta)$

8. $\cos 3\theta \cos 4\theta = \frac{1}{2}(\cos 7\theta + \cos \theta)$

10. $\sin \dfrac{\theta}{2} \cos \dfrac{5\theta}{2} = \frac{1}{2}(\sin 3\theta - \sin 2\theta)$

For Problems 11–18, we are using the formulas:

$$\sin \alpha + \sin \beta = 2 \sin \frac{\alpha + \beta}{2} \cos \frac{\alpha - \beta}{2}$$

$$\sin \alpha - \sin \beta = 2 \sin \frac{\alpha - \beta}{2} \cos \frac{\alpha + \beta}{2}$$

$$\cos \alpha + \cos \beta = 2 \cos \frac{\alpha + \beta}{2} \cos \frac{\alpha - \beta}{2}$$

$$\cos \alpha - \cos \beta = -2 \sin \frac{\alpha + \beta}{2} \sin \frac{\alpha - \beta}{2}$$

12. $\sin 4\theta + \sin 2\theta = 2 \sin 3\theta \cos \theta$

14. $\cos 5\theta - \cos 3\theta = -2 \sin 4\theta \sin \theta$

16. $\cos \theta + \cos 3\theta = 2 \cos 2\theta \cos \theta$

18. $\sin \dfrac{\theta}{2} - \sin \dfrac{3\theta}{2} = 2(-\sin \theta)\cos 2\theta = -2 \sin \theta \cos 2\theta$

20. $\dfrac{\cos \theta + \cos 3\theta}{2 \cos 2\theta} = \dfrac{2 \cos 2\theta \cos \theta}{2 \cos 2\theta} = \cos \theta$

22. $\dfrac{\cos \theta - \cos 3\theta}{\sin 3\theta - \sin \theta} = \dfrac{-2 \sin 2\theta(-\sin \theta)}{2 \sin \theta \cos 2\theta} = \tan 2\theta$

24. $\dfrac{\cos \theta - \cos 5\theta}{\sin \theta + \sin 5\theta} = \dfrac{-2 \sin 3\theta(-\sin 2\theta)}{2 \sin 3\theta \cos 2\theta} = \tan 2\theta$

26. $\sin \theta(\sin 3\theta + \sin 5\theta) = \sin \theta(2 \sin 4\theta \cos \theta) = 2 \sin 4\theta \sin \theta \cos \theta = \cos \theta(2 \sin 4\theta \sin \theta$

$$= \cos \theta \left[2 \cdot \frac{1}{2}(\cos 3\theta - \cos 5\theta) \right] = \cos \theta(\cos 3\theta - \cos 5\theta)$$

28. $\dfrac{\sin 4\theta - \sin 8\theta}{\cos 4\theta - \cos 8\theta} = \dfrac{-2 \sin 2\theta \cos 6\theta}{-2 \sin 6\theta(-\sin 2\theta)} = -\cot 6\theta$

30. $\dfrac{\cos 4\theta - \cos 8\theta}{\cos 4\theta + \cos 8\theta} = \dfrac{-2 \sin 6\theta(-\sin 2\theta)}{2 \cos 6\theta \cos 2\theta} = \tan 2\theta \tan 6\theta$

32.
$$\frac{\cos \alpha + \cos \beta}{\cos \alpha - \cos \beta} = \frac{2 \cos \dfrac{\alpha + \beta}{2} \cos \dfrac{\alpha - \beta}{2}}{-2 \sin \dfrac{\alpha + \beta}{2} \sin \dfrac{\alpha - \beta}{2}} = -\cot \frac{\alpha + \beta}{2} \cot \frac{\alpha - \beta}{2}$$

34.
$$\frac{\sin \alpha - \sin \beta}{\cos \alpha - \cos \beta} = \frac{2 \cos \dfrac{\alpha - \beta}{2} \cos \dfrac{\alpha + \beta}{2}}{-2 \sin \dfrac{\alpha + \beta}{2} \sin \dfrac{\alpha - \beta}{2}} = -\cot \frac{\alpha + \beta}{2}$$

36. $1 - \cos 2\theta + \cos 4\theta - \cos 6\theta = (1 \cos 0\theta - \cos 6\theta) + (\cos 4\theta - \cos 2\theta)$
$$= 2 \sin 3\theta \sin 3\theta - 2 \sin 3\theta \sin \theta = 2 \sin^2 3\theta - 2 \sin 3\theta \sin \theta$$
$$= 2 \sin 3\theta(\sin 3\theta - \sin \theta) = 2 \sin 3\theta(2 \sin \theta \cos 2\theta) = 4 \sin \theta \cos 2\theta \sin 3\theta$$

38. $y = \sin 2\pi(941)t + \sin 2\pi(1477)t = 2 \sin \left[\dfrac{2\pi(941)t + 2\pi(1477)t}{2} \right] \cos \left[\dfrac{2\pi(941)t - 2\pi(1477)t}{2} \right]$
$$= 2 \sin 2418\pi t \cos 536\pi t \text{ since } \cos(-\theta) = \cos \theta$$

40. If $\alpha + \beta + \gamma = \pi$, then

$\tan \alpha + \tan \beta + \tan \gamma$

$$= \frac{\sin \alpha}{\cos \alpha} + \frac{\sin \beta}{\cos \beta} + \frac{\sin \gamma}{\cos \gamma}$$

$$= \frac{\sin \alpha \cos \beta \cos \gamma + \sin \beta \cos \alpha \cos \gamma + \sin \gamma \cos \gamma \cos \beta}{\cos \alpha \cos \beta \cos \gamma}$$

$$= \frac{\cos \gamma(\sin \alpha \cos \beta + \sin \beta \cos \alpha) + \sin \gamma \cos \alpha \cos \beta}{\cos \alpha \cos \beta \cos \gamma}$$

$$= \frac{\cos \gamma \sin(\pi - \gamma) + \sin \gamma \cos \alpha \cos \beta}{\cos \alpha \cos \beta \cos \gamma} = \frac{\cos \gamma \sin \gamma + \sin \gamma \cos \alpha \cos \beta}{\cos \alpha \cos \beta \cos \gamma}$$

$$= \frac{\sin \gamma(\cos \gamma \cos \alpha \cos \beta)}{\cos \alpha \cos \beta \cos \gamma} = \frac{\sin \gamma \{\cos[\pi - (\alpha + \beta)] + \cos \alpha \cos \beta\}}{\cos \alpha \cos \beta \cos \gamma}$$

$$= \frac{\sin \gamma \left\{ \cos \gamma + \dfrac{1}{2}[\cos (\alpha + \beta) + \cos(\alpha - \beta]\right\}}{\cos \alpha \cos \beta \cos \gamma}$$

$$= \frac{\sin \gamma \left\{ \cos [\pi - (\alpha + \beta)] + \dfrac{1}{2}[\cos (\alpha + \beta) + \cos(\alpha - \beta)]\right\}}{\cos \alpha \cos \beta \cos \gamma}$$

$$= \frac{\sin \gamma \left\{ [\cos \pi \cos(\alpha + \beta) + \sin \pi \sin(\alpha + \beta)] + \dfrac{1}{2}[\cos (\alpha + \beta) + \cos(\alpha - \beta)]\right\}}{\cos \alpha \cos \beta \cos \gamma}$$

$$= \frac{\sin \gamma \left[-\cos(\alpha + \beta) + \dfrac{1}{2}\cos(\alpha + \beta) + \dfrac{1}{2}\cos(\alpha - \beta)\right]}{\cos \alpha \cos \beta \cos \gamma}$$

$$= \frac{\sin \gamma \left\{ \dfrac{1}{2}[\cos(\alpha - \beta) - \cos(\alpha + \beta)]\right\}}{\cos \alpha \cos \beta \cos \gamma} = \frac{\sin \gamma \sin \alpha \sin \beta}{\cos \alpha \cos \beta \cos \gamma} = \tan \alpha \tan \beta \tan \gamma$$

42. $2 \sin \dfrac{\alpha - \beta}{2} \cos \dfrac{\alpha + \beta}{2} = 2 \cdot \dfrac{1}{2}\left[\sin\left(\dfrac{\alpha - \beta}{2} + \dfrac{\alpha + \beta}{2}\right) + \sin\left(\dfrac{\alpha - \beta}{2} - \dfrac{\alpha + \beta}{2}\right) \right]$
$$= \sin \frac{2\alpha}{2} + \sin\left(\frac{-2\beta}{2}\right) = \sin \alpha - \sin \beta$$

44. $-2 \sin \dfrac{\alpha + \beta}{2} \sin \dfrac{\alpha - \beta}{2} = -2 \cdot \dfrac{1}{2} \left[\cos \left[\dfrac{\alpha + \beta}{2} - \dfrac{\alpha - \beta}{2} \right] - \cos \left[\dfrac{\alpha - \beta}{2} + \dfrac{\alpha - \beta}{2} \right] \right]$

$$= - \left[\cos \left[\dfrac{2\beta}{2} \right] - \cos \left[\dfrac{2\alpha}{2} \right] \right] = \cos \alpha - \cos \beta$$

9.5 Trigonometric Equations

2. $\tan \theta = 1$

The period of the tangent function is π; and, in the interval $[0, \pi)$, the tangent function has the value

1 at $\dfrac{\pi}{4}$. Therefore,

$$\theta = \dfrac{\pi}{4} + k\pi, \ k \text{ any integer}$$

The solutions on the interval $[0, 2\pi)$, are $\theta = \dfrac{\pi}{4}, \dfrac{5\pi}{4}$

4. $\cos \theta = -\dfrac{\sqrt{3}}{2}$

$$\theta = \dfrac{5\pi}{6} \text{ or } \theta = \dfrac{7\pi}{6} \text{ [on the interval } [0, 2\pi)]$$

6. $\sin \theta = \dfrac{\sqrt{2}}{2}$

$$\theta = \dfrac{\pi}{4} \text{ or } \theta = \dfrac{3\pi}{4} \text{ \{on the interval } [0, 2\pi)]$$

8. $\tan \dfrac{\theta}{2} = \sqrt{3}$

$$\dfrac{\theta}{2} = \dfrac{\pi}{3} + k\pi, \ k \text{ any integer}$$

$$\theta = \dfrac{2\pi}{3} + 2k\pi, \ k \text{ any integer}$$

The only solution on the interval $[0, 2\pi)$ is $\theta = \dfrac{2\pi}{3}$.

10. $\sin \left[3\theta + \dfrac{\pi}{18} \right] = 1$

$$3\theta + \dfrac{\pi}{18} = \dfrac{\pi}{2} + 2k\pi, \ k \text{ any integer}$$

$$3\theta = \dfrac{4\pi}{9} + 2k\pi, \ k \text{ any integer}$$

$$\theta = \dfrac{4\pi}{27} + \dfrac{2}{3}k\pi, \ k \text{ any integer}$$

The solutions on the interval $[0, 2\pi)$ are $\theta = \dfrac{4\pi}{27}, \dfrac{22\pi}{27}$

12. $\cot \dfrac{2\theta}{3} = -\sqrt{3}$

   $\dfrac{2\theta}{3} = \dfrac{5\pi}{4} + k\pi$, k any integer

   $\theta = \dfrac{5\pi}{4} + \dfrac{3}{2}k\pi$, k any integer

The only solution on the interval $[0, 2\pi)$ is $\theta = \dfrac{5\pi}{4}$.

14. $\cos \theta = 0.6$
        $\theta = 0.93$ or $-0.93 + 2\pi$

16. $\cot \theta = 2$. Thus, $\tan \theta = \dfrac{1}{2}$.
        $\theta = .46$ or $\theta = \pi + .46$

18. $\sin \theta = -0.2$
        $\theta = -0.20$ or $\theta = \pi + 0.20$

20. $\csc \theta = -3$. Therefore, $\sin \theta = \dfrac{-1}{3}$
        $\theta = 5.94$ or $\theta = \pi + 0.34$

22. $\sin^2 \theta - 1 = 0$
        $(\sin \theta + 1)(\sin \theta - 1) = 0$
        $\sin \theta = -1$ or $\sin \theta = 1$
            $\theta = \dfrac{3\pi}{2}$ or $\theta = \dfrac{\pi}{2}$

24. $2 \cos^2 \theta + \cos \theta - 1 = 0$
        $(2 \cos \theta - 1)(\cos \theta + 1) = 0$
        $2 \cos \theta - 1 = 0$ or $\cos \theta + 1 = 0$
            $\cos \theta = \dfrac{1}{2}$ or $\cos \theta = -1$
                $\theta = \dfrac{\pi}{3}, \dfrac{5\pi}{3}$ or $\theta = \pi$

26. $(\cot \theta + 1)\left[\csc \theta - \dfrac{1}{2}\right] = 0$

        $\cot \theta + 1 = 0$ or $\csc \theta - \dfrac{1}{2} = 0$

            $\cot \theta = -1$ or $\csc \theta = \dfrac{1}{2}$

            $\theta = \dfrac{3\pi}{4}, \dfrac{7\pi}{4}$ or   No solution.

28. $\cos \theta + \sin \theta = 0$
                $\sin \theta = -\cos \theta$
            $\dfrac{\sin \theta}{\cos \theta} = \dfrac{-\cos \theta}{\cos \theta}$
                $\tan \theta = -1$
                    $\theta = \dfrac{3\pi}{4}$ or $\theta = \dfrac{7\pi}{4}$

30. $\sin 2\theta = \cos \theta$
        $2 \sin \theta \cos \theta = \cos \theta$
            $2 \sin \theta = 1$
                $\sin \theta = \dfrac{1}{2}$
                    $\theta = \dfrac{\pi}{6}$ or $\theta = \dfrac{5\pi}{6}$

32.
$$\tan \theta = \cot \theta$$
$$\tan \theta - \cot \theta = 0$$
$$\frac{\tan^2 \theta - 1}{\tan \theta} = 0$$
$$\tan^2 \theta - 1 = 0$$
$$\tan^2 \theta = 1$$

$$\tan \theta = 1 \qquad \text{or} \qquad \tan \theta = -1$$

$$\theta = \frac{\pi}{4}, \frac{5\pi}{4} \quad \text{or} \qquad \theta = \frac{3\pi}{4}, \frac{7\pi}{4}$$

34.
$$\sin 2\theta \sin \theta = \cos \theta$$
$$2 \sin \theta \cos \theta \sin \theta = \cos \theta$$
$$2 \sin^2 \theta = 1$$
$$\sin^2 \theta = \frac{1}{2}$$
$$\sin \theta = \pm \frac{\sqrt{2}}{2}$$

$$\theta = \frac{\pi}{4}, \frac{3\pi}{4} \qquad \text{or} \qquad \theta = \frac{5\pi}{4}, \frac{7\pi}{4}$$

36.
$$\cos 2\theta + \cos 4\theta = 0$$
$$2 \cos 3\theta \cos \theta = 0$$
$$\cos 3\theta = 0 \qquad \text{or} \qquad \cos \theta = 0$$

$$3\theta = \frac{\pi}{2} + 2k\pi, \; 3\theta = \frac{3\pi}{2} + 2k\pi \qquad \text{or} \qquad \theta = \frac{\pi}{2} + 2k\pi, \; \theta = \frac{3\pi}{2} + 2k\pi, \; k \text{ any integer}$$

$$\theta = \frac{\pi}{6} + \frac{2}{3}k\pi, \; \theta = \frac{\pi}{2} + \frac{2}{3}k\pi \qquad \text{or} \qquad \theta = \frac{\pi}{2} + 2k\pi, \; \theta = \frac{3\pi}{2} + 2k\pi, \; k \text{ any integer}$$

The solutions in $[0, 2\pi)$ are $\theta = \dfrac{\pi}{6}, \dfrac{\pi}{2}, \dfrac{3\pi}{2}, \dfrac{5\pi}{6}, \boxed{\dfrac{9\pi}{6}}, \dfrac{7\pi}{6}, \dfrac{11\pi}{6}$.

$$3\pi/2$$

38.
$$\sin 4\theta - \sin 6\theta = 0$$
$$-2 \sin \theta \cos 5\theta = 0$$
$$\sin \theta = 0 \quad \text{or} \quad \cos 5\theta = 0$$

$$\theta = k\pi \text{ or} \qquad 5\theta = \frac{\pi}{2} + 2k\pi, \; 5\theta = \frac{3\pi}{2} + 2k\pi, \; k \text{ any integer}$$

The solutions in $[0, 2\pi)$ are $\theta = 0, \pi, \; \dfrac{\pi}{10}, \dfrac{\pi}{2}, \dfrac{9\pi}{10}, \dfrac{13\pi}{10}, \dfrac{17\pi}{10}, \dfrac{3\pi}{10} + \dfrac{7\pi}{10}, \dfrac{11\pi}{10}, \dfrac{3\pi}{2}, \dfrac{19\pi}{10}$.

40.
$$\sin^2 \theta = 2 \cos \theta + 2$$
$$1 - \cos^2 \theta = 2 \cos \theta + 2$$
$$\cos^2 \theta + 2 \cos \theta + 1 = 0$$
$$(\cos \theta + 1)^2 = 0$$
$$\cos \theta + 1 = 0$$
$$\cos \theta = -1$$
$$\theta = \pi$$

42.
$$\csc^2 = \cot \theta + 1$$
$$\cot^2 \theta + 1 = \cot \theta + 1$$
$$\cot^2 \theta - \cot \theta = 0$$
$$\cot \theta (\cot \theta - 1) = 0$$
$$\cot \theta = 0 \quad \text{or} \quad \cot \theta = 1$$
$$\theta = \frac{\pi}{2}, \frac{3\pi}{2}, \frac{\pi}{4}, \frac{5\pi}{4}$$

44.
$$\cos 2\theta + 5 \cos \theta + 3 = 0$$
$$2 \cos^2 \theta - 1 + 5 \cos \theta + 3 = 0$$
$$2 \cos^2 \theta + 5 \cos \theta + 2 = 0$$
$$(2 \cos \theta + 1)(\cos \theta + 2) = 0$$
$$\cos \theta = \frac{-1}{2} \text{ or } \cos \theta = -2 \text{ (Impossible,}$$
$$\text{because } |\cos \theta| \leq 1)$$
$$\theta = \frac{2\pi}{3}, \theta = \frac{4\pi}{3}$$

46.
$$\sec \theta = \tan \theta + \cot \theta$$
$$\frac{1}{\cos \theta} = \frac{\sin \theta}{\cos \theta} + \frac{\cos \theta}{\sin \theta}$$
$$\frac{1}{\cos \theta} = \frac{\sin^2 \theta + \cos^2 \theta}{\sin \theta \cos \theta}$$
$$\frac{1}{\cos \theta} = \frac{1}{\sin \theta \cos \theta}$$
$$\frac{\sin \theta \cos \theta}{\cos \theta} = 1$$
$$\sin \theta = 1$$
$$\theta = \frac{\pi}{2} + 2k\pi, \text{ } k \text{ any integer}$$

Since $\sec \dfrac{\pi}{2}$ and $\tan \dfrac{\pi}{2}$ do not exist, there is no real solution.

48. $\sqrt{3} \sin \theta + \cos \theta = 1$

We divide each side of the equation by 2. Then, $\dfrac{\sqrt{3}}{2} \sin \theta + \dfrac{1}{2} \cos \theta = \dfrac{1}{2}$. There is a unique angle, ϕ, $0 \leq \phi < 2\pi$, for which $\cos \phi = \dfrac{\sqrt{3}}{2}$ and $\sin \phi = \dfrac{1}{2}$. The angle ϕ is therefore $\dfrac{\pi}{6}$ and $\cos \phi \sin \theta + \sin \phi \cos \theta = \dfrac{1}{2}$.

$$\sin(\theta + \phi) = \sin\left[\theta + \frac{\pi}{6}\right] = \frac{1}{2}$$
$$\theta + \frac{\pi}{6} = \frac{\pi}{6} \text{ or } \theta + \frac{\pi}{6} = \frac{5\pi}{6}$$
$$\theta = 0 \text{ or } \theta = \frac{2\pi}{3}$$

50.
$$\tan 2\theta + 2 \cos \theta = 0$$
$$\frac{\sin 2\theta}{\cos 2\theta} + 2 \cos \theta = 0$$
$$\frac{2 \sin \theta \cos \theta + 2 \cos \theta \cos 2\theta}{\cos 2\theta} = 0$$
$$2 \cos \theta(\sin \theta + \cos 2\theta) = 0$$
$$2 \cos \theta(\sin \theta + 1 - 2 \sin^2 \theta) = 0$$
$$2 \cos \theta(2 \sin^2 \theta - \sin \theta - 1) = 0$$
$$2 \cos \theta(2 \sin \theta + 1)(\sin \theta - 1) = 0$$
$$2 \cos \theta = 0 \quad \text{or} \quad \sin \theta = -1 \quad \text{or} \quad \sin \theta - 1 = 0$$
$$\cos \theta = 0 \quad \text{or} \quad \sin \theta = \frac{-1}{2} \quad \text{or} \quad \sin \theta = 1$$
$$\theta = \frac{\pi}{2}, \frac{3\pi}{2}, \frac{7\pi}{6}, \frac{11\pi}{6}, \frac{\pi}{2}$$

52.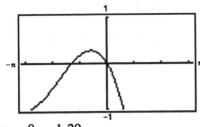

$x = 0; -1.29$

54.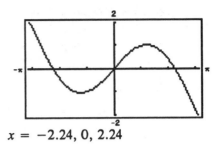

$x = -2.24, 0, 2.24$

56.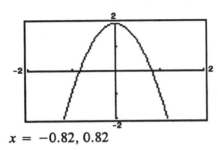

$x = -0.82, 0.82$

60. -0.29

62. -1.25 **64.** $-1.72, 0$ **66.** $-0.61, 0.81$ **68.** 0.31

70. (a) $\sin 2\theta + \cos 2\theta = 0$

Divide each side of the equation by $\sqrt{2}$

$$\frac{1}{\sqrt{2}} \sin 2\theta + \frac{1}{\sqrt{2}} \cos 2\theta = 0$$

$\cos \phi = \dfrac{1}{\sqrt{2}}$ and $\sin \phi = \dfrac{1}{\sqrt{2}}$, so $\phi = 45°$

$\sin(2\theta + 45°) = 0$

$\begin{array}{ll} 2\theta = 0 - 45° \text{ or} & 2\theta = 180° - 45° \\ \theta = -22.5° & \theta = 67.5° \end{array}$

So, the angle θ that maximizes R is $67.5°$. For $\dfrac{\pi}{4} < \theta < \dfrac{\pi}{2}$, only $\theta = \dfrac{3\pi}{8}$ is a solution.

(b) $\begin{aligned} \sin 2\theta + \cos 2\theta &= 0 \\ \tan 2\theta + 1 &= 0 \\ \tan 2\theta &= -1 \\ 2\theta &= 135° \end{aligned}$

$\theta = 67.5°$ Only $\theta = \dfrac{3\pi}{8}$ is a solution.

(c) $R(67.6°) = \dfrac{32^2 \sqrt{2}}{32} (\sin 2 \cdot 67.5° - \cos 2 \cdot 67.5° - 1) = 64 - 32\sqrt{2}$ ft

72. (a) See Figure.

$L = x + y$

$\cos \theta = \dfrac{3}{x}$, so $x = 3 \sec \theta$

$\sin \theta = \dfrac{4}{y}$, so $y = 4 \csc \theta$

Thus, $L(\theta) = 3 \sec \theta + 4 \csc \theta$

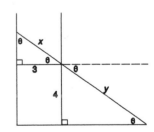

(b) $\quad -3 \sec \theta \tan \theta - 4 \csc \theta \cot \theta = 0$

$$-3 \sec \theta \tan \theta = 4 \csc \theta \cot \theta$$

$$\frac{\sec \theta \tan \theta}{\csc \theta \cot \theta} = \frac{-4}{3}$$

$$\tan^3 \theta = \frac{-4}{3}$$

$$\theta \approx 47.7° \text{ or } 0.83 \text{ radians}$$

(c) $\quad L(\theta) = 3 \sec 47.7° + 4 \csc 47.7° = 9.87 \text{ feet}$

74. $\quad \dfrac{v_1}{v_2} = 1.66, \ \theta_1 = 50°$

Because $\dfrac{\sin \theta_1}{\sin \theta_2} = \dfrac{v_1}{v_2}, \ \dfrac{\sin 50°}{\sin \theta_2} = 1.66$

$$\sin 50° = 1.66 \sin \theta_2$$

$$\frac{\sin 50°}{1.66} = \sin \theta_2$$

$$27.5° = \theta_2$$

76. $\quad \dfrac{v_1}{v_2} = \dfrac{\text{speed of light}}{1.92 \times 10^8} = \dfrac{2.99 \times 10^8}{1.92 \times 10^8}$

The index of refraction of this liquid, with respect to air, for sodium light, is 1.56.

78. $\quad \dfrac{\sin \theta_1}{\sin \theta_2} = 1.52 \qquad$ (The index of refraction of crown glass)

$$\frac{\sin 30°}{\sin \theta_2} = 1.52$$

$$\sin \theta_2 = \frac{\sin 30°}{1.52} \approx 0.3289$$

$$\theta_2 = 19.2°$$

The angle of refraction of a light ray travelling through air that makes an angle of incidence of 30° on a smooth, flat slab of crown glass is 19.2°.

⑨ Chapter Review

2. $\quad \sin \theta \csc \theta - \sin^2 \theta = 1 - \sin^2 \theta = \cos^2 \theta$

4. $\quad (1 - \cos^2 \theta)(1 + \cot^2 \theta) = (\sin^2 \theta)(\csc^2 \theta) = 1$

6. $\quad 4 \sin^2 \theta + 2 \cos^2 \theta = 4(1 - \cos^2 \theta) + 2 \cos^2 \theta = 4 - 4 \cos^2 \theta + 2 \cos^2 \theta = 4 - 2 \cos^2 \theta$

8. $\quad \dfrac{\sin \theta}{1 + \cos \theta} + \dfrac{1 + \cos \theta}{\sin \theta} = \dfrac{\sin^2 \theta + 1 + 2 \cos \theta + \cos^2 \theta}{(1 + \cos \theta)\sin \theta} = \dfrac{1 + 1 + 2 \cos \theta}{(1 + \cos \theta)\sin \theta}$

$$= \dfrac{2(1 + \cos \theta)}{\sin \theta(1 + \cos \theta)} = 2 \csc \theta$$

10. $1 - \dfrac{\cos^2 \theta}{1 + \sin \theta} = \dfrac{1 + \sin \theta - \cos^2 \theta}{1 + \sin \theta} = \dfrac{1 + \sin \theta - (1 - \sin^2 \theta)}{1 + \sin \theta} = \dfrac{\sin^2 \theta + \sin \theta}{1 + \sin \theta}$

$\qquad = \dfrac{\sin \theta(\sin \theta + 1)}{1 + \sin \theta} = \sin \theta$

12. $\dfrac{1 + \sec \theta}{\sec \theta} = \dfrac{1 + \dfrac{1}{\cos \theta}}{\dfrac{1}{\cos \theta}} = \dfrac{\dfrac{\cos \theta + 1}{\cos \theta}}{\dfrac{1}{\cos \theta}} = (\cos \theta + 1) \cdot \left[\dfrac{1 - \cos \theta}{1 - \cos \theta}\right] = \dfrac{1 - \cos^2 \theta}{1 - \cos \theta}$

$\qquad\qquad\qquad\qquad\qquad\qquad\qquad\qquad\qquad\qquad\qquad\qquad = \dfrac{\sin^2 \theta}{1 - \cos \theta}$

14. $\dfrac{\csc \theta}{1 - \cos \theta} = \dfrac{\dfrac{1}{\sin \theta}}{1 - \cos \theta} = \dfrac{1}{\sin \theta(1 - \cos \theta)} \cdot \dfrac{1 + \cos \theta}{1 + \cos \theta} = \dfrac{1 + \cos \theta}{\sin \theta(1 - \cos^2 \theta)} = \dfrac{1 + \cos \theta}{\sin^3 \theta}$

16. $\dfrac{1 - \cos \theta}{1 + \cos \theta} = \dfrac{\dfrac{1 - \cos \theta}{\sin \theta}}{\dfrac{1 + \cos \theta}{\sin \theta}} = \dfrac{\csc \theta - \cot \theta}{\csc \theta + \cot \theta} \cdot \dfrac{\csc \theta - \cot \theta}{\csc \theta - \cot \theta} = \dfrac{(\csc \theta - \cot \theta)^2}{\csc^2 \theta - \cot^2 \theta}$

$\qquad\qquad\qquad\qquad\qquad\qquad\qquad\qquad\qquad\qquad\qquad\qquad = (\csc \theta - \cot \theta)^2$

18. $\dfrac{(2 \sin^2 \theta - 1)^2}{\sin^4 \theta - \cos^4 \theta} = \dfrac{\left[2 \sin^2 \theta - (\cos^2 \theta + \sin^2 \theta)\right]^2}{(\sin^2 \theta - \cos^2 \theta)(\sin^2 \theta + \cos^2 \theta)} = \dfrac{(\sin^2 \theta - \cos^2 \theta)^2}{(\sin^2 \theta - \cos^2 \theta)}$

$\qquad = \sin^2 \theta - \cos^2 \theta = (1 - \cos^2 \theta) - \cos^2 \theta = 1 - 2 \cos^2 \theta$

20. $\dfrac{\sin(\alpha - \beta)}{\sin \alpha \cos \beta} = \dfrac{\sin \alpha \cos \beta - \cos \alpha \sin \beta}{\sin \alpha \cos \beta} = 1 - \cot \alpha \tan \beta$

22. $\dfrac{\cos(\alpha + \beta)}{\sin \alpha \cos \beta} = \dfrac{\cos \alpha \cos \beta - \sin \alpha \sin \beta}{\sin \alpha \cos \beta} = \cot \alpha - \tan \beta$

24. $\sin \theta \tan\dfrac{\theta}{2} = \dfrac{\sin \theta(1 - \cos \theta)}{\sin \theta} = 1 - \cos \theta$

26. $(2 \sin 2\theta)(1 - 2 \sin^2 \theta) = 2 \sin 2\theta \cos 2\theta = \sin 4\theta$

28. $\dfrac{\sin 3\theta \cos \theta - \sin \theta \cos 3\theta}{\sin 2\theta} = \dfrac{\sin(3\theta - \theta)}{\sin 2\theta} = \dfrac{\sin 2\theta}{\sin 2\theta} = 1$

30. $\dfrac{\sin 2\theta + \sin 4\theta}{\sin 2\theta - \sin 4\theta} + \dfrac{\tan 3\theta}{\tan \theta} = \dfrac{2 \sin 3\theta \cos \theta}{-2 \sin \theta \cos 3\theta} + \dfrac{\tan 3\theta}{\tan \theta} = \dfrac{-\tan 3\theta}{\tan \theta} + \dfrac{\tan 3\theta}{\tan \theta} = 0$

32. $\cos 2\theta - \cos 10\theta = 2 \sin 6\theta \sin 4\theta = \dfrac{\sin 4\theta}{\cos 4\theta}(2 \sin 6\theta \cos 4\theta) = (\tan 4\theta)(\sin 2\theta + \sin 10\theta)$

34. $\tan 105° = \tan(60° + 45°) = \dfrac{\tan 60° + \tan 45°}{1 - \tan 60° \tan 45°} = \dfrac{\sqrt{3} + 1}{1 - \sqrt{3}}$

36. $\sin\left[\dfrac{-\pi}{12}\right] = \sin\left[\dfrac{3\pi}{12} - \dfrac{4\pi}{12}\right] = \sin\dfrac{\pi}{4}\cos\dfrac{\pi}{3} - \cos\dfrac{\pi}{4}\sin\dfrac{\pi}{3}$

$$= \dfrac{\sqrt{2}}{2}\cdot\dfrac{1}{2} - \dfrac{\sqrt{2}}{2}\cdot\dfrac{\sqrt{3}}{2} = \dfrac{\sqrt{2}}{4}\left(1 - \sqrt{3}\right)$$

38. $\sin 70°\cos 40° - \cos 70°\sin 40° = \sin(70° - 40°) = \sin 30° = \dfrac{1}{2}$

40. $\sin\dfrac{5\pi}{8} = \sin\dfrac{\frac{5\pi}{4}}{2} = +\sqrt{\dfrac{1 - \cos\dfrac{5\pi}{4}}{2}} = +\sqrt{\dfrac{1 + \dfrac{\sqrt{2}}{2}}{2}} = \dfrac{\sqrt{2 + \sqrt{2}}}{2}$

42. $\cos\alpha = \dfrac{4}{5}, 0 < \alpha < \dfrac{\pi}{2}; \cos\beta = \dfrac{5}{13}, \dfrac{-\pi}{2} < \beta < 0$

Therefore, $\sin\alpha = \dfrac{3}{5}, \tan\alpha = \dfrac{3}{4}, \sin\beta = \dfrac{-12}{13}, \tan\beta = \dfrac{-12}{5}; 0 < \dfrac{\alpha}{2} < \dfrac{\pi}{4}; \dfrac{-\pi}{4} < \dfrac{\beta}{2} < 0$

(a) $\sin(\alpha + \beta) = \sin\alpha\cos\beta + \cos\alpha\sin\beta = \dfrac{3}{5}\cdot\dfrac{5}{13} + \dfrac{4}{5}\cdot\dfrac{-12}{13} = \dfrac{15 - 48}{65} = \dfrac{-33}{65}$

(b) $\cos(\alpha + \beta) = \cos\alpha\cos\beta - \sin\alpha\sin\beta = \dfrac{4}{5}\cdot\dfrac{5}{13} - \dfrac{3}{5}\cdot\dfrac{-12}{13} = \dfrac{20 + 36}{65} = \dfrac{56}{65}$

(c) $\sin(\alpha - \beta) = \sin\alpha\cos\beta - \cos\alpha\sin\beta = \dfrac{3}{5}\cdot\dfrac{5}{13} - \dfrac{4}{5}\cdot\dfrac{-12}{13} = \dfrac{15 + 48}{65} = \dfrac{63}{65}$

(d) $\tan(\alpha + \beta) = \dfrac{\tan\alpha + \tan\beta}{1 - \tan\alpha\tan\beta} = \dfrac{\dfrac{3}{4} + \dfrac{-12}{5}}{1 - \dfrac{3}{4}\cdot\dfrac{-12}{5}} = \dfrac{\dfrac{15 - 48}{20}}{\dfrac{20 + 36}{20}} = \dfrac{-33}{56}$

(e) $\sin 2\alpha = 2\sin\alpha\cos\alpha = 2\cdot\dfrac{3}{5}\cdot\dfrac{4}{5} = \dfrac{24}{25}$

(f) $\cos 2\beta = 2\cos^2\beta - 1 = 2\cdot\left[\dfrac{5}{13}\right]^2 - 1 = \dfrac{-119}{169}$

(g) $\sin\dfrac{\beta}{2} = -\sqrt{\dfrac{1 - \cos\beta}{2}} = -\sqrt{\dfrac{1 - \dfrac{5}{13}}{2}} = -\sqrt{\dfrac{8}{26}} = \dfrac{-2\sqrt{2}}{\sqrt{26}} = \dfrac{-4\sqrt{13}}{26}$

(h) $\cos\dfrac{\alpha}{2} = \sqrt{\dfrac{1 + \cos\alpha}{2}} = \sqrt{\dfrac{1 + \dfrac{4}{5}}{2}} = \sqrt{\dfrac{9}{10}} = \dfrac{3\sqrt{10}}{10}$

44. $\sin\alpha = \dfrac{-4}{5}, \dfrac{-\pi}{2} < \alpha < 0; \cos\beta = \dfrac{-5}{13}, \dfrac{\pi}{2} < \beta < \pi$

Therefore, $\cos\alpha = \dfrac{3}{5}, \tan\alpha = \dfrac{-4}{3}, \sin\beta = \dfrac{12}{13}, \tan\beta = \dfrac{-12}{5}; \dfrac{-\pi}{4} < \dfrac{\alpha}{2} < 0; \dfrac{\pi}{4} < \dfrac{\beta}{2} < \dfrac{\pi}{2}$

(a) $\sin(\alpha + \beta) = \sin\alpha\cos\beta + \cos\alpha\sin\beta = \dfrac{-4}{5}\cdot\dfrac{-5}{13} + \dfrac{3}{5}\cdot\dfrac{12}{13} = \dfrac{20 + 36}{65} = \dfrac{56}{65}$

(b) $\cos(\alpha + \beta) = \cos\alpha\cos\beta - \sin\alpha\sin\beta = \dfrac{3}{5}\cdot\dfrac{-5}{13} - \dfrac{-4}{5}\cdot\dfrac{12}{13} = \dfrac{-15 + 48}{65} = \dfrac{33}{65}$

(c) $\sin(\alpha - \beta) = \sin\alpha\cos\beta - \cos\alpha\sin\beta = \dfrac{-4}{5}\cdot\dfrac{-5}{13} - \dfrac{3}{5}\cdot\dfrac{12}{13} = \dfrac{20 - 36}{65} = \dfrac{-16}{65}$

(d) $\tan(\alpha + \beta) = \dfrac{\tan \alpha + \tan \beta}{1 - \tan \alpha \tan \beta} = \dfrac{\dfrac{-4}{3} + \dfrac{-12}{5}}{1 - \dfrac{-4}{3} \cdot \dfrac{-12}{5}} = \dfrac{\dfrac{-20 - 36}{15}}{\dfrac{5 - 16}{5}} = \dfrac{56}{33}$

(e) $\sin 2\alpha = 2 \sin \alpha \cos \alpha = 2 \cdot \dfrac{4}{5} \cdot \dfrac{3}{5} = \dfrac{-24}{25}$

(f) $\cos 2\beta = 2 \cos^2 \beta - 1 = 2 \cdot \left[\dfrac{-5}{13}\right]^2 - 1 = \dfrac{50 - 169}{169} = \dfrac{-119}{169}$

(g) $\sin \dfrac{\beta}{2} = \sqrt{\dfrac{1 - \cos \beta}{2}} = \sqrt{\dfrac{1 - \dfrac{5}{13}}{2}} = \sqrt{\dfrac{9}{13}} = \dfrac{3\sqrt{13}}{13}$

(h) $\cos \dfrac{\alpha}{2} = \sqrt{\dfrac{1 + \cos \alpha}{2}} = \sqrt{\dfrac{1 + \dfrac{3}{5}}{2}} = \sqrt{\dfrac{4}{5}} = \dfrac{2\sqrt{5}}{5}$

46. $\tan \alpha = \dfrac{-4}{3}$, $\dfrac{\pi}{2} < \alpha < \pi$; $\cot \beta = \dfrac{12}{5}$, $\pi < \beta < \dfrac{3\pi}{2}$

Therefore, $\sin \alpha = \dfrac{4}{5}$, $\cos \alpha = \dfrac{-3}{5}$, $\sin \beta = \dfrac{-5}{13}$, $\cos \beta = \dfrac{-12}{13}$; $\dfrac{\pi}{4} < \dfrac{\alpha}{2} < \dfrac{\pi}{2}$; $\dfrac{\pi}{2} < \dfrac{\beta}{2} < \dfrac{3\pi}{4}$

(a) $\sin(\alpha + \beta) = \sin \alpha \cos \beta + \cos \alpha \sin \beta = \dfrac{4}{5} \cdot \dfrac{-12}{13} + \dfrac{-3}{5} \cdot \dfrac{-5}{13} = \dfrac{-48 + 15}{65} = \dfrac{-33}{65}$

(b) $\cos(\alpha + \beta) = \cos \alpha \cos \beta - \sin \alpha \sin \beta = \dfrac{-3}{5} \cdot \dfrac{-12}{13} - \dfrac{4}{5} \cdot \dfrac{-5}{13} = \dfrac{36 + 20}{65} = \dfrac{56}{65}$

(c) $\sin(\alpha - \beta) = \sin \alpha \cos \beta - \cos \alpha \sin \beta = \dfrac{4}{5} \cdot \dfrac{-12}{13} - \dfrac{-3}{5} \cdot \dfrac{-5}{13} = \dfrac{-48 - 15}{65} = \dfrac{-63}{65}$

(d) $\tan(\alpha + \beta) = \dfrac{\tan \alpha + \tan \beta}{1 - \tan \alpha \tan \beta} = \dfrac{\dfrac{-4}{3} + \dfrac{5}{12}}{1 - \dfrac{-4}{3} \cdot \dfrac{15}{12}} = \dfrac{\dfrac{-16 + 5}{12}}{\dfrac{9 + 5}{9}} = \dfrac{-33}{56}$

(e) $\sin 2\alpha = 2 \sin \alpha \cos \alpha = 2 \cdot \dfrac{4}{5} \cdot \dfrac{-3}{5} = \dfrac{-24}{25}$

(f) $\cos 2\beta = 2 \sin^2 \beta - 1 = \dfrac{144}{169} - \dfrac{25}{169} = \dfrac{119}{169}$

(g) $\sin \dfrac{\beta}{2} = \sqrt{\dfrac{1 - \cos \beta}{2}} = \sqrt{\dfrac{1 + \dfrac{12}{13}}{2}} = \sqrt{\dfrac{25}{26}} = \dfrac{5\sqrt{26}}{26}$

(h) $\cos \dfrac{\alpha}{2} = \sqrt{\dfrac{1 + \cos \alpha}{2}} = \sqrt{\dfrac{1 - \dfrac{3}{5}}{2}} = \sqrt{\dfrac{1}{5}} = \dfrac{\sqrt{5}}{5}$

48. $\csc \alpha = 2$, $\dfrac{\pi}{2} < \alpha < \pi$; $\sec \beta = -3$, $\dfrac{\pi}{2} < \beta < \pi$

Therefore, $\sin \alpha = \dfrac{1}{2}$, $\cos \alpha = \dfrac{-\sqrt{3}}{2}$, $\tan \alpha = \dfrac{-1}{\sqrt{3}}$, $\sin \beta = \dfrac{2\sqrt{2}}{3}$, $\cos \beta = \dfrac{-1}{3}$, $\tan \beta = -2\sqrt{2}$;

$\dfrac{\pi}{4} < \dfrac{\alpha}{2} < \dfrac{\pi}{2}$; $\dfrac{\pi}{4} < \dfrac{\beta}{2} < \dfrac{\pi}{2}$

(a) $\sin(\alpha + \beta) = \sin \alpha \cos \beta + \cos \alpha \sin \beta = \dfrac{1}{2} \cdot \dfrac{-1}{3} + \dfrac{-\sqrt{3}}{2} \cdot \dfrac{2\sqrt{2}}{3} = \dfrac{-1 - 2\sqrt{6}}{6}$

(b) $\cos(\alpha + \beta) = \cos \alpha \cos \beta - \sin \alpha \sin \beta = \dfrac{-\sqrt{3}}{2} \cdot \dfrac{-1}{3} - \dfrac{1}{2} \cdot \dfrac{2\sqrt{2}}{3} = \dfrac{\sqrt{3} - 2\sqrt{2}}{6}$

(c) $\sin(\alpha - \beta) = \sin \alpha \cos \beta - \cos \alpha \sin \beta = \dfrac{1}{2} \cdot \dfrac{-1}{3} - \dfrac{-\sqrt{3}}{3} \cdot \dfrac{2\sqrt{2}}{3} = \dfrac{-1 + 2\sqrt{6}}{6}$

(d) $\tan(\alpha + \beta) = \dfrac{\tan \alpha + \tan \beta}{1 - \tan \alpha \tan \beta} = \dfrac{\dfrac{-\sqrt{3}}{3} - 2\sqrt{2}}{1 - \left[\dfrac{-\sqrt{3}}{3}\right] \cdot \left(-2\sqrt{2}\right)} = \dfrac{\dfrac{-\sqrt{3} - 6\sqrt{2}}{3}}{\dfrac{3 - 2\sqrt{6}}{3}} = \dfrac{-\sqrt{3} - 6\sqrt{2}}{3 - 2\sqrt{6}}$

(e) $\sin 2\alpha = 2 \sin \alpha \cos \alpha = 2 \cdot \dfrac{1}{2} \cdot \dfrac{-\sqrt{3}}{2} = \dfrac{-\sqrt{3}}{2}$

(f) $\cos 2\beta = 2 \cos^2 \beta - 1 = 2 \cdot \left[\dfrac{-1}{3}\right]^2 - 1 = \dfrac{-7}{9}$

(g) $\sin \dfrac{\beta}{2} = \sqrt{\dfrac{1 - \cos \beta}{2}} = \sqrt{\dfrac{1 + \dfrac{1}{3}}{2}} = \sqrt{\dfrac{4}{6}} = \dfrac{2\sqrt{6}}{6} = \dfrac{\sqrt{6}}{3}$

(h) $\cos \dfrac{\alpha}{2} = \sqrt{\dfrac{1 + \cos \alpha}{2}} = \sqrt{\dfrac{1 + \dfrac{-\sqrt{3}}{2}}{2}} = \dfrac{\sqrt{2 - \sqrt{3}}}{2}$

50. $\tan \alpha = -2,\ \dfrac{\pi}{2} < \alpha < \pi;\ \cot \beta = -2,\ \dfrac{\pi}{2} < \beta < \pi$

Therefore, $\sin \alpha = \dfrac{2}{\sqrt{5}},\ \cos \alpha = \dfrac{-1}{\sqrt{5}},\ \sin \beta = \dfrac{1}{\sqrt{5}},\ \cos \beta = \dfrac{-2}{\sqrt{5}},\ \tan \beta = -\dfrac{1}{2};$

$\dfrac{\pi}{4} < \dfrac{\alpha}{2} < \dfrac{\pi}{2};\ \dfrac{\pi}{4} < \dfrac{\beta}{2} < \dfrac{\pi}{2}$

(a) $\sin(\alpha + \beta) = \sin \alpha \cos \beta + \cos \alpha \sin \beta = \dfrac{2}{\sqrt{5}} \cdot \dfrac{-2}{\sqrt{5}} + \dfrac{-1}{\sqrt{5}} \cdot \dfrac{1}{\sqrt{5}} = \dfrac{-4 - 1}{5} = -1$

(b) $\cos(\alpha + \beta) = \cos \alpha \cos \beta - \sin \alpha \sin \beta = \dfrac{-1}{\sqrt{5}} \cdot \dfrac{-2}{\sqrt{5}} - \dfrac{2}{\sqrt{5}} \cdot \dfrac{1}{\sqrt{5}} = \dfrac{2 - 2}{5} = 0$

(c) $\sin(\alpha - \beta) = \sin \alpha \cos \beta - \cos \alpha \sin \beta = \dfrac{2}{\sqrt{5}} \cdot \dfrac{-2}{\sqrt{5}} - \dfrac{-1}{\sqrt{5}} \cdot \dfrac{1}{\sqrt{5}} = \dfrac{-4 + 1}{5} = \dfrac{-3}{5}$

(d) $\tan(\alpha + \beta) = \dfrac{\tan \alpha + \tan \beta}{1 - \tan \alpha \tan \beta} = \dfrac{-2 + \dfrac{-1}{2}}{1 - (-2)\left[\dfrac{-1}{2}\right]} = \text{Undefined}$

(e) $\sin 2\alpha = 2 \sin \alpha \cos \alpha = 2 \cdot \dfrac{2}{\sqrt{5}} \cdot \dfrac{-1}{\sqrt{5}} = \dfrac{-4}{5}$

(f) $\cos 2\beta = 2 \cos^2 \beta - 1 = 2 \cdot \dfrac{4}{5} - 1 = \dfrac{3}{5}$

(g) $\sin \dfrac{\beta}{2} = \sqrt{\dfrac{1 - \cos \beta}{2}} = \sqrt{\dfrac{1 + \dfrac{2\sqrt{5}}{5}}{2}} = \sqrt{\dfrac{5 + 2\sqrt{5}}{10}}$

(h) $\cos \dfrac{\alpha}{2} = \sqrt{\dfrac{1 + \cos \alpha}{2}} = \sqrt{\dfrac{1 + \dfrac{-1}{\sqrt{5}}}{2}} = \dfrac{\sqrt{5 - \sqrt{5}}}{10}$

52. $\sin \theta = \dfrac{-\sqrt{3}}{2}$

$\theta = \dfrac{4\pi}{3}$ or $\theta = \dfrac{5\pi}{3}$

54. $\tan \theta = -\sqrt{3}$

$\theta = \dfrac{2\pi}{3}$ or $\theta = \dfrac{5\pi}{3}$

56. $\cos 2\theta = 0$

$2\theta = \dfrac{\pi}{2} + 2k\pi$ or $2\theta = \dfrac{3\pi}{2} + 2k\pi$, k any integer

$\theta = \dfrac{\pi}{4} + k\pi$ or $\theta = \dfrac{3\pi}{4} + k\pi$, k any integer

The solutions on the interval $[0, 2\pi)$ are $\theta = \dfrac{\pi}{4}, \dfrac{5\pi}{4}, \dfrac{3\pi}{4}, \dfrac{7\pi}{4}$.

58. $\sin 3\theta = 1$

$3\theta = \dfrac{\pi}{2} + 2k\pi$, k any integer

$\theta = \dfrac{\pi}{6} + \dfrac{2k\pi}{3}$, k any integer

The solutions on the interval $[0, 2\pi)$ are $\theta = \dfrac{\pi}{6}, \dfrac{5\pi}{6}, \dfrac{3\pi}{2}$.

60. $\tan \theta = 25$
$\theta = 1.5308$ or $\theta = \pi + 1.5308$

62. $\cos \theta = \sec \theta$

$\cos \theta = \dfrac{1}{\cos \theta}$

$\cos^2 \theta = 1$

$\cos \theta = 1$ or $\cos \theta = -1$

$\theta = 0, \pi$

64.
$$\cos 2\theta = \sin \theta$$
$$1 - 2 \sin^2 \theta = \sin \theta$$
$$2 \sin^2 \theta + \sin \theta - 1 = 0$$
$$(2 \sin \theta - 1)(\sin \theta + 1) = 0$$
$2 \sin \theta - 1 = 0$ or $\sin \theta + 1 = 0$
$2 \sin \theta = 1$ or $\sin \theta = -1$
$\sin \theta = \dfrac{1}{2}$ or $\sin \theta = -1$

$\theta = \dfrac{\pi}{6}, \dfrac{5\pi}{6}, \dfrac{3\pi}{2}$

66.
$$\sin 2\theta - \sin \theta - 2 \cos \theta + 1 = 0$$
$$2 \sin \theta \cos \theta - \sin \theta - 2 \cos \theta + 1 = 0$$
$$\sin \theta(2 \cos \theta - 1) - 1(2 \cos \theta - 1) = 0$$
$$(\sin \theta - 1)(2 \cos \theta - 1) = 0$$
$\sin \theta - 1 = 0$ or $2 \cos \theta - 1 = 0$
$\sin \theta = 1$ or $\cos \theta = \dfrac{1}{2}$

$\theta = \dfrac{\pi}{2}$ or $\theta = \dfrac{\pi}{3}, \dfrac{5\pi}{3}$

68. $2 \cos^2 + \cos \theta - 1 = 0$
$(2 \cos \theta - 1)(\cos \theta + 1) = 0$
$2 \cos \theta - 1 = 0$ or $\cos \theta + 1 = 0$

$\cos \theta = \dfrac{1}{2}$ or $\cos \theta = -1$

$\theta = \dfrac{\pi}{3}, \theta = \dfrac{5\pi}{3}$ or $\theta = \pi$

70.

$$\sin\theta + 2\cos\theta = 1$$
$$2\cos\theta = 1 - \sin\theta$$
$$4\cos^2\theta = (1 - \sin\theta)^2$$
$$4(1 - \sin^2\theta) = 1 - 2\sin\theta + \sin^2\theta$$
$$5\sin^2\theta - 2\sin\theta - 3 = 0$$
$$(5\sin\theta + 3)(\sin\theta - 1) = 0$$

$$5\sin\theta + 3 = 0 \qquad \text{or} \qquad \sin\theta - 1 = 0$$
$$5\sin\theta = -3$$
$$\sin\theta = -3$$

$$\sin\theta = \frac{-3}{5} \qquad \text{or} \qquad \sin\theta = 1$$

$$\theta = \sin^{-1}\left[\frac{-3}{5}\right] \qquad \text{or} \qquad \theta = \frac{\pi}{2}$$

$$\theta \approx -0.6435 \qquad \text{or} \qquad \theta = \frac{\pi}{2}$$

On the interval $\{0, 2\pi\}$, the solutions are $\theta \approx 5.6397$ or $\theta = \dfrac{\pi}{2}$.

Mission Possible

1. There are 19 different distances needed. The angle between the third and first base lines at home plate is 90°. Dividing that by 5° gives 18, but we have to count both endpoints.

2. $\tan 20° = \dfrac{y}{410 + x}$ and $\tan 35° = \dfrac{y}{x}$

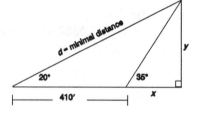

Solve the equations for y:
$$y = (410 + x)\tan 20°;\ y = x\tan 35°$$
Set the equations equal to each other:
$$(410 + x)\tan 20° = x\tan 35°$$
Solve for x:
$$410\tan 20° + x\tan 20° = x\tan 35°$$
$$\frac{410\tan 20°}{\tan 35° - \tan 20°} = x$$
$$x \approx 443.82 \text{ feet}$$
To find the minimal distance, d:
$$\cos 20° = \frac{410 + 443.82}{d}$$
$$d = \frac{410 + 443.82}{\cos 20°} = 908.62 \text{ feet}$$

3. $\tan A = \dfrac{y}{f + x}$ and $\tan B = \dfrac{y}{x}$

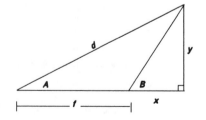

Thus, $y = (f + x)\tan A$ and $y = x\tan B$
So, $(f + x)\tan A = x\tan B$
Solving the above equation for x, we obtain:
$$x = \frac{f \cdot \tan A}{\tan B - \tan A}$$

To find the minimal distance d:

$$\cos A = \frac{f + x}{d}$$

Thus,

$$d = \frac{f + x}{\cos A}$$

$$d = \frac{f + \dfrac{f \tan A}{\tan B - \tan A}}{\cos A}$$

$$d = \frac{f \tan B}{\cos A(\tan B - \tan A)}$$

4. Groups should compare. To find the distance, let $f = 325$, $A = 30°$, $B = 50°$.

$$d = \frac{325 \cdot \tan 50°}{\cos 30°(\tan 50° - \tan 30°)}$$

$$d = 727.92 \text{ feet}$$

5.

$$d = \frac{f \cdot \tan B}{\cos A(\tan B - \tan A)}$$

$$d = \frac{f \cdot \dfrac{\sin B}{\cos B}}{\cos A\dfrac{\sin B}{\cos B} - \sin A}$$

$$d = \frac{f \cdot \sin B}{\cos A \sin B - \sin A \cos B}$$

$$d = \frac{f \cdot \sin B}{\sin(B - A)}$$

6. The path of a projectile ignoring wind resistance and other factors is a parabola. However, the wind and spin of the ball would probably make the path more 3-dimensional.

7. We developed only the straight line distance from home plate to the highest point to be cleared. Any deviation from that straight line, due to trajectory or wind, would increase the distance. We did not consider the distance from the highest point down to the street, which would also add on to the distance.

8. Once you had defined how to measure the length of the home run, it would be possible to designate a "longest home run." However, it would be very difficult to determine a measure, short of building into the ball itself a computer chip that could tell you distance from strike of the bat to first object encountered. Other less accurate measures might involve measuring from home plate to the spot where the ball hit the ground (if that could be determined) although that would not take into account the height the ball had reached or how much the wind had helped or resisted.

Chapter 10

ADDITIONAL APPLICATIONS OF TRIGONOMETRY

10.1 The Law of Sines

2.

$$\alpha + \beta + \gamma = 180°$$
$$45° + 40° + \gamma = 180°$$
$$\gamma = 95°$$

Using the Law of Sines twice,

$$\frac{\sin \alpha}{a} = \frac{\sin \gamma}{c} \qquad\qquad \frac{\sin \beta}{b} = \frac{\sin \gamma}{c}$$

$$\frac{\sin 45°}{a} = \frac{\sin 95°}{4} \qquad\qquad \frac{\sin 40°}{b} = \frac{\sin 95°}{4}$$

$$a = \frac{4 \sin 45°}{\sin 95°} \qquad\qquad b = \frac{4 \sin 40°}{\sin 95°}$$

$$a \approx 2.84 \qquad\qquad b \approx 2.58$$

4.

$$\alpha + \beta + \gamma = 180°$$
$$\alpha + 30° + 125° = 180°$$
$$\alpha = 25°$$

$$\frac{\sin \alpha}{a} = \frac{\sin \beta}{b} \qquad\qquad \frac{\sin \beta}{b} = \frac{\sin \gamma}{c}$$

$$\frac{\sin 25°}{a} = \frac{\sin 30°}{10} \qquad\qquad \frac{\sin 30°}{10} = \frac{\sin 125°}{c}$$

$$a = \frac{10 \sin 25°}{\sin 30°} \qquad\qquad c = \frac{10 \sin 125°}{\sin 30°}$$

$$a \approx 8.45 \qquad\qquad c \approx 16.38$$

6.

$$\alpha + \beta + \gamma = 180°$$
$$10° + 5° + \gamma = 180°$$
$$\gamma = 165°$$

$$\frac{\sin \alpha}{a} = \frac{\sin \gamma}{c} \qquad\qquad \frac{\sin \beta}{b} = \frac{\sin \gamma}{c}$$

$$\frac{\sin 10°}{a} = \frac{\sin 165°}{5} \qquad\qquad \frac{\sin 5°}{b} = \frac{\sin 165°}{5}$$

$$a = \frac{5 \sin 10°}{\sin 165°} \qquad\qquad b = \frac{5 \sin 5°}{\sin 165°}$$

$$a \approx 3.35 \qquad\qquad b \approx 1.68$$

8.
$$\alpha + \beta + \gamma = 180°$$
$$100° + 30° + \gamma = 180°$$
$$\gamma = 50°$$

$$\frac{\sin \alpha}{a} = \frac{\sin \beta}{b} \qquad \frac{\sin \beta}{b} = \frac{\sin \gamma}{c}$$

$$\frac{\sin 100°}{a} = \frac{\sin 30°}{6} \qquad \frac{\sin 30°}{6} = \frac{\sin 50°}{c}$$

$$a = \frac{6 \sin 100°}{\sin 30°} \qquad c = \frac{6 \sin 50°}{\sin 30°}$$

$$a \approx 11.82 \qquad c \approx 9.19$$

10.
$$\alpha + \beta + \gamma = 180°$$
$$50° + \beta + 20° = 180°$$
$$\beta = 110°$$

$$\frac{\sin \alpha}{a} = \frac{\sin \beta}{b} \qquad \frac{\sin \alpha}{a} = \frac{\sin \gamma}{c}$$

$$\frac{\sin 50°}{3} = \frac{\sin 110°}{b} \qquad \frac{\sin 50°}{3} = \frac{\sin 20°}{c}$$

$$b = \frac{3 \sin 110°}{\sin 50°} \qquad c = \frac{3 \sin 20°}{\sin 50°}$$

$$b \approx 3.68 \qquad c \approx 1.34$$

12.
$$\alpha + \beta + \gamma = 180°$$
$$70° + 60° + \gamma = 180°$$
$$\gamma = 50°$$

$$\frac{\sin \alpha}{a} = \frac{\sin \gamma}{c} \qquad \frac{\sin \beta}{b} = \frac{\sin \gamma}{c}$$

$$\frac{\sin 70°}{a} = \frac{\sin 50°}{4} \qquad \frac{\sin 60°}{b} = \frac{\sin 50°}{4}$$

$$a = \frac{4 \sin 70°}{\sin 50°} \qquad b = \frac{4 \sin 60°}{\sin 50°}$$

$$a \approx 4.91 \qquad b \approx 4.52$$

14.
$$\alpha + \beta + \gamma = 180°$$
$$\alpha + 10° + 100° = 180°$$
$$\alpha = 70°$$

$$\frac{\sin \alpha}{a} = \frac{\sin \beta}{b} \qquad \frac{\sin \beta}{b} = \frac{\sin \gamma}{c}$$

$$\frac{\sin 70°}{a} = \frac{\sin 10°}{2} \qquad \frac{\sin 10°}{2} = \frac{\sin 100°}{c}$$

$$a = \frac{2 \sin 70°}{\sin 10°} \qquad c = \frac{2 \sin 100°}{\sin 10°}$$

$$a \approx 10.82 \qquad c \approx 11.34$$

16.
$$\alpha + \beta + \gamma = 180°$$
$$\alpha + 20° + 70° = 180°$$
$$\alpha = 90°$$

$$\frac{\sin \alpha}{a} = \frac{\sin \beta}{b} \qquad\qquad \frac{\sin \beta}{a} = \frac{\sin \gamma}{c}$$

$$\frac{\sin 90°}{1} = \frac{\sin 20°}{b} \qquad\qquad \frac{\sin 90°}{1} = \frac{\sin 70°}{c}$$

$$b = \frac{\sin 20°}{\sin 90°} \qquad\qquad c = \frac{\sin 70°}{\sin 90°}$$

$$b \approx 0.34 \qquad\qquad c \approx 0.94$$

18.   Given that $b = 4$, $c = 3$, and $\beta = 40°$, we use the Law of Sines to find the angle γ:

$$\frac{\sin \beta}{b} = \frac{\sin \gamma}{c}$$

$$\frac{\sin 40°}{4} = \frac{\sin \gamma}{3}$$

$$\sin \gamma = \frac{3 \sin 40°}{4} = 0.4821$$

There are two angles γ, $0° < \gamma < 180°$, for which $\sin \gamma \approx 0.4821$, namely
$$\gamma \approx 28.8° \text{ or } \gamma \approx 151.2°$$
The second possibility is ruled out, because $\beta = 40°$, making $\beta + \gamma \approx 191.2° > 180°$. Now, using $\gamma = 28.8°$, we find $\alpha = 180° - \beta - \gamma \approx 111.2°$. The third side a is determined using the Law of Sines:

$$\frac{\sin \alpha}{a} = \frac{\sin \beta}{b}$$

$$\frac{\sin 111.2°}{a} = \frac{\sin 40°}{4}$$

$$a = \frac{4 \sin 111.2°}{\sin 40°} = 5.80$$

One triangle: $\gamma = 28.8°$, $\alpha = 111.2°$, $a \approx 5.80$

20.   Given $a = 2$, $c = 1$, and $\alpha = 120°$,

$$\frac{\sin \alpha}{a} = \frac{\sin \gamma}{c}$$

$$\frac{\sin 120°}{2} = \frac{\sin \gamma}{1}$$

$$\sin \gamma = \frac{\sin 120°}{2} \approx 0.4330$$

$$\gamma \approx 25.7° \text{ or } \gamma \approx 154.3° \qquad\qquad \text{(Impossible because } \alpha + \gamma \approx 274.3° > 180°)$$

Using $\gamma \approx 25.7°$, $\beta = 180° - \alpha - \gamma \approx 34.3°$,

$$\frac{\sin \alpha}{a} = \frac{\sin \beta}{b}$$

$$\frac{\sin 120°}{2} = \frac{\sin 34.3°}{b}$$

$$b = \frac{2 \sin 34.3°}{\sin 120°} \approx 1.30$$

One triangle: $\gamma = 25.7°$, $\beta = 34.3°$, $b = 1.30$

Chapter 10 Additional Applications of Trigonometry

22. Given $b = 2$, $c = 3$, and $\beta = 40°$,

$$\frac{\sin \beta}{b} = \frac{\sin \gamma}{c}$$

$$\frac{\sin 40°}{2} = \frac{\sin \gamma}{3}$$

$$\sin \gamma = \frac{3 \sin 40°}{2} \approx 0.9642$$

$\gamma_1 \approx 74.6°$ or $\gamma \approx 105.4°$

$\alpha_1 = 180° - \beta - \gamma_1 \approx 65.4°$ or $\alpha_2 = 180° - \beta - \gamma_2 \approx 34.6°$

To find side a,

$$\frac{\sin \alpha}{a} = \frac{\sin \beta}{b}$$

$$\frac{\sin 65.4°}{a_1} = \frac{\sin 40°}{2} \qquad \text{or} \qquad \frac{\sin 34.6°}{a_2} = \frac{\sin 40°}{2}$$

$$a_1 = \frac{2 \sin 65.4°}{\sin 40°} \qquad \text{or} \qquad a_2 = \frac{2 \sin 34.6°}{\sin 40°}$$

$$a_1 \approx 2.83 \qquad \text{or} \qquad a_2 \approx 1.77$$

Two triangles: $\gamma_1 = 74.6°$, $\alpha_1 = 65.4°$, $a_1 = 2.83$

or $\gamma_2 = 105.4°$, $\alpha_2 = 34.6°$, $a_2 = 1.77$

24. Given $a = 3$, $b = 7$, and $\alpha = 70°$,

$$\frac{\sin \alpha}{a} = \frac{\sin \beta}{b}$$

$$\frac{\sin 70°}{3} = \frac{\sin \beta}{7}$$

$$\sin \beta = \frac{7 \sin 70°}{3} \approx 2.1926$$

There is no angle β for which $\sin \beta > 1$. Hence, there can be no triangle having the given measurements.

26. Given $b = 4$, $c = 5$, and $\beta = 95°$,

$$\frac{\sin \beta}{b} = \frac{\sin \gamma}{c}$$

$$\frac{\sin 95°}{4} = \frac{\sin \gamma}{5}$$

$$\sin \gamma = \frac{5 \sin 95°}{4} \approx 1.2452$$

No triangle.

28. Given $b = 4$, $c = 5$, and $\beta = 40°$,

$$\frac{\sin \beta}{b} = \frac{\sin \gamma}{c}$$

$$\frac{\sin 40°}{4} = \frac{\sin \gamma}{5}$$

$$\sin \gamma = \frac{5 \sin 40°}{4} \approx 0.8035$$

$\gamma_1 = 53.5°$ or $\gamma_2 \approx 126.5°$

$\alpha_1 = 180° - \beta - \gamma_1 \approx 86.5°$ or $\alpha_2 = 180° - \beta - \gamma_2 \approx 13.5°$

To find side a,

$$\frac{\sin \alpha}{a} = \frac{\sin \beta}{b}$$

$$\frac{\sin 86.5°}{a_1} = \frac{\sin 40°}{4} \qquad \text{or} \qquad \frac{\sin 13.5°}{a_2} = \frac{\sin 40°}{4}$$

$$a_1 = \frac{4 \sin 86.5°}{\sin 40°} \qquad \text{or} \qquad a_2 = \frac{4 \sin 13.5°}{\sin 40°}$$

$$a_1 \approx 6.21 \qquad\qquad \text{or} \qquad\qquad a_2 \approx 1.45$$

Two triangles: $\quad \gamma_1 = 53.5°$, $\alpha_1 = 86.5°$, $a_1 = 6.21$

$\qquad\qquad\quad$ or $\gamma_2 = 126.5°$, $\alpha_2 = 13.5°$, $a_2 = 1.45$

30.   The angle β is found to be

$$\beta = 180° - 40° - 50° = 90°$$

Let c denote the distance from A to B. Using the Law of Sines,

$$\frac{\sin 90°}{100} = \frac{\sin 50°}{c}$$

$$c = \frac{100 \sin 50°}{\sin 90°} \approx 76.6$$

The distance from the house at A to the house at B is 76.6 feet.

32.   Using information from Problem 31, we know that the distance from A to B is 1490.5 feet. We also know that Angle $ADB = 90°$. Let the height BD of the mountain be denoted as a. To find a, we use the Law of Sines:

$$\frac{\sin 25°}{a} = \frac{\sin 90°}{1490.5}$$

$$a = \frac{1490.5 \sin 25°}{\sin 90°} \approx 629.9$$

The height of the mountain is 629.9 feet.

34. $\gamma = 180° - 69.2° - 65.5° = 45.3°$

$$\frac{\sin 65.5°}{b} = \frac{\sin 45.3°}{880}$$

$$b = \frac{880 \sin 65.5°}{\sin 45.3°} \approx 1126.6$$

Now we have enough information to find the height of the bridge denoted as h.

$$\frac{\sin 69.2°}{h} = \frac{\sin 90°}{1126.6}$$

$$h = \frac{1126.6 \sin 69.2°}{\sin 90°} \approx 1053.2$$

The bridge is 1053.2 feet high.

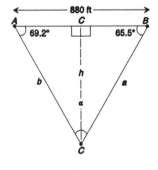

36.   Let distance travelled be 50 miles and 70 miles and rate be 250 mph, then the time, t, it took the aircraft to fly from city A to city B was

$$t = \frac{70 + 50}{250} = 0.48 \text{ hour}$$

Since,

$$\frac{\sin 10°}{70} = \frac{\sin \beta}{50}$$

$$\sin \beta = \frac{50 \sin 10°}{70} \approx 0.1240$$

$$\beta = 7.125°$$

$$\gamma = 180° - 10° - 7.125° = 162.875°$$

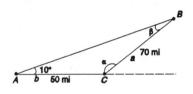

Using the Law of Sines, we now have the necessary information to find c, the distance from A to B.

$$\frac{\sin 10°}{70} = \frac{\sin 162.875°}{c}$$

$$c = \frac{70 \sin 162.875°}{\sin 10°} \approx 118.7$$

$$t = \frac{118.7}{250} = 0.4748 \text{ hour}$$

Thus, it would have taken 0.4748 hours for the correct course. Since $0.48 - 0.4748 = 0.0052$, then about 0.0052 hours or 0.31 minutes were lost due to the error.

38.   Let θ denote Angle AOP.

$$\frac{\sin \theta}{9} = \frac{\sin 15°}{3}$$

$$\sin \theta = \frac{9 \sin 15°}{3} \approx .7765$$

$$\theta \approx 180° - 51° = 129°$$
$$A = 180° - 15° - 129° \approx 36°$$
$$\frac{\sin 36°}{a} = \frac{\sin 15°}{3}$$

$$a = \frac{3 \sin 36°}{\sin 15°}$$
$$\approx 6.8 \text{ inches}$$

The distance from the piston to the center of the crankshaft is 6.8 inches.

40.   Draw a perpendicular line from the shore to $\angle ACB$. Then $\angle ABC = 180° - 90° - 35° = 55°$. Also, $\angle CAB = 180° - 90° - 15° = 75°$.

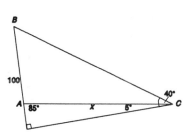

(a)   Using the Law of Sines:

$$\frac{\sin 50°}{3} = \frac{\sin 55°}{x}, \qquad \text{where } x = \text{the distance from the ship to lighthouse } A.$$

$$x = \frac{3 \sin 55°}{\sin 50°} \approx 3.21 \text{ miles}$$

(b)   Using the Law of Sines:

$$\frac{\sin 50°}{3} = \frac{\sin 75°}{y}, \qquad \text{where } y = \text{the distance from the ship to lighthouse } B.$$

$$y = \frac{3 \sin 75°}{\sin 50°} \approx 3.78 \text{ miles}$$

(c)   Using the Law of Sines:

$$\frac{\sin 90°}{3.21} = \frac{\sin 75°}{z}, \qquad \text{where } z = \text{the distance to shore.}$$

$$z = \frac{3.21 \sin 75°}{\sin 90°} \approx 3.10 \text{ miles}$$

42.   The tower forms an angle of 95° with the ground due to the supplementary angles. Thus, $\angle ABC = 180° - 95° - 35° = 50°$ since the angle between the ground and the sight line is $40° - 5° = 35°$. Thus, using the Law of Sines, we have:

$$\frac{\sin 35°}{100} = \frac{\sin 50°}{x}, \qquad \text{where } x \text{ is the distance from the ranger to the fire tower.}$$

$$x = \frac{100 \sin 50°}{\sin 35°} \approx 133.56 \text{ feet}$$

44. $\dfrac{a - b}{c} = \dfrac{a}{c} - \dfrac{b}{c} = \dfrac{\sin \alpha}{\sin \gamma} - \dfrac{\sin \beta}{\sin \gamma} = \dfrac{\sin \alpha - \sin \beta}{\sin \gamma} = \dfrac{2 \sin \dfrac{\alpha - \beta}{2} \cos \dfrac{\alpha + \beta}{2}}{\sin \left[\dfrac{\gamma}{2} + \dfrac{\gamma}{2} \right]}$

$= \dfrac{2 \sin \dfrac{\alpha - \beta}{2} \cos \dfrac{\alpha + \beta}{2}}{\sin \dfrac{\gamma}{2} \cos \dfrac{\gamma}{2} + \cos \dfrac{\gamma}{2} \sin \dfrac{\gamma}{2}} = \dfrac{2 \sin \dfrac{\alpha - \beta}{2} \cos \dfrac{\alpha + \beta}{2}}{2 \sin \dfrac{\gamma}{2} \cos \dfrac{\gamma}{2}} = \dfrac{\sin \dfrac{\alpha - \beta}{2} \cos \left[\dfrac{\pi}{2} - \dfrac{\gamma}{2} \right]}{\sin \dfrac{\gamma}{2} \cos \dfrac{\gamma}{2}}$

$= \dfrac{\sin \dfrac{\alpha - \beta}{2} \sin \dfrac{\gamma}{2}}{\sin \dfrac{\gamma}{2} \cos \dfrac{\gamma}{2}} = \dfrac{\sin \dfrac{\alpha - \beta}{2}}{\cos \dfrac{\gamma}{2}} = \dfrac{\sin \dfrac{1}{2}(\alpha - \beta)}{\cos \dfrac{1}{2} \gamma}$

46. To derive the Law of Tangents, we use Mollweide's Formula:

$$\dfrac{a - b}{a + b} = \dfrac{2 \sin \dfrac{\alpha - \beta}{2} \cos \dfrac{\alpha + \beta}{2}}{2 \sin \dfrac{\alpha + \beta}{2} \cos \dfrac{\alpha - \beta}{2}} = \tan \dfrac{\alpha - \beta}{2} \cot \dfrac{\alpha + \beta}{2} = \dfrac{\tan \dfrac{1}{2}(\alpha - \beta)}{\tan \dfrac{1}{2}(\alpha + \beta)}$$

10.2   The Law of Cosines

2. $a^2 = b^2 + c^2 - 2bc \cos \alpha$
$a^2 = 9 + 16 - 2 \cdot 3 \cdot 4 \cdot \cos 30°$
$a^2 = 25 - 24(0.8660) = 4.216$
$a \approx 2.05$

$\cos \beta = \dfrac{a^2 + c^2 - b^2}{2ac}$

$\cos \beta = \dfrac{4.216 + 16 - 9}{2 \cdot 2.0533 \cdot 4} = \dfrac{11.216}{16.4264} \approx 0.6828$

$\beta \approx 46.9°$

$\cos \gamma = \dfrac{a^2 + b^2 - c^2}{2ab}$

$\cos \gamma = \dfrac{4.216 + 9 - 16}{2 \cdot 2.0533 \cdot 3} = \dfrac{-2.784}{12.3198} \approx -0.2260$

$\gamma \approx 103.1°$

4. $b^2 = a^2 + c^2 - 2ac \cos \beta$
$b^2 = 4 + 25 - 2 \cdot 2 \cdot 5 \cdot \cos 20°$
$b^2 = 29 - 20(0.9397) = 10.206$
$b \approx 3.19$

$\cos \alpha = \dfrac{b^2 + c^2 - a^2}{2bc}$

$\cos \alpha = \dfrac{10.206 + 25 - 4}{2 \cdot 3.1947 \cdot 5} = \dfrac{31.206}{31.947} \approx 0.9768$

$\alpha \approx 12.4°$

$$\cos \gamma = \frac{a^2 + b^2 - c^2}{2ab}$$

$$\cos \gamma = \frac{4 + 10.206 - 25}{2 \cdot 2 \cdot 3.1947} = \frac{-10.794}{12.7788} \approx -0.8447$$

$$\gamma \approx 147.6°$$

6. $\cos \alpha = \dfrac{b^2 + c^2 - a^2}{2bc}$

$$\cos \alpha = \frac{25 + 16 - 64}{2 \cdot 5 \cdot 4} = \frac{-23}{40} = -0.575$$

$$\alpha \approx 125.1°$$

$$\cos \beta = \frac{a^2 + c^2 - b^2}{2ac}$$

$$\cos \beta = \frac{64 + 16 - 25}{2 \cdot 8 \cdot 4} = \frac{55}{64} \approx 0.8594$$

$$\beta \approx 30.8°$$

$$\cos \gamma = \frac{a^2 + b^2 - c^2}{2ab}$$

$$\cos \gamma = \frac{64 + 25 - 16}{2 \cdot 8 \cdot 5} = \frac{73}{80} = 0.9125$$

$$\gamma \approx 24.1°$$

8. $\cos \alpha = \dfrac{b^2 + c^2 - a^2}{2bc}$

$$\cos \alpha = \frac{9 + 16 - 16}{2 \cdot 3 \cdot 4} = \frac{9}{24} = 0.375$$

$$\alpha \approx 68°$$

$$\cos \beta = \frac{a^2 + c^2 - b^2}{2ac}$$

$$\cos \beta = \frac{16 + 16 - 9}{2 \cdot 4 \cdot 4} = \frac{23}{32} \approx 0.7188$$

$$\beta \approx 44°$$

$$\cos \gamma = \frac{a^2 + b^2 - c^2}{2ab}$$

$$\cos \gamma = \frac{16 + 9 - 16}{2 \cdot 4 \cdot 3} = \frac{9}{24} = 0.375$$

$$\gamma \approx 68°$$

10. $b^2 = a^2 + c^2 - 2ac \cos \beta$

$$b^2 = 4 + 1 - 2 \cdot 2 \cdot 1 \cdot \cos 10°$$

$$b^2 = 5 - 4(0.9848) = 1.0608$$

$$b \approx 1.03$$

$$\cos \alpha = \frac{b^2 + c^2 - a^2}{2bc}$$

$$\cos \alpha = \frac{1.0608 + 1 - 4}{2 \cdot 1.03 \cdot 1} = \frac{-1.9392}{2.06} \approx -0.9414$$

$$\alpha \approx 160.3°$$

$$\cos \gamma = \frac{a^2 + b^2 - c^2}{2ab}$$

$$\cos \gamma = \frac{4 + 1.0608 - 1}{2 \cdot 2 \cdot 1.03} = \frac{4.0608}{4.12} \approx 0.9856$$

$$\gamma \approx 9.7°$$

12. $c^2 = a^2 + b^2 - 2ab \cos \gamma$

$$c^2 = 36 + 16 - 2 \cdot 6 \cdot 4 \cdot \cos 60°$$

$$c^2 = 52 - 48(0.5) = 28$$

$$c \approx 5.29$$

$$\cos \alpha = \frac{b^2 + c^2 - a^2}{2bc}$$

$$\cos \alpha = \frac{16 + 28 - 36}{2 \cdot 4 \cdot 5.2915} = \frac{8}{42.332} \approx 0.1890$$

$$\alpha \approx 79.1°$$

$$\cos \beta = \frac{a^2 + c^2 - b^2}{2ac}$$

$$\cos \beta = \frac{36 + 28 - 16}{2 \cdot 6 \cdot 5.2915} = \frac{48}{63.498} \approx 0.7559$$

$$\beta \approx 40.9°$$

14. $a^2 = b^2 + c^2 - 2bc \cos \alpha$
    $a^2 = 16 + 1 - 2 \cdot 4 \cdot 1 \cdot \cos 120°$
    $a^2 = 17 - 8(-0.5) = 21$
    $a \approx 4.58$

$$\cos \beta = \frac{a^2 + c^2 - b^2}{2ac}$$

$$\cos \beta = \frac{21 + 1 - 16}{2 \cdot 4.5826 \cdot 1} = \frac{6}{9.1652} \approx 0.6547$$

$\beta \approx 49.1°$

$$\cos \gamma = \frac{a^2 + b^2 - c^2}{2ab}$$

$$\cos \gamma = \frac{21 + 16 - 1}{2 \cdot 4.5826 \cdot 4} = \frac{36}{36.6608} \approx 0.9820$$

$\gamma \approx 10.9°$

16. $b^2 = a^2 + c^2 - 2ac \cos \beta$
    $b^2 = 9 + 4 - 2 \cdot 3 \cdot 2 \cdot \cos 90°$
    $b^2 = 13 - 12(0) = 13$
    $b \approx 3.61$

$$\cos \alpha = \frac{b^2 + c^2 - a^2}{2bc}$$

$$\cos \alpha = \frac{13 + 4 - 9}{2 \cdot 3.6056 \cdot 2} = \frac{18}{14.4224} \approx 0.5547$$

$\alpha \approx 56.3°$

$$\cos \gamma = \frac{a^2 + b^2 - c^2}{2ab}$$

$$\cos \gamma = \frac{9 + 13 - 4}{2 \cdot 3 \cdot 3.6056} = \frac{18}{21.6336} \approx 0.8320$$

$\gamma \approx 33.7°$

18. $$\cos \alpha = \frac{b^2 + c^2 - a^2}{2bc}$$

$$\cos \alpha = \frac{25 + 9 - 16}{2 \cdot 5 \cdot 3} = \frac{18}{30} \approx 0.6$$

$\alpha \approx 53.1°$

$$\cos \beta = \frac{a^2 + c^2 - b^2}{2ac}$$

$$\cos \beta = \frac{16 + 9 - 25}{2 \cdot 4 \cdot 3} = 0$$

$\beta \approx 90°$

$$\cos \gamma = \frac{a^2 + b^2 - c^2}{2ab}$$

$$\cos \gamma = \frac{16 + 25 - 9}{2 \cdot 4 \cdot 5} = \frac{32}{40} \approx 0.8$$

$\gamma \approx 36.9°$

20. $$\cos \alpha = \frac{b^2 + c^2 - a^2}{2bc}$$

$$\cos \alpha = \frac{9 + 4 - 9}{2 \cdot 3 \cdot 2} = \frac{4}{12} \approx 0.3333$$

$\alpha \approx 70.5°$

$$\cos \beta = \frac{a^2 + c^2 - b^2}{2ac}$$

$$\cos \beta = \frac{9 + 4 - 9}{2 \cdot 3 \cdot 2} = \frac{4}{12} \approx 0.3333$$

$\beta \approx 70.5°$

$$\cos \gamma = \frac{a^2 + b^2 - c^2}{2ab}$$

$$\cos \gamma = \frac{9 + 9 - 4}{2 \cdot 3 \cdot 3} = \frac{14}{18} \approx 0.7778$$

$\gamma \approx 39°$

22. $\cos \alpha = \dfrac{b^2 + c^2 - a^2}{2bc}$

$\cos \alpha = \dfrac{9 + 36 - 16}{2 \cdot 3 \cdot 6} = \dfrac{29}{36} \approx 0.8056$

$\alpha \approx 36.3°$

$\cos \beta = \dfrac{a^2 + c^2 - b^2}{2ac}$

$\cos \beta = \dfrac{16 + 36 - 9}{2 \cdot 4 \cdot 6} = \dfrac{43}{48} \approx 0.8958$

$\beta \approx 26.4°$

$\cos \gamma = \dfrac{a^2 + b^2 - c^2}{2ab}$

$\cos \gamma = \dfrac{16 + 9 - 36}{2 \cdot 4 \cdot 3} = \dfrac{-11}{24} \approx -0.4583$

$\gamma \approx 117.3°$

24. $\cos \alpha = \dfrac{b^2 + c^2 - a^2}{2bc}$

$\cos \alpha = \dfrac{49 + 100 - 81}{2 \cdot 7 \cdot 10} = \dfrac{68}{140} \approx 0.4857$

$\alpha \approx 61°$

$\cos \beta = \dfrac{a^2 + c^2 - b^2}{2ac}$

$\cos \beta = \dfrac{81 + 100 - 49}{2 \cdot 9 \cdot 10} = \dfrac{132}{180} \approx 0.7333$

$\beta \approx 42.8°$

$\cos \gamma = \dfrac{a^2 + b^2 - c^2}{2ab}$

$\cos \gamma = \dfrac{81 + 49 - 100}{2 \cdot 9 \cdot 7} = \dfrac{30}{126} \approx 0.2381$

$\gamma \approx 76.2°$

26. **(a)** Angle ABC, denoted by $\beta = 180° - 50° = 130°$

Now we know two sides and the included angle. We can use Law of Cosines to find the third side, denoted by b.

$$b^2 = a^2 + c^2 - 2ac \cos \beta$$
$$b^2 = (100)^2 + (150)^2 - 2 \cdot 100 \cdot 150 \cdot \cos 130°$$
$$b^2 = 32,500 - 30,000(-0.6428) - 51,784$$
$$b \approx 227.56 \text{ miles}$$

It is 227.56 miles from city A to city C.

(b) Let γ denote the Angle ACB. The angle through which the pilot should turn at city C to return to city A is found by $180° - \gamma$. We use the Law of Cosines to find angle γ.

$$\cos \gamma = \dfrac{a^2 + b^2 - c^2}{2ab}$$
$$\cos \gamma = \dfrac{100,000 + 51,784 - 22,500}{2 \cdot 100 \cdot 227.56} = \dfrac{39,284}{45,512} \approx 0.8632$$
$$\gamma \approx 30.3°$$
$$180° - \gamma \approx 149.7°$$

The pilot should turn through an angle of 149.7°.

28. **(a)** Using the labels provided in this illustration, we are looking for the angle through which the captain should turn to head directly to Barbados, determined by $180° - \gamma$. First, we need to determine the measurement of b using the equation:

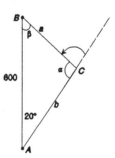

$$\text{Velocity} \times \text{Time} = \text{Distance}$$
$$15 \quad \times \quad 10 \quad = 150$$

Hence, $b = 150$ nautical miles. Before we find γ, we must use the Law of Cosines to determine the third side of the triangle, a.

$$a^2 = b^2 + c^2 - 2bc \cos \alpha$$
$$a^2 = 22,500 + 360,000 - 2 \cdot 150 \cdot 600 \cdot \cos 20°$$
$$a^2 = 382,500 - 180,000(0.9397) = 213,354$$
$$a \approx 461.9$$

Now we use the Law of Cosines to find angle γ:

$$\cos \gamma = \frac{a^2 + b^2 - c^2}{2ab}$$

$$\cos \gamma = \frac{213,354 + 22,500 - 360,000}{2 \cdot 461.9 \cdot 150} = \frac{-124,146}{138,570} \approx -0.8958$$

$$\gamma \approx 153.6°$$
$$180° - \gamma \approx 26.4°$$

The captain should turn through an angle of 77.1°.

(b)   Time $= \dfrac{\text{Distance}}{\text{Velocity}}$

       Time $= \dfrac{b + a}{15}$

       Time $= \dfrac{150 + 461.9}{15} = 40.8$ hours $= 40$ hours 48 minutes

30.   (a) $\alpha = 45°$ because the diagonal of a square bisects the right angle. We know two sides of the triangle and their included angle. Hence, we use the Law of Cosines to find the third side.

$$a^2 = b^2 + c^2 - 2bc \cos \alpha$$
$$a^2 = 2116 + 3600 - 2 \cdot 46 \cdot 60 \cdot \cos 45°$$
$$a^2 = 5716 - 5520(0.7071) = 1812.8$$
$$a \approx 42.6 \text{ feet}$$

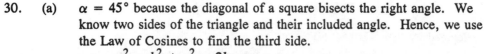

It is 42.6 feet from the pitching rubber to first base on a Little League field.

(b)   Using the Law of Sines, we find angle γ:

$$\frac{\sin 45°}{42.6} = \frac{\sin \gamma}{60}$$
$$\sin \gamma = \frac{60 \sin 45°}{42.6}$$
$$\approx 0.9959$$

Since there are two angles γ, $0° < \gamma < 180°$, for which $\sin \gamma \approx 0.9959$, namely $\gamma = 84.8°$ or $\gamma = 95.2°$. We rule out 84.8° by part (a).

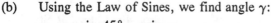

Thus, $\gamma = 95.2°$; $\beta = 180° - \alpha - \gamma \approx 39.8°$

Now we can use the Law of Cosines to find side b:

$$b^2 = a^2 + c^2 - 2ac \cos \beta$$
$$b^2 = 1812.8 + 3600 - 2 \cdot 42.6 \cdot 60 \cdot \cos 39.8°$$
$$b^2 = 5412.8 - 5112(0.7683) = 1487.2946$$
$$b \approx 38.6 \text{ feet}$$

It is 46.3 feet from the pitching rubber to second base.

(c)   From (b), we know that $\gamma = 95.2°$.

$180° - 95.2° = 84.8°$. The pitcher needs to turn an angle of 84.4° to face first base.

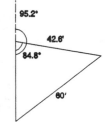

32. $\beta = 85°$ since the inclination to the horizontal is 5°. Hence, we use the Law of Cosines to find b.

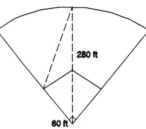

$$b^2 = 500^2 + 100^2 - 2 \cdot 500 \cdot 100 \cdot \cos 85°$$
$$b^2 = 251284.4$$
$$b \approx 502.3 \text{ ft.}$$

$\alpha = 95°$ (supplementary angles). Hence, we use the Law of Cosines to find a.

$$a^2 = 500^2 + 100^2 - 2 \cdot 500 \cdot 100 \cdot \cos 95°$$
$$a^2 = 268715.6$$
$$a \approx 518.4 \text{ ft.}$$

Therefore, the length of the two guy wires will be $a + b = 1019.7$ ft.

34. We can use the Law of Cosines since we know the length of the two sides and the included angle.

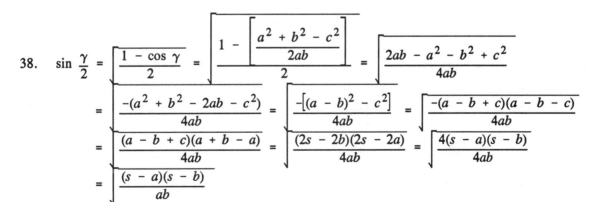

$$c^2 = a^2 + b^2 - 2\,ab \cos \gamma$$
$$c^2 = 280^2 + 60^2 - 2(280)(60)\cos 45°$$
$$c^2 = 58241.2122$$
$$c \approx 241.33 \text{ feet}$$

In Oak Lawn's Little League Field, it is approximately 241.33 feet from dead center to 3rd base.

36. We use the Law of Cosines to find the measurement of side d:

$$d^2 = r^2 + r^2 - 2 \cdot r \cdot r \cdot \cos \theta$$
$$d^2 = 2r^2 - 2r^2 \cos \theta$$
$$d^2 = 2r^2(1 - \cos \theta)$$
$$d^2 = 4r^2 \left[\frac{1 - \cos \theta}{2} \right]$$
$$d = \sqrt{4r^2 \left[\frac{1 - \cos \theta}{2} \right]}$$
$$d = 2r \sin \frac{\theta}{2}$$

Since $d < s$, and $s = r\theta$, we have $2r \sin \dfrac{\theta}{2} < r\theta$

$$\sin \frac{\theta}{2} < \frac{\theta}{2}$$

$\therefore \sin \theta < \theta$ for any angle θ

38.
$$\sin \frac{\gamma}{2} = \sqrt{\frac{1 - \cos \gamma}{2}} = \sqrt{\frac{1 - \left[\dfrac{a^2 + b^2 - c^2}{2ab}\right]}{2}} = \sqrt{\frac{2ab - a^2 - b^2 + c^2}{4ab}}$$

$$= \sqrt{\frac{-(a^2 + b^2 - 2ab - c^2)}{4ab}} = \sqrt{\frac{-[(a - b)^2 - c^2]}{4ab}} = \sqrt{\frac{-(a - b + c)(a - b - c)}{4ab}}$$

$$= \sqrt{\frac{(a - b + c)(a + b - a)}{4ab}} = \sqrt{\frac{(2s - 2b)(2s - 2a)}{4ab}} = \sqrt{\frac{4(s - a)(s - b)}{4ab}}$$

$$= \sqrt{\frac{(s - a)(s - b)}{ab}}$$

2. $A = \frac{1}{2}bc \sin \alpha$

     $A = \frac{1}{2} \cdot 3 \cdot 4 \cdot \sin 30° = 3$

4. $A = \frac{1}{2}ac \sin \beta$

     $A = \frac{1}{2} \cdot 2 \cdot 5 \cdot \sin 20° \approx 1.71$

6. $a = 8, \; b = 5, c = 4$

     $s = \frac{1}{2}(a + b + c) = \frac{1}{2}(8 + 5 + 4) = \frac{17}{2}$

Using Heron's Formula:

$$A = \sqrt{s(s - a)(s - b)(s - c)} = \sqrt{\frac{17}{2} \cdot \frac{1}{2} \cdot \frac{7}{2} \cdot \frac{9}{2}} = \sqrt{\frac{1071}{16}} = \sqrt{66.9375} \approx 8.18$$

8. $s = \frac{1}{2}(a + b + c) = \frac{1}{2}(4 + 3 + 4) = \frac{11}{2}$

$$A = \sqrt{s(s - a)(s - b)(s - c)} = \sqrt{\frac{11}{2} \cdot \frac{3}{2} \cdot \frac{5}{2} \cdot \frac{3}{2}} = \sqrt{\frac{495}{16}} = \sqrt{30.9375} \approx 5.56$$

10. $A = \frac{1}{2}ac \sin \beta$

     $A = \frac{1}{2} \cdot 2 \cdot 1 \cdot \sin 10° \approx 0.17$

12. $A = \frac{1}{2}ab \sin \gamma$

     $A = \frac{1}{2} \cdot 6 \cdot 4 \cdot \sin 60° \approx 10.39$

14. $A = \frac{1}{2}bc \sin \alpha$

     $A = \frac{1}{2} \cdot 4 \cdot 1 \cdot \sin 120° \approx 1.73$

16. $A = \frac{1}{2}ac \sin \beta$

     $A = \frac{1}{2} \cdot 3 \cdot 2 \cdot \sin 90° = 3$

18. $s = \frac{1}{2}(a + b + c) = \frac{1}{2}(4 + 5 + 3) = 6$

     $A = \sqrt{s(s - a)(s - b)(s - c)} = \sqrt{6 \cdot 2 \cdot 1 \cdot 3} = \sqrt{36} = 6$

20. $s = \frac{1}{2}(a + b + c) = \frac{1}{2}(3 + 3 + 2) = 4$

     $A = \sqrt{s(s - a)(s - b)(s - c)} = \sqrt{4 \cdot 1 \cdot 1 \cdot 2} = \sqrt{8} \approx 2.83$

22. $s = \frac{1}{2}(a + b + c) = \frac{1}{2}(4 + 3 + 6) = \frac{13}{2}$

     $A = \sqrt{s(s - a)(s - b)(s - c)} = \sqrt{\frac{13}{2} \cdot \frac{5}{2} \cdot \frac{7}{2} \cdot \frac{1}{2}} = \sqrt{28.4375} \approx 5.33$

24. $s = \frac{1}{2}(a + b + c) = \frac{1}{2}(9 + 7 + 10) = 13$

     $A = \sqrt{s(s - a)(s - b)(s - c)} = \sqrt{13 \cdot 4 \cdot 6 \cdot 3} = \sqrt{936} \approx 30.59$

26. We will find the area of the 3 triangles shown in the diagram. First, we need to find the lengths of sides x and y. Use the Law of Cosines to find the length of side x:

$$x^2 = 35^2 + 80^2 - 2(35)(80)\cos 15°$$
$$x^2 = 2215.8154$$
$$x \approx 47.07 \text{ feet}$$

To find the length of side y, we recognize the angle included between 45' and 20' is $180° - 100° = 80°$. Thus,

$$y^2 = 45^2 + 20^2 - 2(45)(20)\cos 80°$$
$$y^2 = 2112.4333$$
$$y \approx 45.96 \text{ feet}$$

Using Heron's Formula:

Area of triangle 1:

$$\text{Find } s_1 = \frac{1}{2}(35 + 80 + 47.07) = 81.035$$

$$\text{Thus, } A_1 = \sqrt{81.035(81.035 - 35)(81.035 - 80)(81.035 - 47.07)} \approx 362.13 \text{ ft}^2$$

Area of triangle 2:

$$\text{Find } s_2 = \frac{1}{2}(40 + 47.07 + 45.96) = 66.515$$

$$\text{Thus, } A_2 = \sqrt{66.515(66.515 - 40)(66.515 - 47.07)(66.515 - 45.96)} \approx 839.59 \text{ ft}^2$$

Area of triangle 2:

$$\text{Find } s_3 = \frac{1}{2}(45 + 20 + 45.96) = 55.48$$

$$\text{Thus, } A_3 = \sqrt{55.48(55.48 - 45)(55.48 - 20)(55.48 - 45.96)} \approx 443.16 \text{ ft}^2$$

The approximate area of the lake is $A_1 + A_2 + A_3 = 1644.9 \text{ ft}^2$.

28. Using the fact that $A = \frac{1}{2}bc \sin \alpha$ and $A = \frac{1}{2}ac \sin \beta$, and $A = \frac{1}{2}ab \sin \gamma$,

$$A = \frac{1}{2}bc \sin \alpha \qquad\qquad A = \frac{1}{2}ac \sin \beta$$

$$A = \frac{1}{2}b \sin \alpha \left[\frac{b \sin \gamma}{\sin \beta}\right] \qquad\qquad A = \frac{1}{2}c \sin \beta \left[\frac{c \sin \alpha}{\sin \gamma}\right]$$

$$A = \frac{b^2 \sin \alpha \sin \gamma}{2 \sin \beta} \qquad\qquad A = \frac{c^2 \sin \alpha \sin \beta}{2 \sin \gamma}$$

30. $A = \dfrac{a^2 \sin \beta \sin \gamma}{2 \sin \alpha}$ $\qquad\qquad \beta = 180° - \alpha - \gamma = 110°$

$A = \dfrac{9 \sin 110° \sin 20°}{2 \sin 50°} \approx 1.89$

32. $A = \dfrac{c^2 \sin \alpha \sin \beta}{2 \sin \gamma}$ $\qquad\qquad \gamma = 180° - \alpha - \beta = 50°$

$A = \dfrac{16 \sin 70° \sin 60°}{2 \sin 50°} \approx 8.50$

34. $A = \dfrac{b^2 \sin \alpha \sin \gamma}{2 \sin \beta}$ $\qquad\qquad \alpha = 180° - \beta - \gamma = 70°$

$A = \dfrac{4 \sin 70° \sin 100°}{2 \sin 10°} \approx 10.66$

36.　$A = \dfrac{a^2 \sin \beta \sin \gamma}{2 \sin \alpha}$　　　　$\beta = 180° - \alpha - \gamma = 90°$

　　　$A = \dfrac{1 \sin 20° \sin 70°}{2 \sin 90°} \approx 0.16$

10.4　Polar Coordinates

2.
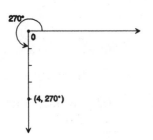

4.
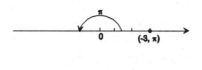

6.
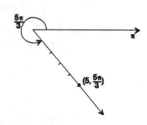

8.
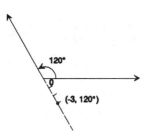

10.
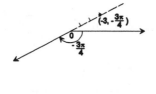

12.
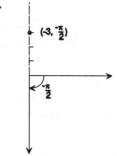

14.　(a)　$\left[4, \dfrac{3\pi}{4} - 2\pi\right] = \left[4, \dfrac{-5\pi}{4}\right]$

　　　(b)　$\left[-4, \dfrac{3\pi}{4} + \pi\right] = \left[-4, \dfrac{7\pi}{4}\right]$

　　　(c)　$\left[4, \dfrac{3\pi}{4} + 2\pi\right] = \left[4, \dfrac{11\pi}{4}\right]$

16.　(a)　$(3, 4\pi - \pi - 4\pi) = (3, -\pi)$
　　　(b)　$(-3, 4\pi - 4\pi) = (-3, 0)$
　　　(c)　$(3, 4\pi - \pi) = (3, 3\pi)$

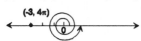

18.　(a)　$(2, \pi - 2\pi) = (2, -\pi)$
　　　(b)　$(-2, \pi - \pi) = (-2, 0)$
　　　(c)　$(2, \pi + 2\pi) = (2, 3\pi)$

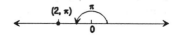

20.　(a)　$\left[2, \dfrac{-2\pi}{3} - \pi\right] = \left[2, \dfrac{-5\pi}{3}\right]$

　　　(b)　$\left[-2, \dfrac{-2\pi}{3} + 2\pi\right] = \left[-2, \dfrac{4\pi}{3}\right]$

　　　(c)　$\left[2, \dfrac{-2\pi}{3} + \pi + 2\pi\right] = \left[2, \dfrac{7\pi}{3}\right]$

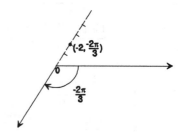

22. $x = r \cos \theta = 4 \cos \dfrac{3\pi}{2} = 0;$

$y = r \sin \theta = 4 \sin \dfrac{3\pi}{2} = -4$

The rectangular coordinates of the point

$\left[4, \dfrac{3\pi}{2}\right]$ are $(0, -4)$.

24. $x = r \cos \theta = -3 \cos \pi = 3;$
$y = r \sin \theta = -3 \sin \pi = 0$
The rectangular coordinates of the point
$(-3, \pi)$ are $(3, 0)$.

26. $x = r \cos \theta = 5 \cos 300° = \dfrac{5}{2};$

$y = r \sin \theta = 5 \sin 300° = \dfrac{-5\sqrt{3}}{2}$

The rectangular coordinates of the point

$\left[5, \dfrac{5\pi}{3}\right]$ are $\left[\dfrac{5}{2}, \dfrac{-5\sqrt{3}}{2}\right]$.

28. $x = r \cos \theta = -3 \cos \dfrac{2\pi}{3} = \dfrac{3}{2};$

$y = r \sin \theta = -3 \sin \dfrac{2\pi}{3} = \dfrac{-3\sqrt{3}}{2}$

The rectangular coordinates of the point

$\left[-3, \dfrac{2\pi}{3}\right]$ are $\left[\dfrac{3}{2}, \dfrac{-3\sqrt{3}}{2}\right]$.

30. $x = r \cos \theta = -3 \cos \dfrac{-3\pi}{4} = \dfrac{-3\sqrt{2}}{2};$

$y = r \sin \theta = -3 \sin \dfrac{-3\pi}{4} = \dfrac{3\sqrt{2}}{2}$

The rectangular coordinates of the point

$\left[-3, \dfrac{-3\pi}{4}\right]$ are $\left[\dfrac{3\sqrt{2}}{2}, \dfrac{3\sqrt{2}}{2}\right]$.

32. $x = r \cos \theta = -3 \cos -90° = 0;$
$y = r \sin \theta = -3 \sin -90° = 3$
The rectangular coordinates of the point
$\left[-3, \dfrac{-90°}{2}\right]$ are $(0, 3)$.

34. $x = r \cos \theta = -3.1 \cos 182° = 3.10;\ y = r \sin \theta = -3.1 \sin 182° = 0.11$

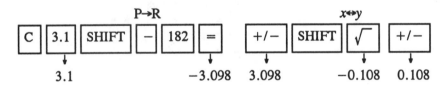

36. $x = 8.1 \cos 5.2 = 3.80;\ y = 8.1 \sin 5.2 = -7.16$

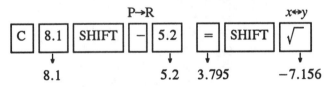

38. Since the point $(0, 2)$ lies on the positive, y-axis, then $r = |y| = |2| = 2$ and $\theta = \dfrac{\pi}{2}$. Thus,

polar coordinates are $\left[2, \dfrac{\pi}{2}\right]$.

40. Since the point $(0, -2)$ lies on the negative y-axis, then $r = |y| = |-2| = 2$ and $\theta = -\frac{\pi}{2}$.

Thus, polar coordinates are $\left[2, \frac{-\pi}{2}\right]$.

42. Since the point $(-3, 3)$ lies in quadrant II, then $r = -\sqrt{9 + 9} = -3\sqrt{2}$ and $\theta = \tan^{-1}\frac{y}{x}$

$= \tan^{-1}(-1) = \frac{-\pi}{4}$. Thus, polar coordinates are $\left[-3\sqrt{2}, \frac{-\pi}{4}\right]$.

44. Since the point $\left(-2, -2\sqrt{3}\right)$ lies in quadrant III, then $r = -\sqrt{4 + 12} = -4$ and

$\theta = \tan^{-1}\left(\sqrt{3}\right) = \frac{\pi}{3}$. Thus, polar coordinates are $\left[-4, \frac{\pi}{3}\right]$.

46. $x = -0.8, y = -2.1$

$r = \sqrt{x^2 + y^2} = \sqrt{5.05} = 2.25$

$\alpha = \tan^{-1}\left|\frac{y}{x}\right| = \tan^{-1}2.625 = 1.21$

$\theta = \pi + \alpha = 4.35$

Polar coordinates of (x, y) are $(2.25, 4.35)$.

48. $x = -2.3, y = 0.2$

$r = \sqrt{x^2 + y^2} = \sqrt{5.33} = 2.31$

$\alpha = \tan^{-1}\left|\frac{y}{x}\right| = \tan^{-1}0.087 = 0.09$

$\theta = \pi - \alpha = 3.05$

Polar coordinates of (x, y) are $(2.31, 3.05)$.

50. $x^2 + y^2 = x$

$r^2 = r\cos\theta$

$r = \cos\theta$

52. $y^2 = 2x$

$r^2\sin^2\theta = 2r\cos\theta$

$r^2\sin^2\theta - 2r\cos\theta = 0$

54. $4x^2y = 1$

$4(r^2\cos^2\theta)(r\sin\theta) = 1$

$4r^3\cos^2\theta\sin\theta - 1 = 0$

56. $y = -3$

$r\sin\theta = -3$

$r\sin\theta + 3 = 0$

58. $r = \sin\theta + 1$

$r^2 = r\sin\theta + r$

$x^2 + y^2 = y \pm \sqrt{x^2 + y^2}$

60. $r = \sin\theta - \cos\theta$

$r^2 = r\sin\theta - r\cos\theta$

$x^2 + y^2 = y - x$

$x^2 + x + y^2 - y = 0$

62. $r = 4$

$r^2 = 16$

$x^2 + y^2 = 16$

64. $r = \dfrac{3}{3 - \cos\theta}$

$3r - r\cos\theta = 3$

$3r - x = 3$

$3r = 3 + x$

$9r^2 = (3 + x)^2$

$9x^2 + 9y^2 = 9 + 6x + x^2$

$8x^2 - 6x + 9y^2 - 9 = 0$

2. $r = 2$

This is of the form $r = a$, $a > 0$. Thus, by Table 7, the graph of $r = 2$ is a circle, center at the pole and radius 2.

If we convert the polar equation to a rectangular equation, then

$$r = 2$$
$$r^2 = 4$$
$$x^2 + y^2 = 4$$

Circle: center at the pole and radius 2.

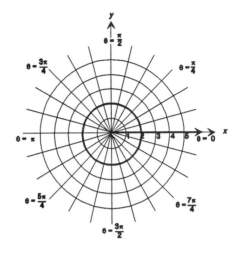

4. $\theta = \dfrac{-\pi}{4}$

This is of the form $\theta = \alpha$. Thus, by Table 7, the graph of $\theta = \dfrac{-\pi}{4}$ is a line passing through the pole making an angle of $\dfrac{-\pi}{4}$ with the polar axis.

If we convert the polar equation to a rectangular equation, then

$$\theta = \frac{-\pi}{4}$$
$$\tan \theta = \tan \frac{-\pi}{4}$$
$$\frac{y}{x} = -1$$
$$y = -x$$

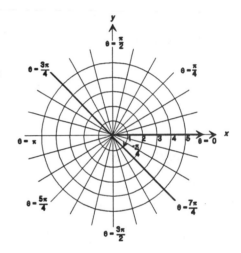

6. $r \cos \theta = 4$

This is of the form $r \cos \theta = b$. Thus, by Table 7, the graph of $r \cos \theta = 4$ is a vertical line.

Since $x = r \cos \theta$, we can write the rectangular equation as $x = 4$. We conclude that the graph is a vertical line 4 units to the right of the pole.

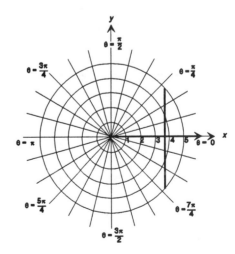

8. $r \sin \theta = -2$

This is of the form $r \sin \theta = a$. Thus, by Table 7, the graph of $r \sin \theta = -2$ is a horizontal line.

Converting to rectangular coordinates, we have
$$r \sin \theta = -2$$
$$y = -2$$
which is the graph of a horizontal line 2 units below the pole.

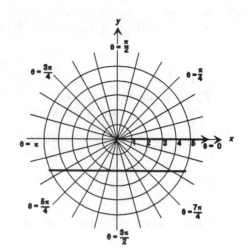

10. $r = 2 \sin \theta$

This is of the form $r = 2a \sin \theta$, $a > 0$. Thus, by Table 7, the graph of $r = 2 \sin \theta$ is a circle, passing through the pole, tangent to the polar axis, center on the line $\theta = \dfrac{\pi}{2}$.

If we convert to a rectangular equation,
$$r = 2 \sin \theta$$
$$r^2 = 2r \sin \theta$$
$$x^2 + y^2 = 2y$$
$$x^2 + y^2 - 2y = 0$$
$$x^2 + (y - 1)^2 = 1$$
This is the equation of a circle, center at $(0, 1)$ in rectangular coordinates and radius 1.

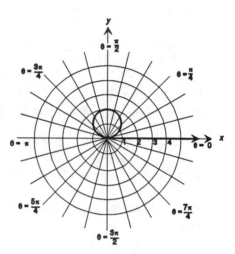

12. $r = -4 \cos \theta$ is the graph of a circle which passes through the pole, tangent to the line $\theta = \dfrac{\pi}{2}$, center on the polar axis. Converting to rectangular coordinates, we have
$$r = -4 \cos \theta$$
$$r^2 = -4r \cos \theta$$
$$x^2 + y^2 = -4x$$
$$x^2 + 4x^2 + y^2 = 0$$
$$(x + 2)^2 + y^2 = 4$$
This is the equation of a circle, center at $(-2, 0)$ in rectangular coordinates and radius 2.

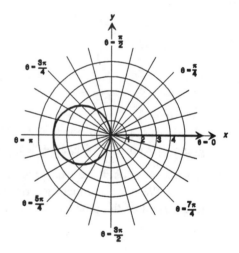

14. We convert the polar equation to a rectangular equation.

$$r \csc \theta = 8$$

$$r \frac{1}{\sin \theta} = 8$$

$$r = 8 \sin \theta$$

$$r^2 = 8 r \sin \theta$$

$$x^2 + y^2 = 8y$$

$$x^2 + y^2 - 8y = 0$$

$$x^2 + (y - 4)^2 = 16$$

This is the equation of a circle, center at $(0, 4)$ in rectangular coordinates and radius 4.

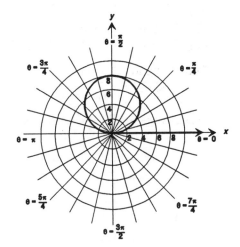

16. We convert to a rectangular equation.

$$r \sec \theta = -4$$

$$r \frac{1}{\cos \theta} = -4$$

$$r = -4 \cos \theta$$

$$r^2 = -4 r \cos \theta$$

$$x^2 + y^2 = -4x$$

$$x^2 + 4x + y^2 = 0$$

$$(x + 2)^2 + y^2 = 4$$

This is the equation of a circle, center $(-2, 0)$ in rectangular coordinates, $(2, \pi)$ in polar coordinates, and radius 2.

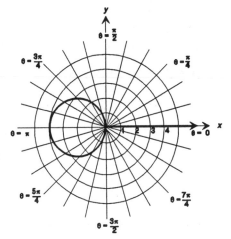

18. $\theta = \dfrac{\pi}{4}$; A

20. $r \cos \theta = 2$; B

22. $r = 2 \sin \theta$; G

24. $r \sin \theta = 2$; C

26. $r = 1 + \sin \theta$. This graph is a cardiod.
We check for symmetry.

Polar axis: Replace θ by $-\theta$. The result is $r = 1 + \sin(-\theta) = 1 - \sin \theta$. The test fails.

The line $\theta = \dfrac{\pi}{2}$:   Replace θ by $\pi - \theta$.

$$r = 1 + \sin(\pi - \theta) = 1 + \sin \pi \cos \theta - \cos \pi \sin \theta = 1 + \sin \theta$$

Thus, the graph is symmetric with respect to the line $\theta = \dfrac{\pi}{2}$.

The pole: Replace r by $-r$. $-r = 1 + \sin \theta$. The test fails.

θ	$\dfrac{-\pi}{2}$	$\dfrac{-\pi}{3}$	$\dfrac{-\pi}{4}$	$\dfrac{-\pi}{6}$	0
$r = 1 + \sin\theta$	0	$1 - \dfrac{\sqrt{3}}{2} \approx 0.1$	$1 - \dfrac{\sqrt{2}}{2} \approx 0.3$	$\dfrac{1}{2}$	1
(r, θ)	$\left[0, \dfrac{-\pi}{2}\right]$	$\left[0.1, \dfrac{-\pi}{3}\right]$	$\left[0.3, \dfrac{-\pi}{4}\right]$	$\left[\dfrac{1}{2}, \dfrac{-\pi}{6}\right]$	$(1, 0)$

θ	$\dfrac{\pi}{6}$	$\dfrac{\pi}{4}$	$\dfrac{\pi}{3}$	$\dfrac{\pi}{2}$
$r = 1 + \sin\theta$	$\dfrac{3}{2}$	$1 + \dfrac{\sqrt{2}}{2} \approx 1.7$	$1 + \dfrac{\sqrt{3}}{2} \approx 1.9$	2
(r, θ)	$\left[\dfrac{3}{2}, \dfrac{\pi}{6}\right]$	$\left[1.7, \dfrac{\pi}{4}\right]$	$\left[1.9, \dfrac{\pi}{3}\right]$	$\left[2, \dfrac{\pi}{2}\right]$

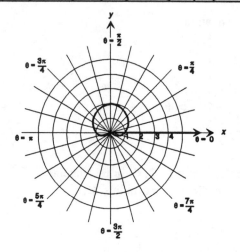

28. $r = 2 - 2\cos\theta$. This is a graph of a cardiod. We check for symmetry.

Polar axis: $r = 2 - 2\cos(-\theta) = 2 - 2\cos\theta$

                     Thus, the graph is symmetric with respect to the polar axis.

The line $\theta = \dfrac{\pi}{2}$: $r = 2 - 2\cos(\pi - \theta) = 2 - 2(\cos\pi\cos\theta + \sin\pi\sin\theta) = 2 + 2\cos\theta$

                     The test fails.

The pole: $-r = 2 - 2\cos\theta$. The test fails.

θ	0	$\dfrac{\pi}{6}$	$\dfrac{\pi}{4}$	$\dfrac{\pi}{3}$	$\dfrac{\pi}{2}$
$r = 2 - 2\cos\theta$	0	$2 - \sqrt{3} \approx 0.3$	$2 - \sqrt{2} \approx 0.6$	1	2
(r, θ)	$(0, 0)$	$\left[0.3, \dfrac{\pi}{6}\right]$	$\left[0.6, \dfrac{\pi}{4}\right]$	$\left[1, \dfrac{\pi}{3}\right]$	$\left[2, \dfrac{\pi}{2}\right]$

θ	$\dfrac{2\pi}{3}$	$\dfrac{3\pi}{4}$	$\dfrac{5\pi}{6}$	π
$r = 2 - 2\cos\theta$	3	$2 + \sqrt{2} \approx 3.4$	$2 + \sqrt{3} \approx 3.7$	4
(r, θ)	$\left[3, \dfrac{2\pi}{3}\right]$	$\left[3.4, \dfrac{3\pi}{4}\right]$	$\left[3.7, \dfrac{5\pi}{6}\right]$	$(4, \pi)$

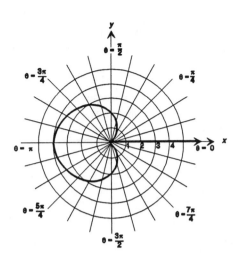

30. $r = 2 - \cos\theta$. This is a graph of a limacon without inner loop. Symmetry.

Polar axis: $r = 2\cos(-\theta) = 2 - \cos\theta$

The graph is symmetric with respect to the polar axis.

The line $\theta = \frac{\pi}{2}$: $r = 2 - \cos(\pi - \theta) = 2 - [\cos\pi\cos\theta + \sin\pi\sin\theta] = 2 + \cos\theta$

The test fails.

The pole: $-r = 2 - \cos\theta$. The test fails.

θ	0	$\dfrac{\pi}{6}$	$\dfrac{\pi}{4}$	$\dfrac{\pi}{3}$	$\dfrac{\pi}{2}$
$r = 2 - \cos\theta$	1	$2 - \dfrac{\sqrt{3}}{2} \approx 1.1$	$2 - \dfrac{\sqrt{2}}{2} \approx 1.3$	$\dfrac{3}{2}$	2
(r, θ)	$(1, 0)$	$\left[1.1, \dfrac{\pi}{6}\right]$	$\left[1.3, \dfrac{\pi}{4}\right]$	$\left[\dfrac{3}{2}, \dfrac{\pi}{3}\right]$	$\left[2, \dfrac{\pi}{2}\right]$

θ	$\dfrac{2\pi}{3}$	$\dfrac{3\pi}{4}$	$\dfrac{5\pi}{6}$	π
$r = 2 - \cos\theta$	$\dfrac{5}{2}$	$2 + \dfrac{\sqrt{2}}{2} \approx 2.7$	$2 + \dfrac{\sqrt{3}}{2} \approx 2.9$	3
(r, θ)	$\left[\dfrac{5}{2}, \dfrac{2\pi}{3}\right]$	$\left[2.7, \dfrac{3\pi}{4}\right]$	$\left[2.9, \dfrac{5\pi}{6}\right]$	$(3, \pi)$

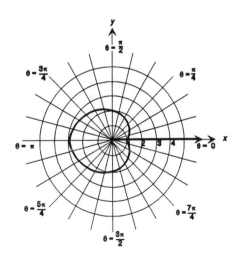

32. $r = 4 + 2 \sin \theta$. This is a graph of a limacon without inner loop. Symmetry.
Polar axis: $r = 4 + 2 \sin(-\theta) = 4 - 2 \sin \theta$. The test fails.

The line $\theta = \frac{\pi}{2}$: $r = 4 + 2 \sin(\pi - \theta) = 4 + 2[\sin \pi \cos \theta - \cos \pi \sin \theta] = 4 + 2 \sin \theta$

The graph is symmetric with respect to the polar axis.

The pole: $-r = 4 + 2 \sin \theta$. The test fails.

θ	$\frac{-\pi}{2}$	$\frac{-\pi}{3}$	$\frac{-\pi}{4}$	$\frac{-\pi}{6}$	0
$r = 4 + 2 \sin \theta$	2	$4 - \sqrt{3} \approx 2.3$	$4 - \sqrt{2} \approx 2.6$	3	4
(r, θ)	$\left(2, \frac{-\pi}{2}\right)$	$\left(2.3, \frac{-\pi}{3}\right)$	$\left(2.6, \frac{-\pi}{4}\right)$	$\left(3, \frac{-\pi}{6}\right)$	$(4, 0)$

θ	$\frac{\pi}{6}$	$\frac{\pi}{4}$	$\frac{\pi}{3}$	$\frac{\pi}{2}$
$r = 4 + 2 \sin \theta$	5	$4 + \sqrt{2} \approx 5.4$	$4 + \sqrt{3} \approx 5.7$	6
(r, θ)	$\left(5, \frac{\pi}{6}\right)$	$\left(5.4, \frac{\pi}{4}\right)$	$\left(5.7, \frac{\pi}{3}\right)$	$\left(6, \frac{\pi}{2}\right)$

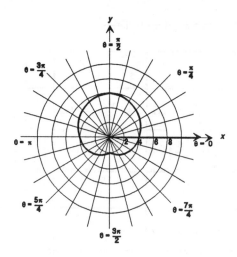

34. $r = 1 - 2 \sin \theta$. This is a graph of a limacon with inner loop. Symmetry.
Polar axis: $r = 1 - 2 \sin(-\theta) = 1 + 2 \sin \theta$. The test fails.

The line $\theta = \frac{\pi}{2}$: $r = 1 - 2 \sin(\pi - \theta) = 1 - [\sin \pi \cos \theta - \cos \pi \sin \theta] = 1 - 2 \sin \theta$

Thus, the graph is symmetric with respect to the line $\theta = \frac{\pi}{2}$.

The pole: $-r = 1 - 2 \sin \theta$. The test fails.

θ	$\dfrac{-\pi}{2}$	$\dfrac{-\pi}{3}$	$\dfrac{-\pi}{4}$	$\dfrac{-\pi}{6}$	0
$r = 1 - 2\sin\theta$	3	$1 + \sqrt{3} \approx 2.7$	$1 + \sqrt{2} \approx 2.4$	2	1
(r, θ)	$\left(3, \dfrac{-\pi}{2}\right)$	$\left(2.7, \dfrac{-\pi}{3}\right)$	$\left(2.4, \dfrac{-\pi}{4}\right)$	$\left(2, \dfrac{-\pi}{6}\right)$	$(1, 0)$

θ	$\dfrac{\pi}{6}$	$\dfrac{\pi}{4}$	$\dfrac{\pi}{3}$	$\dfrac{\pi}{2}$
$r = 1 - 2\sin\theta$	0	$1 - \sqrt{2} \approx -0.4$	$1 + \sqrt{3} \approx -0.7$	-1
(r, θ)	$\left(0, \dfrac{\pi}{6}\right)$	$\left(-0.4, \dfrac{\pi}{4}\right)$	$\left(-0.7, \dfrac{\pi}{3}\right)$	$\left(-1, \dfrac{\pi}{2}\right)$

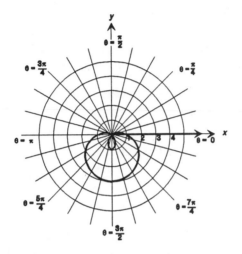

36. $r = 2 + 4\cos\theta$. This is a graph of a limacon with inner loop. Symmetry.

Polar axis: $r = 2 + 4\cos(-\theta) = 2 + 4\cos\theta$

The graph is symmetric with respect to the polar axis.

The line $\theta = \dfrac{\pi}{2}$: $r = 2 + 4\cos(\pi - \theta) = 2 + 4(\cos\pi\cos\theta + \sin\pi\sin\theta) = 2 - 4\cos\pi$

The test fails.

The pole: $-r = 2 + 4\cos\theta$. The test fails.

θ	0	$\dfrac{\pi}{6}$	$\dfrac{\pi}{4}$	$\dfrac{\pi}{3}$	$\dfrac{\pi}{2}$
$r = 2 + 4\cos\theta$	6	$2 + 2\sqrt{3} \approx 5.5$	$2 + 2\sqrt{2} \approx 4.8$	4	2
(r, θ)	$(6, 0)$	$\left(5.5, \dfrac{\pi}{6}\right)$	$\left(4.8, \dfrac{\pi}{4}\right)$	$\left(4, \dfrac{\pi}{3}\right)$	$\left(2, \dfrac{\pi}{2}\right)$

θ	$\dfrac{2\pi}{3}$	$\dfrac{3\pi}{4}$	$\dfrac{5\pi}{6}$	π
$r = 2 + 4\cos\theta$	0	$2 - 2\sqrt{2} \approx -0.8$	$2 - 2\sqrt{3} \approx -1.5$	-2
(r, θ)	$\left(0, \dfrac{2\pi}{3}\right)$	$\left(-0.8, \dfrac{3\pi}{4}\right)$	$\left(-1.5, \dfrac{5\pi}{6}\right)$	$(-2, \pi)$

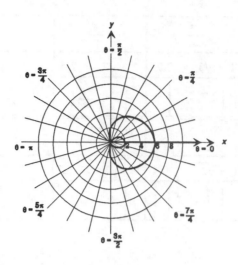

38. $r = 2 \sin 2\,\theta$. This is a graph of a rose with 4 petals. Symmetry.

Polar axis: $r = 2 \sin 2(-\theta) = -2 \sin 2\theta$. The test fails.

The line $\theta = \dfrac{\pi}{2}$: $r = 2 \sin 2(\pi - \theta) = 2 \sin(2\pi - 2\theta) = 2(\sin 2\pi \cos 2\theta + \cos 2\pi \sin 2\theta)$

$$= -2 \sin 2\theta$$

The test fails.

The pole: $-r = 2 \sin 2\theta$. The test fails.

θ	0	$\dfrac{\pi}{6}$	$\dfrac{\pi}{4}$	$\dfrac{\pi}{3}$	$\dfrac{\pi}{2}$
$r = 2 \sin 2\theta$	0	$\sqrt{3} \approx 1.7$	2	$\sqrt{3} \approx 1.7$	0
(r, θ)	$(0, 0)$	$\left[1.7, \dfrac{\pi}{6}\right]$	$\left[2, \dfrac{\pi}{4}\right]$	$\left[1.7, \dfrac{\pi}{3}\right]$	$\left[0, \dfrac{\pi}{2}\right]$

θ	$\dfrac{2\pi}{3}$	$\dfrac{3\pi}{4}$	$\dfrac{5\pi}{6}$	π
$r = 2 \sin 2\theta$	$-\sqrt{3} \approx -1.7$	-2	$-\sqrt{3} \approx -1.7$	0
(r, θ)	$\left[-1.7, \dfrac{2\pi}{3}\right]$	$\left[-2, \dfrac{3\pi}{4}\right]$	$\left[-1.7, \dfrac{5\pi}{6}\right]$	$(0, \pi)$

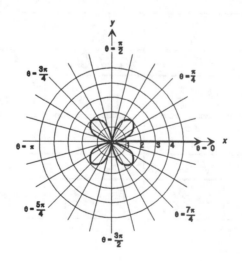

40. $r = 3 \cos 4\theta$. This is a graph of a rose. Symmetry.
Polar axis: $r = 3 \cos 4(-\theta) = 3 \cos 4\theta$
               The graph is symmetric with respect to the polar axis.

The line $\theta = \frac{\pi}{2}$: $r = 3 \cos 4(\pi - \theta) = 3 \cos(4\pi - 4\theta) = 3(\cos 4\pi \cos 4\theta + \sin 4\pi \sin 4\theta)$

                   $= 3 \cos 4\pi$

               Thus, the graph is symmetric with respect to the line $\theta = \frac{\pi}{2}$.

Since it is symmetric with respect to the polar axis and the line $\theta = \frac{\pi}{2}$, it must also be symmetric with respect to the pole.

θ	0	$\frac{\pi}{12}$	$\frac{\pi}{6}$	$\frac{\pi}{4}$	$\frac{\pi}{4}$	$\frac{5\pi}{12}$	$\frac{\pi}{2}$
$r = 3 \cos 4\theta$	3	$\frac{3}{2}$	$\frac{-3}{2}$	-3	$\frac{-3}{2}$	$\frac{3}{2}$	3
(r, θ)	$(3, 0)$	$\left[\frac{3}{2}, \frac{\pi}{12}\right]$	$\left[\frac{-3}{2}, \frac{\pi}{6}\right]$	$\left[-3, \frac{\pi}{4}\right]$	$\left[\frac{-3}{2}, \frac{\pi}{3}\right]$	$\left[\frac{3}{2}, \frac{5\pi}{12}\right]$	$\left[3, \frac{\pi}{2}\right]$

We can use symmetry to complete the graph.

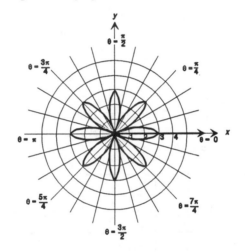

42. $r^2 = \sin 2\theta$. This is a graph of a lemniscate. Symmetry.
Polar axis: $r^2 = \sin 2(-\theta) = -\sin 2\theta$. The test fails.

The line $\theta = \frac{\pi}{2}$: $r^2 = \sin 2(\pi - \theta) = \sin(2\pi - 2\theta) = \sin 2\pi \cos 2\theta - \cos 2\pi \sin 2\theta) = -\sin 2\theta$

               The test fails.
The pole: $(-r)^2 = \sin 2\theta$
                 $r^2 = \sin 2\theta$
               Thus, the graph is symmetric with respect to the pole.

θ	0	$\dfrac{\pi}{6}$	$\dfrac{\pi}{4}$	$\dfrac{\pi}{3}$	$\dfrac{\pi}{2}$
$r^2 = \sin 2\theta$	0	$\dfrac{\sqrt{3}}{2}$	1	$\dfrac{\sqrt{3}}{2}$	0
$r = \pm\sqrt{\sin 2\theta}$	0	$\pm\sqrt{\dfrac{\sqrt{3}}{2}} \approx \pm 0.9$	± 1	$\pm\sqrt{\dfrac{\sqrt{3}}{2}} \approx \pm 0.9$	0
(r, θ)	$(0, 0)$	$\left(0.9, \dfrac{\pi}{6}\right)\ \left(-0.9, \dfrac{\pi}{6}\right)$	$\left(1, \dfrac{\pi}{4}\right)\ \left(-1, \dfrac{\pi}{4}\right)$	$\left(0.9, \dfrac{\pi}{3}\right)\ \left(-0.9, \dfrac{\pi}{3}\right)$	$\left(0, \dfrac{\pi}{2}\right)$

Note there are no points on the graph for $\dfrac{\pi}{2} < \theta < \pi$, since $\sin 2\pi < 0$ for such values.

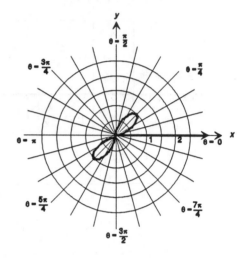

44. $r = 3^{\theta}$. This is a graph of a spiral.
We check for symmetry. All tests fail. There is no number for which $r = 0$. Hence, the graph does not pass through the pole. We observe that r is positive for all θ, that r increases as θ increases, that $r \to 0$ as $\theta \to -\infty$, and that $r \to \infty$ as $\theta \to \infty$.

θ	0	$\dfrac{\pi}{4}$	$\dfrac{\pi}{2}$	π	$\dfrac{3\pi}{2}$	2π
$r = 3^{\theta}$	1	2.4	5.6	31.5	177.2	955.0
(r, θ)	$(0, 1)$	$\left(2.4, \dfrac{\pi}{4}\right)$	$\left(5.6, \dfrac{\pi}{2}\right)$	$(31.5, \pi)$	$\left(177.2, \dfrac{3\pi}{2}\right)$	$(995.0, 2\pi)$

θ	$\dfrac{-\pi}{4}$	$\dfrac{-\pi}{2}$	$-\pi$	$\dfrac{-3\pi}{2}$	-2π
$r = 3^{\theta}$	0.4	0.2	0.0	0.0	0.0
(r, θ)	$\left(0.4, \dfrac{-\pi}{4}\right)$	$\left(0.2, \dfrac{-\pi}{2}\right)$	$(0.0, -\pi)$	$\left(0.0, \dfrac{-3\pi}{2}\right)$	$(0.0, -2\pi)$

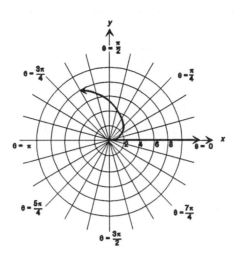

46. $r = 3 + \cos\theta$

This is a graph of a limacon without inner loop.

48. $r = 4 \cos 3\theta$

This is a graph of a rose with 3 petals.

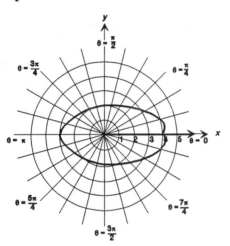

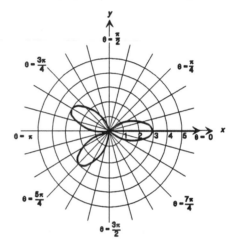

50. $r = \dfrac{2}{1 - 2 \cos\theta}$

We check for symmetry:

Polar axis: $\quad r = \dfrac{2}{1 - 2\cos(-\theta)} = \dfrac{2}{1 - 2\cos\theta}$

Thus, the graph is symmetric with respect to the polar axis.

The line $\theta = \dfrac{\pi}{2}$: $\quad r = \dfrac{2}{1 - 2\cos(\pi - \theta)} = \dfrac{2}{1 - 2(\cos\pi\cos\theta + \sin\pi\sin\theta)} = \dfrac{2}{1 + 2\cos\theta}$

The test fails.

The pole: The test fails.

Due to the symmetry with respect to the polar axis, we only need to assign values to θ from 0 to π.

There is no number θ for which $1 - 2\cos\theta = 0$ or $\cos\theta = \dfrac{1}{2}$, $\theta = \dfrac{\pi}{3} \pm 2k\pi$ or $\theta = \dfrac{5\pi}{3} \pm 2k\pi$,

k any integer.

θ	0	$\dfrac{\pi}{6}$	$\dfrac{\pi}{4}$	$\dfrac{\pi}{2}$
$r = \dfrac{2}{1 - 2\cos\theta}$	-2	$\dfrac{2}{1 - \sqrt{3}} \approx -2.7$	$\dfrac{2}{1 - \sqrt{2}} \approx -4.8$	2
(r, θ)	$(-2, 0)$	$\left(-2.7, \dfrac{\pi}{6}\right)$	$\left(-4.8, \dfrac{\pi}{4}\right)$	$\left(2, \dfrac{\pi}{2}\right)$
θ	$\dfrac{2\pi}{3}$	$\dfrac{3\pi}{4}$	$\dfrac{5\pi}{6}$	π
$r = \dfrac{2}{1 - 2\cos\theta}$	1	$\dfrac{2}{1 + \sqrt{2}} \approx 0.8$	$\dfrac{2}{1 + \sqrt{3}} \approx 0.7$	$\dfrac{2}{3}$
(r, θ)	$\left(1, \dfrac{2\pi}{3}\right)$	$\left(0.8, \dfrac{3\pi}{4}\right)$	$\left(0.7 \ \dfrac{5\pi}{6}\right)$	$\left(\dfrac{2}{3}, \pi\right)$

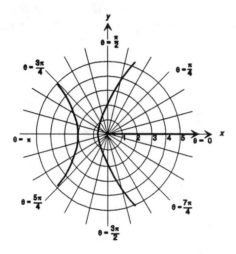

52. $r = \dfrac{1}{1 - \cos\theta}$

We check for symmetry:

Polar axis: $r = \dfrac{1}{1 - \cos(-\theta)} = \dfrac{1}{1 - \cos\theta}$

Thus, the graph is symmetric with respect to the polar axis.

The line $\theta = \dfrac{\pi}{2}$: $r = \dfrac{1}{1 - \cos(\pi - \theta)} = \dfrac{1}{1 + \cos\theta}$. The test fails.

The pole: $-r = \dfrac{1}{1 - \cos\theta}$. The test fails.

Due to symmetry with respect to the polar axis, we only need to assign values to θ from 0 to π. Since $1 - \cos\theta = 0$, $\theta \neq 0 \pm 2k\pi$, k any integer.

θ	$\dfrac{\pi}{6}$	$\dfrac{\pi}{4}$	$\dfrac{\pi}{3}$	$\dfrac{\pi}{2}$
$r = \dfrac{1}{1 - \cos\theta}$	$\dfrac{1}{1 - \frac{\sqrt{3}}{2}} \approx 7.5$	$\dfrac{1}{1 - \frac{\sqrt{2}}{2}} \approx 0.3$	2	1
(r, θ)	$\left(7.5, \dfrac{\pi}{6}\right)$	$\left(3.4, \dfrac{\pi}{4}\right)$	$\left(2, \dfrac{\pi}{3}\right)$	$\left(1, \dfrac{\pi}{2}\right)$

θ	$\dfrac{2\pi}{3}$	$\dfrac{3\pi}{4}$	$\dfrac{5\pi}{6}$	π
$r = \dfrac{1}{1 - \cos\theta}$	$\dfrac{2}{3}$	$\dfrac{1}{1 + \frac{\sqrt{2}}{2}} \approx 0.6$	$\dfrac{1}{1 + \frac{\sqrt{3}}{2}} \approx 0.5$	$\dfrac{1}{2}$
(r, θ)	$\left(\dfrac{2}{3}, \dfrac{2\pi}{3}\right)$	$\left(0.6, \dfrac{3\pi}{4}\right)$	$\left(0.5, \dfrac{5\pi}{6}\right)$	$\left(\dfrac{1}{2}, \pi\right)$

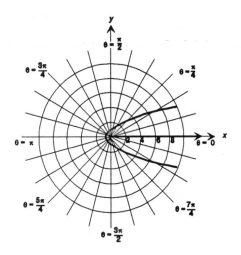

54. $r = \dfrac{3}{\theta}$

All tests for symmetry fail. We observe that r decreases as θ increases. Also θ cannot be 0.

θ	$\dfrac{\pi}{6}$	$\dfrac{\pi}{4}$	$\dfrac{\pi}{3}$	$\dfrac{\pi}{2}$	π	$\dfrac{3\pi}{2}$	2π
$r = \dfrac{3}{\theta}$	$\dfrac{18}{\pi} \approx 5.7$	$\dfrac{12}{\pi} \approx 3.8$	$\dfrac{9}{\pi} \approx 2.9$	$\dfrac{6}{\pi} \approx 1.9$	$\dfrac{3}{\pi} \approx 1.0$	$\dfrac{2}{\pi} \approx 0.6$	$\dfrac{3}{2\pi} \approx 0.5$
(r, θ)	$\left(5.7, \dfrac{\pi}{6}\right)$	$\left(3.8, \dfrac{\pi}{4}\right)$	$\left(2.9, \dfrac{\pi}{3}\right)$	$\left(1.9, \dfrac{\pi}{2}\right)$	$(1.0, \pi)$	$\left(0.6, \dfrac{3\pi}{2}\right)$	$(0.5, 2\pi)$

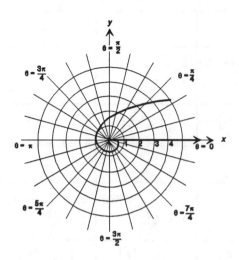

56. $r = \sin \theta \tan \theta$

θ	0	$\dfrac{\pi}{6}$	$\dfrac{\pi}{4}$	$\dfrac{\pi}{3}$
$r = \sin \theta \tan \theta$	0	$\dfrac{1}{2\sqrt{3}} \approx 0.3$	$\dfrac{\sqrt{2}}{2} \approx 0.7$	$\dfrac{3}{2}$
(r, θ)	$(0, 0)$	$\left(0.3, \dfrac{\pi}{6}\right)$	$\left(0.7, \dfrac{\pi}{4}\right)$	$\left(\dfrac{3}{2}, \dfrac{\pi}{3}\right)$
θ	$\dfrac{3\pi}{4}$	π	$\dfrac{5\pi}{4}$	2π
$r = \sin \theta \tan \theta$	$\dfrac{-\sqrt{2}}{2} \approx -0.7$	0	$\dfrac{-\sqrt{2}}{2} \approx -0.7$	0
(r, θ)	$\left(-0.7, \dfrac{3\pi}{4}\right)$	$(0, \pi)$	$\left(-0.7, \dfrac{5\pi}{4}\right)$	$(0, 2\pi)$

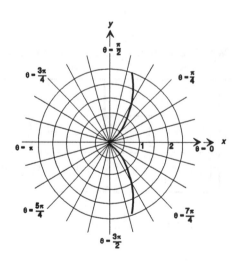

58. $\cos \dfrac{\theta}{2}$

Symmetry:

Polar axis: $\qquad r = \cos\dfrac{(-\theta)}{2} = \cos\dfrac{\theta}{2}$

Thus, symmetric with respect to the polar axis.

The line $\theta = \dfrac{\theta}{2}$**:** $\quad r = \cos\dfrac{1}{2}(\pi - \theta) = \cos\left[\dfrac{\pi}{2} - \dfrac{\theta}{2}\right] = \cos\dfrac{\pi}{2}\cos\dfrac{\theta}{2} + \sin\dfrac{\pi}{2}\sin\dfrac{\theta}{2}$

$\qquad\qquad\qquad = 0 + \sin\dfrac{\theta}{2} = \sin\dfrac{\theta}{2}$

The test fails.

The pole: $\qquad$ The test fails.

θ	0	$\dfrac{\pi}{6}$	$\dfrac{\pi}{4}$	$\dfrac{\pi}{3}$	$\dfrac{\pi}{2}$
$r = \cos\dfrac{\theta}{2}$	1	0.96	0.92	$\dfrac{\sqrt{3}}{2} \approx 0.9$	$\dfrac{\sqrt{2}}{2} \approx 0.7$
(r, θ)	$(1, 0)$	$\left[0.96, \dfrac{\pi}{6}\right]$	$\left[0.92, \dfrac{\pi}{4}\right]$	$\left[0.9, \dfrac{\pi}{3}\right]$	$\left[0.7, \dfrac{\pi}{2}\right]$

θ	$\dfrac{2\pi}{3}$	$\dfrac{3\pi}{4}$	$\dfrac{5\pi}{6}$	π
$r = \cos\dfrac{\theta}{2}$	$\dfrac{1}{2}$	0.4	0.3	0
(r, θ)	$\left[\dfrac{1}{2}, \dfrac{2\pi}{3}\right]$	$\left[0.4, \dfrac{3\pi}{4}\right]$	$\left[0.3, \dfrac{5\pi}{6}\right]$	$(0, \pi)$

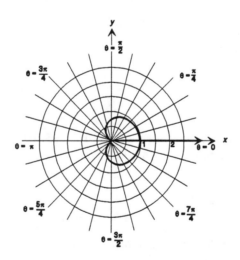

60. We convert the polar equation to a rectangular equation.

$\qquad r\cos\theta = a$

$\qquad\quad x = a$

Thus, the graph $r\cos\theta = a$ is a vertical line a units to the right of the pole if $a > 0$, and a units to the left of the pole if $a < 0$.

62. We convert the polar equation to a rectangular equation.

$$r = -2a \sin \theta, \ a > 0$$
$$r^2 = -2ar \sin \theta$$
$$x^2 + y^2 = -2ay$$
$$x^2 + y^2 + 2y = 0$$
$$x^2 + (y + a)^2 = a^2$$

Circle: radius a, center at $(0, -a)$ in rectangular coordinates.

64. We convert the polar equation to a rectangular equation.

$$r = -2a \cos \theta, \ a > 0$$
$$r^2 = -2ar \cos \theta$$
$$x^2 + y^2 = -2ax$$
$$x^2 + 2ax + y^2 = 0$$
$$(x + a)^2 + y^2 = a^2$$

Circle: radius a, center at $(-a, 0)$ in rectangular coordinates.

66. Symmetry with respect to the Pole: In a polar equation, replace θ by $\pi + \theta$. If an equivalent equation results, the graph is symmetric with respect to the pole.

(a) $r^2 = \sin \theta$: $\quad r^2 = \sin(\pi + \theta)$
$\qquad\qquad\qquad r^2 = -\sin \theta$
$\qquad\qquad\qquad$ New test fails.

$\qquad\qquad\qquad (-r)^2 = \sin \theta$
$\qquad\qquad\qquad\quad r^2 = \sin \theta$
$\qquad\qquad\qquad$ Test works.

(b) $r = \cos^2 \theta$: $\quad -r = \cos^2 \theta$
$\qquad\qquad\qquad\quad$ Test fails.

$\qquad\qquad\qquad r = \cos^2(\pi + \theta)$
$\qquad\qquad\qquad r = (-\cos \theta)^2$
$\qquad\qquad\qquad r = \cos^2 \theta$
$\qquad\qquad\qquad$ New test works.

10.6 The Complex Plane; Demoivre's Theorem

2. $r = \sqrt{x^2 + y^2} = \sqrt{(-1)^2 + (1)^2} = \sqrt{2}$ and

$\tan \theta = \dfrac{y}{x} = \dfrac{1}{-1} = -1$

Thus, $\theta = 135°$ and $r = \sqrt{2}$, so the polar form of $z = -1 + i$ is

$z = r(\cos \theta + i \sin \theta) = \sqrt{2}\,(\cos 135° + i \sin 135°)$

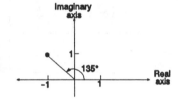

4. $r = \sqrt{x^2 + y^2} = \sqrt{(1)^2 + \left(-\sqrt{3}\right)^2} = 2$ and

$\tan \theta = \dfrac{y}{x} = \dfrac{-\sqrt{3}}{1} = -\sqrt{3}$

Thus, $\theta = 300°$ and $r = 2$, so the polar form of $z = 1 - \sqrt{3}\,i$ is
$z = r(\cos \theta + i \sin \theta) = 2(\cos 300° + i \sin 300°)$

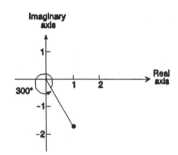

6. $r = \sqrt{x^2 + y^2} = \sqrt{(-2)^2 + (0)^2} = 2$ and $\tan \theta = \dfrac{y}{x} = 0.$

Thus, $\theta = 180°$ and $r = 2$, so the polar form of $z = -2$ is
$z = r(\cos \theta + i \sin \theta) = 2(\cos 180° + i \sin 180°)$

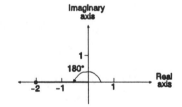

Chapter 10 Additional Applications of Trigonometry

8. $r = \sqrt{x^2 + y^2} = \sqrt{\left(9\sqrt{3}\right)^2 + (9)^2} = \sqrt{324} = 18$ and

$\tan \theta = \dfrac{y}{x} = \dfrac{9}{9\sqrt{3}} = \dfrac{1}{\sqrt{3}}$

Thus, $\theta = 30°$ and $r = 18$, so the polar form of $z = 9\sqrt{3} + 9i$ is
$z = r(\cos \theta + i \sin \theta) = 18(\cos 30° + i \sin 30°)$

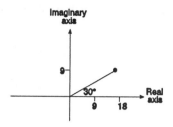

10. $r = \sqrt{x^2 + y^2} = \sqrt{(2)^2 + \left(\sqrt{3}\right)^2} = \sqrt{7}$

$\tan \theta = \dfrac{y}{x} = \dfrac{\sqrt{3}}{2} \approx 0.8660$

$\theta = 40.9°$

The polar form of $z = 2 + \sqrt{3}\, i$ is $z = \sqrt{7}\,(\cos 40.9° + i \sin 40.9°)$

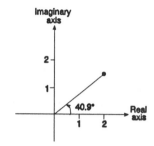

12. $r = \sqrt{x^2 + y^2} = \sqrt{\sqrt{5})^2 + (-1)^2} = \sqrt{6}$

$\tan \theta = \dfrac{y}{x} = \dfrac{-1}{\sqrt{5}} \approx -0.4472$

$\theta = 335.9°$

The polar form of $z = \sqrt{5} - i$ is $z = \sqrt{6}\,(\cos 335.9° + i \sin 335.9°)$

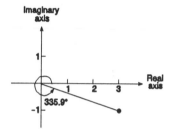

14. $3(\cos 210° + i \sin 210°) = 3\left[-\dfrac{\sqrt{3}}{2} + \dfrac{-1}{2}i\right] = \dfrac{-3\sqrt{3}}{2} - \dfrac{3}{2}i$

16. $2\left[\cos \dfrac{5\pi}{6} + i \sin \dfrac{5\pi}{6}\right] = 2\left[-\dfrac{\sqrt{3}}{2} + \dfrac{1}{2}i\right] = -\sqrt{3} + i$

18. $4\left[\cos\dfrac{\pi}{2} + i \sin\dfrac{\pi}{2}\right] = 4(0 + i) = 4i$

20. $0.4(\cos 200° + i \sin 200°) = 0.4(-0.9397 + -0.3420i) = -0.376 - 0.137i$

22. $3\left[\cos \dfrac{\pi}{10} + i \sin \dfrac{\pi}{10}\right] = 3(0.9511 + 03090i) = 2.853 + 0.927i$

24. $zw = \cos(120° + 100°) + i \sin(120° + 100°)$
$zw = \cos 220° + i \sin 220°$

$\dfrac{z}{w} = \cos(120° - 100°) + i \sin(120° - 100°)$

$\dfrac{z}{w} = \cos 20° + i \sin 20°$

26.
$$zw = 2 \cdot 6[\cos(80° + 200°) + i \sin(80° + 200°)]$$
$$zw = 12(\cos 280° + i \sin 280°)$$
$$\frac{z}{w} = \frac{2}{6}[\cos(80° - 200°) + i \sin(80° - 200°)]$$
$$\frac{z}{w} = \frac{1}{3}[\cos(-120°) + i \sin(-120°)]$$
$$\frac{z}{w} = \frac{1}{3}(\cos 240° + i \sin 240°)$$

28.
$$zw = 4 \cdot 2\left[\cos\left(\frac{3\pi}{8} + \frac{9\pi}{16}\right) + i \sin\left(\frac{3\pi}{8} + \frac{9\pi}{16}\right)\right]$$
$$zw = 8\left[\cos\frac{15\pi}{16} + i \sin\frac{15\pi}{16}\right]$$
$$\frac{z}{w} = \frac{4}{2}\left[\cos\left(\frac{3\pi}{8} - \frac{9\pi}{16}\right) + i \sin\left(\frac{3\pi}{8} - \frac{9\pi}{16}\right)\right]$$
$$\frac{z}{w} = 2\left[\cos\left(\frac{-3\pi}{16}\right) + i \sin\left(\frac{-3\pi}{16}\right)\right]$$
$$\frac{z}{w} = 2\left[\cos\frac{29\pi}{16} + i \sin\frac{29\pi}{16}\right]$$

30.
$$r_1 = \sqrt{x^2 + y^2} = \sqrt{2}$$
$$\tan\theta_1 = \frac{y}{x} = -1, \text{ so } \theta = 315°$$
$$z = \sqrt{2}\,(\cos 315° + i \sin 315°)$$
$$r_2 = \sqrt{x^2 + y^2} = \sqrt{4} = 2, \ \tan\theta_2 = \frac{y}{x} = -\sqrt{3}, \text{ so } \theta = 300°$$
$$w = 2(\cos 300° + i \sin 300°)$$
$$zw = 2\sqrt{2}\left[\cos(315° + 300° - 360°) + i \sin(315° + 300° - 360°)\right]$$
$$zw = 2\sqrt{2}\,(\cos 255° + i \sin 255°)$$
$$\frac{z}{w} = \frac{\sqrt{2}}{2}\left[\cos(315° - 300°) + i \sin(315° - 300°)\right]$$
$$\frac{z}{w} = \frac{\sqrt{2}}{2}(\cos 15° + i \sin 15°)$$

32.
$$[3(\cos 80° + i \sin 80°)]^3 = 3^3[\cos(3 \cdot 80°) + i \sin(3 \cdot 80°)] = 27(\cos 240° + i \sin 240°)$$
$$= 27\left[\frac{-1}{2} + \frac{-\sqrt{3}}{2}i\right] = \frac{-27}{2} - \frac{27\sqrt{3}}{2}i$$

34.
$$\left[\sqrt{2}\left(\cos\frac{5\pi}{16} + i \sin\frac{5\pi}{16}\right)\right]^4 = \sqrt{2}^4\left[\cos\left(4 \cdot \frac{5\pi}{16}\right) + i \sin\left(4 \cdot \frac{5\pi}{16}\right)\right]$$
$$= 4\left[\cos\frac{5\pi}{4} + i \sin\frac{5\pi}{4}\right] = 4\left[\frac{-\sqrt{2}}{2} + \frac{-\sqrt{2}}{2}i\right] = -2\sqrt{2} - 2\sqrt{2}\,i$$

36. $\left[\dfrac{1}{2}(\cos 72° + i \sin 72°)\right]^5 = \left(\dfrac{1}{2}\right)^5\left[\cos(5 \cdot 72°) + i \sin(5 \cdot 72°)\right]$

$\qquad\qquad = \dfrac{1}{32}(\cos 360° + i \sin 360°) = \dfrac{1}{32}(1) = \dfrac{1}{32}$

38. $\left[\sqrt{3}\left(\cos \dfrac{5\pi}{18} + i \sin \dfrac{5\pi}{18}\right)\right]^6 = \sqrt{3}^6\left[\cos\left(6 \cdot \dfrac{5\pi}{18}\right) + i \sin\left(6 \cdot \dfrac{5\pi}{18}\right)\right]$

$\qquad\qquad = 27\left[\cos \dfrac{5\pi}{3} + i \sin \dfrac{5\pi}{3}\right] = 27\left[\dfrac{1}{2} + \dfrac{-\sqrt{3}}{2}i\right] = \dfrac{27}{2}\left(1 - \sqrt{3}\,i\right)$

40. $\qquad \sqrt{3} - i = 2(\cos 330° + i \sin 330°)$

$\qquad \left(\sqrt{3} - i\right)^6 = 2^6\left[\cos(6 \cdot 330°) + i \sin(6 \cdot 330°)\right] = 64(\cos 1980° + i \sin 1980°)$
$\qquad\qquad = 64(\cos 180° + i \sin 180°) = 64(-1 + 0i) = -64$

42. $\quad r = \sqrt{1^2 + \left(-\sqrt{5}\right)^2} = \sqrt{6},\ \tan \theta = -\sqrt{5},\ \theta = 294.1°$

$\qquad \left(1 - \sqrt{5}\,i\right)^8 = \sqrt{6}^8\left[\cos(8 \cdot 294.1°) + i \sin(8 \cdot 294.1°)\right] = 1296(\cos 2352.8° + i \sin 2352.8°)$
$\qquad\qquad = 1296(\cos 192.8° + i \sin 192.8°) = 1296(-0.9751 + -0.2215i)$
$\qquad\qquad = -1263.7 - 287.1i$

44. $\quad \sqrt{3} - i = 2(\cos 330° + i \sin 330°)$

$\qquad z_k = \sqrt[4]{2}\left[\cos\left(\dfrac{330°}{4} + \dfrac{360°k}{4}\right) + i \sin\left(\dfrac{330°}{4} + \dfrac{360°k}{4}\right)\right],\ k = 0, 1, 2, 3$

$\qquad$ Thus, $z_0 = \sqrt[4]{2}(\cos 82.5° + i \sin 82.5°)$

$\qquad\qquad z_1 = \sqrt[4]{2}(\cos 172.5° + i \sin 172.5°)$

$\qquad\qquad z_2 = \sqrt[4]{2}(\cos 262.5° + i \sin 262.5°)$

$\qquad\qquad z_3 = \sqrt[4]{2}(\cos 352.5° + i \sin 352.5°)$

46. $\quad r = \sqrt{(-8)^2 + (-8)^2} = 8\sqrt{2},\ \tan \theta = 1,\ \theta = 225°$

$\qquad -8 - 8i = 8\sqrt{2}\,(\cos 225° + i \sin 225°)$

$\qquad z_k = \sqrt[3]{8\sqrt{2}}\left[\cos\left(\dfrac{225°}{3} + \dfrac{360°k}{3}\right) + i \sin\left(\dfrac{225°}{3} + \dfrac{360°k}{3}\right)\right],\ k = 0, 1, 2$

$\qquad$ Thus, $z_0 = 2\sqrt[6]{2}(\cos 75° + i \sin 75°)$

$\qquad\qquad z_1 = 2\sqrt[6]{2}(\cos 195° + i \sin 195°)$

$\qquad\qquad z_2 = 2\sqrt[6]{2}(\cos 315° + i \sin 315°)$

48. $\quad -8 = 8(\cos 180° = i \sin 180°)$

$\qquad z_k = \sqrt[3]{8}\left[\cos\left(\dfrac{180°}{3} + \dfrac{360°k}{3}\right) + i \sin\left(\dfrac{180°}{3} + \dfrac{360°k}{3}\right)\right],\ k = 0, 1, 2$

$\qquad$ Thus, $z_0 = 2(\cos 60° + i \sin 60°)$

$\qquad\qquad z_1 = 2(\cos 180° + i \sin 180°)$

$\qquad\qquad z_2 = 2(\cos 300° + i \sin 300°)$

50. $-i = 1(\cos 270° + i \sin 270°),$

$$z_k = \sqrt[5]{1}\left[\cos\left(\frac{270°}{5} + \frac{360°k}{5}\right) + i \sin\left(\frac{270°}{5} + \frac{360°k}{5}\right)\right], \; k = 0, 1, 2, 3, 4$$

Thus, $z_0 = \cos 54° + i \sin 54°$
$z_1 = \cos 126° + i \sin 126°$
$z_2 = \cos 198° + i \sin 198°$
$z_3 = \cos 270° + i \sin 270°$
$z_4 = \cos 342° + i \sin 342°$

52. $1 = (\cos 0° + i \sin 0°)$

$$z_k = \left[\cos\left(\frac{0°}{6} + \frac{360°k}{6}\right) + i \sin\left(\frac{0°}{6} + \frac{360°k}{6}\right)\right], \; k = 0, 1, 2, 3, 4, 5$$

Thus, $z_0 = \cos 0° + i \sin 0°$
$z_1 = \cos 60° + i \sin 60°$
$z_2 = \cos 120° + i \sin 120°$
$z_3 = \cos 180° + i \sin 180°$
$z_4 = \cos 240° + i \sin 240°$
$z_5 = \cos 300° + i \sin 300°$

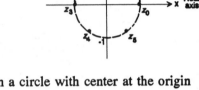

54. Since $|z_k| = \sqrt[n]{r}$ for all k, each of the complex n^{th} roots lies on a circle with center at the origin and radius $\sqrt[n]{r}$.

56. Let $z_1 = r_1(\cos \theta_1 + i \sin \theta_1)$ and $z_2 = r_2(\cos \theta_2 + i \sin \theta_2)$

$$\frac{z_1}{z_2} = \frac{r_1(\cos \theta_1 + i \sin \theta_1)}{r_2(\cos \theta_2 + i \sin \theta_2)} = \frac{r_1}{r_2} \cdot \frac{(\cos \theta_1 + i \sin \theta_1)}{(\cos \theta_2 + i \sin \theta_2)} \cdot \frac{(\cos \theta_2 - i \sin \theta_2)}{(\cos \theta_2 - i \sin \theta_2)}$$

$$= \frac{r_1}{r_2} \cdot \frac{\cos \theta_1 \cos \theta_2 - i \cos \theta_2 \sin \theta_2 + i \sin \theta_1 \cos \theta_2 + \sin \theta_1 \sin \theta_2}{\cos^2 \theta_2 + \sin^2 \theta_2}$$

$$= \frac{r_1}{r_2} \cdot \frac{(\cos \theta_1 \cos \theta_2 + \sin \theta_1 \sin \theta_2) + i(\sin \theta_1 \cos \theta_2 - \cos \theta_1 \sin \theta_2)}{1}$$

$$= \frac{r_1}{r_2}\left[\cos(\theta_1 - \theta_2) + i \sin(\theta_1 - \theta_2)\right]$$

10   Chapter Review

2.
$$\frac{\sin \alpha}{a} = \frac{\sin \gamma}{c}$$
$$\frac{\sin 10°}{a} = \frac{\sin 40°}{2}$$
$$a = \frac{2 \sin 10°}{\sin 40°} \approx 0.54$$
$$\beta = 180° - \alpha - \gamma = 130°$$
$$\frac{\sin 130°}{b} = \sin \frac{40°}{2}$$
$$b = \frac{2 \sin 130°}{\sin 40°} \approx 2.38$$

4.
$$\frac{\sin \alpha}{a} = \frac{\sin \gamma}{c}$$
$$\frac{\sin 60°}{2} = \frac{\sin \gamma}{5}$$
$$\sin \gamma = \frac{5 \sin 60°}{2} \approx 2.1651$$

Because $\sin \gamma$ cannot be greater than 1, this is impossible. No triangle.

6. $$\frac{\sin \alpha}{a} = \frac{\sin \gamma}{c}$$
$$\frac{\sin \alpha}{3} = \frac{\sin 20°}{1}$$
$$\sin \alpha = 3 \sin 20° \approx 1.0261$$
Impossible! No triangle.

8. $$\frac{\sin \alpha}{a} = \frac{\sin \beta}{b}$$
$$\frac{\sin \alpha}{3} = \frac{\sin 80°}{5}$$
$$\sin \alpha = \frac{3 \sin 80°}{5} \approx 0.5909$$
$$\alpha \approx 36.2°$$
$$\gamma \approx 180° - \alpha - \beta \approx 63.8°$$
$$c^2 = a^2 + b^2 - 2ab \cos \gamma$$
$$c^2 = 9 + 25 - 2(3)(5) \cos 63.8°$$
$$\approx 20.7548$$
$$c \approx 4.56$$

10. $$\cos \alpha = \frac{b^2 + c^2 - a^2}{2bc}$$
$$\cos \alpha = \frac{49 + 64 - 100}{2(7)(8)} \approx 0.1161$$
$$\alpha \approx 83.3°$$
$$\cos \beta = \frac{a^2 + c^2 - b^2}{2ac}$$
$$\cos \beta = \frac{100 + 64 - 49}{2(10)(8)} = 0.71875$$
$$\beta = 44.0°$$
$$\gamma \approx 180° - \alpha - \beta \approx 52.7°$$

12. $$c^2 = a^2 + b^2 - 2ab \cos \gamma$$
$$c^2 = 16 + 1 - 2(4)(1) \cos 100°$$
$$\approx 18.3892$$
$$c \approx 4.29$$
$$\cos \alpha = \frac{b^2 + c^2 - a^2}{2bc}$$
$$\cos \alpha = \frac{1 + 18.3892 - 16}{2(1)(4.2883)} \approx 0.3952$$
$$\alpha \approx 66.7°$$
$$\beta \approx 180° - \alpha - \gamma \approx 13.3°$$

14. $$\frac{\sin \alpha}{a} = \frac{\sin \beta}{b}$$
$$\frac{\sin 20°}{2} = \frac{\sin \beta}{3}$$
$$\sin \beta = \frac{3 \sin 20°}{2} \approx 0.5130$$
(Two triangles)

$$\beta_1 \approx 30.9°$$ $$\beta_2 = 149.1°$$
$$\gamma_1 = 180° - \alpha - \beta_1 = 129.1°$$ $$\gamma_2 = 180° - \alpha - \beta_2 \approx 10.9°$$
$$c_1^2 = a^2 + b^2 - 2ab \cos \gamma_1$$ $$c_2^2 = a^2 + b^2 - 2ab \cos \gamma_2$$
$$c_1^2 = 4 + 9 - 2(2)(3)\cos 129.1°$$ $$c_2^2 = 4 + 9 - 2(2)(3) \cos 10.9°$$
$$c_1^2 \approx 20.5681$$ $$c_2^2 = 1.2165$$
$$c_1 \approx 4.54$$ $$c_2 \approx 1.10$$

16. $$\cos \alpha = \frac{b^2 + c^2 - a^2}{2bc}$$
$$\cos \alpha = \frac{4 + 4 - 9}{2(2)(2)} = -0.125$$
$$\alpha = 97.2°$$
$$\cos \beta = \frac{a^2 + c^2 - b^2}{2ac}$$
$$\cos \beta = \frac{9 + 4 - 4}{2(3)(2)} = 0.75$$
$$\beta = 41.4°$$
$$\cos \gamma = \frac{a^2 + b^2 - c^2}{2ab}$$
$$\cos \gamma = \frac{9 + 4 - 4}{2(3)(2)} = 0.75$$
$$\gamma = 41.4°$$

18. $$\gamma = 180° - \alpha - \beta = 60°$$
$$\frac{\sin \alpha}{a} = \frac{\sin \beta}{b} = \frac{\sin \gamma}{c}$$
$$\frac{\sin 20°}{4} = \frac{\sin 100°}{b}$$
$$b = \frac{4 \sin 100°}{\sin 20°}$$
$$b \approx 11.52$$

$$\frac{\sin 20°}{4} = \frac{\sin 60°}{c}$$
$$c = \frac{4 \sin 60°}{\sin 20°}$$
$$c \approx 10.13$$

20.
$$c^2 = a^2 + b^2 - 2ac \cos \gamma$$
$$c^2 = 1 + 4 - 2(1)(2)\cos 60° = 3$$
$$c \approx 1.73$$
$$\frac{\sin \alpha}{a} = \frac{\sin \gamma}{c}$$
$$\frac{\sin \alpha}{1} = \frac{\sin 60°}{1.7321}$$
$$\sin \alpha \approx 0.5000$$
$$\alpha \approx 30°$$
$$\beta \approx 180° - \alpha - \gamma = 90°$$

22. $A = \dfrac{1}{2}bc \sin \alpha$

  $A = \dfrac{1}{2} \cdot 5 \cdot 4 \cdot \sin 20° \approx 3.42$

24. $A = \dfrac{1}{2}ab \sin \gamma$

  $A = \dfrac{1}{2} \cdot 2 \cdot 1 \cdot \sin 100° \approx 0.98$

26. $s = \dfrac{1}{2}(a + b + c)$

  $s = \dfrac{1}{2}(10 + 7 + 8) = \dfrac{25}{2}$

  $A = \sqrt{s(s - a)(s - b)(s - c)} = \sqrt{\left[\dfrac{25}{2}\right]\left[\dfrac{5}{2}\right]\left[\dfrac{11}{2}\right]\left[\dfrac{9}{2}\right]} = \dfrac{\sqrt{12,375}}{4} \approx 27.81$

28. $s = \dfrac{1}{2}(a + b + c)$

  $s = \dfrac{1}{2}(3 + 2 + 2) = \dfrac{7}{2}$

  $A = \sqrt{s(s - a)(s - b)(s - c)} = \sqrt{\left[\dfrac{7}{2}\right]\left[\dfrac{1}{2}\right]\left[\dfrac{3}{2}\right]\left[\dfrac{3}{2}\right]} = \dfrac{\sqrt{63}}{4} \approx 1.98$

30. $A = \dfrac{c^2 \sin \alpha \sin \beta}{2 \sin \gamma}$
  $\beta = 180° - \alpha - \gamma = 130°$
  $A = \dfrac{9 \sin 10° \sin 130°}{2 \sin 40°} = \dfrac{4.5(.1736)(0.766)}{(0.6428)} \approx 0.93$

32. $x = r \cos \theta \qquad y = r \sin \theta$

  $x = 4 \cos \dfrac{2\pi}{3} \qquad y = 4 \sin \dfrac{2\pi}{3}$

  $x = -2 \qquad\quad y = 2\sqrt{3}$

  $\left(-2, \ 2\sqrt{3}\right)$

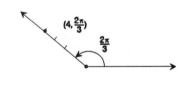

34. $x = r \cos \theta \qquad\qquad y = r \sin \theta$

  $x = -1 \cos \dfrac{5\pi}{3} \qquad y = -1 \sin \dfrac{5\pi}{4}$

  $x = \dfrac{\sqrt{2}}{2} \qquad\qquad y = \dfrac{\sqrt{2}}{2}$

  $\left[\dfrac{\sqrt{2}}{2}, \ \dfrac{\sqrt{2}}{2}\right]$

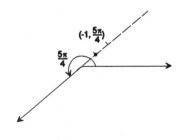

36.

$$x = r \cos \theta \qquad\qquad y = r \sin \theta$$

$$x = -4 \cos \frac{-\pi}{4} \qquad y = -4 \sin \frac{-\pi}{4}$$

$$x = -2\sqrt{2} \qquad\qquad y = 2\sqrt{2}$$

$$\left(-2\sqrt{2},\ 2\sqrt{2}\right)$$

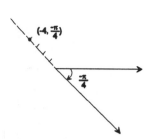

38. Since the point $(1, -1)$ lies in quadrant IV, then $r = \sqrt{1 + 1} = \sqrt{2}$ and $\theta = \tan^{-1}(-1) = \frac{-\pi}{4}$. Hence, two pairs of polar coordinates are $\left[\sqrt{2},\ \frac{-\pi}{4}\right]$ and $\left[-\sqrt{2},\ \frac{3\pi}{4}\right]$.

40. Since the point $(2, 0)$ lies on the positive x-axis, then $r = \sqrt{4 + 0} = 2$ and $\theta = \tan^{-1} 0 = 0$. Hence, two pairs of polar coordinates are $(2, 0)$ and $(-2, \pi)$.

42. Since the point $(-5, 12)$ lies in quadrant II, $-\sqrt{25 + 144} = -13$ and $\theta = \tan{-1} \left[\frac{-12}{5}\right] \approx -1.18$. Hence, two pairs of polar coordinates are $(-13, -1.18)$ and $(13, 1.97)$.

44.

$$2x^2 - 2y^2 = 5y$$
$$2r^2 \cos^2 \theta - 2r^2 \sin^2 \theta = 5r \sin \theta$$
$$2r^2(\cos^2 \theta - \sin^2 \theta) = 5r \sin \theta$$
$$\frac{2}{5} r(\cos 2\theta) = \sin \theta$$
$$2r \cos 2\theta = 5 \sin \theta$$

46.

$$x^2 + 2y^2 = \frac{y}{x}$$
$$r^2 \cos^2 \theta + 2r^2 \sin^2 \theta = \tan \theta$$
$$r^2(\cos^2 \theta + 2 \sin^2 \theta) = \tan \theta$$
$$r^2(1 + \sin^2 \theta) = \tan \theta$$

48.

$$y(x^2 - y^2) = 3$$
$$r \sin \theta (r^2 - 2r^2 \sin \theta) = 3$$
$$r^3 \sin \theta (1 - 2 \sin^2 \theta) = 3$$
$$r^3 \sin \theta \cos 2\theta = 3$$

50.

$$3r = \sin \theta$$
$$3r^2 = r \sin \theta$$
$$3(x^2 + y^2) = y$$
$$3x^2 + 3y^2 - y = 0$$

52.

$$\theta = \frac{\pi}{4}$$
$$\tan \theta = \tan \frac{\pi}{4}$$
$$\frac{y}{x} = 1$$
$$y = x$$
$$y - x = 0$$

54.

$$r^2 \tan \theta = 1$$
$$(x^2 + y^2)\frac{y}{x} = 1$$
$$xy + \frac{y^3}{x} = 1$$
$$y^3 + x^2 y - x = 0$$

56. $r = 3 \sin \theta$
Test for symmetry:

   With respect to the Pole: Test fails.
   With respect to the Polar Axis: $r = 3 \sin(\theta) = -3 \sin \theta$. Test fails.

   With respect to the Line $\theta = \frac{\pi}{2}$: $r = 3 \sin(\pi - \theta) = 3(\sin \pi \cos \theta - \cos \pi \sin \theta) = 3 \sin \theta$

   Thus, the graph is symmetric with respect to the line $\theta = \frac{\pi}{2}$.

θ	0	$\dfrac{\pi}{6}$	$\dfrac{\pi}{4}$	$\dfrac{\pi}{3}$	$\dfrac{\pi}{2}$	$\dfrac{2\pi}{3}$	$\dfrac{5\pi}{6}$
$r = 3\sin\theta$	0	$\dfrac{3}{2}$	$\dfrac{3\sqrt{2}}{2}$	$\dfrac{3\sqrt{3}}{2}$	3	$\dfrac{3\sqrt{3}}{2}$	$\dfrac{3}{2}$
(r, θ)	$(0, 0)$	$\left[\dfrac{3}{2}, \dfrac{\pi}{6}\right]$	$\left[\dfrac{3\sqrt{2}}{2}, \dfrac{\pi}{4}\right]$	$\left[\dfrac{3\sqrt{3}}{2}, \dfrac{\pi}{3}\right]$	$\left[3, \dfrac{\pi}{2}\right]$	$\left[\dfrac{3\sqrt{3}}{2}, \dfrac{2\pi}{3}\right]$	$\left[\dfrac{3}{2}, \dfrac{5\pi}{6}\right]$
θ	π	$\dfrac{7\pi}{6}$	$\dfrac{4\pi}{3}$	$\dfrac{3\pi}{2}$	$\dfrac{5\pi}{3}$	$\dfrac{11\pi}{6}$	2π
$r = 3\sin\theta$	0	$\dfrac{-3}{2}$	$\dfrac{-3\sqrt{3}}{2}$	-3	$\dfrac{-3\sqrt{3}}{2}$	$\dfrac{-3}{2}$	0
(r, θ)	$(0, \pi)$	$\left[\dfrac{-3}{2}, \dfrac{7\pi}{6}\right]$	$\left[\dfrac{-3\sqrt{3}}{2}, \dfrac{4\pi}{3}\right]$	$\left[-3, \dfrac{3\pi}{2}\right]$	$\left[\dfrac{-3\sqrt{3}}{2}, \dfrac{5\pi}{3}\right]$	$\left[\dfrac{-3}{2}, \dfrac{11\pi}{6}\right]$	$(0, 2\pi)$

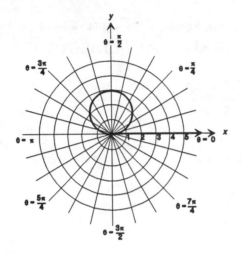

58. $r = 2 + \cos\theta$

This is a limacon without inner loop.

Test for symmetry:

With respect to the Pole: Test fails.

With respect to the Polar Axis: $r = 2 + \cos(-\theta) = 2 + \cos\theta$. Thus, the graph is symmetric with respect to the polar axis.

With respect to the line $\theta = \dfrac{\pi}{2}$: Test fails.

θ	0	$\dfrac{\pi}{6}$	$\dfrac{\pi}{4}$	$\dfrac{\pi}{3}$
$r = 2 + \cos\theta$	3	$\dfrac{4+\sqrt{3}}{2}$	$\dfrac{4+\sqrt{2}}{2}$	$\dfrac{5}{2}$
(r, θ)	$(3, 0)$	$\left(\dfrac{4+\sqrt{3}}{2}, \dfrac{\pi}{6}\right)$	$\left(\dfrac{4+\sqrt{2}}{2}, \dfrac{\pi}{4}\right)$	$\left(\dfrac{5}{2}, \dfrac{\pi}{3}\right)$

θ	$\dfrac{\pi}{2}$	$\dfrac{2\pi}{3}$	$\dfrac{5\pi}{6}$	π
$r = 2 + \cos\theta$	2	$\dfrac{3}{2}$	$\dfrac{4-\sqrt{3}}{2}$	1
(r, θ)	$\left(2, \dfrac{\pi}{2}\right)$	$\left(\dfrac{3}{2}, \dfrac{2\pi}{3}\right)$	$\left(\dfrac{4-\sqrt{3}}{2}, \dfrac{5\pi}{6}\right)$	$(1, \pi)$

The remaining points on the graph can be found by using symmetry with respect to the polar axis.

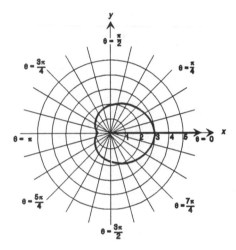

60. $r = 1 - 2\sin\theta$
This is a limacon with inner loop.
Test for symmetry:

   With respect to the Pole: Test fails.
   With respect to the Polar Axis: $r = 1 - 2\sin(-\theta) = 1 + \sin\theta$. Thus, the test fails.

   With respect to the Line $\theta = \dfrac{\pi}{2}$: The result is $r = 1 - 2\sin(\pi - \theta) = 1 - 2\sin\theta$.

The graph is symmetric with respect to the line $\theta = \dfrac{\pi}{2}$.

θ	0	$\dfrac{\pi}{6}$	$\dfrac{\pi}{4}$	$\dfrac{\pi}{3}$	$\dfrac{\pi}{2}$
$r = 1 - 2\sin\theta$	1	0	$1 - \sqrt{2}$	$1 - \sqrt{3}$	-1
(r, θ)	$(1, 0)$	$\left[0, \dfrac{\pi}{6}\right]$	$\left[1 - \sqrt{2}, \dfrac{\pi}{4}\right]$	$\left[1 - \sqrt{3}, \dfrac{\pi}{3}\right]$	$\left[-1, \dfrac{\pi}{2}\right]$

θ	$\dfrac{2\pi}{3}$	$\dfrac{5\pi}{6}$	π	$\dfrac{7\pi}{6}$
$r = 1 - 2\sin\theta$	$1 - \sqrt{3}$	0	1	2
(r, θ)	$\left[1 - \sqrt{3}, \dfrac{2\pi}{3}\right]$	$\left[0, \dfrac{5\pi}{6}\right]$	$(1, \pi)$	$\left[2, \dfrac{7\pi}{6}\right]$

θ	$\dfrac{3\pi}{2}$	$\dfrac{5\pi}{3}$	$\dfrac{11\pi}{6}$	2π
$r = 1 - 2\sin\theta$	3	$1 + \sqrt{3}$	-2	1
(r, θ)	$\left[3, \dfrac{3\pi}{2}\right]$	$\left[1 + \sqrt{3}, \dfrac{5\pi}{3}\right]$	$\left[-2, \dfrac{11\pi}{6}\right]$	$(1, 2\pi)$

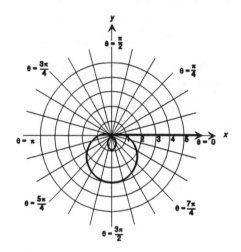

62. $r = \sqrt{x^2 + y^2} = \sqrt{3 + 1}$, $= 2$, $\tan\theta = \dfrac{-1}{\sqrt{3}}$

$\theta = 150°$

Thus, $-\sqrt{3} + i = 2(\cos 150° + i \sin 150°)$

64. $r = \sqrt{x^2 + y^2} = \sqrt{(3)^2 + (-2)^2} = \sqrt{13}$

$\tan\theta = \dfrac{-2}{3}$, $\theta = 326.3°$

Thus, $3 - 2i = \sqrt{13}\,(\cos 326.3° + i \sin 326.3°)$

66. $3(\cos 60° + i \sin 60°) = 3\left[\dfrac{1}{2} + i\,\dfrac{\sqrt{3}}{2}\right] = \dfrac{3}{2} + \dfrac{3\sqrt{3}}{2}i$

68. $4\left[\cos\dfrac{3\pi}{4} + i \sin\dfrac{3\pi}{4}\right] = 4\left[\dfrac{-\sqrt{2}}{2} + \dfrac{\sqrt{2}}{2}i\right] = -2\sqrt{2}\,(1 - i)$

70. $0.5(\cos 160° + i \sin 160°) = 0.5(-0.9397 + 0.3420i) = -0.47 + 0.17i$

72. $zw = \cos(205° + 85°) + i \sin(205° + 85°)$
$zw = \cos 290° + i \sin 290°$

$\dfrac{z}{w} = \cos(205° - 85°) + i \sin(205° - 85°)$

$\dfrac{z}{w} = \cos 120° + i \sin 120°$

74. $zw = 2 \cdot 3\left[\cos\left(\dfrac{5\pi}{3} + \dfrac{\pi}{3}\right) + i \sin\left(\dfrac{5\pi}{3} + \dfrac{\pi}{3}\right)\right] = 6(\cos 2\pi + i \sin 2\pi)$

$\dfrac{z}{w} = \dfrac{2}{3}\left[\cos\left(\dfrac{5\pi}{3} - \dfrac{\pi}{3}\right) + i \sin\left(\dfrac{5\pi}{3} - \dfrac{\pi}{3}\right)\right] = \dfrac{2}{3}\left[\cos\dfrac{4\pi}{3} + i \sin\dfrac{4\pi}{3}\right]$

76. $zw = 4 \cdot 1[\cos(50° + 340°) + i \sin(50° + 340°)]$
$= 4[\cos(390° - 360°) + i \sin(390° - 360°)] = 4(\cos 30° + i \sin 30°)$

$\dfrac{z}{w} = \dfrac{4}{1}[\cos(50° - 340°) + i \sin(50° - 340°)] = 4(\cos 70° + i \sin 70°)$

78. $[2(\cos 50° + i \sin 50°)]^3 = 2^3[3 \cdot 50°) + i \sin(3 \cdot 50°)] = 8(\cos 150° + i \sin 150°)$

$$= 8\left[\dfrac{-\sqrt{3}}{2} + \dfrac{1}{2}i\right] = 4\left(-\sqrt{3} + i\right)$$

80. $\left[2\left(\cos\dfrac{5\pi}{16} + i \sin\dfrac{5\pi}{16}\right)\right]^4 = 2^4\left[\cos\left(4 \cdot \dfrac{5\pi}{16}\right) + i \sin\left(4 \cdot \dfrac{5\pi}{16}\right)\right]$

$$= 16\left[\cos\dfrac{5\pi}{4} + i \sin\dfrac{5\pi}{4}\right] = 16\left[\dfrac{-\sqrt{2}}{2} + \dfrac{-\sqrt{2}}{2}i\right] = -8\sqrt{2}(1 + i)$$

82. $2 - 2i = 2\sqrt{2}(\cos 315° + i \sin 315°)$

$(2 - 2i)^8 = \left(2\sqrt{2}\right)^8[\cos(8 \cdot 315°) + i \sin(86 \cdot 315°)]$
$= 4096[\cos(2520° - 2520°) + i \sin(2520° - 2520°)]$
$= 4096(\cos 0° + i \sin 0°) = 4096(1) = 4096$

84. $(1 - 2i) = \sqrt{5}(\cos 296.6° + i \sin 296.6°)$

$(1 - 2i)^4 = \left(\sqrt{5}\right)^4[\cos(4 \cdot 296.6°) + i \sin(4 \cdot 296.6°)]]$
$= 25(\cos(1186.4 - 1080) + i \sin(1186.4 - 1080)]$
$= 25(\cos 106.4° + i \sin 106.4°) = 25(-0.2823 + 0.9593i) = -7.06 + 23.98i$

86. $-16 = 16(\cos 180° + i \sin 180°)$

$z_k = \sqrt[4]{16}\left[\cos\left(\dfrac{180°}{4} + \dfrac{360°k}{4}\right) + i \sin\left(\dfrac{180°}{4} + \dfrac{360°k}{4}\right)\right], k = 0, 1, 2, 3$
Thus, $z_0 = 2(\cos 45° + i \sin 45°)$
$z_1 = 3(\cos 135° + i \sin 135°)$
$z_2 = 2(\cos 225° + i \sin 225°)$
$z_3 = 2(\cos 315° + i \sin 315°)$

88. (a)   420 miles per hour × 10 minutes = 420 miles per hour × $\frac{1}{6}$ hour = 70 miles

$$a^2 = 300^2 + 70^2 - 2 \cdot 300 \cdot 70 \cos 5°$$
$$a^2 = 53059.8$$
$$a \approx 230.3 \text{ miles}$$

$$\frac{\sin 5°}{230.3} = \frac{\sin C}{300}$$
$$\sin C \approx 0.1135$$
$$C \approx 173.5$$

The pilot should turn through an angle of 180° − 173.5 = 6.5° to correct the course.

(b)   From city A to city B it takes $\frac{300 \text{ miles}}{420 \text{ miles/hour}} = \frac{5}{7}$ hour. The distance from the point of the error to city B is 230.3 miles. From city A to the point of the error, the time required is $\frac{70 \text{ miles}}{420 \text{ miles/hour}} = \frac{1}{6}$ hour. Because of the error, the pilot must increase his speed so that the 230.3 miles is covered in $\frac{5}{7} - \frac{1}{6} = \frac{23}{42}$ hour. This will require a speed of $\frac{230.3}{\frac{23}{42}} = 420.55$ mph.

90.   By supplementary angles, $\angle CAB = 180° - 120° = 60°$ and $\angle ABC = 180° - 115° = 65°$. Therefore, $\angle ACB = 180° - 60° - 65° = 55°$. To find the length of $\overline{AC}$, use the Law of Sines:

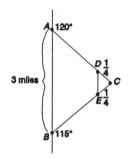

3 miles

$$\frac{\sin 55°}{3} = \frac{\sin 65°}{\overline{AC}}$$
$$\overline{AC} = \frac{3 \sin 65°}{\sin 55°} \approx 3.32 \text{ miles}$$

Thus, the length $\overline{AD} = 3.32 - 0.25 = 3.07$ miles
To find the length of $\overline{BC}$, use the Law of Sines:

$$\frac{\sin 55°}{3} = \frac{\sin 60°}{\overline{BC}}$$
$$\overline{BE} = \frac{3 \sin 60°}{\sin 55°} \approx 3.17 \text{ miles}$$

Thus, the length of $\overline{BE} = 3.17 - 0.25 = 2.92$ miles.
Since $\overline{DE}$ is parallel to $\overline{AB}$, $\angle ABC = \angle DEC$; therefore, $\angle DEC = 65°$. We use the Law of Sines to find $\overline{DE}$:

$$\frac{\sin 65°}{0.25} = \frac{\sin 55°}{\overline{DE}}$$
$$\overline{DE} = \frac{0.25 \sin 55°}{\sin 65°} \approx 0.23 \text{ miles}$$

The length of the highway is 3.07 + 2.92 + 0.23 = 6.22 miles

92.   We can use the Law of Cosines to find the distance between the two homes.
$$c^2 = 50^2 + 60^2 - 2(50)(60)\cos 80°$$
$$c^2 \approx 5058.1109$$
$$c \approx 71 \text{ feet}$$

94. We divide the irregular parcel of land into two triangles. First, we
find x using the Law of Cosines.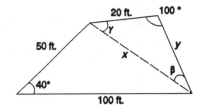

$$x^2 = (100)^2 + (50)^2 - 2(50)(100) \cos 40°$$
$$x^2 = 4839.56$$
$$x = 69.567$$

Now we can find the area of one triangle:

$$A = \frac{1}{2}(50)(100)\sin 40° = 1606.97$$

Using the Law of Sines, we can find β:

$$\frac{\sin 100°}{x} = \frac{\sin \beta}{20}$$
$$\frac{\sin 100°}{69.567} = \frac{\sin \beta}{20}$$
$$\sin \beta = \frac{20 \sin 100°}{69.567} \approx 0.2831$$
$$\beta = 16.447°$$
$$\gamma = 180° - 100° - \beta \approx 63.553°$$

Using the Law of sines, we find γ:

$$\frac{\sin 63.553°}{y} = \frac{\sin 100°}{69.567}$$
$$y = \frac{69.567 \sin 63.533°}{\sin 100°} = 63.247$$

Now we can find the area of the second triangle:

$$A = \frac{1}{2}(20(63.247)\sin 100° = 622.86.$$

Adding the areas of the two triangles, we find the area of the irregular parcel of land:
Area = $1606.97 + 622.86 \approx 2{,}229.8$

If the land is being sold for $100 per square foot, the cost of the parcel is
$2229.8 \times \$100 = \$222{,}980.00$.

Mission Possible

1. By tracing another circle around the red rock, this one with a radius of 70 yards, you will be able to walk around the circle until you find the point where the red rock appears to be 40° clockwise from the highest hill. That will be the location of the lost palm tree.

2.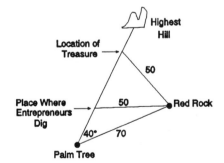

3. This triangle is an example of the SSA problem. There are actually two triangles with sides 50 and 70 and with a 40° angle across from the 50 yd. side. Along the sight line from the palm tree to the highest hill, there would be two possible vertices. The entrepreneurs went to the first one, which was approximately 32 yds. from the palm tree. When that proved fruitless, they gave up, but you knew there was another possibility. Going 75 yds. along the sight line, you come to the other possible vertex. That is where your team found the treasure.

4. Amateur anthropologists or treasure hunters, whether on land or on the ocean depths, have to be very careful about what they retrieve. Often claims against the treasures found are registered with various governments. It is not inconceivable that the descendants of those from whom the pirates stole their treasure have liens against the jewelry or coins or that certain governments or clans believe they have a right to some or all of the treasure. And at the very least, you will have to pay taxes. And I'd be a little nervous about the revenge of the 5 entrepreneurs; a legal sharing of the proceeds might head off a less civilized response to their realization that you had recovered the booty.

ANALYTIC GEOMETRY

11.2 The Parabola

2. G   4. D   6. A   8. F

10. Focus at $(0, 2)$; vertex at $(0, 0)$
$x^2 = 8y$

Letting $y = 1$, $x^2 = 8$ or $x = \pm 2\sqrt{2}$. The
points $(-2\sqrt{2}, 1)$ and $(2\sqrt{2}, 1)$ define the
latus rectum.

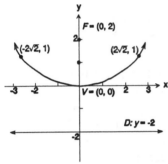

12. Focus at $(-4, 0)$; vertex at $(0, 0)$
$y^2 = -16x$
Letting $x = -4$, $y^2 = 64$, or $y = \pm 8$. The
points $(-4, 8)$ and $(-4, -8)$ define the
latus rectum.

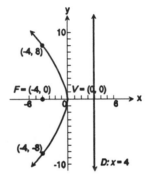

14. Focus at $(0, -1)$; directrix the line $y = 1$
$x^2 = -4y$
Letting $x = -1$, $x^2 = 4$, or $x = \pm 2$. The
points $(-2, -1)$ and $(2, -1)$ define the latus
rectum.

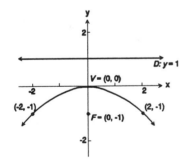

16. Directrix the line $x = \dfrac{-1}{2}$; vertex at $(0, 0)$

$y^2 = 2x$

Letting $x = \dfrac{1}{2}$, $y^2 = 1$, or $y = \pm 1$. The

points $\left[\dfrac{1}{2}, -1\right]$ and $\left[\dfrac{1}{2}, 1\right]$ define the
latus rectum.

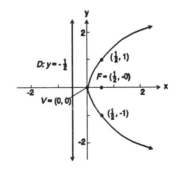

18. Focus at $(4, -2)$; vertex at $(6, -2)$

$(y + 2)^2 = 8(x - 4)$

Letting $x = 6$, $(y + 2)^2 = 16$, or $y + 2 = \pm 4$ so that $y = 2$ or $y = -6$. The points $(6, 2)$ and $(6, -6)$ define the latus rectum.

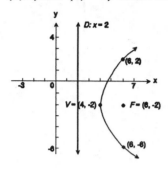

20. Vertex at $(0, 0)$; axis of symmetry the x-axis; containing the point $(2, 3)$

$$y^2 = 4ax \text{ or } y^2 = -4ax$$
$$9 = 8a \text{ or } 9 = -8a$$
$$a = \frac{9}{8} \text{ since } a > 0$$
$$y^2 = \frac{9}{2}x$$

Letting $x = \frac{9}{8}$, $y^2 = \frac{9}{2} \cdot \frac{9}{8} = \frac{81}{16}$ or

$y = \pm\frac{9}{4}$. The points $\left[\frac{9}{8}, \frac{-9}{4}\right]$ and

$\left[\frac{9}{8}, \frac{9}{4}\right]$ define the latus rectum.

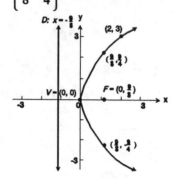

22. Focus at $(2, 4)$; directrix the line $x = -4$.

$(y - 4)^2 = 12(x + 1)$

Letting $x = 2$, $(y - 4)^2 = 36$ or $y - 4 = \pm 6$ so that $y = 10$ or $y = -2$. The points $(2, 10)$ and $(2, -2)$ define the latus rectum.

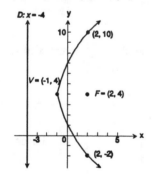

24. Focus at $(-4, 4)$; directrix the line $y = -2$.

$(x + 4)^2 = 12(y - 1)$

Letting $y = 4$, $(x + 2)^2 = 36$ or $x + 4 = \pm 6$ so that $x = 2$ or $x = -10$. The points $(2, 4)$ and $(-10, 4)$ define the latus rectum.

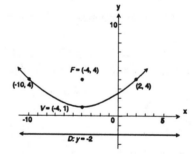

26. $y^2 = 8x$
Vertex $(0, 0)$; focus $(2, 0)$;
directrix the line $x = -2$

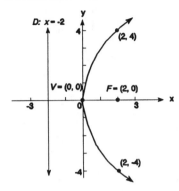

28. $x^2 = -4y$
Vertex $(0, 0)$; focus $(0, -1)$;
directrix the line $y = 1$

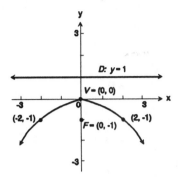

30. $(x + 4)^2 = 16(y + 2)$
Vertex $(-4, -2)$; focus $(-4, 2)$;
directrix the line $y = -6$

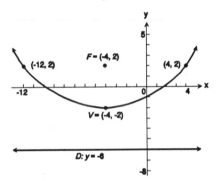

32. $(y + 1)^2 = -4(x - 2)$
Vertex $(2, -1)$; focus $(1, -1)$;
directrix the line $x = 3$

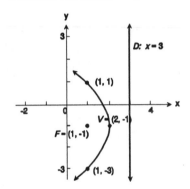

34. $(x - 2)^2 = 4(y - 3)$
Vertex $(2, 3)$; focus $(2, 4)$;
directrix the line $y = 2$

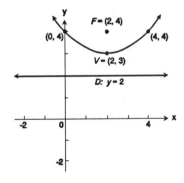

36. $x^2 + 6x - 4y + 1 = 0$
$$x^2 + 6x = 4y - 1$$
$$x^2 + 6x + 9 = 4y + 8$$
$$(x + 3)^2 = 4(y + 2)$$
Vertex $(-3, -2)$; focus $(-3, -1)$;
directrix the line $y = -3$

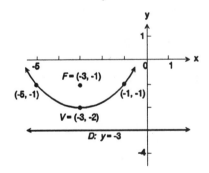

38.
$$y^2 - 2y = 8x - 1$$
$$y^2 - 2y + 1 = 8x$$
$$(y - 1)^2 = 8(x - 0)$$
Vertex (0, 1); focus (2, 1);
directrix the line $x = -2$.

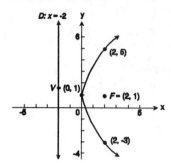

40.
$$x^2 - 4x = 2y$$
$$x^2 - 4x + 4 = 2y + 4$$
$$(x - 2)^2 = 2(y + 2)$$
Vertex (2, -2); focus $\left[2, \dfrac{-3}{2}\right]$;
directrix the line $y = \dfrac{-5}{2}$

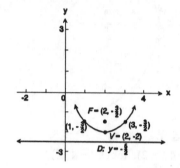

42.
$$y^2 + 12y = -x + 1$$
$$y^2 + 12y + 36 = -x + 37$$
$$(y + 6)^2 = -(x - 37)$$
Vertex (37, -6); focus $\left[\dfrac{147}{4}, -6\right]$; directrix the line $x = \dfrac{149}{4}$

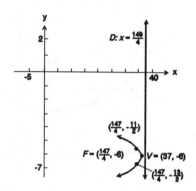

44.
$$(x - 1)^2 = c(y - 2)$$
$$1 = c(-1)$$
$$c = -1$$
$$(x - 1)^2 = -(y - 2)$$

46.
$$(x - 0)^2 = c(y + 1)$$
$$x^2 = c(y + 1)$$
$$4 = c$$
$$x^2 = 4(y + 1)$$

48.
$$(x - 1)^2 = c(y + 1)$$
$$1 = c(2)$$
$$c = \dfrac{1}{2}$$
$$(x - 1)^2 = \dfrac{1}{2}(y + 1)$$

50.
$$y^2 = c(x - 1)$$
$$1 = c(-1)$$
$$c = -1$$
$$y^2 = -(x - 1)$$

52. Situate the parabola so that its vertex is at (0, 0), and it opens up. Then, we know
$$x^2 = 4ay$$
Since the parabola is 6 feet across and 2 feet deep, the points (3, 2) and (−3, 2) must satisfy the equation:
$$x^2 = 4ay$$
$$9 = 8a \text{ (Letting } x = 3 \text{ and } y = 2)$$
$$a = \frac{9}{8} = 1.125$$
But a is, by definition, the distance from the vertex to the focus. Therefore, the receiver is 1.125 feet or 13.5 inches from the vertex of the dish.

$x^2 = c(x-10)$

56. Vertex: (0, 10); $(x - 10)^2 = cy$
Point on the parabola: (200, 100):
$$(200 - 10)^2 = c(100)$$
$$c = \frac{(190)^2}{100} = 361$$
For $x = 150$, $(150 - 10)^2 = 361y$
$$y = \frac{(140)^2}{361}$$
$$y \approx 54.29 \text{ feet} \quad 60.25$$

58. Situate the parabola so that its vertex is at (0, 0), and it opens up. Then, we know:
$$x^2 = 4ay$$
The light source is located 2 feet from the base, so $a = 2$.
The depth is 4 feet, so $y = 4$. Thus, we have:
$$x^2 = 4(2)(4)$$
$$x^2 = 32$$
$$x = \pm 4\sqrt{2}$$
Thus, the width of the opening should be
$$2(4\sqrt{2}) = 8\sqrt{2} \text{ feet.}$$

62. Situate the parabola so that its vertex is at (0, 0) and it opens down. Then, we know:
$$x^2 = -4ay$$
Since the bridge has a height of 10 feet when it is 40 feet from the center, we know two points on the parabola are (−40, −10) and (40, −10). See Figure.
The form of the equation of this parabola is
$$x^2 = -4ay$$

54. Situate the parabola so that its vertex is at (0, 0) and it opens up. Then, we know
$$x^2 = 4ay$$
Since the focus is 1 inch from the vertex, we have $a = +1$. The depth is 2 inches, so $y = 2$. Thus, we have:
$$x^2 = 4(1)(2)$$
$$x^2 = 8$$
$$x = \pm 2\sqrt{2}$$
Thus, the diameter of the opening is
$$2(2\sqrt{2}) = 4\sqrt{2} \text{ inches.}$$

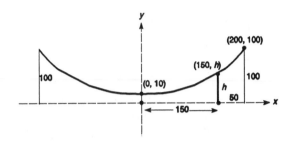

60. Situate the parabola so that its vertex is at (0, 0), and it opens up. Then, we know:
$$x^2 = 4ay$$
Since the parabola is 4 inches across and 36 inches deep, the points (2, 36) and (−2, 36) must satisfy the equation:
$$x^2 = 4ay$$
$$4 = 4a(36)$$
$$a = \frac{1}{36}$$
$$a \approx 0.0278 \text{ inches}$$
But a is, by definition, the distance from the vertex to the focus. Therefore, the light collected will be concentrated 0.0278 inches from the vertex.

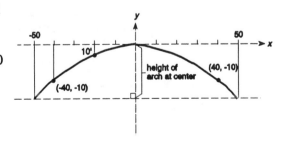

Therefore, letting $x = 40$ and $y = -10$, we find a.

$$40^2 = -4a(-10)$$
$$a = 40$$

So, the equation of the parabola is:

$$x^2 = -160y$$

The height of the arch at the center is found by letting $x = 50$ since the value of y when $x = 50$ represents the height of the arch at the center. So,

$$50^2 = -160y$$
$$y = -15.625$$

Therefore, the height of the arch at the center is 15.625 feet.

64. $Cy^2 + Dx = 0 \qquad C \neq 0, D \neq 0$

$$Cy^2 = -Dx$$

$$y^2 = \frac{-D}{C}x$$

Parabola; vertex at $(0, 0)$; axis of symmetry the x-axis; focus at $\left[\frac{-D}{4C}, 0\right]$; directrix the line

$$x = \frac{D}{4C}$$

66. $Cy^2 + Dx + Ey + F = 0 \qquad C \neq 0$

(a) If $D \neq 0$, then

$$Cy^2 + Ey = -Dx - F$$

$$C\left[y^2 + \frac{E}{C}y\right] = -Dx - F$$

$$C\left[y^2 + \frac{E}{C}y + \frac{E^2}{4C^2}\right] = -Dx - F + \frac{E^2}{4C}$$

$$\left[y + \frac{E}{2C}\right]^2 = \frac{-D}{C}\left[x + \frac{F}{D} - \frac{E^2}{4DC}\right]$$

Parabola, vertex at $\left[\frac{E^2}{4DC} - \frac{F}{D}, \frac{-E}{2C}\right]$; axis of symmetry parallel to the x-axis.

(b) If $D = 0$, then

$$Cy^2 + Ey + F = 0$$

$$y = \frac{-E \pm \sqrt{E^2 - 4CF}}{2C}$$

If $E^2 - 4CF = 0$, then

$$y = \frac{-E}{2C}$$

Horizontal line

(c) If $D = 0$, then

$$C^2 + Ey + F = 0$$

$$y = \frac{-E \pm \sqrt{E^2 - 4CF}}{2C}$$

If $E^2 - 4CF > 0$, then

$$y = \frac{-E + \sqrt{E^2 - 4CF}}{2C} \text{ or }$$

$$y = \frac{-E - \sqrt{E^2 - 4CF}}{2C}$$

Two horizontal lines

(d) If $D = 0$, then

$$Cy^2 + Ey + F = 0$$

$$y = \frac{-E \pm \sqrt{E^2 - 4CF}}{2C}$$

If $E^2 - 4CF < 0$, there is no real solution. Hence, the graph contains no points.

2. D

4. A

6. $\dfrac{x^2}{9} + \dfrac{y^2}{4} = 1$
    $c^2 = 9 - 4 = 5$
    Foci $\left(\sqrt{5},\ 0\right),\ \left(-\sqrt{5},\ 0\right)$
    Vertices $(3, 0),\ (-3, 0)$

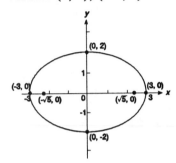

8. $\dfrac{x^2}{4} + \dfrac{y^2}{16} = 1$
    $c^2 = 16 - 4 = 12$
    Foci $\left(0,\ 2\sqrt{3}\right),\ \left(0,\ -2\sqrt{3}\right)$
    Vertices $(0, 4),\ (0, -4)$

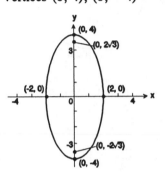

10. $x^2 + 9y^2 = 18$
     $\dfrac{x^2}{18} + \dfrac{y^2}{2} = 1$
     $c^2 = 18 - 2 = 16$
     Foci $(4, 0),\ (-4, 0)$
     Vertices $\left(3\sqrt{2},\ 0\right),\ \left(-3\sqrt{2},\ 0\right)$

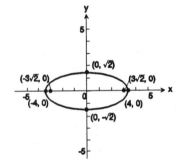

12. $4y^2 + 9x^2 = 36$
     $\dfrac{y^2}{9} + \dfrac{x^2}{4} = 1$
     $c^2 = 9 - 4 = 5$
     Foci $\left(0,\ \sqrt{5}\right),\ \left(0,\ -\sqrt{5}\right)$
     Vertices $(0, 3),\ (0, -3)$

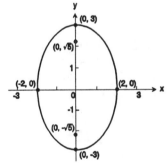

14. $x^2 + y^2 = 4$

$$\frac{x^2}{4} + \frac{y^2}{4} = 1$$

Foci (0, 0)
Vertices (2, 0)(−2, 0), (0, 2)(0, −2)

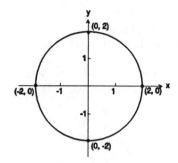

16. Center (0, 0); focus (−1, 0);
vertex (3, 0)
$b^2 = 9 − 1 = 8$

$$\frac{x^2}{9} + \frac{y^2}{8} = 1$$

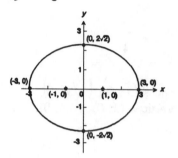

18. Center (0, 0); focus (0, 1);
vertex (0, −2)
$b^2 = 4 − 1 = 3$

$$\frac{x^2}{3} + \frac{y^2}{4} = 1$$

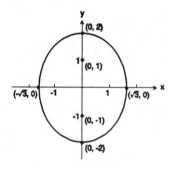

20. Focus (0, −4);
vertices (0, ±8)
$b^2 = 64 − 16 = 48$

$$\frac{x^2}{48} + \frac{y^2}{64} = 1$$

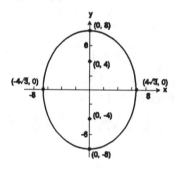

22. Foci (0, ±2); length of major axis is 8
$b^2 = 16 − 4 = 12$

$$\frac{x^2}{12} + \frac{y^2}{16} = 1$$

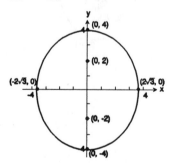

24. Vertices (±5, 0); $c = 2$
$b^2 = 25 − 4 = 21$

$$\frac{x^2}{25} + \frac{y^2}{21} = 1$$

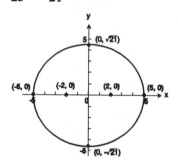

26. $\dfrac{(x + 1)^2}{1} + \dfrac{(y + 1)^2}{4} = 1$

28. $\dfrac{x^2}{4} + (y − 1)^2 = 1$

30. $\dfrac{(x + 4)^2}{9} + \dfrac{(y + 2)^2}{4} = 1$

$c^2 = 9 - 4 = 5$

Center $(-4, -2)$; foci $\left(-4 \pm \sqrt{5}, -2\right)$; vertices $(-4 \pm 3, -2)$

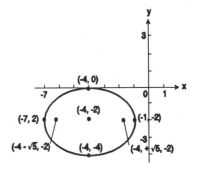

32. $9(x - 3)^2 + (y + 2)^2 = 18$

$\dfrac{(x - 3)^2}{2} + \dfrac{(y + 2)^2}{18} = 1$

$c^2 = 18 - 2 = 16$

Center $(3, -2)$; foci $(3, -2 \pm 4)$; vertices $\left(3, -2, \pm 3\sqrt{2}\right)$

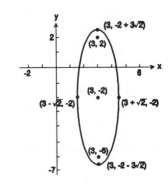

34. $x^2 + 3y^2 - 12y + 9 = 0$

$x^2 + 3(y^2 - 4y + 4) = -9 + 12$

$x^2 + 3(y - 2)^2 = 3$

$\dfrac{x^2}{3} + \dfrac{(y - 2)^2}{1} = 1$

$c^2 = 3 - 1 = 2$

Center $(0, 2)$; foci $\left(\pm\sqrt{2}, 2\right)$; vertices $(\pm\sqrt{3}, 2)$

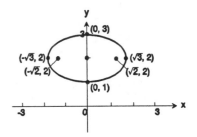

36. $4x^2 + 3y^2 + 8x - 6y = 5$

$4(x^2 + 2x + 1) + 3(y^2 - 2y + 1) = 5 + 4 + 3$

$4(x + 1)^2 + 3(y - 1)^2 = 12$

$\dfrac{(x + 1)^2}{3} + \dfrac{(y - 1)^2}{4} = 1$

$c^2 = 4 - 3 = 1$

Center $(-1, 1)$; foci $(-1, 1 \pm 1)$; vertices $(-1, 1 \pm 2)$

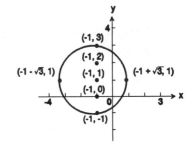

38. $x^2 + 9y^2 + 6x - 18y + 9 = 0$

$(x^2 + 6x + 9) + 9(y^2 - 2y + 1) = -9 + 9 + 9$

$(x + 3)^2 + 9(y - 1)^2 = 9$

$\dfrac{(x + 3)^2}{9} + \dfrac{(y - 1)^2}{1} = 1$

$c^2 = 9 - 1 = 8$

Center $(-3, 1)$; foci $\left(-3 \pm \sqrt{8}, 1\right)$; vertices $(-3 \pm 3, 1)$

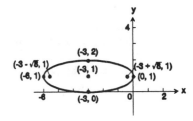

40.

$$9x^2 + y^2 - 18x = 0$$
$$9(x^2 - 2x + 1) + y^2 = 0 + 9$$
$$9(x - 1)^2 + y^2 = 9$$
$$(x - 1)^2 + \frac{y^2}{9} = 1$$

$$c^2 = 9 - 1 = 8$$

Center $(1, 0)$; foci $\left(1, \pm\sqrt{8}\right)$; vertices $(1, \pm 3)$

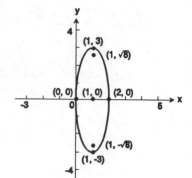

42.
Center at $(-3, 1) \Rightarrow h = -3, k = 1$
Vertex at $(-3, 3) \Rightarrow a = 2$
Focus at $(-3, 0) \Rightarrow c = 1$
$b^2 = 4 - 1 = 3$

The form of the equation is:
$$\frac{(x - h)^2}{b^2} + \frac{(y - k)^2}{a^2} = 1$$
$$\frac{(x + 3)^2}{3} + \frac{(y - 1)^2}{4} = 1$$

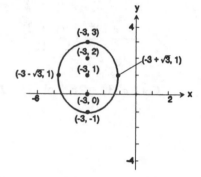

44.
Foci at $(1, 2)$ and $(-3, 2) \Rightarrow$ center is
$(-1, 2)$ and $c = 2$.
Vertex at $(-4, 2) \Rightarrow a = 3$
$b^2 = 9 - 4 = 5$

The form of the equation is:
$$\frac{(x - h)^2}{a^2} + \frac{(y - k)^2}{b^2} = 1$$
$$\frac{(x + 1)^2}{9} + \frac{(y - 2)^2}{5} = 1$$

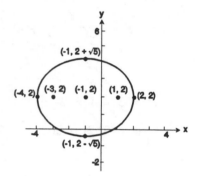

46.
Vertices at $(2, 5)$ and $(2, -1) \Rightarrow a = 3$ and the center is $(2, 2)$; $c = 2$.
Thus, $b^2 = a^2 - c^2 = 9 - 4 = 5$

The form of the equation is:
$$\frac{(x - h)^2}{b^2} + \frac{(y - k)^2}{a^2} = 1$$
$$\frac{(x - 2)^2}{5} + \frac{(y - 2)^2}{9} = 1$$

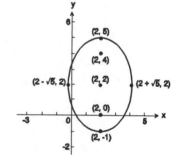

48.
Center at $(1, 2)$ and focus at $(1, 4) \Rightarrow c = 2$.

The form of the equation is:
$$\frac{(x - h)^2}{b^2} + \frac{(y - k)^2}{a^2} = 1$$
$$\frac{(x - 1)^2}{b^2} + \frac{(y - 2)^2}{a^2} = 1$$

Contains the point $(2, 2) \Rightarrow \dfrac{(2 - 1)^2}{b^2} + \dfrac{(2 - 2)^2}{a^2} = 1$
or $b^2 = 1$ and $a^2 = c^2 + b^2 = 4 + 1 = 5$

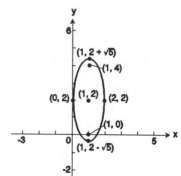

Thus, $\dfrac{(x-1)^2}{1} + \dfrac{(y-2)^2}{5} = 1$

50. Center at (1, 2) and vertex at (1, 4) $\Rightarrow a = 2$.

The form of the equation is: $\dfrac{(x-h)^2}{b^2} + \dfrac{(y-k)^2}{a^2} = 1$

Contains the point (2, 2) $\Rightarrow \dfrac{(x-1)^2}{b^2} + \dfrac{(y-2)^2}{4} = 1$

$$\dfrac{(2-1)^2}{b^2} + \dfrac{(2-2)^2}{4} = 1$$

or $b^2 = 1$ and $c^2 = a^2 - b^2 = 4 - 1 = 3$

Thus, $\dfrac{(x-1)^2}{1} + \dfrac{(y-2)^2}{4} = 1$

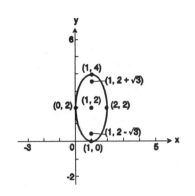

52.
$$y = \sqrt{9 - 9x^2}$$
$$y^2 = 9 - 9x^2, \quad y \geq 0$$
$$9x^2 + y^2 = 9, \quad y \geq 0$$
$$x^2 + \dfrac{y^2}{9} = 1, \quad y \geq 0$$

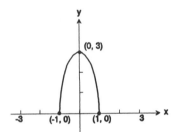

54.
$$y = -\sqrt{4 - 4x^2}$$
$$y^2 = 4 - 4x^2, \quad y \leq 0$$
$$4x^2 + y^2 = 4$$
$$x^2 + \dfrac{y^2}{4} = 1, \quad y \leq 0$$

56. $\dfrac{x^2}{225} + \dfrac{y^2}{100} = 1$

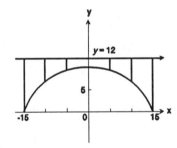

When $x = -15$, $1 + \dfrac{y^2}{100} = 1$ so that $y = 0$

Distance from roadway to arch is $12 - y = 12 - 0 = 12$ feet.

When $x = -10$, $\dfrac{100}{225} + \dfrac{y^2}{100} = 1$

$$y^2 = 100 \left[\dfrac{125}{225} \right] \approx 55.56$$
$$y \approx 7.45$$

Distance from roadway to arch is $12 - 7.45 = 4.55$ feet.

When $x = -5$, $\dfrac{25}{225} + \dfrac{y^2}{100} = 1$

$$y^2 = 100 \left[\dfrac{200}{225} \right] \approx 88.95$$
$$y \approx 9.43$$

Distance from roadway to arch is $12 - 9.43 = 2.57$ feet.
When $x = 0$, distance is $12 - 10 = 2$ feet

By symmetry,   at $x = 5$, distance is 2.57 feet;
at $x = 10$, distance is 4.55 feet;
at $x = 15$, distance is 12 feet.

58.   The distance from the nearest wall to one focus is 6 feet. The distance between foci is 100 feet. Therefore, the major axis is 112 feet long $(6' + 6' + 100')$. c represents the distance from the center to a focus; therefore, $c = 50$ feet since the distance between foci is 100 feet. a is the distance from the center to a vertex; therefore, a is $\dfrac{112}{2} = 56$ feet. Since $b^2 = a^2 - c^2$, $b^2 = 56^2 - 50^2 = 636$, so $b \approx 25.2$ feet. Since b is the height from the center to the top of the ellipse on the minor axis, the elliptical ceiling is 25.2 feet at the center.

60.   Set up a rectangular coordinate system so that the center of the ellipse is at the origin and the major axis is along the x-axis. The equation of the ellipse is:

$$\frac{x^2}{a^2} + \frac{y^2}{b^2} = 1$$

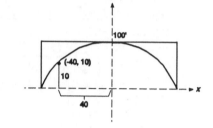

Since the bridge spans 100 feet, $a = 50$.
We know a point on the ellipse is $(-40, 10)$; therefore, we could solve for b:

$$\frac{(-40)^2}{50^2} + \frac{10^2}{b^2} = 1$$

$$\frac{1600}{2500} + \frac{100}{b^2} = 1$$

$$1600b^2 + 250000 = 2500b^2$$

$$900b^2 = 250000$$

$$b \approx 16.67 \text{ feet}$$

Therefore, the height of the arch at its center is 16.67 feet.

62.   Let $a = $ half the length of the major axis.

Then $\dfrac{x^2}{a^2} + \dfrac{y^2}{400} = 1$

Since $(28, 13)$ is a point on the graph, we know that

$$\frac{28^2}{a^2} + \frac{13^2}{400} = 1$$

$$\frac{784}{a^2} = 1 - \frac{169}{400}$$

$$\frac{784}{a^2} = \frac{231}{400}$$

$$231a^2 = 313600$$

$$a^2 \approx 1357.57$$

$$a \approx 36.84$$

The span of the bridge is 2 times a, or approximately 73.68 feet.

64. If the mean distance is 142 million miles, then $a = 142$. Therefore, the length of the major axis is 284. So, the aphelion is $284 - 128.5 = 155.5$ million miles. The distance from the center of the ellipse to the sun (a focus) is $142 - 128.5 = 13.5$ million miles. Therefore, $c = 13.5$. We use $b^2 = a^2 - c^2$ to find b^2.

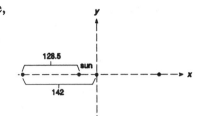

$$b^2 = a^2 - c^2$$
$$b^2 = 142^2 - 13.5^2$$
$$b^2 = 19981.75$$

So, the equation of the orbit of Mars is

$$\frac{x^2}{(142)^2} + \frac{y^2}{19981.75} = 1$$

66. The aphelion is found by adding the mean distance and the distance from the center to the sun. The mean distance is the sum of the perihelion and the distance from the sun to the center. Therefore, the mean distance is $4551 + 897.5 = 5448.5$ million miles. The aphelion is $5448.5 + 897.5 = 6346$ million miles. We need b^2 to write the equation for the orbit of Pluto. Since $b^2 = a^2 - c^2$ with $a = 5448.5$ and $c = 897.5$, we have

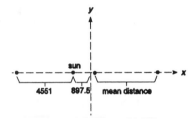

$$b^2 = 5448.5^2 - 897.5^2$$
$$= 28880646$$

Therefore, the equation for the orbit of Pluto is:

$$\frac{x^2}{5448.5^2} + \frac{y^2}{28880646} = 1$$

68. Place the x-axis along the 80 foot portion and the y-axis along the 40 foot portion. An equation of the ellipse is

$$\frac{x^2}{(40)^2} + \frac{y^2}{(20)^2} = 1$$

When $x = 34$,

$$\frac{(30)^2}{(40)^2} + \frac{(y)^2}{(20)^2} = 1$$

$$\frac{y^2}{(20)^2} = 1 - \left[\frac{3}{4}\right]^2 = \frac{7}{16}$$

$$y^2 = (20)^2\left[\frac{7}{16}\right]$$

$$y = \frac{20\sqrt{7}}{4} = 5\sqrt{7}$$

The width 10 feet from the side is $2\left(5\sqrt{7}\right) = 10\sqrt{7}$ feet.

70. $$Ax^2 + Cy^2 + Dx + Ey + F = 0 \quad A \neq 0, C \neq 0$$
$$Ax^2 + Dx + Cy^2 + Ey = -F$$
$$A\left[x^2 + \frac{D}{A}x\right] + C\left[y^2 + \frac{E}{C}y\right] = -F$$
$$A\left[x + \frac{D}{2A}\right]^2 + C\left[y + \frac{E}{2C}\right]^2 = -F + \frac{D^2}{4A} + \frac{E^2}{4C}$$

(a)   If $\dfrac{D^2}{4A} + \dfrac{E^2}{4C} - F$ is of the same sign as A (and C), this is the equation of an ellipse with center at $\left[\dfrac{-D}{2A}, \dfrac{-E}{2C}\right]$.

(b)   If $\dfrac{D^2}{4A} + \dfrac{E^2}{4C} - F = 0$, the graph is the single point $\left[\dfrac{-D}{2A}, \dfrac{-E}{2C}\right]$.

(c)   If $\dfrac{D^2}{4A} + \dfrac{E^2}{4C} - F$ is of the opposite sign to A (and C), the graph contains no points since, in this case, the left side has a sign opposite that of the right side.

11.4   The Hyperbola

2.   C

4.   D

6.   Center $(0, 0)$; focus $(0, 5)$; vertex $(0, 3)$
$a^2 = 9, c^2 = 25, b^2 = 25 - 9 = 16$

$$\frac{y^2}{9} - \frac{x^2}{16} = 1$$

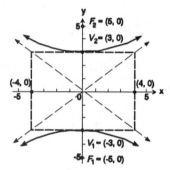

8.   Center $(0, 0)$; focus $(-3, 0)$; vertex $(2, 0)$
$a^2 = 4, c^2 = 9, b^2 = 9 - 4 = 5$

$$\frac{y^2}{4} - \frac{x^2}{5} = 1$$

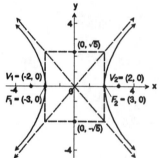

10.   Focus $(0, 6)$; vertices $(0, -2), (0, 2)$
$a^2 = 4, c^2 = 36, b^2 = 36 - 4 = 32$

$$\frac{y^2}{4} - \frac{x^2}{32} = 1$$

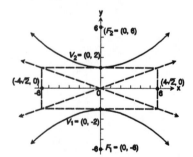

12.   Vertices $(-4, 0), (4, 0)$;
asymptote the line $y = 2x$

$$y = \frac{b}{a}x = 2x$$
$$a = 4 \Rightarrow b = 8 \Rightarrow c^2 = 16 + 64 = 80$$

Focus $\left(4\sqrt{5}, 0\right), \left(-4\sqrt{5}, 0\right)$

$$\frac{x^2}{16} - \frac{y^2}{64} = 1$$

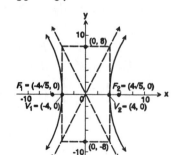

14. Foci $(0, -2)$, $(0, 2)$;
asymptote the line $y = -x$

$$y = \frac{a}{b}x = -x$$
$$a = b,$$
$$a^2 + a^2 = 4$$
$$2a^2 = 4 \Rightarrow a^2 = 2$$
$$\Rightarrow b^2 = 2$$
$$\frac{y^2}{2} - \frac{x^2}{2} = 1$$

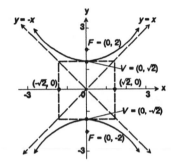

16. $\dfrac{y^2}{16} - \dfrac{x^2}{4} = 1$

The transverse axis is the y-axis.
Center $(0, 0)$
Vertices $(0, 4)$, $(0, -4)$

Foci $\left(0, 2\sqrt{5}\right) \left(0, -2\sqrt{5}\right)$
Asymptotes: $y = 2x$ and $y = -2x$

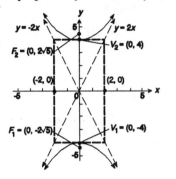

18. $y^2 - 4x^2 = 16$

$$\frac{y^2}{16} - \frac{x^2}{4} = 1$$

The transverse axis is the y-axis.
Center $(0, 0)$
Vertices $(0, 4)$, $(0, -4)$

Foci $\left(0, 2\sqrt{5}\right) \left(0, -2\sqrt{5}\right)$
Asymptotes: $y = 2x$ and $y = -2x$

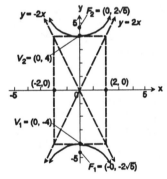

20. $x^2 - y^2 = 4$

$$\frac{x^2}{4} - \frac{y^2}{4} = 1$$

The transverse axis is the y-axis.
Center $(0, 0)$
Vertices $(2, 0)$, $(-2, 0)$

Foci $\left(2\sqrt{2}, 0\right) \left(-2\sqrt{2}, 0\right)$
Asymptotes: $y = x$ and $y = -x$

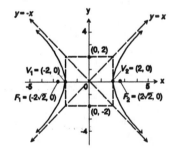

22. $2x^2 - y^2 = 4$

$\dfrac{x^2}{2} - \dfrac{y^2}{4} = 1$

The transverse axis is the x-axis.
Center $(0, 0)$

Vertices $\left(\sqrt{2}, 0\right), \left(-\sqrt{2}, 0\right)$

Foci $\left(\sqrt{6}, 0\right) \left(-\sqrt{6}, 0\right)$

Asymptotes: $y = \sqrt{2}\,x$ or $y = -\sqrt{2}\,x$

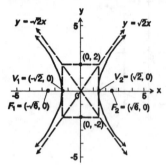

24. $y^2 - x^2 = 1$

26. $\dfrac{x^2}{4} - \dfrac{y^2}{16} = 1$

28. Center $(-3, 1)$, focus $(-3, 6)$
Vertex $(-3, 4)$;
$(h, k) = (-3, 1)$, $a = 3$, $c = 5$
$b^2 = 25 - 9 = 16$

$$\dfrac{(y - 1)^2}{9} - \dfrac{(x + 3)^2}{16} = 1$$

Asymptotes: $y - 1 = \pm\dfrac{3}{4}(x + 3)$

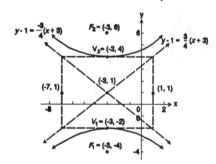

30. Center $(1, 4)$, focus $(-2, 4)$
Vertex $(0, 4)$;
$(h, k) = (1, 4)$, $a = 1$, $c = 3$
$b^2 = 9 - 1 = 8$

$$\dfrac{(x - 1)^2}{1} - \dfrac{(y - 4)^2}{8} = 1$$

Asymptotes: $y - 4 = \pm 2\sqrt{2}\,(x - 1)$

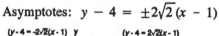
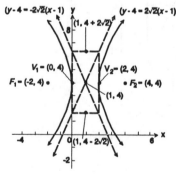

     Chapter 11 Analytic Geometry

32. Focus $(-4, 0)$, vertices $(-4, 4)$ and $(-4, 2)$

$(h, k) = (-4, 3)$, $a = 1$, $c = 3$

$b^2 = 9 - 1 = 8$

$$\frac{(y - 3)^2}{1} - \frac{(x + 4)^2}{8} = 1$$

Asymptotes: $y - 3 = \pm\dfrac{\sqrt{2}}{4}(x + 4)$

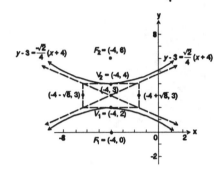

34. Vertices $(1, -3)$ $(1, 1)$

Asymptote: $\dfrac{(x - 1)}{2} = \dfrac{(y + 1)}{3}$

$(h, k) = (1, -1)$, $a = 2$

Asymptote: $y + 1 = \pm\dfrac{2}{b}(x - 1)$

and $y + 1 = \dfrac{3}{2}(x - 1)$

$\Rightarrow \dfrac{2}{b} = \dfrac{3}{2}$

$b = \dfrac{4}{3}$

$c^2 = 4 + \dfrac{16}{9} = \dfrac{52}{9}$

foci $\left[1, -1 \pm \dfrac{\sqrt{52}}{3}\right]$

$$\frac{(y + 1)^2}{4} - \frac{9(x - 1)^2}{16} = 1$$

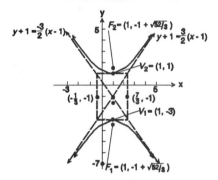

36. $\dfrac{(y + 3)^2}{4} - \dfrac{(x - 2)^2}{9} = 1$

Transverse axis parallel to y-axis.

Center $(2, -3)$, foci $\left(2, -3 \pm \sqrt{13}\right)$
Vertices $(2, -5)$, $(2, -1)$

Asymptotes: $y + 3 = \pm\dfrac{2}{3}(x - 2)$

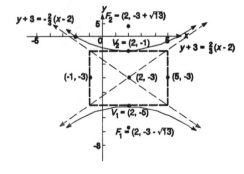

38. $(x + 4)^2 - 9(y - 3)^2 = 9$

$\dfrac{(x + 4)^2}{9} - (y - 3)^2 = 1$

Transverse axis parallel to x-axis.

Center $(-4, 3)$, foci $\left(-4 \pm \sqrt{10}, 3\right)$
Vertices $(-7, 3)$ and $(-1, 3)$

Asymptotes: $y - 3 = \pm\dfrac{1}{3}(x + 4)$

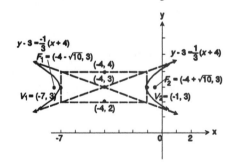

40. $(y - 3)^2 - (x + 2)^2 = 4$

$$\frac{(y - 3)^2}{4} - \frac{(x + 2)^2}{4} = 1$$

Transverse axis parallel to y-axis.

Center $(-2, 3)$, foci $\left(-2, 3 \pm 2\sqrt{2}\right)$
Vertices $(-2, 1)$ and $(-2, 5)$
Asymptotes: $y - 3 = \pm(x + 2)$

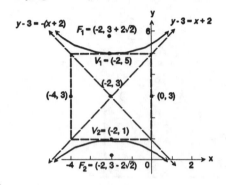

42. $y^2 - x^2 - 4y + 4x - 1 = 0$

$(y^2 - 4y + 4) - (x^2 - 4x + 4) = 1 + 4 - 4$

$$(y - 2)^2 - (x - 2)^2 = 1$$

Transverse axis parallel to y-axis.

Center $(2, 2)$, foci $\left(2, 2 \pm \sqrt{2}\right)$
Vertices $(2, 1)$ and $(2, 3)$
Asymptotes: $y - 2 = \pm(x - 2)$

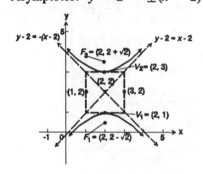

44.

$$2x^2 - y^2 + 4x + 4y - 4 = 0$$
$$2(x^2 + 2x + 1) - (y^2 - 4y + 4) = 4 + 2 - 4$$
$$2(x + 1)^2 - (y - 2)^2 = 2$$
$$(x + 1)^2 - \frac{(y - 2)^2}{2} = 1$$

Transverse axis parallel to x-axis.

Center $(-1, 2)$, foci $\left(-1 \pm \sqrt{3}, 2\right)$
Vertices $(0, 2)$ and $(-2, 2)$

Asymptotes: $y - 2 = \pm\sqrt{2}\,(x + 1)$

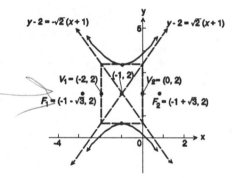

46.

$$2y^2 - x^2 + 2x + 8y + 3 = 0$$
$$2(y^2 + 4y + 4) - (x^2 - 2x + 1) = -3 + 8 - 1$$
$$2(y + 2)^2 - (x - 1)^2 = 4$$
$$\frac{(y + 2)^2}{2} - \frac{(x - 1)^2}{4} = 1$$

Transverse axis parallel to y-axis.

Center $(1, -2)$, foci $\left(1, -2 \pm \sqrt{6}\right)$

Vertices $\left(1, -2 \pm \sqrt{2}\right)$

Asymptotes: $y + 2 = \pm\frac{\sqrt{2}}{2}(x - 1)$

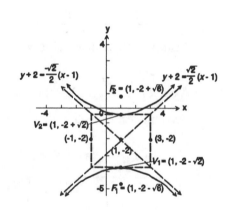

48.
$$x^2 - 3y^2 + 8x - 6y + 4 = 0$$
$$x^2 + 8x + 16 - 3(y^2 + 2y + 1) = -4 + 16 - 3$$
$$(x + 4)^2 - 3(y + 1)^2 = 9$$

$$\frac{(x + 4)^2}{9} - \frac{(y + 1)^2}{3} = 1$$

Transverse axis parallel to x-axis.

Center $(-4, -1)$; foci $\left(-4 \pm 2\sqrt{3}, -1\right)$
Vertices $(-4 \pm 3, -1)$

Asymptotes: $y + 1 = \pm \frac{\sqrt{3}}{3}(x + 4)$

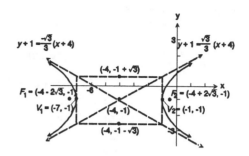

50.
$$y = -\sqrt{9 + 9x^2}$$
$$y^2 = 9 + 9x^2, \quad y \le 0$$
$$y^2 - 9x^2 = 9, \qquad y \le 0$$

$$\frac{y^2}{9} - x^2 = 1, \qquad y \le 0$$

Transverse axis parallel to y-axis

Center $(0, 0)$; Foci $(0, \pm\sqrt{10})$
Vertices $(0, \pm 3)$
Asymptotes: $y = \pm 3x$

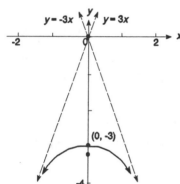

52.
$$y = \sqrt{-1 + x^2}$$
$$y^2 = -1 + x^2, \qquad y \ge 0$$
$$y^2 - x^2 = -1, \qquad y \ge 0$$
$$x^2 - y^2 = 1, \qquad y \ge 0$$

Transverse axis parallel to x-axis.

Center $(0, 0)$; Foci $(\pm\sqrt{2}, 0)$
Vertices $(\pm 1, 0)$
Asymptotes: $y = \pm 1x$

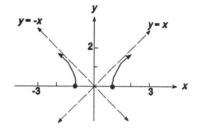

54. (a) Set up a rectangular coordinate system so the two stations lie on the x-axis and the origin is midway between them. The ship lies on a hyperbola whose foci are the locations of the two stations. Since the time difference is .00032 seconds and the speed of the signal is 186,000 miles per second, the difference of the distances from the ship to each station (foci) is

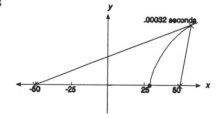

distance $= (186000)(.00032) = 59.52$ miles

The difference of the distances from the ship to each station, 59.52, equals $2a$, so $a = 29.76$ and the vertex of the corresponding hyperbola is at $(29.76, 0)$. Since the focus is at $(50, 0)$, following this hyperbola the ship would reach shore 20.24 miles from the master station.

(b) The ship should follow a hyperbola with vertex at $(40, 0)$. For this hyperbola, $a = 40$, so the constant difference of the distances from the ship to each station is 80. The time difference the ship should look for is:

$$\text{time} = \frac{80}{186,000} = 0.0004301 \text{ seconds}$$

(c) We need to find the equation of the hyperbola with vertex at (40, 0) and a focus at (50, 0). The form of the equation of this hyperbola is $\dfrac{x^2}{a^2} - \dfrac{y^2}{b^2} = 1$

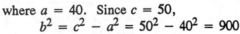
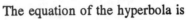

where $a = 40$. Since $c = 50$,
$$b^2 = c^2 - a^2 = 50^2 - 40^2 = 900$$
The equation of the hyperbola is
$$\frac{x^2}{1600} - \frac{y^2}{900} = 1$$
Since the ship is 20 miles off shore, we use $y = 20$ and solve the above equation for x.
$$\frac{x^2}{1600} - \frac{20^2}{900} = 1$$
$$\frac{x^2}{1600} = 1 + \frac{20^2}{900} = 1.4444$$
$$x^2 = 1600(1.4444)$$
$$x \approx 48$$
The ship is at the position (48, 20).

58. $e = \dfrac{c}{a}$; if $a = b$, then $c^2 = 2a^2$.

Thus $\dfrac{c^2}{a^2} = 2$ or $e = \dfrac{c}{a} = \sqrt{2}$

60. $\dfrac{y^2}{a^2} - \dfrac{x^2}{b^2} = 1$

$\dfrac{y^2}{a^2} = 1 + \dfrac{x^2}{b^2}$

$y^2 = a^2 \left[1 + \dfrac{x^2}{b^2} \right]$

$y^2 = \dfrac{a^2 x^2}{b^2} \left[\dfrac{b^2}{x^2} + 1 \right]$

$y = \pm \dfrac{ax}{b} \sqrt{\dfrac{b^2}{x^2} + 1}$

As $x \to -\infty$ or as $x \to \infty$, the term $\dfrac{b^2}{x^2}$ gets close to 0, so the expression under the radical gets closer to 1. Thus, the graph of the hyperbola gets closer to the lines
$$y = \frac{-a}{b}x \text{ and } y = \frac{a}{b}x$$
These lines are asymptotes of the hyperbola.

62. $Ax^2 + Cy^2 + Dx + Ey + F = 0$ where A and C are of opposite sign
$$Ax^2 + Dx + Cy^2 + Ey = -F$$
$$A\left[x^2 + \frac{D}{A}x \right] + C\left[y^2 + \frac{E}{C}y \right] = -F$$
$$A\left[x + \frac{D}{2A} \right]^2 + C\left[y + \frac{E}{2C} \right]^2 = -F + \frac{D^2}{4A} + \frac{E^2}{4C}$$

(a) If $\dfrac{D^2}{4A} + \dfrac{E^2}{4C} - F \neq 0$, this is the equation of a hyperbola with center at $\left[\dfrac{-D}{2A}, \dfrac{-E}{2C} \right]$ if

$C < 0$ or $\left[\dfrac{-E}{2C}, \dfrac{-D}{2A} \right]$ if $A < 0$.

(b) If $\dfrac{D^2}{4A} + \dfrac{E^2}{4C} - F = 0$, the graph is the intersecting lines $x = \dfrac{-D}{2A}$ and $y = \dfrac{-E}{2C}$.

2. $2y^2 - 3y + 3x = 0$
    Here $A = 0$ and $C = 0$, so that $AC = (0)(2) = 0$, which means the equation defines a parabola.

4. $2x^2 + y^2 - 8x + 4y + 2 = 0$
    Here $A = 2$ and $C = 1$, so that $AC = (2)(1) = 2 > 0$ which means the equation defines an ellipse.

6. $4x^2 - 3y^2 - 8x + 6y + 1 = 0$
    Here $A = 4$ and $C = -3$, so that $AC = (4)(-3) = -12 < 0$ which means the equation defines a hyperbola.

8. $y^2 - 8x^2 - 2x - y = 0$
    Here $A = -8$ and $C = 1$, so that $AC = (-8)(1) = -8 < 0$ which means the equation defines a hyperbola.

10. $2x^2 + 2y^2 - 8x + 8y = 0$
    Here $A = 2$ and $C = 2$, so that $AC = (2)(2) = 4 > 0$ which means the equation defines an ellipse, specifically a circle.

For Problems 12-32, we use the formulas $\cot 2\theta = \dfrac{A - C}{B}$, $y = x' \cos \theta - y' \sin \theta$, *and*
$y = x' \sin \theta + y' \cos \theta$.

12. $A = 1$, $B = -4$, $C = 1$, $\cot 2\theta = 0$ so that $\theta = \dfrac{\pi}{4}$

    $x = \dfrac{\sqrt{2}}{2}(x' - y')$; $y = \dfrac{\sqrt{2}}{2}(x' + y')$

14. $A = 3$, $B = -10$, $C = 3$, $\cot 2\theta = 0$ so that $\theta = \dfrac{\pi}{4}$

    $x = \dfrac{\sqrt{2}}{2}(x' - y')$; $y = \dfrac{\sqrt{2}}{2}(x' + y')$

16. $A = 11$, $B = 10\sqrt{3}$, $C = 1$, $\cot 2\theta = \dfrac{10}{10\sqrt{3}} = \dfrac{1}{\sqrt{3}}$; $\cos 2\theta = \dfrac{-1}{2}$

    $\cos 2\theta = \dfrac{1}{2}$, $2\theta = 60°$, $\theta = 30°$

    $\sin \theta = \sqrt{\dfrac{1 - \dfrac{1}{2}}{2}} = \dfrac{1}{2}$; $\cos \theta = \sqrt{\dfrac{1 + \dfrac{1}{2}}{2}} = \dfrac{\sqrt{3}}{2}$

    $x = \dfrac{\sqrt{3}}{2}x' - \dfrac{1}{2}y' = \dfrac{1}{2}\left(\sqrt{3}x' - y'\right)$; $y = \dfrac{1}{2}x' + \dfrac{\sqrt{3}}{2}y' = \dfrac{1}{2}\left(x' + \sqrt{3}y'\right)$

18. $A = 1, B = 4, C = 4, \cot 2\theta = \dfrac{-3}{4}; \cos 2\theta = \dfrac{-3}{5}$

$$\sin \theta = \sqrt{\dfrac{1 + \dfrac{3}{5}}{2}} = \dfrac{2}{\sqrt{5}} = \dfrac{2\sqrt{5}}{5}; \quad \cos \theta = \sqrt{\dfrac{1 - \dfrac{3}{5}}{2}} = \dfrac{1}{\sqrt{5}} = \dfrac{\sqrt{5}}{5}$$

$$x = \dfrac{\sqrt{5}}{5}x' - \dfrac{2\sqrt{5}}{5}y' = \dfrac{\sqrt{5}}{5}(x' - 2y'); \quad y = \dfrac{2\sqrt{5}}{5}x' + \dfrac{\sqrt{5}}{5}y' = \dfrac{\sqrt{5}}{5}(2x' + y')$$

20. $A = 34, B = -24, C = 41, \cot 2\theta = \dfrac{-7}{-24} = \dfrac{7}{24}; \cos 2\theta = \dfrac{7}{25}$

$$\sin \theta = \sqrt{\dfrac{1 - \dfrac{7}{25}}{2}} = \dfrac{3}{5}, \quad \cos \theta = \sqrt{\dfrac{1 + \dfrac{7}{25}}{2}} = \dfrac{4}{5}$$

$$x = \dfrac{4}{5}x' - \dfrac{3}{5}y' = \dfrac{1}{5}(4x' - 3y'); \quad y = \dfrac{3}{5}x' + 4 + 5y' = \dfrac{1}{5}(3x' + 4y')$$

22. $\theta = \dfrac{\pi}{4}$ (see Problem 12)

$x^2 - 4xy + y^2 - 3 = 0$

$$\left[\dfrac{\sqrt{2}}{2}(x' - y')\right]^2 - 4\left[\dfrac{\sqrt{2}}{2}(x' - y')\right]\left[\dfrac{\sqrt{2}}{2}(x' + y')\right] + \left[\dfrac{\sqrt{2}}{2}(x' + y')\right]^2 - 3 = 0$$

$$\dfrac{1}{2}(x'^2 - 2x'y' + y'^2) - \dfrac{4}{2}(x'^2 - y'^2) + \dfrac{1}{2}(x'^2 + 2x'y' + y'^2) = 3$$

$$-2x'^2 + 6y'^2 = 6$$

$$y'^2 - \dfrac{x'^2}{3} = 1$$

Hyperbola; center at origin; transverse axis the y'-axis; vertices at $(0, \pm 1)$.

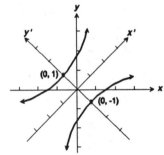

24. $\theta = \dfrac{\pi}{4}$ (see Problem 14)

$3x^2 - 10xy + 3y^2 - 32 = 0$

$$3\left[\dfrac{\sqrt{2}}{2}(x' - y')\right]^2 - 10\left[\dfrac{\sqrt{2}}{2}(x' - y')\right]\left[\dfrac{\sqrt{2}}{2}(x' + y')\right] + 3\left[\dfrac{\sqrt{2}}{2}(x' + y')\right]^2 = 32$$

$$\dfrac{3}{2}(x'^2 - 2x'y' + y'^2) - \dfrac{10}{2}(x'^2 - y'^2) + \dfrac{3}{2}(x'^2 + 2x'y' + y'^2) = 32$$

$$-4x'^2 + 16y'^2 = 64$$

$$\dfrac{y'^2}{4} - \dfrac{x'^2}{16} = 1$$

Hyperbola; center at origin; transverse axis the y'-axis; vertices at $(0, \pm 2)$.

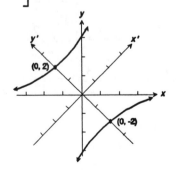

26. $\theta = 30°$ (see Problem 16)

$11x^2 + 10\sqrt{3}\,xy + y^2 - 4 = 0$

$11\left[\frac{1}{2}\left(\sqrt{3}\,x' - y'\right)\right]^2 + 10\sqrt{3}\left[\frac{1}{2}\left(\sqrt{3}\,x' - y'\right)\right]\left[\frac{1}{2}\left(x' + \sqrt{3}\,y'\right)\right] + \left[\frac{1}{2}\left(x' + \sqrt{3}\,y'\right)\right]^2 = 4$

$11\left(3x'^2 - 2\sqrt{3}\,x'y' + y'^2\right) + 10\sqrt{3}\left(\sqrt{3}\,x'^2 + 2x'y' - \sqrt{3}\,y'^2\right) + \left(x'^2 + 2\sqrt{3}\,x'y' + 3y'^2\right) = 16$

$64x'^2 - 16y'^2 = 16$

$4x'^2 - y'^2 = 1$

Hyperbola; center at $(0, 0)$; transverse axis the x'-axis; vertices at

$\left[\pm\frac{1}{2}, 0\right]$

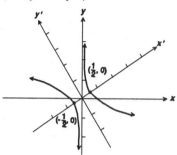

28. $\theta = 63°$ (see Problem 18)

$x^2 + 4xy + 4y^2 + 5\sqrt{5}\,y + 5 = 0$

$\left[\frac{\sqrt{5}}{5}\left(x' - 2y'\right)\right]^2 + 4\left[\frac{\sqrt{5}}{5}\left(x' - 2y'\right)\right]\left[\frac{\sqrt{5}}{5}\left(2x' + y'\right)\right] + \left[\frac{\sqrt{5}}{5}\left(2x' + y'\right)\right]^2$

$+ 5\sqrt{5}\left[\frac{\sqrt{5}}{5}\left(x' - 2y'\right)\right] + 5 = 0$

$\frac{1}{5}\left(x'^2 - 4x'y' + 4y'^2\right) + \frac{4}{5}\left(2x'^2 - 3x'y' - 2y'^2\right) + \frac{4}{5}\left(4x'^2 + 4x'y' + y'^2\right) + 5\left(2x' + y'\right) = -5$

$5x'^2 + 10x' + 5y' = -5$

$x'^2 + 2x' + y' = -1$

$x'^2 + 2x' + 1 + y' = 0$

$\left(x' + 1\right)^2 + y' = 0$

$y' = -\left(x' + 1\right)^2$

Parabola; vertex at $(-1, 0)$; axis of symmetry parallel to the y'-axis.

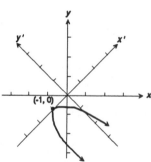

30. $\theta \approx 37°$ (see Problem 20)

$34x^2 - 24xy + 41y^2 - 25 = 0$

$34\left[\frac{1}{5}\left(4x' - 3y'\right)\right]^2 - 24\left[\frac{1}{5}\left(4x' - 3y'\right)\right]\left[\frac{1}{5}\left(3x' + 4y'\right)\right] + 41\left[\frac{1}{5}\left(3x' + 4y'\right)\right]^2 = 25$

$\frac{34}{25}\left(16x'^2 - 24x'y' + 9y'^2\right) - \frac{24}{25}\left(12x'^2 + 7x'y' - 12y'^2\right) + \frac{41}{25}\left(9x'^2 + 24x'y' + 16y'^2\right) = 25$

$\frac{625}{25}x'^2 + \frac{1250}{25}y'^2 = 25$

$25x'^2 + 50y'^2 = 0$

$x'^2 + 2y'^2 = 1$

Ellipse; center at $(0, 0)$; major axis the x'-axis; vertices at $(0, \pm 1)$.

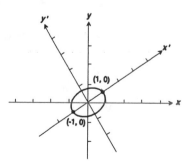

32. $A = 16$, $B = 24$, $C = 9$, $\cot 2\theta = \dfrac{7}{24}$; $\cos 2\theta = \dfrac{7}{25}$

$\sin \theta = \sqrt{\dfrac{1 - \dfrac{7}{25}}{2}} = \dfrac{3}{5}$; $\cos \theta = \sqrt{\dfrac{1 + \dfrac{7}{25}}{2}} = \dfrac{4}{5}$; $\theta \approx 37°$

$x = \dfrac{4}{5}x' - \dfrac{3}{5}y' = \dfrac{1}{5}(4x' - 3y')$; $y = \dfrac{3}{5}x' + \dfrac{4}{5}y' = \dfrac{1}{5}(3x' + 4y')$

$16x^2 + 24xy + 9y^2 - 60x + 80y = 0$

$16\left[\dfrac{1}{5}(4x' - 3y')\right]^2 + 24\left[\dfrac{1}{5}(4x' - 3y')\right]\left[\dfrac{1}{5}(3x' + 4y')\right] + 9\left[\dfrac{1}{5}(3x' + 4y')\right]^2$

$\qquad + 60\left[\dfrac{1}{5}(4x' - 3y')\right] + 80\left[\dfrac{1}{5}(3x' + 4y')\right] = 0$

$\dfrac{16}{25}(16x'^2 - 24x'y' + 9y'^2) + \dfrac{24}{25}(12x'^2 + 7x'y' - 12y'^2) + \dfrac{9}{25}(9x'^2 + 24x'y' + 16y'^2)$

$\qquad - 12(4x' - 3y') + 16(3x' + 4y') = 0$

$\dfrac{625}{25}x'^2 + 100y' = 0$

$\qquad\qquad 25x'^2 = -100y'$

$\qquad\qquad\quad x'^2 = -4y'$

Parabola; vertex at $(0, 0)$; focus at $(0, -1)$.

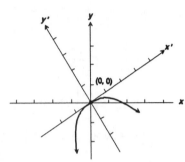

34. $B^2 - 4AC = 9 - 32 = -23 < 0$; ellipse   **36.** $B^2 - 4AC = 9 - 16 = -7 < 0$; ellipse

38. $B^2 - 4AC = 144 - 160 = -16 < 0$;   **40.** $B^2 - 4AC = 144 - 144 = 0$; parabola
ellipse

42. $B^2 - 4AC = 4 - 12 = -8 < 0$; ellipse

44. $A' + C' = [A \cos^2 \theta + B \sin \theta \cos \theta + C \sin^2 \theta] + [A \sin^2 \theta - B \sin \theta \cos \theta + C \cos^2 \theta]$

$\qquad\qquad = A(\cos^2 \theta + \sin^2 \theta) + \sin \theta \cos \theta(B - B) + C(\sin^2 \theta + \cos^2 \theta)$

$\qquad\qquad = A(1) + \sin \theta \cos \theta(0) + C(1) = A + C$

46.   Since $B^2 - 4AC = B'^2 - 4A'C'$ (see Problem 35), then if $\cot 2\theta = \dfrac{A - C}{B}$, then $B' = 0$ and

$B^2 = 4AC = -4A'C'$. Hence,

$\qquad Ax^2 + Bxy + Cy^2 + Dx + Ey + F = 0$ now has the form

$\qquad A'x'^2 + C'^2y'^2 + D'x' + E'y' + F' = 0$. Thus,

    (a)   If $B^2 - 4AC = -4A'C' = 0$, then $A'C' = 0$. Using the theorem for identifying conics without completing any squares, the equation is a parabola.

    (b)   If $B^2 - 4AC = -4A'C' < 0$, then $A'C' > 0$. Hence, the equation is an ellipse (or a circle).

    (c)   If $B^2 - 4AC = -4A'C' > 0$, then $A'C' < 0$. Hence, the equation is a hyperbola.

48.

$$x^{1/2} + y^{1/2} = a^{1/2}$$
$$y^{1/2} = a^{1/2} - x^{1/2}$$
$$y = (a^{1/2} - x^{1/2})^2$$
$$y = a - 2a^{1/2}x^{1/2} + x$$
$$2a^{1/2}x^{1/2} = (a + x) - y$$
$$4ax = (a + x) - 2y(a + x) + y^2$$
$$4ax = a^2 + 2ax + x^2 - 2ay + 2xy + y^2$$
$$0 = x^2 + y^2 + 2ax - 2xy + 2ay + a^2$$

$B^2 - 4AC \; 4 - 4 = 0$

The equation is a parabola.

11.6 Polar Equations of Conics

For Problems 2-16, use Table 5.

2. $r = \dfrac{3}{1 - \sin \theta}$; $e = 1$; $p = 3$; parabola; directrix is parallel to the polar axis 3 units below the pole.

4. $r = \dfrac{2}{1 + 2 \cos \theta}$; $ep = 2$, $e = 2$; $p = 1$; hyperbola; directrix is perpendicular to the polar axis 1 unit to the right of the pole.

6. $r = \dfrac{6}{8 \left[1 + \dfrac{1}{4} \sin \theta \right]} = \dfrac{\dfrac{3}{4}}{1 + \dfrac{1}{4} \sin \theta}$; $ep = \dfrac{3}{4}$, $e = \dfrac{1}{4}$; $p = 3$

Ellipse; directrix is parallel to the polar axis 3 units above the pole.

8. $r = \dfrac{3}{1 - \sin \theta}$; $ep = 3$, $e = 1$, $p = 3$

Parabola; directrix is parallel to the polar axis 3 units below the pole.

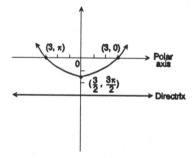

10. $r = \dfrac{10}{5 \left[1 + \dfrac{4}{5} \cos \theta \right]} = \dfrac{2}{1 + \dfrac{4}{5} \cos \theta}$; $ep = 2$, $e = \dfrac{4}{5}$, $p = \dfrac{5}{2}$

Ellipse; directrix is perpendicular to the polar axis $\dfrac{5}{2}$ units to the

right of the pole. Vertices are at $\left[\dfrac{10}{9}, 0 \right]$ and $(10, \pi)$.

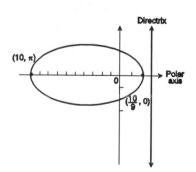

12. $r = \dfrac{12}{4(1 + 2 \sin \theta)} = \dfrac{3}{1 + 2 \sin \theta}$; $ep = 3$, $e = 2$, $p = \dfrac{3}{2}$

Hyperbola, directrix is parallel to the polar axis $\dfrac{3}{2}$ units above the

pole. Vertices are at $\left[1, \dfrac{\pi}{2}\right]$ and $\left[-3, \dfrac{3\pi}{2}\right]$.

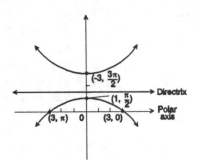

14. $r = \dfrac{8}{2 + 4 \cos \theta} = \dfrac{8}{2(1 + 2 \cos \theta)} = \dfrac{4}{1 + 2 \cos \theta}$

$e = 2$; $ep = 4$, so $p = \dfrac{4}{2} = 2$

The conic is a hyperbola; the transverse axis is perpendicular to the directrix. The directrix is perpendicular to the polar axis 1 unit to the right of the pole.

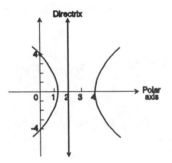

16. $r = \dfrac{2}{2 - \cos \theta} = \dfrac{1}{1 - \dfrac{1}{2} \cos \theta}$; $ep = 1$, $e = \dfrac{1}{2}$, $p = 2$

Ellipse; directrix is perpendicular to the polar axis 2 units to the left of the pole.

Vertices are at $(2, 0)$ and $\left[\dfrac{2}{3}, \pi\right]$.

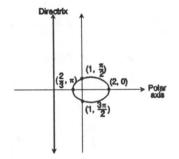

18. $r = \dfrac{3 \csc \theta}{\csc \theta - 1} = \dfrac{3}{1 - \sin \theta}$; $ep = 3$, $e = 1$, $p = 3$

Parabola; directrix is parallel to the polar axis 3 units below the

pole. Vertices are at $\left[\dfrac{3}{2}, \dfrac{3\pi}{2}\right]$.

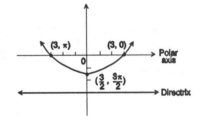

20. $r = \dfrac{3}{1 - \sin \theta}$
$r - r \sin \theta = 3$
$r = 3 + r \sin \theta$
$r^2 = (3 + r \cos \theta)^2$
$x^2 + y^2 = (3 + y)^2$
$x^2 + y^2 = 9 + 6y + y^2$
$x^2 - 6y - 9 = 0$

22. $r = \dfrac{10}{5 + 4 \cos \theta}$
$5r + 4r \cos \theta = 10$
$5r = 10 - 4r \cos \theta$
$25r^2 = (10 - 4r \cos \theta)^2$
$25(x^2 + y^2) = (10 - 4x)^2$
$25x^2 + 25y^2 = 100 - 80x + 16x^2$
$9x^2 + 25y^2 + 80x - 100 = 0$

24. $r = \dfrac{12}{4 + 8 \sin \theta}$

$4r + 8r \sin \theta = 12$

$4r = 12 - 8r \sin \theta$

$r = 3 - 2r \sin \theta$

$r^2 = (3 - 2r \sin \theta)^2$

$x^2 + y^2 = (3 - 2y)^2$

$x^2 + y^2 = 9 - 12y + 4y^2$

$x^2 - 3y^2 + 12y - 9 = 0$

26. $r = \dfrac{8}{2 + 4 \cos \theta}$

$r(2 + 4 \cos \theta) = 8$

$2r + 4r \cos \theta = 8$

$2r = 8 - 4r \cos \theta$

$r = 4 - 2r \cos \theta$

$r^2 = (4 - 2r \cos \theta)^2$

$x^2 + y^2 = (4 - 2x)^2$

$x^2 + y^2 = 16 - 16x + 4x^2$

$3x^2 - y^2 - 16x + 16 = 0$

28. $2r - r \cos \theta = 2$

$2r = 2 + r \cos \theta$

$4r^2 = (2 + r \cos \theta)^2$

$4(x^2 + y^2) = (2 + x)^2$

$4x^2 + 4y^2 = 4 + 4x + x^2$

$3x^2 + 4y^2 - 4x - 4 = 0$

30. $r = \dfrac{3 \csc \theta}{\csc \theta - 1}$

$r = \dfrac{3}{1 - \sin \theta}$

$r - r \sin \theta = 3$

$r = 3 + r \sin \theta$

$r^2 = (3 + r \sin \theta)^2$

$x^2 + y^2 = (3 + y)^2$

$x^2 + y^2 = 9 + 6y + y^2$

$x^2 - 6y - 9 = 0$

32. $r = \dfrac{ep}{1 - e \sin \theta}$

$e = 1, p = 2$

$r = \dfrac{2}{1 - \sin \theta}$

34. $r = \dfrac{ep}{1 + e \cos \theta}$

$e = \dfrac{2}{3}, p = 3$

$r = \dfrac{2}{1 + \dfrac{2}{3} \sin \theta}$

$r = \dfrac{6}{3 + 2 \sin \theta}$

36. $r = \dfrac{ep}{1 + e \cos \theta}$

$e = 5, p = 5$

$r = \dfrac{25}{1 + 5 \cos \theta}$

38. $d(F, P) = ed(D, P)$

$d(D, P) = p - r \sin \theta$

$\therefore r = e(p - r \sin \theta)$

$r = ep - er \sin \theta$

$r + er \sin \theta = ep$

$r = \dfrac{ep}{1 + e \sin \theta}$

40. $r = \dfrac{(3.442)10^7}{1 - 0.206 \cos \theta}$

At aphelion, the greatest distance from the sun, $\cos \theta = +1$. Therefore,

$$r = \frac{(3.442)10^7}{1 - 0.206(+1)} = \frac{(3.442)10^7}{.794} = 4.335 \times 10^7 \text{ miles}$$

At perihelion, the shortest distance from the sun, $\cos \theta = -1$. Therefore,

$$r = \frac{(3.442)10^7}{1 - 0.206(-1)} = \frac{(3.442)10^7}{1.206} = 2.854 \times 10^7 \text{ miles}$$

2. $x = t - 3, y = 2t + 4$
$y = 2(x + 3) + 4$
$y = 2x + 10$

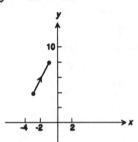

4. $x = \sqrt{2t}, y = 4t$

$y = 4\left[\dfrac{x^2}{2}\right]$

$y = 2x^2$

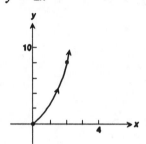

6. $x = \sqrt{t} + 4, y = \sqrt{t} - 4$
$x = (y + 4) + 4$
$x = y + 8$

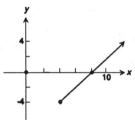

8. $x = 2t - 4, y = 4t^2$

$y = 4\left[\dfrac{x + 4}{2}\right]^2$
$y = (x + 4)^2$

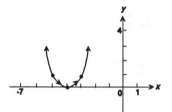

10. $x = e^t, y = e^{-t}$

$x = e^t = \dfrac{1}{e^{-t}} = \dfrac{1}{y}$

$xy = 1$

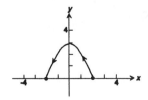

12. $x = t^{3/2} + 1, y = \sqrt{t}$
$x = (y^2)^{3/2} + 1$
$x = y^3 + 1$

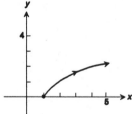

14. $x = 2\cos t, y = 3\sin t$

$\left[\dfrac{x}{2}\right]^2 + \left[\dfrac{y}{3}\right]^2 = \cos^2 t + \sin^2 t$

$\dfrac{x^2}{4} + \dfrac{y^2}{9} = 1$

16. $x = 2\cos t, y = \sin t$

$\left[\dfrac{x}{2}\right]^2 + y^2 = \cos^2 t + \sin^2 t$

$\dfrac{x^2}{4} + y^2 = 1$

18. $x = \csc t, y = \cot t$
$$\cot^2 t + 1 = \csc^2 t$$
$$y^2 + 1 = x^2$$
$$x^2 - y^2 = 1$$

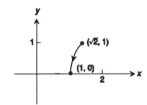

20. $x = t^2, y = \ln t$
$$y = \ln\sqrt{x}$$
$$y = \frac{1}{2}\ln x$$

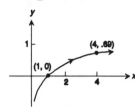

22. $x = t, y = t^4 + 1$
$x = t^3, y = t^{12} + 1$

24. $x = t, y = t^2$
$x = \sqrt{t}, y = t$

26. $x = 2\sin \omega t, y = 3\cos \omega t$
$$\frac{2\pi}{\omega} = 1, \omega = 2\pi$$
$$x = 2\sin 2\pi t, y = 3\cos 2\pi t, \quad 0 \le t \le 1$$

28. $x = 2\cos \omega t, y = 3\sin \omega t$
$$\frac{2\pi}{\omega} = 3, \omega = \frac{2\pi}{3}$$
$$x = 2\cos\frac{2\pi}{3}t, y = 3\sin\frac{2\pi}{3}t, 0 \le t \le 3$$

30.

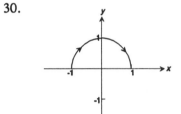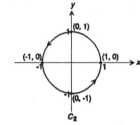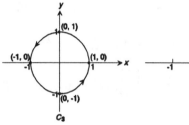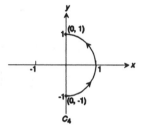

32. (a) $x = (v_0\cos\theta)t, y = (v_0\sin\theta)t - 16t^2$
$$t = \frac{x}{v_0\cos\theta}$$

$$y = v_0\sin\theta\left[\frac{x}{v_0\cos\theta}\right] - 16 \cdot \frac{x^2}{v_0^2\cos^2\theta}$$

$$y = (\tan\theta)x - \frac{16}{v_0^2\cos^2\theta}x^2$$

Thus, y is a quadratic function of x; its graph is a parabola.

(b) $y = 0 \Rightarrow (v_0\sin\theta)t - 16t^2 = 0$
$$\Rightarrow t[(v_0\sin\theta) - 16t] = 0$$
$$\Rightarrow t = 0 \text{ or } v_0\sin\theta = 16t$$
$$\Rightarrow t = \frac{1}{16}v_0\sin\theta$$

(c) $t = \frac{1}{16}v_0\sin\theta \Rightarrow x = v_0\cos\theta\left[\frac{1}{16}v_0\sin\theta\right]$

$$x = \frac{v_0^2\sin 2\theta}{32} \text{ feet}$$

(d) $x = y$

$$(V_0 \cos \theta)t = (V_0 \sin \theta)t - 16t^2$$

$$16t^2 + (V_0 \cos \theta)t - (V_0 \sin \theta)t = 0$$

$$t(16t + V_0 \cos \theta - V_0 \sin \theta) = 0$$

$$t = 0 \text{ or } 16t + V_0 \cos \theta - V_0 \sin \theta = 0$$

$$t = \frac{V_0 \sin \theta - V_0 \cos \theta}{16}$$

$$t = \frac{V_0}{16}(\sin \theta - \cos \theta)$$

At $t = \dfrac{V_0}{16}(\sin \theta - \cos \theta)$:

$$x = V_0 \cos \theta \left[\frac{V_0}{16}(\sin \theta - \cos \theta) \right] = \frac{V_0^2}{16}\cos \theta(\sin \theta - \cos \theta)$$

$$y = V_0 \sin \theta \left[\frac{V_0}{16}(\sin \theta - \cos \theta) \right] - 16\left[\frac{V_0^2}{16^2}(\sin \theta - \cos \theta)^2 \right]$$

$$= \frac{V_0^2}{16}\sin \theta(\sin \theta - \cos \theta) - \frac{V_0^2}{16}(\sin^2 \theta - 2\sin \theta \cos \theta + \cos^2\theta)$$

$$= \frac{V_0^2}{16}\sin \theta(\sin \theta - \cos \theta) - \frac{V_0^2}{16}(1 - 2\sin \theta \cos \theta)$$

Now, we must find $\sqrt{x^2 + y^2}$.

$$x^2 = \left[\frac{V_0^2}{16}\cos \theta(\sin \theta - \cos \theta) \right]^2 = \frac{V_0^4}{16^2}\cos^2 \theta(1 - 2\sin \theta \cos \theta)$$

$$y^2 = \left[\frac{V_0^2}{16}\sin \theta(\sin \theta - \cos \theta) - \frac{V_0^2}{16}(1 - 2\sin \theta \cos \theta) \right]^2$$

$$= \frac{V_0^4}{16^2}\Big[\sin^2 \theta(1 - 2\sin \theta \cos \theta) - 2\sin \theta(\sin \theta - \cos \theta)$$

$$(1 - 2\sin \theta \cos \theta) + (1 - 2\sin \theta \cos)^2\Big]$$

$$\sqrt{x^2 + y^2} = \sqrt{\begin{array}{l} \dfrac{V_0^4}{16^2}\cos^2 \theta(1 - 2\sin \theta \cos \theta) + \dfrac{V_0^4}{16^2}\sin^2 \theta(1 - 2\sin \theta \cos \theta) \\[2mm] - \dfrac{2V_0^3}{16}\sin \theta(\sin \theta - \cos \theta)(1 - 2\sin \theta \cos \theta) + V_0^2(1 - 2\sin \theta \cos \theta)^2 \end{array}}$$

$$= \sqrt{\begin{array}{l} (1 - 2\sin \theta \cos \theta)\dfrac{V_0^4}{16^2}\Big[(\sin^2 \theta + \cos^2 \theta) - 2\sin \theta(\sin \theta - \cos \theta) \\[2mm] + V_0^2(1 - 2\sin \theta \cos \theta)\Big] \end{array}}$$

$$= \sqrt{(1 - 2\sin \theta \cos \theta)\dfrac{V_0^4}{16^2}\Big[1 - 2(\sin^2 \theta - \sin \theta \cos \theta) + 1 - 2\sin \theta \cos \theta\Big]}$$

$$= \sqrt{(1 - 2 \sin \theta \cos \theta)\frac{v_0^4}{16^2}\left[1 - 2 \sin^2 \theta + 2 \sin \theta \cos \theta + 1 - 2 \sin \theta \cos \theta\right]}$$

$$= \sqrt{(1 - 2 \sin \theta \cos \theta)\frac{v_0^4}{16^2}\left[2 - 2 \sin^2 \theta\right]} = \sqrt{2}\frac{v_0^2}{16}\sqrt{(1 - 2 \sin \theta \cos \theta)(\cos^2 \theta)}$$

$$= \frac{v_0^2}{16}\sqrt{2} \cos \theta\sqrt{(\sin \theta - \cos \theta)^2} = \frac{v_0^2}{16}\sqrt{2} \cos \theta(\sin \theta - \cos \theta)$$

34.

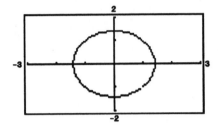

36.

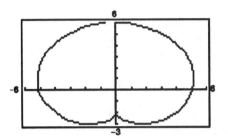

11 Chapter Review

2. $16x^2 = y \Rightarrow x^2 = \frac{1}{16}y$

Parabola
Vertex $(0, 0)$

Focus $\left[0, \frac{1}{64}\right]$

Directrix: $y = \frac{-1}{64}$

4. $\frac{y^2}{25} - x^2 = 1$

Hyperbola
Center $(0, 0)$
Vertices $(0, \pm 5)$

Foci $\left(0, \pm\sqrt{26}\right)$

Asymptotes: $y = \pm\frac{1}{5}x$

6. $\frac{x^2}{9} + \frac{y^2}{16} = 1$

Ellipse
Center $(0, 0)$
Vertices $(0, \pm 4)$

Foci $\left(0, \pm\sqrt{7}\right)$

8. $3y^2 - x^2 = 9$

$\frac{y^2}{3} - \frac{x^2}{9} = 1$

Hyperbola
Center $(0, 0)$

Vertices $\left(0, \pm\sqrt{3}\right)$

Foci $\left(0, \pm 2\sqrt{3}\right)$

Asymptotes: $y = \pm\frac{\sqrt{3}}{3}x$

10. $9x^2 + 4y^2 = 36$

$\dfrac{x^2}{4} + \dfrac{y^2}{9} = 1$

Ellipse

Center $(0, 0)$

Vertices $(0, \pm 3)$

Foci $\left(0, \pm\sqrt{5}\right)$

12. $2y^2 - 4y = x - 2$

$2(y^2 - 2y + 1) = x$

$(y - 1)^2 = \dfrac{1}{2}x$

Parabola

Vertex $(0, 1)$

Focus $\left[\dfrac{1}{8}, 1\right]$

Directrix: $y = \dfrac{-1}{8}$

14. $4x^2 + y^2 + 8x - 4y + 4 = 0$

$4(x^2 + 2x + 1) + (y^2 - 4y + 4) = -4 + 4 + 4$

$4(x + 1)^2 + (y - 2)^2 = 4$

$(x + 1)^2 + \dfrac{(y - 2)^2}{4} = 1$

Ellipse

Center $(-1, 2)$

Vertices $(-1, 0)$ and $(-1, 4)$

Foci $\left(-1, 2 \pm \sqrt{3}\right)$

16. $4x^2 + 9y^2 - 16x + 18y = 11$

$4(x^2 - 4x + 4) + 9(y^2 + 2y + 1) = 11 + 16 + 9$

$4(x - 2)^2 + 9(y + 1)^2 = 36$

$\dfrac{(x - 2)^2}{9} + \dfrac{(y + 1)^2}{4} = 1$

Ellipse

Center $(2, -1)$

Vertices $(5, -1)$ and $(-1, -1)$

Foci $\left(2 \pm \sqrt{5}, -1\right)$

18. $4y^2 + 3x - 16y + 19 = 0$

$4(y^2 - 4y + 4) = -3x - 19 + 16$

$(y - 2)^2 = \dfrac{1}{4}(-3x - 3)$

$(y - 2)^2 = \dfrac{-3}{4}(x + 1)$

Parabola

Vertex $(-1, 2)$

Focus $\left[-\dfrac{19}{16}, 2\right]$

Directrix: $x = -\dfrac{13}{16}$

20.
$$x^2 - y^2 - 2x - 2y = 1$$
$$(x^2 - 2x + 1) - (y^2 + 2y + 1) = 1 + 1 - 1$$
$$(x - 1)^2 - (y + 1)^2 = 1$$

Hyperbola
Center $(1, -1)$
Vertices $(0, -1)$ and $(2, -1)$

Foci $\left(1 \pm \sqrt{2}, -1\right)$
Asymptotes: $y + 1 = \pm(x - 1)$

22. Ellipse; center $(0, 0)$; focus $(0, 3)$;
vertex $(0, 5)$
$c^2 = 9, a^2 = 25 \Rightarrow$
$b^2 = 25 - 9 = 16$

$$\frac{x^2}{16} + \frac{y^2}{25} = 1$$

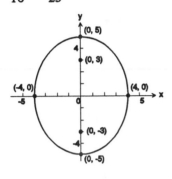

24. Parabola; vertex $(0, 0)$; directrix: $y = -3$
$a = 3$
$x^2 = 12y$

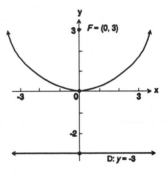

26. Hyperbola; vertices $(\pm 2, 0)$; focus $(4, 0)$
$a^2 = 4, c^2 = 16 \Rightarrow$
$b^2 = 16 - 4 = 12$

$$\frac{x^2}{4} - \frac{y^2}{12} = 1$$

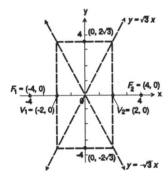

28. Ellipse; center $(-1, 2)$; focus $(0, 2)$;
vertex $(2, 2)$
$a^2 = 9, c^2 = 1, b^2 = 9 - 1 = 8$

$$\frac{(x + 1)^2}{9} + \frac{(y - 2)^2}{8} = 1$$

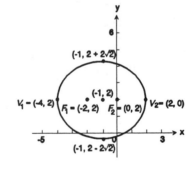

30. Parabola; focus $(3, 6)$; directrix: $y = 8$
$(x - 3)^2 = -4(y - 7)$

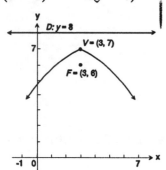

32. Hyperbola; vertices $(-3, 3)$, $(5, 3)$;
focus $(7, 3)$
$a = 4$, $c = 6 \Rightarrow b^2 = 36 - 16 = 20$
Center $(1, 3)$

Asymptotes: $y - 3 = \pm\dfrac{\sqrt{5}}{2}(x - 1)$

$$\dfrac{(x - 1)^2}{16} - \dfrac{(y - 3)^2}{20} = 1$$

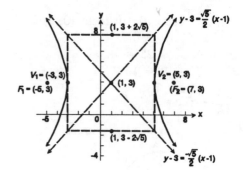

34. Center $(4, -2)$, $a = 1$, $c = 4$
Transverse axis parallel to y-axis.

Asymptotes: $y + 2 = \pm\dfrac{\sqrt{15}}{15}(x - 4)$

Hyperbola: $b^2 = 16 - 1 = 15$

$$(y + 2)^2 - \dfrac{(x - 4)^2}{15} = 1$$

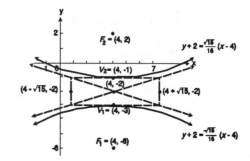

36. Vertices $(4, 0)$, $(4, 4)$;
asymptote: $y + 2x - 10 = 0$
Hyperbola: center $(4, 2)$

$a = 2$, $y - 2 = -\dfrac{2}{b}(x - 4)$

$y + \dfrac{2}{b}x - \dfrac{8}{b} - 2 = 0$

Since $y + 2x - 10 = 0$, then $b = 1$.
$c^2 = 1 + 4 = 5$

$$\dfrac{(y - 2)^2}{4} - \dfrac{(x - 4)^2}{1} = 1$$

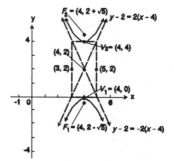

38. $2x^2 - y + 8x = 0$
$y = 2x^2 - 8x$
Here $A = 2$ and $C = 0$ so that $AC = 0$. The equation is a parabola.

40. $x^2 - 8y^2 - x - 2y = 0$
Here $A = 1$ and $C = -8$ so that $AC = -8$. The equation is a hyperbola.

42. $4x^2 + 4xy + y^2 - 8\sqrt{5}x + 16\sqrt{5}y = 0$
Here $A = 4$, $B = 4$, and $C = 1$ so that $B^2 - 4AC = 0$. The equation is a parabola.

44. $4x^2 - 10xy + 4y^2 - 9 = 0$
Here $A = 4$, $B = -10$, and $C = 4$ so that $B^2 - 4AC = 36 > 0$. The equation is a hyperbola.

Chapter 11 Analytic Geometry

46. $4x^2 + 12xy - 10y^2 + x + y - 10 = 0$
     Here $A = 4$, $B = 12$, and $C = -10$ so that $B^2 - 4AC = 304 > 0$. The equation is a hyperbola.

48. $A = 2$, $B = -5$, $C = 2$, $\cot 2\theta = 0$; $\theta = 45°$

   $$x = \frac{\sqrt{2}}{2}(x' - y'); \quad y = \frac{\sqrt{2}}{2}(x' + y')$$

   $$2\left[\frac{\sqrt{2}}{2}(x' - y')\right]^2 - 5\left[\frac{\sqrt{2}}{2}(x' - y')\right]\left[\frac{\sqrt{2}}{2}(x' + y')\right] + 2\left[\frac{\sqrt{2}}{2}(x' + y')\right]^2 - \frac{9}{2} = 0$$

   $$2 \cdot \frac{1}{2}(x'^2 - 2x'y' + y'^2) - 5 \cdot \frac{1}{2}(x'^2 - y'^2) + 2 \cdot \frac{1}{2}(x'^2 + 2x'y' + y'^2) = \frac{9}{2}$$

   $$-\frac{1}{2}x'^2 + \frac{9}{2}y'^2 = \frac{9}{2}$$

   $$y'^2 - \frac{x'^2}{9} = 1$$

   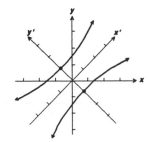

   Hyperbola; center at origin; transverse axis is the y'-axis; Vertices at $(0, \pm 1)$.

50. $A = 1$, $B = 4$, $C = 4$, $\cot 2\theta = \frac{-3}{4}$; $\cos 2\theta = \frac{-3}{5}$

   $$\sin\theta = \sqrt{\frac{1 + \dfrac{3}{5}}{2}} = \frac{2}{\sqrt{5}} = \frac{2\sqrt{5}}{5}; \quad \cos\theta = \sqrt{\frac{1 - \dfrac{3}{5}}{2}} = \frac{1}{\sqrt{5}} = \frac{\sqrt{5}}{5}$$

   $\theta = 63°$

   $$x = \frac{\sqrt{5}}{5}(x' - 2y'), \quad y = \frac{\sqrt{5}}{5}(2x' + y')$$

   $$\left[\frac{\sqrt{5}}{5}(x' - 2y')\right]^2 + 4\left[\frac{\sqrt{5}}{5}(x' - 2y')\right]\left[\frac{\sqrt{5}}{5}(2x' + y')\right] + 4\left[\frac{\sqrt{5}}{5}(2x' + y')\right]^2$$

   $$+ 8\sqrt{5}\left[\frac{\sqrt{5}}{5}(x' - 2y')\right] - 16\sqrt{5}\left[\frac{\sqrt{5}}{5}(2x' + y')\right] = 0$$

   $$\frac{1}{5}(x'^2 - 4x'y' + 4y'^2) + \frac{4}{5}(2x'^2 - 3x'y' - 2y'^2) + \frac{4}{5}(4x'^2 + 4x'y' + y'^2) + 8(x' - 2y') - 16(2x' + y') = 0$$

   $$5x'^2 - 24x' - 32y' = 0$$
   $$x'^2 - \frac{24}{5}x' = \frac{32}{5}y'$$
   $$x'^2 - \frac{24}{5}x' + \frac{144}{25} = \frac{32}{5}y' + \frac{144}{25}$$
   $$\left[x' - \frac{12}{5}\right]^2 = \frac{32}{5}\left[y' + \frac{9}{10}\right]$$

   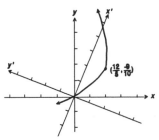

   Parabola; vertex at $\left[\dfrac{12}{5}, \dfrac{-9}{10}\right]$; focus parallel to the y'-axis at

   $\left[\dfrac{12}{5}, \dfrac{7}{10}\right]$.

52. $A = 9, B = -24, C = 16, \cot 2\theta = \dfrac{7}{24}; \cos 2\theta = \dfrac{7}{25}; \theta = 74°$

$$\sin \theta = \sqrt{\dfrac{1 - \dfrac{7}{25}}{2}} = \dfrac{3}{5}; \cos \theta = \sqrt{\dfrac{1 + \dfrac{7}{25}}{2}} = \dfrac{4}{5}$$

$$x = \dfrac{4}{5}x' - \dfrac{3}{5}y' = \dfrac{1}{5}(4x' - 3y'); \; y = \dfrac{3}{5}x' + \dfrac{4}{5}y' = \dfrac{1}{5}(3x' + 4')$$

$$9\left[\dfrac{1}{5}(4x' - 3y')\right]^2 - 24\left[\dfrac{1}{5}(4x' - 3y')\right]\left[\dfrac{1}{5}(3x' + 4y')\right] + 16\left[\dfrac{1}{5}(3x' + 4y')\right]^2$$

$$+ 30\left[\dfrac{1}{5}(4x' - 3y')\right] + 40\left[\dfrac{1}{5}(3x' + 4y')\right] = 0$$

$$\dfrac{9}{25}(16x'^2 - 24x'y' + 9y'^2) - \dfrac{24}{25}(12x'^2 + 7x'y' - 12y'^2) + \dfrac{16}{25}(9x'^2 + 24x'y' + 16y'^2)$$

$$+ 6(4x' - 3y') + 8(3x' + 4y') = 0$$

$$25y'^2 + 48x' + 14y' = 0$$

$$y'^2 + \dfrac{14}{25}y' = \dfrac{-48}{25}x'$$

$$y'^2 + \dfrac{14}{25}y' + \dfrac{49}{625} = \dfrac{-48}{25}x' + \dfrac{49}{625}$$

$$\left[y' + \dfrac{7}{25}\right]^2 = \dfrac{-48}{25}\left[x' - \dfrac{49}{1200}\right]$$

Focus parallel to the x'-axis at $\left[\dfrac{-527}{1200}, \dfrac{-7}{25}\right]$.

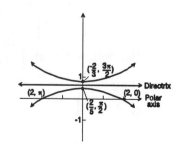

54. $ep = 6, e = 1, p = 6$
Parabola; directrix is parallel to the polar axis 6 units above the pole.

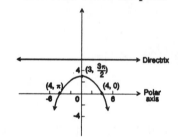

56. $r = \dfrac{2}{3 + 2 \cos \theta} = \dfrac{\dfrac{2}{3}}{1 + \dfrac{2}{3} \cos \theta};$

$ep = \dfrac{2}{3}, e = \dfrac{2}{3}, p = 1$

Ellipse; directrix is perpendicular to the polar axis 1 unit to the right of the pole. Vertices are at $\left[\dfrac{2}{5}, 0\right]$ and $(2, \pi)$.

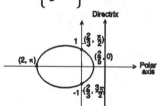

58. $r = \dfrac{10}{5 + 20 \sin \theta} = \dfrac{2}{1 + 4 \sin \theta}; ep = 2, e = 4, p = \dfrac{1}{2}$

Hyperbola, directrix is parallel to the polar axis $\dfrac{1}{2}$ unit above the pole.

Vertices are at $\left[\dfrac{2}{5}, \dfrac{\pi}{2}\right]$ and $\left[\dfrac{-2}{3}, \dfrac{3\pi}{2}\right]$.

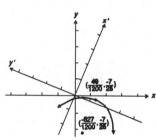

60. $r = \dfrac{6}{2 - \sin \theta}$

$r(2 - \sin \theta) = 6$

$2r - r \sin \theta = 6$

$2r = 6 + r \sin \theta$

$4r^2 = (6 + r \sin \theta)^2$

$4(x^2 + y^2) = (6 + y)^2$

$4x^2 + 4y^2 = 36 + 12y + y^2$

$4x^2 + 3y^2 - 12y - 36 = 0$

62. $r = \dfrac{2}{3 + 2 \cos \theta}$

$r(3 + 2 \cos \theta) = 2$

$3r + 2r \cos \theta = 2$

$3r = 2 - 2r \cos \theta$

$9r^2 = (2 - 2r \cos \theta)^2$

$9(x^2 + y^2) = (2 - 2x)^2$

$9x^2 + 9y^2 = 4 - 8x + 4x^2$

$5x^2 + 9y^2 + 8x - 4 = 0$

64. $x = 2t^2 + 6, \; y = 5 - t$

$x = 2(5 - y)^2 + 6$

$x = 2(25 - 10y + y^2) + 6$

$x = 56 - 20y + 2y^2$

$\dfrac{1}{2}x = 28 - 10y + y^2$

$y^2 - 10y = \dfrac{1}{2}x - 28$

$y^2 - 10y + 25 = \dfrac{1}{2}x - 3$

$(y - 5)^2 = \dfrac{1}{2}(x - 6)$

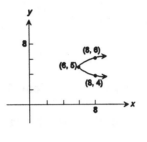

66. $x = \ln t, \; y = t^3$

$x = \ln y^{1/3}$

$x = \dfrac{1}{3}\ln y$

68. $x = t^{3/2}, \; y = 2t + 4$

$y = 2x^{2/3} + 4$

70. $x^2 - 4y^2 = 16$

$\dfrac{x^2}{16} - \dfrac{y^2}{4} = 1$

Vertices $(0, 4)(0, -4)$; foci $\left(0, \, 2\sqrt{5}\right), \left(0, \, -2\sqrt{5}\right)$

Thus, the ellipse with foci $(0, 4)$, $(0, -4)$ and vertices $\left(0, \, 2\sqrt{5}\right)$, $\left(0, \, -2\sqrt{5}\right)$ is $\dfrac{x^2}{20} + \dfrac{y^2}{4} = 1$

$(b^2 = 20 - 16 = 4)$

72. Let $P = (x, y)$ be a point in the plane, $Q = (5, 0)$ and $R = \left[\dfrac{16}{5}, 0\right]$

$d(P, Q) = \dfrac{5}{4} \cdot d(P, R)$. When $y = 0$,

$$\sqrt{(x - 5)^2 + y^2} = \dfrac{5}{4}\sqrt{\left[x - \dfrac{16}{5}\right]^2}$$

$$(x - 5)^2 + y^2 = \dfrac{25}{16}\left[x - \dfrac{16}{5}\right]^2$$

$$16(x^2 - 10x + 25) + 16y^2 = 25\left[x^2 - \dfrac{32}{5}x + \dfrac{256}{25}\right]$$

$$16y^2 - 9x^2 = -144$$

$$\dfrac{x^2}{16} - \dfrac{y^2}{9} = 1$$

This is a hyperbola, center at $(0, 0)$, transverse axis along the x-axis.

74. Situate the parabola so that its vertex is at $(0, 0)$ and it opens down. Then, we know
$$x^2 = 4ay$$
Since the bridge has a span of 60 feet, we know that it spans 30 feet on either side of the vertex. The maximum height of the parabolic arch, 20 feet, is also the height of the bridge 30 feet on either side of the vertex. Therefore, we know two points on the parabola, $(30, -20)$ and $(-30, -20)$. See figure.
Letting $x = 30$, $y = -20$, we find a:
$$(30)^2 = 4a(-20)$$
$$a = -11.25$$
Therefore, the equation of the parabola is
$$x^2 = -45y$$

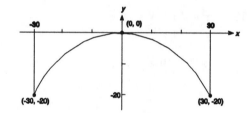

For $x = 5$ feet from the center, height $= 20 - 0.56 = 19.44$ feet
For $x = 10$ feet from the center, height $= 20 - 2.22 = 17.78$ feet
For $x = 20$ feet from the center, height $= 20 - 8.89 = 11.11$ feet

76. The major axis is 80 feet; therefore, $a = 40$. The minor axis is 50 feet; therefore, $b = 25$. Since $b^2 = a^2 - c^2$, $c^2 = a^2 - b^2$; so $c^2 = 40^2 - 25^2 = 975$, which implies $c \approx 31.2$ feet. Therefore, the foci are located $40 - 31.2 = 8.8$ feet in from each wall.

Mission Possible

1. a) Choose the top of the arch as the origin, or b) choose the point at water level below the top of the arch.

2. Then the equation will be of the form a) $y = ax^2$ or b) $y = ax^2 + 350$, with $a = -\dfrac{2}{1575}$ in either case.

3. A clearance of 280 feet would be represented by a y-value of a) -70 or b) 280. Corresponding x-values would be ± 234.8, so the width of the channel would be 469.6 feet.

4. Choose the point at water level below the top of the arch as the origin.

Chapter 11 Analytic Geometry

5. Then the equation of the ellipse would be $\dfrac{x^2}{525^2} + \dfrac{y^2}{350^2} = 1$. Solving for y gives

$y = 350 \cdot \sqrt{\left(1 - \dfrac{x^2}{525^2}\right)}$. Setting $y - 280$ gives corresponding x-values of $\pm\,315$, so the width of the channel would be 630 feet.

6. The ellipse definitely gives the wider channel at **every** height, but we may not be able to afford the expense of keeping the channel dredged to that width. On the other hand, with a great deal of channel traffic, the wider channel is important. In flood time, the ellipse continues to give the wider channel; 587.6' vs. 437.2' if the water level rises 10 feet.

SYSTEMS OF EQUATIONS AND INEQUALITIES

12.1 Systems of Linear Equations: Substitution; Elimination

2. $3(-2) + 2(4) = 2$ and
$(-2) - 7(4) = -30$

4. $2\left[\dfrac{-1}{2}\right] + \dfrac{1}{2}(2) = 0$ and

$3\left[\dfrac{-1}{2}\right] - 4(2) = \dfrac{-19}{2}$

6. $(-2)^2 - (-1)^2 = 3$ and
$(-2)(-1) = 2$

8. $\dfrac{2}{2-1} + 3 = 5$ and $3(2) - 3 = 3$

10. $4(2) - 1 = 7$ and $8(2) + 5(-3) - 1 = 0$
and $-2 - (-3) + 5(1) = 6$

12. $\begin{cases} x + 2y = 5 \\ x + y = 3 \end{cases}$
$x = 3 - y$
$\quad (3 - y) + 2y = 5$
$\qquad\qquad\qquad y = 2$
$x = 3 - 2 = 1$
The solution is $x = 1$, $y = 2$.

14. $\begin{cases} x + 3y = 5 \\ 2x - 3y = -8 \end{cases}$
$x = 5 - 3y$
$\quad 2(5 - 3y) - 3y = -8$
$\qquad 10 - 6y - 3y = -8$
$\qquad\qquad\qquad -9y = -18$
$\qquad\qquad\qquad\quad y = 2$
$x = 5 - 3(2) = -1$
The solution is $x = -1$, $y = 2$.

16. $\begin{cases} 4x + 5y = -3 \\ -2y = -4 \end{cases}$
$y = 2$
$\quad 4x + 5(2) = -3$
$\qquad\qquad 4x = -13$
$\qquad\qquad\quad x = \dfrac{-13}{4}$
The solution is $x = \dfrac{-13}{4}$, $y = 2$.

18. $\begin{cases} 2x + 4y = \dfrac{2}{3} \\ 3x - 5y = -10 \end{cases} \rightarrow \begin{cases} 6x + 12y = 2 \\ -6x + 10y = 20 \end{cases} \rightarrow \begin{cases} 6x + 12y = 2 \\ 22y = 22 \end{cases} \rightarrow \begin{cases} 6x + 12y = 2 \\ y = 1 \end{cases}$

$\rightarrow \begin{cases} 6x + 12(1) = 2 \\ y = 1 \end{cases} \rightarrow \begin{cases} 6x = -10 \\ y = 1 \end{cases} \rightarrow \begin{cases} x = \dfrac{-5}{3} \\ y = 1 \end{cases}$

The solution is $x = \dfrac{-5}{3}$, $y = 1$.

20. $\begin{cases} x - y = 5 \\ -3x + 3y = 2 \end{cases}$
$x = 5 + y$
$\quad -3(5 + y) + 3y = 2$
$\quad -15 - 3y + 3y = 2$
$\quad\qquad 0 \cdot y = 17$

The system is inconsistent.

22. $\begin{cases} 3x + 3y = -1 \\ 4x + y = \dfrac{8}{3} \end{cases} \rightarrow \begin{cases} 3x + 3y = -1 \\ -12x - 3y = -8 \end{cases} \rightarrow \begin{cases} 3x + 3y = -1 \\ -9x = -9 \end{cases} \rightarrow \begin{cases} 3x + 3y = -1 \\ x = 1 \end{cases} \rightarrow \begin{cases} y = -\dfrac{4}{3} \\ x = 1 \end{cases}$

The solution is $x = 1, y = -\dfrac{4}{3}$.

24. $\begin{cases} 3x - y = 7 \\ 9x - 3y = 21 \end{cases}$
$y = 3x - 7$
$\quad 9x - 3(3x - 7) = 21$
$\quad 9x - 9x + 21 = 21$
$\quad\qquad\qquad 21 = 21$

The solutions are $y = 3x - 7$, where x can be any real number. In other words, the system has infinitely many solutions.

26. $\begin{cases} 3x - 2y = 0 \\ 5x + 10y = 4 \end{cases} \rightarrow \begin{cases} 15x - 10y = 0 \\ 5x + 10y = 4 \end{cases} \rightarrow \begin{cases} 20x = 4 \\ 5x + 10y = 4 \end{cases}$

$\rightarrow \begin{cases} x = \dfrac{1}{5} \\ 5\left(\dfrac{1}{5}\right) + 10y = 4 \end{cases} \rightarrow \begin{cases} x = \dfrac{1}{5} \\ 10y = 3 \end{cases} \rightarrow \begin{cases} x = \dfrac{1}{5} \\ y = \dfrac{3}{10} \end{cases}$

The solution is $x = \dfrac{1}{5}, y = \dfrac{3}{10}$.

28. $\begin{cases} \dfrac{1}{2}x + y = -2 \\ x - 2y = 8 \end{cases}$
$x = 8 + 2y$
$\quad \dfrac{1}{2}(8 + 2y) + y = -2$
$\quad\qquad 4 + y + y = -2$
$\quad\qquad\qquad 2y = -6$
$\quad\qquad\qquad\quad y = -3$
$x = 8 + 2(-3) = 2$

The solution is $x = 2, y = -3$.

30. $\begin{cases} \dfrac{1}{3}x - \dfrac{3}{2}y = -5 \\ \dfrac{3}{4}x + \dfrac{1}{3}y = 11 \end{cases}$ $\rightarrow$ $\begin{cases} 2x - 9y = -30 \\ 9x + 4y = 132 \end{cases}$ $\rightarrow$ $\begin{cases} -18x + 81y = 270 \\ 18x + 8y = 264 \end{cases}$ $\rightarrow$ $\begin{cases} -18x + 81y = 270 \\ 89y = 534 \end{cases}$

$\rightarrow$ $\begin{cases} -18x + 81y = 270 \\ y = 6 \end{cases}$ $\rightarrow$ $\begin{cases} -18x + 81(6) = 270 \\ y = 6 \end{cases}$ $\rightarrow$ $\begin{cases} x = 12 \\ y = 6 \end{cases}$

The system is $x = 12$, $y = 6$.

32. $\begin{cases} 2x - y = -1 \\ x + \dfrac{1}{2}y = \dfrac{3}{2} \end{cases}$

$y = 2x + 1$

$x + \dfrac{1}{2}(2x + 1) = \dfrac{3}{2}$

$x + x + \dfrac{1}{2} = \dfrac{3}{2}$

$2x = 1$

$x = \dfrac{1}{2}$

$y = 2\left[\dfrac{1}{2}\right] + 1 = 2$

The solution is $x = \dfrac{1}{2}$, $y = 2$.

34. $\begin{cases} 2x + y = -4 \\ -2y + 4z = 0 \\ 3x - 2z = -11 \end{cases}$ $\rightarrow$ $\begin{cases} 6x + 3y = -12 \\ -2y + 4z = 0 \\ -6x + 4z = 22 \end{cases}$ $\rightarrow$ $\begin{cases} 6x + 3y = -12 \\ -2y + 4z = 0 \\ 3y + 4z = 10 \end{cases}$

$\rightarrow$ $\begin{cases} 6x + 3y = -12 \\ -5y = -10 \\ 3y + 4z = 10 \end{cases}$ $\rightarrow$ $\begin{cases} 6x + 3(2) = -12 \\ y = 2 \\ 3(2) + 4z = 10 \end{cases}$ $\rightarrow$ $\begin{cases} x = -3 \\ y = 2 \\ z = 1 \end{cases}$

The solution is $x = -3$, $y = 2$, $z = 1$.

36. $\begin{cases} 2x + y - 3z = 0 \\ -2x + 2y + z = -7 \\ 3x - 4y - 3z = 7 \end{cases}$

$y = -2x + 3z$

$\begin{cases} -2x + 2(-2x + 3z) + z = -7 \\ 3x - 4(-2x + 3z) - 3z = 7 \end{cases}$

$\begin{cases} -6x + 7z = -7 \\ 11x - 15z = 7 \end{cases}$

$7z = -7 + 6x$

$z = -1 + \dfrac{6}{7}x$

$11x - 15\left[-1 + \dfrac{6}{7}x\right] = 7$

$11x + 15 - \dfrac{90}{7}x = 7$

$\dfrac{-13}{7}x = -8$

$x = \dfrac{56}{13}$

$z = -1 + \dfrac{6}{7}\left[\dfrac{56}{13}\right] = \dfrac{35}{13}$

$y = -2\left[\dfrac{56}{13}\right] + 3\left[\dfrac{35}{13}\right] = \dfrac{-7}{13}$

The solution is $x = \dfrac{56}{13}$, $y = \dfrac{-7}{13}$, $z = \dfrac{35}{13}$.

38. $\begin{cases} 2x - 3y - z = 0 \\ -x + 2y + z = 5 \\ 3x - 4y - z = 1 \end{cases}$

$z = 2x - 3y$

$\begin{cases} -x + 2y + (2x - 3y) = 5 \\ 3x - 4y - (2x - 3y) = 1 \end{cases}$

$\begin{cases} x - y = 5 \\ x - y = 1 \end{cases}$

$x = 5 + y$

$5 + y - y = 0$

$0 \cdot y = -5$

The system is inconsistent.

40. $\begin{cases} 2x - 3y - z = 0 \\ 3x + 2y + 2z = 2 \\ x + 5y + 3z = 2 \end{cases}$

$x = 2 - 5y - 3z$

$\begin{cases} 2(2 - 5y - 3z) - 3y - z = 0 \\ 3(2 - 5y - 3z) + 2y + 2z = 2 \end{cases}$

$\begin{cases} -13y - 7z = -4 \\ -13y - 7z = -4 \end{cases}$

Thus, the solution can be written:

$$y = \frac{-7}{13}z + \frac{4}{13}$$

$$x = \frac{-4}{13}z + \frac{6}{13}$$

where z is any real number.

42. $\begin{cases} 3x - 2y + 2z = 6 \\ 7x - 3y + 2z = -1 \\ 2x - 3y + 4z = 0 \end{cases} \rightarrow \begin{cases} 6x - 4y + 4z = 12 \\ 7x - 3y + 2z = -1 \\ -6x + 9y - 12z = 0 \end{cases} \rightarrow \begin{cases} 6x - 4y + 4z = 12 \\ 7x - 3y + 2z = -1 \\ 5y - 8z = 12 \end{cases}$

$\rightarrow \begin{cases} 42x - 28y + 28z = 84 \\ -42x + 18y - 12z = 6 \\ 5y - 8z = 12 \end{cases} \rightarrow \begin{cases} 42x - 28y + 28z = 84 \\ -10y + 16z = 90 \\ 5y - 8z = 12 \end{cases}$

$\rightarrow \begin{cases} 42x - 28y + 28z = 84 \\ -10y + 16z = 90 \\ 10y - 16z = 24 \end{cases} \rightarrow \begin{cases} 42x - 28y + 28z = 84 \\ -10y + 16z = 90 \\ 0 = 114 \end{cases}$

The system is inconsistent.

44. $\begin{cases} x - y + z = -4 \\ 2x - 3y + 4z = -15 \\ 5x + y - 2z = 12 \end{cases} \rightarrow \begin{cases} -2x + 2y - 2z = 8 \\ 2x - 3y + 4z = -15 \\ 5x + y - 2z = 12 \end{cases} \rightarrow \begin{cases} -2x + 2y - 2z = 8 \\ -y + 2z = -7 \\ 5x + y - 2z = 12 \end{cases}$

$\rightarrow \begin{cases} -10x + 10y - 10z = 40 \\ -y + 2z = -7 \\ 10x + 2y - 4z = 24 \end{cases} \rightarrow \begin{cases} -10x + 10y - 10z = 40 \\ -y + 2z = -7 \\ 12y - 14z = 64 \end{cases}$

$\rightarrow \begin{cases} -10x + 10y - 10z = 40 \\ -12y + 24z = -84 \\ 12y - 14z = 64 \end{cases} \rightarrow \begin{cases} -10x + 10y - 10z = 40 \\ -12y + 24z = -84 \\ 10z = -20 \end{cases}$

$\rightarrow \begin{cases} -10x + 10y - 10z = 40 \\ -12y + 24z = -84 \\ z = -2 \end{cases} \rightarrow \begin{cases} -10x + 10y - 10(-2) = 40 \\ -12y + 24(-2) = -84 \\ z = -2 \end{cases}$

$\rightarrow \begin{cases} -10x + 10y = 20 \\ -12y = -36 \\ z = -2 \end{cases} \rightarrow \begin{cases} -10x + 10y = 20 \\ y = 3 \\ z = -2 \end{cases} \rightarrow \begin{cases} -10x = -10 \\ y = 3 \\ z = -2 \end{cases} \rightarrow \begin{cases} x = 1 \\ y = 3 \\ z = -2 \end{cases}$

The solution is $x = 1$, $y = 3$, $z = -2$.

46.
$$\begin{cases} x + 4y - 3z = -8 \\ 3x - y + 3z = 12 \\ x + y + 6z = 1 \end{cases}$$
$$x = -8 - 4y + 3z$$
$$\begin{cases} 3(-8 - 4y + 3z) - y + 3z = 12 \\ (-8 - 4y + 3z) + y + 6z = 1 \end{cases}$$
$$\begin{cases} -13y + 12z = 36 \\ -3y + 9z = 9 \end{cases}$$
$$-3y = 9 - 9z$$
$$y = -3 + 3z$$
$$-13(-3 + 3z) + 12z = 36$$
$$-27z = -3$$
$$z = \frac{1}{9}$$
$$y = -3 + 3\left[\frac{1}{9}\right] = -\frac{8}{3}$$
$$x = -8 - 4\left[\frac{-8}{3}\right] + 3\left[\frac{1}{9}\right] = 3$$

The solution is $x = 3$, $y = \dfrac{-8}{3}$, $z = \dfrac{1}{9}$.

48.
$$\begin{cases} \dfrac{4}{x} - \dfrac{3}{y} = 0 \\ \dfrac{6}{x} + \dfrac{3}{2y} = 2 \end{cases}$$
Let $u = \dfrac{1}{x}$ and $v = \dfrac{1}{y}$, then
$$\begin{cases} 4u - 3v = 0 \\ 6u + \dfrac{3}{2}v = 2 \end{cases}$$
$$4u = 3v$$
$$u = \frac{3}{4}v$$
$$6\left[\frac{3}{4}v\right] + \frac{3}{2}v = 2$$
$$6v = 2$$
$$v = \frac{2}{6} = \frac{1}{3}$$
$$u = \frac{3}{4}\left[\frac{1}{3}\right] = \frac{1}{4}$$
Thus, $\quad x = \dfrac{1}{\frac{1}{4}} = 4$
$$y = \frac{1}{\frac{1}{3}} = 3$$

The solution is $x = 4$, $y = 3$.

50.
$$\begin{cases} y = -\sqrt{3}x + 100 \\ y = 0.2x + \sqrt{19} \end{cases}$$

Using the ZOOM and TRACE features, we approximate the point of intersection at (49.50, 14.25).

52.
$$\begin{cases} \sqrt{5}x - \sqrt{6}y + 60 = 0 \\ 0.2x + 0.3y + \sqrt{5} = 0 \end{cases}$$

Using the ZOOM and TRACE features, we approximate the point of intersection at (−20.22, 6.03).

54.
$$\begin{cases} \sqrt{6}x - \sqrt{5}y + \sqrt{1.1} = 0 \\ y = -0.2x + 0.1 \end{cases}$$

Using the ZOOM and TRACE feathers, we approximate the point of intersection at (−0.28, 0.15).

56.
$$\begin{cases} x - y = 40 \\ 6y - x = 5 \end{cases}$$
$$x - y = 40$$
$$-x + 6y = 5$$
$$5y = 45$$
$$y = 9$$
$$x - 9 = 40$$
$$x = 49$$

The two numbers are 49 and 9.

58. Let ℓ = length and w = width, then
$$\begin{cases} 2\ell + 2w = 3000 \\ \ell - w = 50 \end{cases}$$
$$\ell = 50 + w$$
$$2(50 + w) + 2w = 3000$$
$$100 + 4w = 3000$$
$$4w = 2900$$
$$w = 725$$
$$\ell = 50 + 725 = 775$$

The length is 775 meters and the width is 725 meters.

Chapter 12 Systems of Equations and Inequalities

60. Let x be the number of adults who paid and y be the number of senior citizens who paid.

$$\begin{cases} x + y = 325 \\ 9x + 7y = 2495 \end{cases}$$

$x = 325 - y$

$9(325 - y) + 7y = 2495$

$\qquad\qquad -2y = -430$

$\qquad\qquad\quad y = 215$

$x = 325 - 215 = 110$

Thus, 110 were adults and 215 were senior citizens.

62. Let w be the wind speed, then

$2(150 + w) = 3(150 - w)$

$300 + 2w = 450 - 3w$

$\qquad 5w = 150$

$\qquad\; w = 30$

The wind speed is 30 miles per hour.

64. Let x be amount invested in AA bonds and let y be amount invested in a bank certificate.

(a)
$$\begin{cases} x + y = 150,000 \\ .10x + .05y = 12,000 \end{cases}$$

$x = 150,000 - y$

$10(150,000 - y) + 5y = 1,200,000$

$1,500,000 - 10y + 5y = 1,200,000$

$\qquad\qquad\qquad -5y = -300,000$

$\qquad\qquad\qquad\quad y = 60,000$

$x = 150,000 - 60,000 = 90,000$

Thus, the retired couple should invest $90,000 in AA bonds and $60,000 in a bank certificate to realize exactly $12,000.

(b)
$$\begin{cases} x + y = 150,000 \\ .10x + .05y = 14,000 \end{cases}$$

$x = 150,000 - y$

$\qquad\qquad 10x + 5y = 1,400,000$

$10(150,000 - y) + 5y = 1,400,000$

$1,500,000 - 10y + 5y = 1,400,000$

$\qquad\qquad\qquad -5y = -100,000$

$\qquad\qquad\qquad\quad y = 20,000$

$x = 150,000 - 20,000 = 130,000$

Thus, the retired couple should invest $130,000 in AA bonds and $20,000 in a bank certificate to realize exactly $14,000 in 2 years.

66. Let x be the cost of a hot dog and y be the cost of a soft drink, then

$$\begin{cases} 10x + 5y = 12.50 \\ 7x + 4y = 9.00 \end{cases} \rightarrow \begin{cases} 40x + 20y = 50 \\ -35x - 20y = -45 \end{cases}$$

$5x = 5$

$x = 1$

Thus, $\qquad 7(1) + 4y = 9$

$\qquad\qquad\quad 4y = 2$

$\qquad\qquad\quad\; y = \dfrac{1}{2} = .50$

Thus, the hot dog costs $1.00 and the soft drink costs $.50.

68. Let x be the speed of the swimmer and let y be the speed of the current, then

$$\begin{cases} (x + y)3 = 15 \\ (x - y)5 = 15 \end{cases} \rightarrow \begin{cases} 3x + 3y = 15 \\ 5x - 5y = 15 \end{cases} \rightarrow \begin{cases} 15x + 15y = 75 \\ 15x - 15y = 45 \end{cases} \rightarrow \begin{cases} 15x + 15y = 75 \\ 30x = 120 \end{cases}$$

$$\rightarrow \begin{cases} 15x + 15y = 75 \\ x = 4 \end{cases} \rightarrow \begin{cases} 15(4) + 15y = 75 \\ x = 4 \end{cases} \rightarrow \begin{cases} y = 1 \\ x = 4 \end{cases}$$

The average speed of the swimmer is 4 miles per hour and the speed of the current is 1 mile per hour.

70. Let x be the number of nickels, y be the number of dimes and z be the number of quarters, then

$$\begin{cases} x + y + z = 37 \\ 5x + 10y + 25z = 325 \\ y = x + 5 \end{cases} \rightarrow \begin{cases} -5x - 5y - 5z = -185 \\ 5x + 10y + 25z = 325 \\ -x + y = 5 \end{cases} \rightarrow \begin{cases} -5x - 5y - 5z = -185 \\ 5y + 20z = 140 \\ -x + y = 5 \end{cases}$$

$$\rightarrow \begin{cases} -5x - 5y - 5z = -185 \\ 5y + 20z = 140 \\ 5x - 5y = -25 \end{cases} \rightarrow \begin{cases} -5x - 5y - 5z = -185 \\ 5y + 20z = 140 \\ -10y - 5z = -210 \end{cases} \rightarrow \begin{cases} -5x - 5y - 5z = -185 \\ 10y + 40z = 280 \\ -10y - 5z = -210 \end{cases}$$

$$\rightarrow \begin{cases} -5x - 5y - 5z = -185 \\ 10y + 40z = 280 \\ 35z = 70 \end{cases} \rightarrow \begin{cases} x = 15 \\ y = 20 \\ z = 2 \end{cases}$$

72.

$$\begin{cases} I_3 = I_1 + I_2 \\ 8 = 4(I_1 + I_2) + 6I_2 \\ 8I_1 - 6I_2 = 4 \end{cases}$$

$$\begin{cases} I_3 = I_1 + I_2 \\ 4I_1 + 10I_2 = 8 \\ 8I_1 - 6I_2 = 4 \end{cases}$$

$$\begin{cases} I_3 = I_1 + I_2 \\ 8I_1 + 20I_2 = 16 \\ -8I_1 + 6I_2 = -4 \end{cases}$$

$$26I_2 = 12$$

$$I_2 = \frac{12}{26} = \frac{6}{13}$$

$$4I_1 + 10\left[\frac{6}{13}\right] = 8$$

$$4I_1 = 8 - \frac{60}{13} = \frac{104 - 60}{13} = \frac{44}{13}$$

$$I_1 = \frac{11}{13}$$

$$I_3 = I_1 + I_2 = \frac{4}{13} + \frac{6}{13} = \frac{17}{13}$$

74. Let x = number of 2-person work stations, and y = number of 3-person work stations.
Then: $\quad 2x + 3y = 38$ (The lab can be used by 38 students at one time.)
$\qquad\quad x + y = 16$ (The total number of work stations is 16.)
Now solve for x and y:

$$\begin{cases} 2x + 3y = 38 & (1) \\ x + y = 16 & (2) \end{cases}$$

$$\begin{cases} 2x + 3y = 38 & (1) \\ -2x - 2y = -32 & (2) \quad \text{Multiply both sides by } -2. \end{cases}$$

$$\begin{cases} 2x + 3y = 38 & (1) \\ y = 6 & (2) \quad \text{Replace (2) by (1) + (2).} \end{cases}$$

$$\begin{cases} 2x + 18 = 38 & (1) \quad \text{Back-substitute; } y = 6. \\ y = 6 & (2) \end{cases}$$

$$\begin{cases} x = 10 & (1) \\ y = 6 & (2) \end{cases}$$

Therefore, the lab contains $x = 10$ work stations for 2 students, and $y = 6$ work stations for 3 students.

78. $y = x^2 + bx + c$

passing through $(1, 3)$: $3 = 1 + b + c$
passing through $(3, 5)$: $5 = 9 + 3b + c$

$$\begin{cases} b + c = 2 & (1) \\ 3b + c = -4 & (2) \end{cases}$$

$$\begin{cases} c = 2 - b & (1) \quad \text{Solve for } c. \\ 3b + c = -4 & (2) \end{cases}$$

$$\begin{cases} c = 2 - b & (1) \\ 3b + (2 - b) = -4 & (2) \quad \text{Substitute.} \end{cases}$$

$$\begin{cases} c = 2 - b & (1) \\ b = -3 & (2) \quad \text{Solve for } b. \end{cases}$$

$$\begin{cases} c = 5 & (1) \quad \text{Substitute and solve.} \\ b = -3 & (2) \end{cases}$$

The equation for the parabola is $y = x^2 - 3x + 5$, so $b = -3, c = 5$.

80.　$y = x^2 + bx + c$

$$\begin{cases} x_1^2 + bx_1 + c = y_1 \\ x_2^2 + bx_2 + c = y_2 \end{cases}$$

$$c = y_1 - x_1^2 - bx_1$$

$$x_2^2 + bx_2 + y_1 - x_1^2 - bx_1 = y_2$$

$$b(x_2 - x_1) = y_2 - y_1 + x_1^2 - x_2^2$$

$$b = \frac{y_2 - y_1 + x_1^2 - x_2^2}{(x_2 - x_1)}$$

$$c = y_1 - x_1^2 - \frac{(y_2 - y_1 + x_1^2 - x_2^2)}{(x_2 - x_1)} x_1$$

82.　$y = ax^2 + bx + c$

　　passing through $(-1, 2)$　　$2 = a - b + c$
　　passing through $(1, 4)$　　$4 = a + b + c$
　　passing through $(2, 4)$　　$4 = 4a + 2b + c$

$$\begin{cases} a - b + c = 2 & (1) \\ a + b + c = 4 & (2) \\ 4a + 2b + c = 4 & (3) \end{cases}$$

$$\begin{cases} a - b + c = 2 & (1) \\ 2a + 2c = 6 & \text{Replacing (2) by (1) + (2).} \\ 4a + 2b + c = 4 & (3) \end{cases}$$

$$\begin{cases} 2a - 2b + 2c = 4 & \text{Multiplying (1) by 2.} \\ 2a + 2c = 6 & (2) \\ 4a + 2b + c = 4 & (3) \end{cases}$$

$$\begin{cases} 2a - 2b + 2c = 4 & (1) \\ 2a + 2c = 6 & (2) \\ 6a + 3c = 8 & \text{Add (1) and (3).} \end{cases}$$

$$\begin{cases} 2a - 2b + 2c = 4 & (1) \\ a = -c + 3 & (2) \quad \text{Solve for } a. \\ 6a + 3c = 8 & (3) \end{cases}$$

$$\begin{cases} 2a - 2b + 2c = 4 & (1) \\ a = -c + 3 & (2) \\ 6(-c + 3) + 3c = 8 & (3) \quad \text{Substitute (2) into (3).} \end{cases}$$

$$\begin{cases} 2a - 2b + 2c = 4 & (1) \\ a = -c + 3 & (2) \\ c = \dfrac{10}{3} & (3) \quad \text{Solve for } c. \end{cases}$$

$$\begin{cases} 2a - 2b + 2c = 4 & (1) \\ a = \dfrac{-1}{3} & (2) \quad \text{Solve for } a. \\ c = \dfrac{10}{3} & (3) \end{cases}$$

$$\begin{cases} b = 1 & (1) \quad \text{Substitute for } a. \\ a = \dfrac{-1}{3} & (2) \\ c = \dfrac{10}{3} & (3) \end{cases}$$

The equation is $y = \dfrac{-1}{3}x^2 + x + \dfrac{10}{3}$, so $a = -\dfrac{1}{3}$, $b = 1$, $c = \dfrac{10}{3}$.

84. $y = m_1 x + b_1$
$y = m_2 x + b_2$
where $m_1 = m_2 = m$ and $b_1 \neq b_2$
$mx + b_1 = mx + b_2 \Rightarrow b_1 = b_2$
Thus, if there were a solution $b_1 = b_2$, since $b_1 \neq b_2$ there is no solution.

12.2 Systems of Linear Equations: Matrices

2. $\begin{bmatrix} 3 & 4 & | & 7 \\ 4 & -2 & | & 5 \end{bmatrix}$

4. $\begin{bmatrix} 9 & -1 & | & 0 \\ 3 & -1 & | & 4 \end{bmatrix}$

6. $\begin{bmatrix} \dfrac{4}{3} & \dfrac{-3}{2} & | & \dfrac{3}{4} \\ \dfrac{-1}{4} & \dfrac{1}{3} & | & \dfrac{2}{3} \end{bmatrix}$

8. $\begin{bmatrix} 5 & -1 & -1 & | & 0 \\ 1 & 1 & 0 & | & 5 \\ 2 & 0 & -3 & | & 2 \end{bmatrix}$

10. $\begin{bmatrix} 2 & 3 & -4 & | & 0 \\ 1 & 0 & -5 & | & -2 \end{bmatrix}$

12. $\begin{bmatrix} 1 & -3 & -3 & | & -3 \\ 2 & -5 & 2 & | & -4 \\ -3 & 2 & 4 & | & 6 \end{bmatrix} \rightarrow \begin{bmatrix} 1 & -3 & -3 & | & -3 \\ 0 & 1 & 8 & | & 2 \\ -3 & 2 & 4 & | & 6 \end{bmatrix} \rightarrow \begin{bmatrix} 1 & -3 & -3 & | & -3 \\ 0 & 1 & 8 & | & 2 \\ 0 & -7 & -5 & | & -3 \end{bmatrix} \rightarrow \begin{bmatrix} 1 & -3 & -3 & | & -3 \\ 0 & 1 & 8 & | & 2 \\ 0 & 0 & 51 & | & 11 \end{bmatrix}$

(a) $R_2 = -2r_1 + r_2$ (b) $R_3 = 3r_1 + r_3$ (c) $R_3 = 7r_2 + r_3$

14. $\begin{bmatrix} 1 & -3 & 3 & | & -5 \\ 2 & -5 & -3 & | & -5 \\ -3 & -2 & 4 & | & 6 \end{bmatrix} \rightarrow \begin{bmatrix} 1 & -3 & 3 & | & -5 \\ 0 & 1 & -9 & | & 5 \\ -3 & -2 & 4 & | & 6 \end{bmatrix} \rightarrow \begin{bmatrix} 1 & -3 & 3 & | & -5 \\ 0 & 1 & -9 & | & 5 \\ 0 & -11 & 13 & | & -9 \end{bmatrix} \rightarrow \begin{bmatrix} 1 & -3 & 3 & | & -5 \\ 0 & 1 & -9 & | & 5 \\ 0 & 0 & -86 & | & 46 \end{bmatrix}$

          (a) $R_2 = -2r_1 + r_2$ (b) $R_3 = 3r_1 + r_3$ (c) $R_3 = 11r_2 + r_3$

16. $\begin{bmatrix} 1 & -3 & -4 & | & -6 \\ 2 & -5 & 6 & | & -6 \\ -3 & 1 & 4 & | & 6 \end{bmatrix} \rightarrow \begin{bmatrix} 1 & -3 & -4 & | & -6 \\ 0 & 1 & 14 & | & 6 \\ -3 & 1 & 4 & | & 6 \end{bmatrix} \rightarrow \begin{bmatrix} 1 & -3 & -4 & | & -6 \\ 0 & 1 & 14 & | & 6 \\ 0 & -8 & -8 & | & -12 \end{bmatrix} \rightarrow \begin{bmatrix} 1 & -3 & -4 & | & -6 \\ 0 & 1 & 14 & | & 6 \\ 0 & 0 & 104 & | & 36 \end{bmatrix}$

          (a) $R_2 = -2r_1 + r_2$ (b) $R_3 = 3r_1 + r_3$ (c) $R_3 = 8r_2 + r_3$

18. $\begin{bmatrix} 1 & -3 & -1 & | & 2 \\ 2 & -5 & 2 & | & 6 \\ -3 & -6 & 4 & | & 6 \end{bmatrix} \rightarrow \begin{bmatrix} 1 & -3 & -1 & | & 2 \\ 0 & 1 & 4 & | & 2 \\ -3 & -6 & 4 & | & 6 \end{bmatrix} \rightarrow \begin{bmatrix} 1 & -3 & -1 & | & 2 \\ 0 & 1 & 4 & | & 2 \\ 0 & -15 & 1 & | & 12 \end{bmatrix} \rightarrow \begin{bmatrix} 1 & -3 & -1 & | & 2 \\ 0 & 1 & 4 & | & 2 \\ 0 & 0 & 61 & | & 42 \end{bmatrix}$

          (a) $R_2 = -2r_1 + r_2$ (b) $R_3 = 3r_1 + r_3$ (c) $R_3 = 15r_2 + r_3$

20. $\begin{bmatrix} 1 & -3 & 5 & | & -3 \\ 2 & -5 & 1 & | & -4 \\ -3 & 3 & 4 & | & 6 \end{bmatrix} \rightarrow \begin{bmatrix} 1 & -3 & 5 & | & -3 \\ 0 & 1 & -9 & | & 2 \\ -3 & 3 & 4 & | & 6 \end{bmatrix} \rightarrow \begin{bmatrix} 1 & -3 & 5 & | & -3 \\ 0 & 1 & -9 & | & 2 \\ 0 & -6 & 19 & | & -3 \end{bmatrix} \rightarrow \begin{bmatrix} 1 & -3 & 5 & | & -3 \\ 0 & 1 & -9 & | & 2 \\ 0 & 0 & -35 & | & 9 \end{bmatrix}$

          (a) $R_2 = -2r_1 + r_2$ (b) $R_3 = 3r_1 + r_3$ (c) $R_3 = 6r_2 + r_3$

22. $\begin{cases} x = -4 \\ y = 0 \end{cases}$ consistent $x = -4, y = 0$

24. $\begin{cases} x = 0 \\ y = 0 \\ 0 = 2 \end{cases}$ inconsistent

26. $\begin{cases} x + 4z = 4 \\ y + 3z = 2 \\ 0 = 0 \end{cases}$ consistent $x = 4 - 4z, y = 2 - 3z$ z any real number

28. $\begin{cases} x_1 = 1 \\ x_2 + 2x_4 = 2 \\ x_3 + 3x_4 = 0 \end{cases}$ consistent $x_1 = 1, x_2 = 2 - 2x_4, x_3 = -3x_4$ x_4 any real number

30. $\begin{cases} x_1 = 1 \\ x_2 = 2 \\ x_3 + 2x_4 = 3 \end{cases}$ consistent $x_1 = 1, x_2 = 2, x_3 = 3 - 2x_4$ x_4 any real number

32. $\begin{bmatrix} 1 & 2 & | & 5 \\ 1 & 1 & | & 3 \end{bmatrix} \rightarrow \begin{bmatrix} 1 & 2 & | & 5 \\ 0 & -1 & | & -2 \end{bmatrix} \rightarrow \begin{bmatrix} 1 & 2 & | & 5 \\ 0 & 1 & | & 2 \end{bmatrix} \rightarrow \begin{bmatrix} 1 & 0 & | & 1 \\ 0 & 1 & | & 2 \end{bmatrix}$

$\qquad\qquad\quad \uparrow \qquad\qquad\quad \uparrow \qquad\qquad\quad \uparrow$

$\qquad\quad R_2 = -r_1 + r_2 \quad R_2 = -r_2 \quad R_1 = -2r_2 + r_1$

The solution is $x = 1, y = 2$.

34. $\begin{bmatrix} 1 & 3 & | & 5 \\ 2 & -3 & | & -8 \end{bmatrix} \rightarrow \begin{bmatrix} 1 & 3 & | & 5 \\ 0 & -9 & | & -18 \end{bmatrix} \rightarrow \begin{bmatrix} 1 & 3 & | & 5 \\ 0 & 1 & | & 2 \end{bmatrix} \rightarrow \begin{bmatrix} 1 & 0 & | & 1 \\ 0 & 1 & | & 2 \end{bmatrix}$

$\qquad\quad \uparrow \qquad\qquad\qquad\quad \uparrow \qquad\qquad\qquad \uparrow$

$\qquad\quad R_2 = -2r_1 + r_2 \quad R_2 = \dfrac{-1}{9}r_2 \quad R_1 = -3r_2 + r_1$

The solution is $x = -1, y = 2$.

36. $\begin{bmatrix} 2 & 4 & | & 16 \\ 3 & -5 & | & -9 \end{bmatrix} \rightarrow \begin{bmatrix} 1 & 2 & | & 8 \\ 3 & -5 & | & -9 \end{bmatrix} \rightarrow \begin{bmatrix} 1 & 2 & | & 8 \\ 0 & -11 & | & -33 \end{bmatrix} \rightarrow \begin{bmatrix} 1 & 2 & | & 8 \\ 0 & 1 & | & 3 \end{bmatrix} \rightarrow \begin{bmatrix} 1 & 0 & | & 2 \\ 0 & 1 & | & 3 \end{bmatrix}$

$\qquad\qquad \uparrow \qquad\qquad\qquad \uparrow \qquad\qquad\qquad\quad \uparrow \qquad\qquad\quad \uparrow$

$\qquad\quad R_1 = \dfrac{1}{2}r_1 \qquad R_2 = -3r_1 + r_2 \quad R_2 = \dfrac{-1}{11}r_2 \quad R_1 = -2r_2 + r_1$

The solution is $x = 2, y = 3$.

38. $\begin{bmatrix} 1 & -1 & | & 5 \\ -3 & 3 & | & 2 \end{bmatrix} \rightarrow \begin{bmatrix} 1 & -1 & | & 5 \\ 0 & 0 & | & 17 \end{bmatrix}$

$\qquad\qquad\qquad \uparrow$

$\qquad\qquad R_2 = 3r_1 + r_2$

The system is inconsistent.

40. $\begin{bmatrix} 3 & 3 & | & 3 \\ 4 & 2 & | & \dfrac{8}{3} \end{bmatrix} \rightarrow \begin{bmatrix} 1 & 1 & | & 1 \\ 4 & 2 & | & \dfrac{8}{3} \end{bmatrix} \rightarrow \begin{bmatrix} 1 & 1 & | & 1 \\ 0 & -2 & | & \dfrac{-4}{3} \end{bmatrix} \rightarrow \begin{bmatrix} 1 & 1 & | & 1 \\ 0 & 1 & | & \dfrac{2}{3} \end{bmatrix} \rightarrow \begin{bmatrix} 1 & 0 & | & \dfrac{1}{3} \\ 0 & 1 & | & \dfrac{2}{3} \end{bmatrix}$

$\qquad\qquad \uparrow \qquad\qquad\qquad \uparrow \qquad\qquad\qquad \uparrow \qquad\qquad\quad \uparrow$

$\qquad\quad R_1 = \dfrac{1}{3}r_1 \qquad R_2 = -4r_1 + r_2 \quad R_2 = \dfrac{-1}{2}r_2 \qquad R_1 = -r_2 + r_1$

The solution is $x = \dfrac{1}{3}, y = \dfrac{2}{3}$.

42. $\begin{bmatrix} 3 & -1 & | & 7 \\ 9 & -3 & | & 21 \end{bmatrix} \rightarrow \begin{bmatrix} 1 & \dfrac{-1}{3} & | & \dfrac{7}{3} \\ 9 & -3 & | & 21 \end{bmatrix} \rightarrow \begin{bmatrix} 1 & \dfrac{-1}{3} & | & \dfrac{7}{3} \\ 0 & 0 & | & 0 \end{bmatrix}$

$\qquad\qquad \uparrow \qquad\qquad\qquad\qquad \uparrow$

$\qquad\quad R_1 = \dfrac{1}{3}r_1 \qquad\qquad R_2 = -9r_1 + r_2$

We write the solutions as $y = 3x - 7$, where x can be any real number or $x = \dfrac{1}{3}y + \dfrac{7}{3}$, where y can be any real number.

44. $\begin{bmatrix} \dfrac{1}{2} & 1 & \bigm| & -2 \\ 1 & -2 & \bigm| & 8 \end{bmatrix} \rightarrow \begin{bmatrix} 1 & -2 & \bigm| & 8 \\ \dfrac{1}{2} & 1 & \bigm| & -2 \end{bmatrix} \rightarrow \begin{bmatrix} 1 & -2 & \bigm| & 8 \\ 0 & 2 & \bigm| & -6 \end{bmatrix} \rightarrow \begin{bmatrix} 1 & -2 & \bigm| & 8 \\ 0 & 1 & \bigm| & -3 \end{bmatrix} \rightarrow \begin{bmatrix} 1 & 0 & \bigm| & 2 \\ 0 & 1 & \bigm| & -3 \end{bmatrix}$

$\qquad\qquad$ Interchange r_1 and r_2 $\qquad R_2 = \dfrac{-1}{2}r_1 + r_2$ $\quad R_2 = \dfrac{1}{2}r_2$ $\qquad R_1 = 2r_2 + r_1$

The solution is $x = 2$, $y = -3$.

46. $\begin{bmatrix} 2 & -1 & \bigm| & -1 \\ 1 & \dfrac{1}{2} & \bigm| & \dfrac{3}{2} \end{bmatrix} \rightarrow \begin{bmatrix} 1 & \dfrac{1}{2} & \bigm| & \dfrac{3}{2} \\ 2 & -1 & \bigm| & -1 \end{bmatrix} \rightarrow \begin{bmatrix} 1 & \dfrac{1}{2} & \bigm| & \dfrac{3}{2} \\ 0 & -2 & \bigm| & -4 \end{bmatrix} \rightarrow \begin{bmatrix} 1 & \dfrac{1}{2} & \bigm| & \dfrac{3}{2} \\ 0 & 1 & \bigm| & 2 \end{bmatrix} \rightarrow \begin{bmatrix} 1 & 0 & \bigm| & \dfrac{1}{2} \\ 0 & 1 & \bigm| & 2 \end{bmatrix}$

$\qquad\qquad$ Interchange r_1 and r_2 $\qquad R_2 = -2r_1 + r_2$ $\quad R_2 = \dfrac{-1}{2}r_2$ $\qquad R_1 = \dfrac{-1}{2}r_2 + r_1$

The solution is $x = \dfrac{1}{2}$, $y = 2$.

48. $\begin{bmatrix} 2 & 1 & 0 & \bigm| & -4 \\ 0 & -2 & 4 & \bigm| & 0 \\ 3 & 0 & -2 & \bigm| & -11 \end{bmatrix} \rightarrow \begin{bmatrix} 1 & \dfrac{1}{2} & 0 & \bigm| & -2 \\ 0 & -2 & 4 & \bigm| & 0 \\ 3 & 0 & -2 & \bigm| & -11 \end{bmatrix} \rightarrow \begin{bmatrix} 1 & \dfrac{1}{2} & 0 & \bigm| & -2 \\ 0 & -2 & 4 & \bigm| & 0 \\ 3 & \dfrac{-3}{2} & -2 & \bigm| & -5 \end{bmatrix} \rightarrow \begin{bmatrix} 1 & \dfrac{1}{2} & 0 & \bigm| & -2 \\ 0 & 1 & -2 & \bigm| & 0 \\ 0 & \dfrac{-3}{2} & -2 & \bigm| & -5 \end{bmatrix}$

$\qquad\qquad R_1 = \dfrac{1}{2}r_1$ $\qquad\qquad R_3 = -3r_1 + r_3$ $\qquad R_2 = \dfrac{-1}{2}r_2$

$\rightarrow \begin{bmatrix} 1 & 0 & 1 & \bigm| & -2 \\ 0 & 1 & -2 & \bigm| & 0 \\ 0 & 0 & -5 & \bigm| & -5 \end{bmatrix} \rightarrow \begin{bmatrix} 1 & 0 & 1 & \bigm| & -2 \\ 0 & 1 & -2 & \bigm| & 0 \\ 0 & 0 & 1 & \bigm| & 1 \end{bmatrix} \rightarrow \begin{bmatrix} 1 & 0 & 0 & \bigm| & -3 \\ 0 & 1 & 0 & \bigm| & 2 \\ 0 & 0 & 1 & \bigm| & 1 \end{bmatrix}$

$\qquad R_1 = \dfrac{-1}{2}r_2 + r_1$ $\quad R_3 = \dfrac{-1}{5}r_3$ $\qquad\quad R_2 = 2r_3 + r_2$

$\qquad R_3 = \dfrac{3}{2}r_2 + r_3$ $\qquad\qquad\qquad\qquad R_1 = -r_3 + r_1$

The solution is $x = -3$, $y = 2$, $z = 1$.

50. $\begin{bmatrix} 2 & 1 & -3 & | & 0 \\ -2 & 2 & 1 & | & -7 \\ 3 & -4 & -3 & | & 7 \end{bmatrix} \rightarrow \begin{bmatrix} 1 & \frac{1}{2} & \frac{-3}{2} & | & 0 \\ -2 & 2 & 1 & | & -7 \\ 3 & -4 & -3 & | & 7 \end{bmatrix} \rightarrow \begin{bmatrix} 1 & \frac{1}{2} & \frac{-3}{2} & | & 0 \\ 0 & 3 & -2 & | & -7 \\ 3 & \frac{-11}{2} & \frac{3}{2} & | & 7 \end{bmatrix} \rightarrow \begin{bmatrix} 0 & \frac{1}{2} & \frac{-3}{2} & | & 0 \\ 0 & 1 & \frac{-2}{3} & | & \frac{-7}{3} \\ 0 & \frac{-11}{2} & \frac{3}{2} & | & 7 \end{bmatrix}$

$\uparrow$ $\qquad$ $\uparrow$ $\qquad$ $\uparrow$

$R_1 = \frac{1}{2}r_1$ $\qquad$ $R_2 = 2r_1 + r_2$ $\qquad$ $R_2 = \frac{1}{3}r_2$

$\qquad\qquad\qquad R_3 = -3r_1 + r_3$

$\rightarrow \begin{bmatrix} 1 & 0 & \frac{-7}{6} & | & \frac{7}{6} \\ 0 & 1 & \frac{-2}{3} & | & \frac{-7}{3} \\ 0 & 0 & \frac{-13}{6} & | & \frac{-35}{6} \end{bmatrix} \rightarrow \begin{bmatrix} 1 & 0 & \frac{-7}{6} & | & \frac{7}{6} \\ 0 & 1 & \frac{-2}{3} & | & \frac{-7}{3} \\ 0 & 0 & 1 & | & \frac{35}{13} \end{bmatrix} \rightarrow \begin{bmatrix} 1 & 0 & 0 & | & \frac{56}{13} \\ 0 & 1 & 0 & | & \frac{-7}{13} \\ 0 & 0 & 1 & | & \frac{35}{13} \end{bmatrix}$

$\uparrow$ $\qquad\qquad$ $\uparrow$ $\qquad\qquad$ $\uparrow$

$R_1 = \frac{-1}{2}r_2 + r_1$ $\qquad$ $R_3 = \frac{-6}{13}r_3$ $\qquad$ $R_2 = \frac{2}{3}r_3 + r_2$

$R_3 = \frac{11}{2}r_2 + r_3$ $\qquad\qquad\qquad$ $R_1 = \frac{7}{6}r_3 + r_1$

The solution is $x = \frac{56}{13}$, $y = \frac{-7}{13}$, $z = \frac{35}{13}$.

52. $\begin{bmatrix} 2 & -3 & -1 & | & 0 \\ -1 & 2 & 1 & | & 5 \\ 3 & -4 & -1 & | & 1 \end{bmatrix} \rightarrow \begin{bmatrix} 1 & -2 & -1 & | & -5 \\ 2 & -3 & -1 & | & 0 \\ 3 & -4 & -1 & | & 1 \end{bmatrix} \rightarrow \begin{bmatrix} 1 & -2 & -1 & | & -5 \\ 0 & 1 & 1 & | & 10 \\ 0 & 2 & 2 & | & 16 \end{bmatrix} \rightarrow \begin{bmatrix} 1 & 0 & 1 & | & 15 \\ 0 & 1 & 1 & | & 10 \\ 0 & 0 & 0 & | & -4 \end{bmatrix}$

$\uparrow$ $\qquad\qquad$ $\uparrow$ $\qquad$ $\uparrow$

Interchange r_1 and $-r_2$ $\qquad$ $R_2 = -2r_1 + r_2$ $\quad$ $R_1 = 2r_2 + r_1$

$\qquad\qquad\qquad\qquad\qquad R_3 = -3r_1 + r_3$ $\quad$ $R_3 = -2r_2 + r_3$

The system is inconsistent.

54. $\begin{bmatrix} 2 & -3 & -1 & | & 0 \\ 3 & 2 & 2 & | & 2 \\ 1 & 5 & 3 & | & 2 \end{bmatrix} \rightarrow \begin{bmatrix} 1 & 5 & 3 & | & 2 \\ 3 & 2 & 2 & | & 2 \\ 2 & -3 & -1 & | & 0 \end{bmatrix} \rightarrow \begin{bmatrix} 1 & 5 & 3 & | & 2 \\ 0 & -13 & -7 & | & -4 \\ 0 & -13 & -7 & | & -4 \end{bmatrix}$

$\uparrow$ $\qquad\qquad$ $\uparrow$

Interchange r_1 and r_2 $\qquad$ $R_2 = -3r_1 + r_2$

$\qquad\qquad\qquad\qquad\qquad R_3 = -2r_1 + r_3$

Thus, the solution can be written:

$y = \frac{-7}{13}z + \frac{4}{13}$

$\qquad\qquad$ where z is any number.

$x = \frac{-4}{13}z + \frac{6}{13}$

56. $\begin{bmatrix} 3 & -2 & 2 & | & 6 \\ 7 & -3 & 2 & | & -1 \\ 2 & -3 & 4 & | & 0 \end{bmatrix} \rightarrow \begin{bmatrix} 1 & \frac{-2}{3} & \frac{2}{3} & | & 2 \\ 7 & -3 & 2 & | & -1 \\ 2 & -3 & 4 & | & 0 \end{bmatrix} \rightarrow \begin{bmatrix} 1 & \frac{-2}{3} & \frac{2}{3} & | & 2 \\ 0 & \frac{5}{3} & \frac{-8}{3} & | & -15 \\ 0 & \frac{-5}{3} & \frac{8}{3} & | & -2 \end{bmatrix}$

$\uparrow$ $\uparrow$

$R_1 = \frac{1}{3}r_1$ $R_2 = -7r_1 + r_2$

$R_3 = -2r_1 + r_3$

$\rightarrow \begin{bmatrix} 1 & \frac{-2}{3} & \frac{2}{3} & | & 2 \\ 0 & 1 & \frac{-8}{5} & | & -9 \\ 0 & \frac{-5}{3} & \frac{8}{3} & | & -2 \end{bmatrix} \rightarrow \begin{bmatrix} 1 & 0 & \frac{-5}{15} & | & -4 \\ 0 & 1 & \frac{-8}{5} & | & -9 \\ 0 & 0 & 0 & | & -17 \end{bmatrix}$

$\uparrow$ $\uparrow$

$R_2 = \frac{3}{5}r_2$ $R_1 = \frac{2}{3}r_2 + r_1$

$R_3 = \frac{5}{3}r_2 + r_3$

The system is inconsistent.

58. $\begin{bmatrix} 1 & -1 & 1 & | & -4 \\ 2 & -3 & 4 & | & -15 \\ 5 & 1 & -2 & | & 12 \end{bmatrix} \rightarrow \begin{bmatrix} 1 & -1 & 1 & | & -4 \\ 0 & -1 & 2 & | & -7 \\ 0 & 6 & -7 & | & 32 \end{bmatrix} \rightarrow \begin{bmatrix} 1 & -1 & 1 & | & -4 \\ 0 & 1 & -2 & | & 7 \\ 0 & 6 & -7 & | & 32 \end{bmatrix}$

$\uparrow$ $\uparrow$

$R_2 = -2r_1 + r_2$ $R_2 = -r_2$

$R_3 = -5r_1 + r_2$

$\rightarrow \begin{bmatrix} 1 & 0 & -1 & | & 3 \\ 0 & 1 & -2 & | & 7 \\ 0 & 0 & 5 & | & -10 \end{bmatrix} \rightarrow \begin{bmatrix} 1 & 0 & -1 & | & 3 \\ 0 & 1 & -2 & | & 7 \\ 0 & 0 & 1 & | & -2 \end{bmatrix} \rightarrow \begin{bmatrix} 1 & 0 & 0 & | & 1 \\ 0 & 1 & 0 & | & 3 \\ 0 & 0 & 1 & | & -2 \end{bmatrix}$

$\uparrow$ $\uparrow$ $\uparrow$

$R_1 = r_2 + r_1$ $R_3 = \frac{1}{5}r_3$ $R_1 = r_3 + r_2$

$R_3 = -6r_2 + r_3$ $R_2 = 2r_3 + r_2$

The solution is $x = 1, y = 3, z = -2$.

60. $\begin{bmatrix} 1 & 4 & -3 & | & -8 \\ 3 & -1 & 3 & | & 12 \\ 1 & 1 & 6 & | & 1 \end{bmatrix} \rightarrow \begin{bmatrix} 1 & 4 & -3 & | & -8 \\ 0 & -13 & 12 & | & 36 \\ 0 & -3 & 9 & | & 9 \end{bmatrix} \rightarrow \begin{bmatrix} 1 & 4 & -3 & | & -8 \\ 0 & 1 & \frac{-12}{13} & | & \frac{-36}{13} \\ 0 & -3 & 9 & | & 9 \end{bmatrix}$

$\uparrow$ $\qquad\qquad\qquad\qquad$ $\uparrow$

$R_2 = -3r_1 + r_2 \qquad R_2 = \frac{-1}{13}r_2$

$R_3 = -r_1 + r_3$

$\rightarrow \begin{bmatrix} 1 & 0 & \frac{9}{13} & | & \frac{40}{13} \\ 0 & 1 & \frac{-12}{13} & | & \frac{-36}{13} \\ 0 & 0 & \frac{81}{13} & | & \frac{9}{13} \end{bmatrix} \rightarrow \begin{bmatrix} 1 & 0 & \frac{9}{13} & | & \frac{40}{13} \\ 0 & 1 & \frac{-12}{13} & | & \frac{-36}{13} \\ 0 & 0 & 1 & | & \frac{1}{9} \end{bmatrix} \rightarrow \begin{bmatrix} 1 & 0 & 0 & | & 3 \\ 0 & 1 & 0 & | & \frac{-8}{3} \\ 0 & 0 & 1 & | & \frac{1}{9} \end{bmatrix}$

$\uparrow$ $\qquad\qquad\qquad\qquad$ $\uparrow$ $\qquad\qquad\qquad$ $\uparrow$

$R_1 = -4r_2 + r_1 \qquad R_3 = \frac{13}{81}r_3 \qquad R_1 = \frac{-9}{13}r_3 + r_1$

$R_3 = 3r_2 + r_3 \qquad\qquad\qquad\qquad\qquad R_2 = \frac{12}{13}r_3 + r_2$

The solution is $x = 3$, $y = \frac{-8}{3}$, $z = \frac{1}{9}$.

62. $\begin{bmatrix} 1 & 1 & 0 & | & 1 \\ 2 & -1 & 1 & | & 1 \\ 1 & 2 & 1 & | & \frac{8}{3} \end{bmatrix} \rightarrow \begin{bmatrix} 1 & 1 & 0 & | & 1 \\ 0 & -3 & 1 & | & -1 \\ 0 & 1 & 1 & | & \frac{5}{3} \end{bmatrix} \rightarrow \begin{bmatrix} 1 & 1 & 0 & | & 1 \\ 0 & 1 & 1 & | & \frac{5}{3} \\ 0 & -3 & 1 & | & -1 \end{bmatrix}$

$\uparrow$ $\qquad\qquad\qquad\qquad$ $\uparrow$

$R_2 = -2r_1 + r_2 \qquad$ Interchange $r_2 + r_3$

$R_3 = -r_1 + r_3$

$\rightarrow \begin{bmatrix} 1 & 0 & -1 & | & -\frac{2}{3} \\ 0 & 1 & 1 & | & \frac{5}{3} \\ 0 & 0 & 4 & | & 4 \end{bmatrix} \rightarrow \begin{bmatrix} 1 & 0 & -1 & | & -\frac{2}{3} \\ 0 & 1 & 1 & | & \frac{5}{3} \\ 0 & 0 & 1 & | & 1 \end{bmatrix} \rightarrow \begin{bmatrix} 1 & 0 & 0 & | & \frac{1}{3} \\ 0 & 1 & 0 & | & \frac{2}{3} \\ 0 & 0 & 1 & | & 1 \end{bmatrix}$

$\uparrow$ $\qquad\qquad\qquad\qquad$ $\uparrow$ $\qquad\qquad\qquad$ $\uparrow$

$R_1 = -r_2 + r_1 \qquad R_3 = \frac{1}{4}r_3 \qquad R_1 = r_3 + r_1$

$R_3 = 3r_2 + r_3 \qquad\qquad\qquad\qquad\qquad R_2 = -r_3 + r_2$

The solution is $x = \frac{1}{3}$, $y = \frac{2}{3}$, $z = 1$.

64.

$$\begin{bmatrix} 1 & 1 & 1 & 1 & | & 4 \\ -1 & 2 & 1 & 0 & | & 0 \\ 2 & 3 & 1 & -1 & | & 6 \\ -2 & 1 & -2 & 2 & | & -1 \end{bmatrix} \rightarrow \begin{bmatrix} 1 & 1 & 1 & 1 & | & 4 \\ 0 & 3 & 2 & 1 & | & 4 \\ 0 & 1 & -1 & -3 & | & -2 \\ 0 & 3 & 0 & 4 & | & 7 \end{bmatrix} \rightarrow \begin{bmatrix} 1 & 1 & 1 & 1 & | & 4 \\ 0 & 1 & -1 & -3 & | & -2 \\ 0 & 3 & 2 & 1 & | & 4 \\ 0 & 3 & 0 & 4 & | & 7 \end{bmatrix} \rightarrow \begin{bmatrix} 1 & 0 & 2 & 4 & | & 6 \\ 0 & 1 & -1 & -3 & | & -2 \\ 0 & 0 & 5 & 10 & | & 10 \\ 0 & 0 & 3 & 13 & | & 13 \end{bmatrix}$$

$$\begin{array}{ccc} \uparrow & \uparrow & \uparrow \\ \begin{array}{l} R_2 = r_1 + r_2 \\ R_3 = -2r_1 + r_3 \\ R_4 = 2r_1 + r_4 \end{array} & \text{Interchange } r_2 + r_3 & \begin{array}{l} R_1 = -2r_3 + r_1 \\ R_3 = -3r_2 + r_3 \\ R_4 = -3r_2 + r_4 \end{array} \end{array}$$

$$\rightarrow \begin{bmatrix} 1 & 0 & 2 & 4 & | & 6 \\ 0 & 1 & -1 & -3 & | & -2 \\ 0 & 0 & 1 & 2 & | & 2 \\ 0 & 0 & 3 & 13 & | & 13 \end{bmatrix} \rightarrow \begin{bmatrix} 1 & 0 & 0 & 0 & | & 2 \\ 0 & 1 & 0 & -1 & | & 0 \\ 0 & 0 & 1 & 2 & | & 2 \\ 0 & 0 & 0 & 7 & | & 7 \end{bmatrix} \rightarrow \begin{bmatrix} 1 & 0 & 0 & 0 & | & 2 \\ 0 & 1 & 0 & -1 & | & 0 \\ 0 & 0 & 1 & 2 & | & 2 \\ 0 & 0 & 0 & 1 & | & 1 \end{bmatrix} \rightarrow \begin{bmatrix} 1 & 0 & 0 & 0 & | & 2 \\ 0 & 1 & 0 & 0 & | & 1 \\ 0 & 0 & 1 & 0 & | & 0 \\ 0 & 0 & 0 & 1 & | & 1 \end{bmatrix}$$

$$\begin{array}{cccc} \uparrow & \uparrow & \uparrow & \uparrow \\ R_3 = \dfrac{1}{5}r_3 & \begin{array}{l} R_1 = -2r_3 + r_1 \\ R_2 = r_3 + r_2 \\ R_4 = -3r_3 + r_4 \end{array} & R_4 = \dfrac{1}{7}r_4 & \begin{array}{l} R_2 = r_4 + r_2 \\ R_3 = -2r_4 + r_3 \end{array} \end{array}$$

The solution is $x = 2$, $y = 1$, $z = 0$, $w = 1$.

66.

$$\begin{bmatrix} 1 & 2 & -1 & | & 3 \\ 2 & -1 & 2 & | & 6 \\ 1 & -3 & 3 & | & 4 \end{bmatrix} \rightarrow \begin{bmatrix} 1 & 2 & -1 & | & 3 \\ 0 & -5 & 4 & | & 0 \\ 0 & -5 & 4 & | & 1 \end{bmatrix} \rightarrow \begin{bmatrix} 1 & 2 & -1 & | & 3 \\ 0 & 1 & -\dfrac{4}{5} & | & 0 \\ 0 & -5 & 4 & | & 1 \end{bmatrix} \rightarrow \begin{bmatrix} 1 & 0 & \dfrac{3}{5} & | & 3 \\ 0 & 1 & -\dfrac{4}{5} & | & 0 \\ 0 & 0 & 0 & | & 1 \end{bmatrix}$$

$$\begin{array}{ccc} \uparrow & \uparrow & \uparrow \\ \begin{array}{l} R_2 = -2r_1 + r_2 \\ R_3 = -r_1 + r_3 \end{array} & R_2 = \dfrac{-1}{5}r_2 & \begin{array}{l} R_1 = -2r_2 + r_1 \\ R_3 = 5r_2 + r_3 \end{array} \end{array}$$

The system is inconsistent.

68.

$$\begin{bmatrix} 2 & 1 & -1 & | & 4 \\ -1 & 1 & 3 & | & 1 \end{bmatrix} \rightarrow \begin{bmatrix} 1 & -1 & -3 & | & -1 \\ 2 & 1 & -1 & | & 4 \end{bmatrix} \rightarrow \begin{bmatrix} 1 & -1 & -3 & | & -1 \\ 0 & 3 & 5 & | & 6 \end{bmatrix}$$

$$\begin{array}{cc} \uparrow & \uparrow \\ \text{Interchange } r_1 \text{ and } -r_2 & R_2 = -2r_1 + r_2 \end{array}$$

$$\rightarrow \begin{bmatrix} 1 & -1 & -3 & | & -1 \\ 0 & 1 & \dfrac{5}{3} & | & 2 \end{bmatrix} \rightarrow \begin{bmatrix} 1 & 0 & \dfrac{4}{3} & | & 1 \\ 0 & 1 & \dfrac{5}{3} & | & 2 \end{bmatrix}$$

$$\begin{array}{cc} \uparrow & \uparrow \\ R_2 = \dfrac{1}{3}r_2 & R_1 = r_2 + r_1 \end{array}$$

The solution is: $\quad x = 1 + \dfrac{4}{3}z$

$$y = 2 - \dfrac{5}{3}z$$

where z can be any real number.

70.

$$\begin{bmatrix} 1 & -3 & 1 & | & 1 \\ 2 & -1 & -4 & | & 0 \\ 1 & -3 & 2 & | & 1 \\ 1 & -2 & 0 & | & 5 \end{bmatrix} \rightarrow \begin{bmatrix} 1 & -3 & 1 & | & 1 \\ 0 & 5 & -6 & | & -2 \\ 0 & 0 & 1 & | & 0 \\ 0 & 1 & -1 & | & 4 \end{bmatrix} \rightarrow \begin{bmatrix} 1 & -3 & 1 & | & 1 \\ 0 & 1 & -1 & | & 4 \\ 0 & 0 & 1 & | & 0 \\ 0 & 5 & -6 & | & -2 \end{bmatrix} \rightarrow \begin{bmatrix} 1 & 0 & -2 & | & 13 \\ 0 & 1 & -1 & | & 4 \\ 0 & 0 & 1 & | & 0 \\ 0 & 0 & -1 & | & -22 \end{bmatrix}$$

$$\begin{aligned} R_2 &= -2r_1 + r_2 \\ R_3 &= -r_1 + r_3 \\ R_4 &= -r_1 + r_4 \end{aligned} \qquad \text{Interchange } r_2 + r_4 \qquad \begin{aligned} R_1 &= 3r_2 + r_1 \\ R_4 &= -5r_2 + r_4 \end{aligned}$$

$$\rightarrow \begin{bmatrix} 1 & 0 & -2 & | & 13 \\ 0 & 1 & -1 & | & 4 \\ 0 & 0 & 1 & | & 0 \\ 0 & 0 & 1 & | & 22 \end{bmatrix} \rightarrow \begin{bmatrix} 1 & 0 & -2 & | & 13 \\ 0 & 1 & -1 & | & 4 \\ 0 & 0 & 0 & | & -22 \\ 0 & 0 & 1 & | & 22 \end{bmatrix}$$

$$R_4 = -r_4 \qquad\qquad R_3 = -r_4 + r_3$$

The system is inconsistent.

72.

$$\begin{bmatrix} -4 & 1 & 0 & 0 & | & 5 \\ 2 & -1 & 1 & -1 & | & 5 \\ 0 & 0 & 1 & 1 & | & 4 \end{bmatrix} \rightarrow \begin{bmatrix} 1 & \frac{-1}{4} & 0 & 0 & | & \frac{-5}{4} \\ 2 & -1 & 1 & -1 & | & 5 \\ 0 & 0 & 1 & 1 & | & 4 \end{bmatrix} \rightarrow \begin{bmatrix} 1 & \frac{-1}{4} & 0 & 0 & | & \frac{-5}{4} \\ 0 & \frac{-1}{2} & 1 & -1 & | & \frac{15}{2} \\ 0 & 0 & 1 & 1 & | & 4 \end{bmatrix}$$

$$R_1 = \frac{-1}{4}r_1 \qquad\qquad R_2 = -2r_1 + r_2$$

$$\rightarrow \begin{bmatrix} 1 & \frac{-1}{4} & 0 & 0 & | & \frac{-5}{4} \\ 0 & 1 & -2 & 2 & | & -15 \\ 0 & 0 & 1 & 1 & | & 4 \end{bmatrix} \rightarrow \begin{bmatrix} 1 & 0 & \frac{-1}{2} & \frac{1}{2} & | & -5 \\ 0 & 1 & -2 & 2 & | & -15 \\ 0 & 0 & 1 & 1 & | & 4 \end{bmatrix} \rightarrow \begin{bmatrix} 1 & 0 & 0 & 1 & | & -3 \\ 0 & 1 & 0 & 4 & | & -7 \\ 0 & 0 & 1 & 1 & | & 4 \end{bmatrix}$$

$$R_2 = -2r_2 \qquad\qquad R_1 = \frac{1}{4}r_2 + r_1 \qquad\qquad \begin{aligned} R_1 &= \frac{1}{2}r_3 + r_1 \\ R_2 &= 2r_3 + r_2 \end{aligned}$$

The solution is: $\quad x = -3 - w$
$\qquad\qquad\qquad y = -7 - 4w \qquad$ where w can be any real number.
$\qquad\qquad\qquad z = 4 - w$

74. $y = ax^2 + bx + c$

$$\begin{bmatrix} 1 & 1 & 1 & | & -1 \\ 9 & 3 & 1 & | & -1 \\ 4 & -2 & 1 & | & 14 \end{bmatrix} \rightarrow \begin{bmatrix} 1 & 1 & 1 & | & -1 \\ 0 & -6 & -8 & | & 8 \\ 0 & -6 & -3 & | & 18 \end{bmatrix} \rightarrow \begin{bmatrix} 1 & 1 & 1 & | & -1 \\ 0 & 1 & \frac{4}{3} & | & \frac{-4}{3} \\ 0 & -6 & -3 & | & 18 \end{bmatrix}$$

$$\uparrow \qquad\qquad\qquad \uparrow$$

$$R_2 = -9r_1 + r_2 \quad R_2 = \frac{-1}{6}r_2$$

$$R_3 = -4r_1 + r_2$$

$$\rightarrow \begin{bmatrix} 1 & 0 & \frac{-1}{3} & | & \frac{1}{3} \\ 0 & 1 & \frac{4}{3} & | & \frac{-4}{3} \\ 0 & 0 & 5 & | & 10 \end{bmatrix} \rightarrow \begin{bmatrix} 1 & 0 & \frac{-1}{3} & | & \frac{1}{3} \\ 0 & 1 & \frac{4}{3} & | & \frac{-4}{3} \\ 0 & 0 & 1 & | & 2 \end{bmatrix} \rightarrow \begin{bmatrix} 1 & 0 & 0 & | & 1 \\ 0 & 1 & 0 & | & -4 \\ 0 & 0 & 1 & | & 2 \end{bmatrix}$$

$$\uparrow \qquad\qquad\qquad \uparrow \qquad\qquad\qquad \uparrow$$

$$R_1 = -r_2 + r_1 \quad R_3 = \frac{1}{5}r_3 \qquad R_2 = \frac{-4}{3}r_3 + r_2$$

$$R_3 = 6r_2 + r_3 \qquad\qquad\qquad R_1 = \frac{1}{2}r_3 + r_1$$

The parabola is $y = x^2 - 4x + 2$.

76. $f(x) = ax^3 + bx^2 + cx + d$

$$\begin{bmatrix} -8 & 4 & -2 & 1 & | & -10 \\ -1 & 1 & -1 & 1 & | & 3 \\ 1 & 1 & 1 & 1 & | & 5 \\ 27 & 9 & 3 & 1 & | & 15 \end{bmatrix} \rightarrow \begin{bmatrix} 1 & 1 & 1 & 1 & | & 5 \\ -1 & 1 & -1 & 1 & | & 3 \\ -8 & 4 & -2 & 1 & | & -10 \\ 27 & 9 & 3 & 1 & | & 15 \end{bmatrix} \rightarrow \begin{bmatrix} 1 & 1 & 1 & 1 & | & 5 \\ 0 & 2 & 0 & 2 & | & 8 \\ 0 & 12 & 6 & 9 & | & 30 \\ 0 & -18 & -24 & -26 & | & -120 \end{bmatrix}$$

$$\uparrow \qquad\qquad\qquad\qquad \uparrow$$

$$\text{Interchange } r_1 + r_2 \qquad R_2 = r_1 + r_2$$

$$R_3 = 8r_1 + r_3$$

$$R4 = -27r_1 + r_4$$

$$\rightarrow \begin{bmatrix} 1 & 1 & 1 & 1 & | & 5 \\ 0 & 1 & 0 & 1 & | & 4 \\ 0 & 12 & 6 & 9 & | & 30 \\ 0 & -18 & -24 & -26 & | & -120 \end{bmatrix} \rightarrow \begin{bmatrix} 1 & 0 & 1 & 0 & | & 1 \\ 0 & 1 & 0 & 1 & | & 4 \\ 0 & 0 & 6 & -3 & | & -18 \\ 0 & 0 & -24 & -8 & | & -48 \end{bmatrix} \rightarrow \begin{bmatrix} 1 & 0 & 1 & 0 & | & 1 \\ 0 & 1 & 0 & 1 & | & 4 \\ 0 & 0 & 1 & \frac{-1}{2} & | & -3 \\ 0 & 0 & -24 & -8 & | & -48 \end{bmatrix}$$

$$\uparrow \qquad\qquad\qquad\qquad \uparrow \qquad\qquad\qquad\qquad \uparrow$$

$$R_2 = \frac{1}{2}r_2 \qquad\qquad R_1 = -r_2 + r_1 \qquad\qquad R_3 = \frac{1}{6}r_3$$

$$R_3 = -12r_2 + r_3$$

$$R_4 = 18r_2 + r_3$$

$$\rightarrow \begin{bmatrix} 1 & 0 & 0 & \frac{1}{2} & \Big| & 4 \\ 0 & 1 & 0 & 1 & \Big| & 4 \\ 0 & 0 & 1 & \frac{-1}{2} & \Big| & -3 \\ 0 & 0 & 0 & -20 & \Big| & -120 \end{bmatrix} \rightarrow \begin{bmatrix} 1 & 0 & 0 & \frac{1}{2} & \Big| & 4 \\ 0 & 1 & 0 & 1 & \Big| & 4 \\ 0 & 0 & 1 & \frac{-1}{2} & \Big| & -3 \\ 0 & 0 & 0 & 1 & \Big| & 6 \end{bmatrix} \rightarrow \begin{bmatrix} 1 & 0 & 0 & 0 & \Big| & 1 \\ 0 & 1 & 0 & 0 & \Big| & -2 \\ 0 & 0 & 1 & 0 & \Big| & 0 \\ 0 & 0 & 0 & 1 & \Big| & 6 \end{bmatrix}$$

$$R_1 = -r_3 + r_1 \qquad R_4 = \frac{1}{20} r_4 \qquad R_1 = \frac{-1}{2} r_4 + r_1$$
$$R_4 = 24r_3 + r_4 \qquad\qquad\qquad R_2 = -r_4 + r_2$$
$$\qquad\qquad\qquad\qquad\qquad\qquad R_3 = \frac{1}{2} r_4 + r_3$$

Thus, $f(x) = x^3 - 2x^2 + 6$.

78. $\dfrac{1}{M} + \dfrac{1}{D} + \dfrac{1}{K} = \dfrac{1}{10}$

$\dfrac{1}{K} + \dfrac{1}{D} = \dfrac{1}{15}$

	Hours to do job	Part of job done in 1 hr.
Mike	M	$\dfrac{1}{M}$
Dan	D	$\dfrac{1}{D}$
Katy	K	$\dfrac{1}{K}$
Together	10	$\dfrac{1}{10}$

$$\frac{1}{M} + \frac{1}{D} + \frac{1}{K} = \frac{1}{10}$$

	Hours to do job	Part of job done in 1 hr.
Dan	D	$\dfrac{1}{D}$
Katy	K	$\dfrac{1}{K}$
Together	15	$\dfrac{1}{15}$

$$\frac{1}{D} + \frac{1}{K} = \frac{1}{15}$$

	Hours to do job	Part of job done in 1 hr.
Mike	M	$\dfrac{1}{M}$
Dan	D	$\dfrac{1}{D}$
Katy	K	$\dfrac{1}{K}$
All 3 together	4	$\dfrac{1}{4}$
Mike & Dan	8	$\dfrac{1}{8}$

$$4\left[\frac{1}{M} + \frac{1}{D} + \frac{1}{K}\right] + 8\left[\frac{1}{M} + \frac{1}{D}\right] = 1$$

$$\frac{1}{M} + \frac{1}{D} + \frac{1}{K} = \frac{1}{4}$$

$$\frac{1}{M} + \frac{1}{D} = \frac{1}{8}$$

$$\frac{1}{M} + \frac{1}{15} = \frac{1}{10}$$

$$\frac{1}{M} = \frac{1}{30}$$

$$4\left[\frac{1}{10}\right] + 8\left[\frac{1}{30} + \frac{1}{D}\right] = 1$$

$$\frac{1}{D} = \frac{5}{120} = \frac{1}{24}$$

$$\frac{1}{K} = \frac{1}{10} - \frac{1}{24} - \frac{1}{10} = \frac{1}{40}$$

$$M = 30 \text{ hr.}; \quad D = 24 \text{ hr.}; \quad K = 40 \text{ hr.}$$

80. x = price of hamburger, y = price of fries, z = price of cola

$$\begin{cases} 8x + 6y + 6z = 26.10 \\ 10x + 6y + 8z = 31.60 \\ 3x + 2y + 4z = 10.95 \end{cases}$$

$$\begin{bmatrix} 8 & 6 & 6 & | & 26.10 \\ 10 & 6 & 8 & | & 31.60 \\ 3 & 2 & 4 & | & 10.95 \end{bmatrix}$$

$$\begin{bmatrix} 4 & 3 & 3 & | & 13.05 \\ 5 & 3 & 4 & | & 15.80 \\ 3 & 2 & 4 & | & 10.95 \end{bmatrix}$$

$$\begin{bmatrix} 4 & 3 & 3 & | & 13.05 \\ 1 & 0 & 1 & | & 2.75 \\ 3 & 2 & 4 & | & 10.95 \end{bmatrix}$$

$$\begin{bmatrix} 1 & 0 & 1 & | & 2.75 \\ 0 & 3 & -1 & | & 2.05 \\ 0 & 2 & 1 & | & 2.70 \end{bmatrix}$$

$$\begin{bmatrix} 1 & 0 & 1 & | & 2.75 \\ 0 & 1 & -2 & | & -.65 \\ 0 & 2 & 1 & | & 2.70 \end{bmatrix}$$

$$\begin{bmatrix} 1 & 0 & 1 & | & 2.75 \\ 0 & 1 & -2 & | & -.65 \\ 0 & 0 & 5 & | & 4.00 \end{bmatrix}$$

$$\begin{bmatrix} 1 & 0 & 1 & | & 2.75 \\ 0 & 1 & -2 & | & -.65 \\ 0 & 0 & 1 & | & .80 \end{bmatrix}$$

$$\begin{bmatrix} 1 & 0 & 0 & | & 1.95 \\ 0 & 1 & 0 & | & .95 \\ 0 & 0 & 1 & | & .80 \end{bmatrix}$$

Yes!   hamburgers 1.95
       fries   .95
       cola   .80

82. x = amount invested at 7%
     y = amount invested at 9%
     z = amount invested at 11%
     I = income

$$\begin{cases} x + y + z = 25{,}000 \\ .07x + .09y + .11z = I \end{cases}$$

$$\begin{cases} x + y + z = 25{,}000 \\ 7x + 9y + 11z = 100I \end{cases}$$

$$\begin{cases} x + y + z = 25{,}000 \\ 2y + 4z = 100I - 175{,}000 \end{cases}$$

$$\begin{cases} x + y + z = 25{,}000 \\ y + 2z = 50I - 87{,}500 \end{cases}$$

(a) $I = 1500$

$$\begin{cases} x + y + z = 25{,}000 \\ y + 2z = -12{,}500 \end{cases}$$

$$\begin{cases} x = z + 37{,}500 \\ y = -2z - 12{,}500 \end{cases}$$

Investing all the money at 7% yields more than $1500

(b) $I = 2000$

$$\begin{cases} x + y + z = 25{,}000 \\ y + 2z = 12{,}500 \end{cases}$$

$$\begin{cases} x = z + 12{,}500 \\ y = -2z + 12{,}500 \end{cases}$$

7%	9%	11%
12,500	12,500	0
15,500	6,500	3000
18,750	0	6250

(c) $I = 2500$

$$\begin{cases} x + y + z = 25{,}000 \\ y + 2z = 37{,}500 \end{cases}$$

$$\begin{cases} x = z - 12{,}500 \\ y = -2z + 37{,}500 \end{cases}$$

7%	9%	11%
0	12,500	12,500
1,000	10,500	13,500
6,250	0	18,750

84. $\begin{cases} 2I_2 = 4 \\ I_1 + 5I_4 = 8 \\ I_1 + 3I_3 = 4 \end{cases}$

$I_1 = I_3 + I_4$

$$\begin{bmatrix} 0 & 2 & 0 & 0 & | & 4 \\ 1 & 0 & 0 & 5 & | & 8 \\ 1 & 0 & 3 & 0 & | & 4 \\ 1 & 0 & -1 & -1 & | & 0 \end{bmatrix}$$

$$\begin{bmatrix} 1 & 0 & 0 & 5 & | & 4 \\ 0 & 2 & 0 & 0 & | & 4 \\ 1 & 0 & 3 & 0 & | & 4 \\ 1 & 0 & -1 & -1 & | & 0 \end{bmatrix}$$

$$\begin{bmatrix} 1 & 0 & 0 & 5 & | & 4 \\ 0 & 1 & 0 & 0 & | & 2 \\ 1 & 0 & 3 & -5 & | & 0 \\ 0 & 0 & -1 & -6 & | & -4 \end{bmatrix}$$

$$\begin{bmatrix} 1 & 0 & 0 & 5 & | & 4 \\ 0 & 1 & 0 & 0 & | & 2 \\ 0 & 0 & 1 & 6 & | & 4 \\ 0 & 0 & 3 & -5 & | & 0 \end{bmatrix}$$

$$\begin{bmatrix} 1 & 0 & 0 & 5 & | & 4 \\ 0 & 1 & 0 & 0 & | & 2 \\ 0 & 0 & 1 & 6 & | & 4 \\ 0 & 0 & 0 & -23 & | & -12 \end{bmatrix}$$

$$\begin{bmatrix} 1 & 0 & 0 & 5 & | & 4 \\ 0 & 1 & 0 & 0 & | & 2 \\ 0 & 0 & 1 & 6 & | & 4 \\ 0 & 0 & 0 & 1 & | & \frac{12}{23} \end{bmatrix}$$

$I_4 = \dfrac{12}{23},\ I_3 = 4 - 6\left[\dfrac{12}{23}\right]$

$\qquad = \dfrac{92 - 72}{23} = \dfrac{20}{23}$

$I_2 = 2,\ I_1 = 4 - 5\left[\dfrac{12}{23}\right]$

$\qquad = \dfrac{92 - 60}{23} = \dfrac{32}{23}$

90. If $a_1 \neq 0$, $\begin{bmatrix} a_1 & b_1 & | & c_1 \\ a_2 & b_2 & | & c_2 \end{bmatrix} \rightarrow \begin{bmatrix} 1 & \dfrac{b_1}{a_1} & | & \dfrac{c_1}{a_1} \\ a_2 & b_2 & | & c_2 \end{bmatrix}$

$$\rightarrow \begin{bmatrix} 1 & \dfrac{b_1}{a_1} & \bigg| & \dfrac{c_1}{a_1} \\ 0 & b_2 - \dfrac{a_2 b_1}{a_1} & \bigg| & c_2 - \dfrac{a_2 c_1}{a_1} \end{bmatrix} \rightarrow \begin{bmatrix} 1 & \dfrac{b_1}{a_1} & \bigg| & \dfrac{c_1}{a_1} \\ 0 & \dfrac{b_2 a_1 - a_2 b_1}{a_1} & \bigg| & \dfrac{c_2 a_1 - a_2 c_1}{a_1} \end{bmatrix} \rightarrow \begin{bmatrix} 1 & \dfrac{b_1}{a_1} & \bigg| & \dfrac{c_1}{a_1} \\ 0 & 0 & \bigg| & \dfrac{c_2 a_1 - a_2 c_1}{a_1} \end{bmatrix}$$

The system is inconsistent if $a_1 c_2 \neq a_2 c_1$, and has infinitely many solutions if $a_1 c_2 = a_2 c_1$.

If $a_1 = 0$, then either $a_2 = 0$ or $b_1 = 0$.
If $a_2 = 0$, the system
$$b_1 y = c_1$$
$$b_2 y = c_2$$
is consistent if $b_1 c_2 \neq b_2 c_1$ and has infinitely many solutions if $b_1 c_2 = b_2 c_1$.
If $b_1 = 0$, then the system is inconsistent if $c_1 \neq 0$ and has infinitely many solutions if $c_2 = 0$.

12.3 Systems of Linear Equations: Determinants

2. $\begin{vmatrix} 6 & 1 \\ 5 & 2 \end{vmatrix} = 12 - 5 = 7$

4. $\begin{vmatrix} 8 & -3 \\ 4 & 2 \end{vmatrix} = 16 - (-12) = 28$

6. $\begin{vmatrix} -4 & 2 \\ -5 & 3 \end{vmatrix} = -12 - (-10) = -2$

8. $\begin{vmatrix} 1 & 3 & -2 \\ 6 & 1 & -5 \\ 8 & 2 & 3 \end{vmatrix} = 1 \begin{vmatrix} 1 & -5 \\ 2 & 3 \end{vmatrix} - 3 \begin{vmatrix} 6 & -5 \\ 8 & 3 \end{vmatrix} + (-2) \begin{vmatrix} 6 & 1 \\ 8 & 2 \end{vmatrix} = 1(13) - 3(58) + (-2)(4) = -169$

10. $\begin{vmatrix} 3 & -9 & 4 \\ 1 & 4 & 0 \\ 8 & -3 & 1 \end{vmatrix} = 3 \begin{vmatrix} 4 & 0 \\ -3 & 1 \end{vmatrix} - (-9) \begin{vmatrix} 1 & 0 \\ 8 & 1 \end{vmatrix} + 4 \begin{vmatrix} 1 & 4 \\ 8 & -3 \end{vmatrix} = 3(4) + 9(1) + 4(-35) = -119$

12. $\begin{cases} x + 2y = 5 \\ x + y = 3 \end{cases}$

$D = \begin{vmatrix} 1 & 2 \\ 1 & 1 \end{vmatrix} = 1 - 2 = -1$, $x = \dfrac{D_x}{D} = \dfrac{\begin{vmatrix} 5 & 2 \\ 3 & 1 \end{vmatrix}}{-1} = \dfrac{-1}{-1} = 1$, $y = \dfrac{D_y}{D} = \dfrac{\begin{vmatrix} 1 & 5 \\ 1 & 3 \end{vmatrix}}{-1} = \dfrac{-2}{-1} = 2$

The solution is $x = 1$, $y = 2$.

14. $\begin{cases} x + 3y = 5 \\ 2x - 3y = -8 \end{cases}$

$D = \begin{vmatrix} 1 & 3 \\ 2 & -3 \end{vmatrix} = -3 - 6 = -9$

$x = \dfrac{D_x}{D} = \dfrac{\begin{vmatrix} 5 & 3 \\ -8 & -3 \end{vmatrix}}{-9} = \dfrac{9}{-9} = -1,$

$y = \dfrac{\begin{vmatrix} 1 & 5 \\ 2 & -8 \end{vmatrix}}{-9} = \dfrac{-18}{-9} = 2$

The solution is $x = -1, y = 2.$

16. $\begin{cases} 4x + 5y = -3 \\ -2y = -4 \end{cases}$

$D = \begin{vmatrix} 4 & 5 \\ 0 & -2 \end{vmatrix} = -8$

$x = \dfrac{D_x}{D} = \dfrac{\begin{vmatrix} -3 & 5 \\ -4 & -2 \end{vmatrix}}{-8} = \dfrac{26}{-8} = \dfrac{-13}{4},$

$y = \dfrac{\begin{vmatrix} 4 & -3 \\ 0 & -4 \end{vmatrix}}{-8} = \dfrac{-16}{-8} = 2$

The solution is $x = \dfrac{-13}{4}, y = 2.$

18. $\begin{cases} 2x + 4y = 16 \\ 3x - 5y = -9 \end{cases}$

$D = \begin{vmatrix} 2 & 4 \\ 3 & -5 \end{vmatrix} = -22$

$x = \dfrac{D_x}{D} = \dfrac{\begin{vmatrix} 16 & 4 \\ -9 & -5 \end{vmatrix}}{-22} = \dfrac{-44}{-22} = 2,$

$y = \dfrac{D_y}{D} = \dfrac{\begin{vmatrix} 2 & 16 \\ 3 & -9 \end{vmatrix}}{-22} = \dfrac{-66}{-22} = 3$

The solution is $x = 2, y = 3.$

20. $\begin{cases} -x + 2y = 5 \\ 4x - 8y = 6 \end{cases}$

$D = \begin{vmatrix} -1 & 2 \\ 4 & -8 \end{vmatrix} = 0$

Thus, Cramer's Rule is not applicable.

22. $\begin{cases} 3x + 3y = 3 \\ 4x + 2y = \dfrac{8}{3} \end{cases}$

$D = \begin{vmatrix} 3 & 3 \\ 4 & 2 \end{vmatrix} = -6$

$x = \dfrac{D_x}{D} = \dfrac{\begin{vmatrix} 3 & 3 \\ \frac{8}{3} & 2 \end{vmatrix}}{-6} = \dfrac{-2}{-6} = \dfrac{1}{3},$

$y = \dfrac{D_y}{D} = \dfrac{\begin{vmatrix} 3 & 3 \\ 4 & \frac{8}{3} \end{vmatrix}}{-6} = \dfrac{-4}{-6} = \dfrac{2}{3}$

The solution is $x = \dfrac{1}{3}, y = \dfrac{2}{3}.$

24. $\begin{cases} 3x - 2y = 0 \\ 5x + 10y = 4 \end{cases}$

$D = \begin{vmatrix} 3 & -2 \\ 5 & 10 \end{vmatrix} = 40$

$x = \dfrac{D_x}{D} = \dfrac{\begin{vmatrix} 0 & -2 \\ 4 & 10 \end{vmatrix}}{40} = \dfrac{8}{40} = \dfrac{1}{5},$

$y = \dfrac{D_y}{D} = \dfrac{\begin{vmatrix} 3 & 0 \\ 5 & 4 \end{vmatrix}}{40} = \dfrac{12}{40} = \dfrac{3}{10}$

The solution is $x = \dfrac{1}{5}, y = \dfrac{3}{10}.$

26.
$$\begin{cases} \dfrac{1}{2}x + y = -2 \\ x - 2y = 8 \end{cases}$$

$$D = \begin{vmatrix} \dfrac{1}{2} & 1 \\ 1 & -2 \end{vmatrix} = -2$$

$$x = \frac{D_x}{D} = \frac{\begin{vmatrix} -2 & 1 \\ 8 & -2 \end{vmatrix}}{-2} = \frac{-4}{-2} = 2,$$

$$y = \frac{D_y}{D} = \frac{\begin{vmatrix} \dfrac{1}{2} & -2 \\ 1 & 8 \end{vmatrix}}{-2} = \frac{6}{-2} = -3$$

The solution is $x = 2$, $y = -3$.

28.
$$\begin{cases} 2x - y = -1 \\ x + \dfrac{1}{2}y = \dfrac{3}{2} \end{cases}$$

$$D = \begin{vmatrix} 2 & -1 \\ 1 & \dfrac{1}{2} \end{vmatrix} = 2$$

$$x = \frac{D_x}{D} = \frac{\begin{vmatrix} -1 & -1 \\ \dfrac{3}{2} & \dfrac{1}{2} \end{vmatrix}}{2} = \frac{1}{2} = 2,$$

$$y = \frac{D_y}{D} = \frac{\begin{vmatrix} 2 & -1 \\ 1 & \dfrac{3}{2} \end{vmatrix}}{2} = \frac{4}{2} = 2$$

The solution is $x = \dfrac{1}{2}$, $y = 2$.

30.
$$\begin{cases} x - y + z = -4 \\ 2x - 3y + 4z = -15 \\ 5x + y - 2z = 12 \end{cases}$$

$$D = \begin{vmatrix} 1 & -1 & 1 \\ 2 & -3 & 4 \\ 5 & 1 & -2 \end{vmatrix} = 1\begin{vmatrix} -3 & 4 \\ 1 & -2 \end{vmatrix} - (-1)\begin{vmatrix} 2 & 4 \\ 5 & -2 \end{vmatrix} + 1\begin{vmatrix} 2 & -3 \\ 5 & 1 \end{vmatrix} = 1(2) + (-24) + 17 = -5$$

$$D_x = \begin{vmatrix} -4 & -1 & 1 \\ -15 & -3 & 4 \\ 12 & 1 & -2 \end{vmatrix} = -4\begin{vmatrix} -3 & 4 \\ 1 & -2 \end{vmatrix} - (-1)\begin{vmatrix} -15 & 4 \\ 12 & -2 \end{vmatrix} + 1\begin{vmatrix} -15 & -3 \\ 12 & 1 \end{vmatrix}$$

$$= -4(2) + (-18) + 21 = -5$$

$$D_y = \begin{vmatrix} 1 & -4 & 1 \\ 2 & -15 & 4 \\ 5 & 12 & -2 \end{vmatrix} = 1\begin{vmatrix} -15 & 4 \\ 12 & -2 \end{vmatrix} - (-4)\begin{vmatrix} 2 & 4 \\ 5 & -2 \end{vmatrix} + 1\begin{vmatrix} 2 & -15 \\ 5 & 12 \end{vmatrix}$$

$$= -18 + 4(-24) + 99 = -15$$

$$D_z = \begin{vmatrix} 1 & -1 & -4 \\ 2 & -3 & -15 \\ 5 & 1 & 12 \end{vmatrix} = 1\begin{vmatrix} -3 & -15 \\ 1 & 12 \end{vmatrix} - (-1)\begin{vmatrix} 2 & -15 \\ 5 & 12 \end{vmatrix} - 4\begin{vmatrix} 2 & -3 \\ 5 & 1 \end{vmatrix}$$

$$= -21 + 99 - 4(17) = 10$$

$$x = \frac{D_x}{D} = \frac{-5}{-5} = 1, \quad y = \frac{D_y}{D} = \frac{-15}{-5} = 3, \quad z = \frac{D_z}{D} = \frac{10}{-5} = -2$$

The solution is $x = 1$, $y = 3$, $z = -2$.

32. $\begin{cases} x + 4y - 3z = -8 \\ 3x - y + 3z = 12 \\ x + y + 6z = 1 \end{cases}$

$$D = \begin{vmatrix} 1 & 4 & -3 \\ 3 & -1 & 3 \\ 1 & 1 & 6 \end{vmatrix} = 1\begin{vmatrix} -1 & 3 \\ 1 & 6 \end{vmatrix} - 4\begin{vmatrix} 3 & 3 \\ 1 & 6 \end{vmatrix} + (-3)\begin{vmatrix} 3 & -1 \\ 1 & 1 \end{vmatrix}$$

$$= -9 - 4(15) - 13(4) = -81$$

$$D_x = \begin{vmatrix} -8 & 4 & -3 \\ 12 & -1 & 3 \\ 1 & 1 & 6 \end{vmatrix} = -8\begin{vmatrix} -1 & 3 \\ 1 & 6 \end{vmatrix} - 4\begin{vmatrix} 12 & 3 \\ 1 & 6 \end{vmatrix} - 3\begin{vmatrix} 12 & -1 \\ 1 & 1 \end{vmatrix}$$

$$= -8(-9) - 4(69) - 3(13) = -243$$

$$D_y = \begin{vmatrix} 1 & -8 & -3 \\ 3 & 12 & 3 \\ 1 & 1 & 6 \end{vmatrix} = 1\begin{vmatrix} 12 & 3 \\ 1 & 6 \end{vmatrix} - (-8)\begin{vmatrix} 3 & 3 \\ 1 & 6 \end{vmatrix} - 3\begin{vmatrix} 3 & 12 \\ 1 & 1 \end{vmatrix}$$

$$= 69 + 8(15) - 3(-9) = 216$$

$$D_z = \begin{vmatrix} 1 & 4 & -8 \\ 3 & -1 & 12 \\ 1 & 1 & 1 \end{vmatrix} = 1\begin{vmatrix} -1 & 12 \\ 1 & 1 \end{vmatrix} - 4\begin{vmatrix} 3 & 12 \\ 1 & 1 \end{vmatrix} - 8\begin{vmatrix} 3 & -1 \\ 1 & 1 \end{vmatrix}$$

$$= -13 - 4(-9) - 8(4) = -9$$

$$x = \frac{D_x}{D} = \frac{-243}{-81} = 3, \; y = \frac{D_y}{D} = \frac{216}{-81} = \frac{-8}{3}, \; z = \frac{D_z}{D} = \frac{-9}{-81} = \frac{1}{9}$$

The solution is $x = 3$, $y = \dfrac{-8}{3}$, $z = \dfrac{1}{9}$.

34. $\begin{cases} x - y + 2z = 5 \\ 3x + 2y = 4 \\ -2x + 2y - 4z = -10 \end{cases}$

$$D = \begin{vmatrix} 1 & -1 & 2 \\ 3 & 2 & 0 \\ -2 & 2 & -4 \end{vmatrix} = 1\begin{vmatrix} 2 & 0 \\ 2 & -4 \end{vmatrix} - (-1)\begin{vmatrix} 3 & 0 \\ -2 & -4 \end{vmatrix} + 2\begin{vmatrix} 3 & 2 \\ -2 & 2 \end{vmatrix} = -8 + (-12) + 2(10) = 0$$

Thus, Cramer's Rule is not applicable.

36. $\begin{cases} x + 4y - 3z = 0 \\ 3x - y + 3z = 0 \\ x + y + 6z = 0 \end{cases}$

$$D = \begin{vmatrix} 1 & 4 & -3 \\ 3 & -1 & 3 \\ 1 & 1 & 1 \end{vmatrix} = 1\begin{vmatrix} -1 & 3 \\ 1 & 6 \end{vmatrix} - 4\begin{vmatrix} 3 & 3 \\ 1 & 6 \end{vmatrix} - 3\begin{vmatrix} 3 & -1 \\ 1 & 1 \end{vmatrix} = -9 - 4(15) - 3(4) = -81$$

$$D_x = \begin{vmatrix} 0 & 4 & -3 \\ 0 & -1 & 3 \\ 0 & 1 & 6 \end{vmatrix} = 0 \;\; \text{(by Theorem 12)}$$

$$D_y = \begin{vmatrix} 1 & 0 & -3 \\ 3 & 0 & 3 \\ 1 & 0 & 6 \end{vmatrix} = 0 \;\; \text{(by Theorem 12)}$$

$$D_z = \begin{vmatrix} 1 & 4 & 0 \\ 3 & -1 & 0 \\ 1 & 1 & 0 \end{vmatrix} = 0 \quad \text{(by Theorem 12)}$$

$$x = \frac{D_x}{D} = \frac{0}{-81} = 0, \quad y = \frac{D_y}{D} = \frac{0}{-81} = 0, \quad z = \frac{D_z}{D} = \frac{0}{-81} = 0$$

The solution is $x = 0$, $y = 0$, $z = 0$.

38. $\begin{cases} x - y + 2z = 5 \\ 3x + 2y = 0 \\ -2x + 2y - 4z = 0 \end{cases}$

$$D = \begin{vmatrix} 1 & -1 & 2 \\ 3 & 2 & 0 \\ -2 & 2 & -4 \end{vmatrix} = 1\begin{vmatrix} 2 & 0 \\ 2 & -4 \end{vmatrix} - (-1)\begin{vmatrix} 3 & 0 \\ -2 & -4 \end{vmatrix} + 2\begin{vmatrix} 3 & 2 \\ -2 & 2 \end{vmatrix} = -8 + (-12) + 2(10) = 0$$

Thus, Cramer's Rule is not applicable.

40. $\begin{cases} \dfrac{4}{x} - \dfrac{3}{y} = 0 \\[2mm] \dfrac{6}{x} + \dfrac{3}{2y} = 2 \end{cases}$

Let $u = \dfrac{1}{x}$ and $v = \dfrac{1}{y}$, then $\quad 4u - 3v = 0$

$$6u + \frac{3}{2}v = 2$$

$$D = \begin{vmatrix} 4 & -3 \\ 6 & \dfrac{3}{2} \end{vmatrix} = 30$$

$$u = \frac{D_u}{D} = \frac{\begin{vmatrix} 0 & -3 \\ 2 & \dfrac{3}{2} \end{vmatrix}}{24} = \frac{6}{24} = \frac{1}{4}, \quad \frac{1}{x} = \frac{1}{4} \rightarrow x = 4$$

$$v = \frac{D_v}{D} = \frac{\begin{vmatrix} 4 & 0 \\ 6 & 2 \end{vmatrix}}{24} = \frac{8}{24} = \frac{1}{3}, \quad \frac{1}{y} = \frac{1}{3} \rightarrow y = 3$$

The solution is $x = 4$, $y = 3$.

42. $\begin{vmatrix} x & 1 \\ 3 & x \end{vmatrix} = -2$

$$x^2 - 3 = 2$$
$$x^2 = 1$$
$$x = \pm 1$$

44. $\begin{vmatrix} 3 & 2 & 4 \\ 1 & x & 5 \\ 0 & 1 & -2 \end{vmatrix} = 0$

$$3 \begin{vmatrix} x & 5 \\ 1 & -2 \end{vmatrix} - 2 \begin{vmatrix} 1 & 5 \\ 0 & -2 \end{vmatrix} + 4 \begin{vmatrix} 1 & x \\ 0 & 1 \end{vmatrix} = 0$$
$$3(-2x - 5) - 2(-2) + 4(1) = 0$$
$$-6x = 7$$
$$x = \frac{-7}{6}$$

46. $\begin{vmatrix} x & 1 & 2 \\ 1 & x & 3 \\ 0 & 1 & 2 \end{vmatrix} = -4x$

$$x \begin{vmatrix} x & 3 \\ 1 & 2 \end{vmatrix} - 1 \begin{vmatrix} 1 & 3 \\ 0 & 2 \end{vmatrix} + 2 \begin{vmatrix} 1 & x \\ 0 & 1 \end{vmatrix} = -4x$$
$$x(2x - 3) - 2 + 2 = -4x$$
$$2x^2 + x = 0$$
$$x(2x + 1) = 0$$
$$x = 0 \text{ or } x = \frac{-1}{2}$$

48. $\begin{vmatrix} x & y & z \\ u & v & w \\ 2 & 4 & 6 \end{vmatrix} = 2 \begin{vmatrix} x & y & z \\ u & v & w \\ 1 & 2 & 3 \end{vmatrix} = 2(4) = 8 \quad \text{(by Theorem 13)}$

50. $\begin{vmatrix} 1 & 2 & 3 \\ x - u & y - v & z - w \\ u & v & w \end{vmatrix} = 4$

By Theorem 11, rows one and three were interchanged to change the sign of the determinant to negative, but rows two and three were also interchanged which changes the sign of the determinant back to positive. Multiplying -1 times row 3 and adding to row 2 does not change the value of the determinant.

52. $\begin{vmatrix} x & y & z - x \\ u & v & w - u \\ 1 & 2 & 2 \end{vmatrix} = 4$

Column 1 is multiplied by -1 and the result is added to column 3. Thus, the value of the determinant remains unchanged.

54. $\begin{vmatrix} x + 3 & y + 6 & z + 9 \\ 3u - 1 & 3v - 2 & 3w - 3 \\ 1 & 2 & 3 \end{vmatrix} = 12$

Row 3 is multiplied by 3 and the result is added to row 1. This does not change the value of the determinant. Multiplying row 3 by -1 and adding to row 2 does not change the value of the determinant. Row 2 is multiplied by 3, so the value of the determinant is multiplied by 3.

56. Any point (x, y) on the line containing (x_2, y_2) and (x_3, y_3) obeys

$$\begin{vmatrix} x & y & 1 \\ x_2 & y_2 & 1 \\ x_3 & y_3 & 1 \end{vmatrix} = 0$$

If (x_1, y_1) is also on this line, so that (x_1, y_1), (x_2, y_2), (x_3, y_3) are collinear, then

$$\begin{vmatrix} x_1 & y_1 & 1 \\ x_2 & y_2 & 1 \\ x_3 & y_3 & 1 \end{vmatrix} = 0$$

Conversely, if $\begin{vmatrix} x_1 & y_1 & 1 \\ x_2 & y_2 & 1 \\ x_3 & y_3 & 1 \end{vmatrix} = 0$, then (x_1, y_1) is a point on the line containing (x_2, y_2) and (x_3, y_3), so the three points are collinear.

58. Cramer's Rule for two equations containing two variables asserts that the solution to the system of equations

$$ax + by = s$$
$$cx + dy = t$$

is given by $x = \dfrac{\begin{vmatrix} s & b \\ t & d \end{vmatrix}}{\begin{vmatrix} a & b \\ c & d \end{vmatrix}}$, $y = \dfrac{\begin{vmatrix} a & s \\ c & t \end{vmatrix}}{\begin{vmatrix} a & b \\ c & d \end{vmatrix}}$

provided that $D = \begin{vmatrix} a & b \\ c & d \end{vmatrix} = ad - bc \neq 0$

Case 1: If $a = 0$, then $b \neq 0$ and $c \neq 0$, so $D = -bc \neq 0$.

Then, $x = \dfrac{sd - bt}{-bc}$, $y = \dfrac{-sc}{-bc}$ is a solution.

Case 2: If $b = 0$, then $a \neq 0$ and $d \neq 0$, so $D = ad \neq 0$.

Then, $x = \dfrac{sd}{ad}$, $y = \dfrac{at - sc}{ad}$ is a solution.

Case 3: If $c = 0$, then $a \neq 0$ and $d \neq 0$, so $D = ad \neq 0$.

Then, $x = \dfrac{sd - bt}{ad}$, $y = \dfrac{at}{ad}$ is a solution.

Case 4: If $d = 0$, then $b \neq 0$ and $c \neq 0$, so $D = -bc \neq 0$.

Then, $x = \dfrac{-bt}{-bc}$, $y = \dfrac{at - sc}{-bc}$ is a solution.

60.

$$\begin{vmatrix} a_{11} & a_{12} & a_{13} \\ ka_{21} & ka_{22} & ka_{23} \\ a_{31} & a_{32} & a_{33} \end{vmatrix}$$

$= a_{11}(ka_{22}a_{33} - ka_{23}a_{32}) - a_{12}(ka_{21}a_{33} - ka_{23}a_{31}) + a_{13}(ka_{21}a_{32} - ka_{22}a_{31})$

$= ka_{11}(a_{22}a_{32} - a_{23}a_{32}) - ka_{12}(a_{21}a_{33} - ka_{23}a_{31}) + ka_{13}(a_{21}a_{32} - a_{22}a_{31})$

$= k[a_{11}(a_{22}a_{32} - a_{23}a_{32}) - a_{12}(a_{21}a_{33} - a_{23}a_{31}) + a_{13}(a_{21}a_{32} - a_{22}a_{31})]$

$$= k\begin{vmatrix} a_{11} & a_{12} & a_{13} \\ a_{21} & a_{22} & a_{23} \\ a_{31} & a_{32} & a_{33} \end{vmatrix}$$

62.

$$\begin{vmatrix} a_{11} + ka_{21} & a_{12} + ka_{22} & a_{31} + ka_{23} \\ a_{21} & a_{22} & a_{23} \\ a_{31} & a_{32} & a_{33} \end{vmatrix}$$

$= (a_{11} + ka_{21})(a_{22}a_{33} - a_{23}a_{32}) - (a_{12} + ka_{22})(a_{21}a_{33} - a_{23}a_{31}) + (a_{13} + ka_{23})(a_{21}a_{32} - a_{22}a_{31})$

$= a_{11}(a_{22}a_{33} - a_{23}a_{32}) + ka_{21}(a_{22}a_{33} - a_{23}a_{32}) - a_{12}(a_{21}a_{33} - a_{23}a_{31}) - ka_{22}(a_{21}a_{33} - a_{23}a_{31})$
$\quad + a_{13}(a_{21}a_{32} - a_{22}a_{31}) + ka_{23}(a_{21}a_{32} - a_{22}a_{31})$

$= a_{11}(a_{22}a_{33} - a_{23}a_{32}) - a_{12}(a_{21}a_{33} - a_{23}a_{31}) + a_{13}(a_{21}a_{32} - a_{22}a_{31}) + ka_{21}a_{22}a_{33} - ka_{21}a_{23}a_{32}$
$\quad - ka_{22}a_{21}a_{33} + ka_{22}a_{23}a_{31} + ka_{23}a_{21}a_{32} - k_{23}a_{22}a_{31}$

$= a_{11}(a_{22}a_{33} - a_{23}a_{32}) - a_{12}(a_{21}a_{33} - a_{23}a_{31}) + a_{13}(a_{21}a_{32} - a_{22}a_{31})$

$$= \begin{vmatrix} a_{11} & a_{12} & a_{13} \\ a_{21} & a_{22} & a_{23} \\ a_{31} & a_{32} & a_{33} \end{vmatrix}$$

2. $\begin{cases} x^2 + y^2 = 8 \\ x^2 + y^2 + 4y = 0 \end{cases}$

$$8 + 4y = 0$$
$$4y = -8$$
$$y = -2$$
$$x^2 + (-2)^2 = 8$$
$$x^2 + 4 = 8$$
$$x^2 = 4$$
$$x = \pm 2$$
$$x = -2, y = -2; x = 2, -2$$

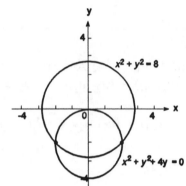

4. $\begin{cases} x^2 + y^2 = 10 \\ y = x + 2 \end{cases}$

$$x^2 + (x + 2)^2 = 10$$
$$x^2 + x^2 + 4x + 4 = 10$$
$$2x^2 + 4x - 6 = 0$$
$$2(x^2 + 2x - 3) = 0$$
$$2(x + 3)(x - 1) = 0$$
$$x = 1, \qquad x = -3$$
$$y = 1 + 2 = 3, y = -3 + 2 = -1$$
$$x = 1, y = 3; x = -3, y = -1$$

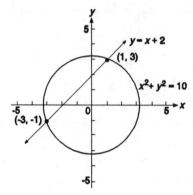

6. $\begin{cases} x^2 + y^2 = 16 \\ x^2 - 2y = 8 \end{cases}$

$$y^2 + 2y = 8$$
$$y^2 + 2y - 8 = 0$$
$$(y + 4)(y - 2) = 0$$
$$y = -4 \text{ or } y = 2$$

Thus, there are three points of intersection:

$$x = 0, y = -4; x = \sqrt{12}, y = 2;$$
$$x = -\sqrt{12}, y = 2$$

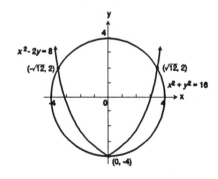

8. $\begin{cases} x^2 = y \\ xy = 1 \end{cases}$

$$y = x^2$$
$$x(x^2) = 1$$
$$x^3 = 1$$
$$x = 1$$

Thus, there is one point of intersection:
$$x = 1, y = 1$$

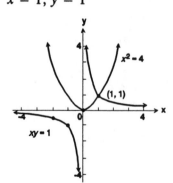

10. $\begin{cases} xy = 1 \\ y = 2x + 1 \end{cases}$

$$x(2x + 1) = 1$$
$$2x^2 + x - 1 = 0$$
$$(2x - 1)(x + 1) = 0$$
$$x = \frac{1}{2}, \qquad x = -1$$
$$y = 2, \qquad y = -1$$
$$x = \frac{1}{2}, y = 2; x = -1, y = -1$$

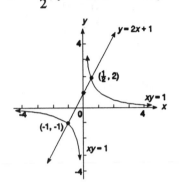

12. $\begin{cases} x^2 + y^2 = 10 \\ xy = 3 \end{cases}$

$$x = \frac{3}{y}$$
$$\left[\frac{3}{y}\right]^2 + y^2 = 10$$
$$y^2 \left[\frac{9}{y^2} + y^2\right] = 10y^2$$
$$9 + y^4 = 10y^2$$
$$y^4 - 10y^2 + 9 = 0$$
$$(y^2 - 9)(y^2 - 1) = 0$$
$$y^2 = 9 \qquad y^2 = 1$$
$$y = \pm 3 \qquad y = \pm 1$$
$$x = -3, y = -1; x = 3, y = 1;$$
$$x = -1, y = -3; x = 1, y = 3$$

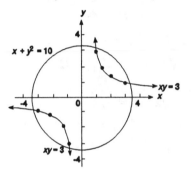

14. $\begin{cases} x^2 - y^2 = 21 \\ x + y = 7 \end{cases}$

$$y = 7 - x$$
$$x^2 - (7 - x)^2 = 21$$
$$x^2 - 49 + 14x - x^2 = 21$$
$$14x = 70$$
$$x = \frac{70}{14}$$
$$= 5$$
$$y = 7 - 5 = 2$$
Hence, $x = 5, y = 2$

16. $\begin{cases} x^2 - 4y^2 = 16 \\ 2y - x = 2 \end{cases}$

$$x = 2y - 2$$
$$(2y - 2)^2 - 4y^2 = 16$$
$$4y^2 - 8y + 4 - 4y^2 = 16$$
$$-8y = 12$$
$$y = \frac{-12}{8}$$
$$= \frac{-3}{2}$$
$$x = 2\left[\frac{-3}{2}\right] - 2 = -5$$
Hence, $x = -5, y = \frac{-3}{2}$

18. $\begin{cases} 2x^2 - xy + y^2 = 8 \\ \qquad\qquad xy = 4 \end{cases}$

$x = \dfrac{4}{y}$

$2\left[\dfrac{4}{y}\right]^2 - \dfrac{4}{y}(y) + y^2 = 8$

$\dfrac{32}{y^2} + y^2 = 12$

$32 + y^4 = 12y^2$

$y^4 - 12y^2 + 32 = 0$

Let $u = y^2$, then $u^2 = y^4$

$u^2 - 12u + 32 = 0$

$(u - 8)(u - 4) = 0$

$u = 8 \qquad$ or $\quad u = 4$

$y^2 = 8 \qquad$ or $\quad y^2 = 4$

$y = \pm2\sqrt{2}$, or $y = \pm2$

If $y = 2\sqrt{2}$, $x = \dfrac{4}{2\sqrt{2}} = \sqrt{2}$

If $y = -2\sqrt{2}$, $x = -\sqrt{2}$

If $y = 2$, $x = \dfrac{4}{2} = 2$

If $y = -2$, $x = -2$

20. $\begin{cases} 2y^2 - 3xy + 6y + 2x + 4 = 0 \\ \qquad\qquad\quad 2x - 3y + 4 = 0 \end{cases}$

$2y^2 - 3xy + 9y = 0$

$x = \dfrac{3y - 4}{2}$

$2y^2 - 3\left[\dfrac{3y - 4}{2}\right]y + 9y = 0$

$4y^2 - 9y^2 + 12y + 18y = 0$

$-5y^2 + 30y = 0$

$y(-5y + 30) = 0$

$y = 0$ or $y = 6$

If $y = 0$, $x = -2$

If $y = 6$, $x = 7$

22. $\begin{cases} 3x^2 - 2y^2 + 5 = 0 \\ 2x^2 - y^2 + 2 = 0 \end{cases}$

$\begin{cases} 3x^2 - 2y^2 + 5 = 0 \\ -4x^2 - 2y^2 - 4 = 0 \end{cases}$

$-x^2 + 1 = 0$

$-x^2 = -1$

$x = \pm1$

$3(1)^2 - 2y^2 + 5 = 0$

$-2y^2 = -8$

$y^2 = 4$

$y = \pm2$

$2(1)^2 - y^2 + 2 = 0$

$-y^2 = -4$

$y = \pm2$

The solutions are $x = 1$, $y = 2$; $x = -1$,
$y = 2$; $x = 1$, $y = -2$; $x = -1$, $y = -2$

24. $\begin{cases} x^2 - 3y^2 + 1 = 0 \\ 2x^2 - 7y^2 + 5 = 0 \end{cases}$

$\begin{cases} -2x^2 + 6y^2 - 2 = 0 \\ 2x^2 - 7y^2 + 5 = 0 \end{cases}$

$-y^2 + 3 = 0$

$y^2 = 3$

$y = \pm\sqrt{3}$

If $y = \sqrt{3}$, $x^2 - 3\left(\sqrt{3}\right)^2 + 1 = 0$

$x^2 = 8$

$x = \pm2\sqrt{2}$

If $y = -\sqrt{3}$, $x^2 - 3\left(-\sqrt{3}\right)^2 + 1 = 0$

$x = \pm2\sqrt{2}$

26. $\begin{cases} 5xy + 13y^2 + 36 = 0 \\ \quad\ \ xy + 7y^2 = 6 \end{cases}$

$xy = 6 - 7y^2$

$5(6 - 7y^2) + 13y^2 + 36 = 0$

$30 - 35y^2 + 13y^2 + 36 = 0$

$\qquad\qquad\qquad -22y^2 = -66$

$\qquad\qquad\qquad\quad y^2 = \dfrac{66}{22} = 3$

$\qquad\qquad\qquad\quad\ y = \pm\sqrt{3}$

If $y = \sqrt{3}$, $x\left(\sqrt{3}\right) + 7\left(\sqrt{3}\right)^2 = 6$

$\qquad\qquad\quad \sqrt{3}\,x + 21 = 6$

$\qquad\qquad\qquad \sqrt{3}\,x = -15$

$\qquad\qquad\qquad\quad\ x = \dfrac{-15}{\sqrt{3}}$

$\qquad\qquad\qquad\qquad = -5\sqrt{3}$

If $y = -\sqrt{3}$, $x = 5\sqrt{3}$

28. $\begin{cases} y^2 - x^2 + 4 = 0 \\ 2x^2 + 3y^2 = 6 \end{cases}$

$y^2 = x^2 - 4$

$2x^2 + 3(x^2 - 4) = 6$

$2x^2 + 3x^2 - 12 = 6$

$\qquad\quad 5x^2 - 18 = 0$

$\qquad\qquad\quad x^2 = \dfrac{18}{5}$

$\qquad\qquad\quad\ x = \pm\sqrt{\dfrac{18}{5}} = \dfrac{\pm 3\sqrt{2}}{5}$

$y^2 = \dfrac{18}{5} - 4$

$y^2 = \dfrac{-2}{5}$

No real solution.

The system is inconsistent.

30. $\begin{cases} 4x^2 + 3y^2 = 4 \\ 2x^2 - 6y^2 = -3 \end{cases}$

$\begin{cases} \ \ 4x^2 + 3y^2 = 4 \\ -4x^2 + 12y^2 = 6 \end{cases}$

$15y^2 = 10$

$\quad y^2 = \dfrac{10}{15} = \dfrac{2}{3}$

$\quad\ y = \pm\dfrac{\sqrt{6}}{3}$

If $y = \pm\dfrac{\sqrt{6}}{3}$, $\quad 4x^2 + 3\left[\dfrac{6}{9}\right] = 4$

$\qquad\qquad\qquad\qquad 4x^2 = 2$

$\qquad\qquad\qquad\qquad\ x^2 = \dfrac{1}{2}$

$\qquad\qquad\qquad\qquad\ x = \pm\dfrac{\sqrt{2}}{2}$

32. $\begin{cases} \dfrac{2}{x^2} - \dfrac{3}{y^2} + 1 = 0 \\ \dfrac{6}{x^2} - \dfrac{7}{y^2} + 2 = 0 \end{cases}$

$\begin{cases} 2y^2 - 3x^2 + x^2y^2 = 0 \\ 6y^2 - 7x^2 + 2x^2y^2 = 0 \end{cases}$

$\begin{cases} -6y^2 + 9x^2 - 3 + x^2y^2 = 0 \\ \ \ 6y^2 - 7x^2 + 2x^2y^2 = 0 \end{cases}$

$2x^2 - x^2y^2 = 0$

$y^2 = 2, x \neq 0$

$y = \pm\sqrt{2}$

If $y = \pm\sqrt{2}$, then

$\qquad \dfrac{2}{x^2} - \dfrac{3}{\left(\sqrt{2}\right)^2} + 1 = 0$

$\qquad\qquad\qquad \dfrac{2}{x^2} = \dfrac{1}{2}$

$\qquad\qquad\qquad\ x^2 = 4$

$\qquad\qquad\qquad\ x = \pm 2$

34.

$$\begin{cases} \dfrac{1}{x^4} - \dfrac{1}{y^4} = 1 \\[2mm] \dfrac{1}{x^4} + \dfrac{1}{y^4} = 4 \end{cases}$$

$$\begin{cases} y^4 - x^4 = x^4 y^4 \\ y^4 + x^4 = 4x^4 y^4 \end{cases}$$

$$2y^4 = 5x^4 y^4$$

$$x^4 = \frac{2}{5},\ y \neq 0$$

$$x = \pm \sqrt[4]{\frac{2}{5}}$$

If $x = \pm \sqrt[4]{\dfrac{2}{5}}$, then

$$\frac{1}{\left[\sqrt[4]{\dfrac{2}{5}}\right]^4} - \frac{1}{y^4} = 1$$

$$\frac{5}{2} - \frac{1}{y^4} = 1$$

$$\frac{-1}{y^4} = \frac{-3}{2}$$

$$3y^4 = 2$$

$$y^4 = \frac{2}{3}$$

$$y = \pm \sqrt[4]{\frac{2}{3}}$$

38.

$$\begin{cases} 5x^2 + 4xy + 3y^2 = 36 \\ x^2 + xy + y^2 = 9 \end{cases}$$

$$\begin{cases} 5x^2 + 4xy + 3y^2 = 36 \\ -4x^2 - 4xy + 4y^2 = -36 \end{cases}$$

$$x^2 - y^2 = 0$$
$$x^2 = y^2$$
$$x = \pm y$$

If $x = y$, $\ x^2 + x^2 + x^2 = 9$
$$3x^2 = 9$$
$$x^2 = 3$$
$$x = \pm\sqrt{3}$$

Thus, if $x = \sqrt{3}$, $y = \sqrt{3}$
and if $x = -\sqrt{3}$, $y = -\sqrt{3}$.

36.

$$\begin{cases} x^2 - xy - 2y^2 = 0 \\ xy + x + 6 = 0 \end{cases}$$

$$(x - 2y)(x + y) = 0$$

$$x - 2y = 0 \quad \text{or} \quad x + y = 0$$
$$x = 2y \quad \text{or} \quad x = -y$$

If $x = 2y$,
$$(2y)y + 2y + 6 = 0$$
$$2y^2 + 2y + 6 = 0$$
$$2(y^2 + y + 3) = 0$$
$$\text{(No real solution)}$$

or if $x = -y$,
$$(-y)y + (-y) + 6 = 0$$
$$-y^2 - y + 6 = 0$$
$$y^2 + y - 6 = 0$$
$$(y + 3)(y - 2) = 0$$
$$y = -3 \quad \text{or} \quad y = 2$$

Thus, $x = 3$ or $x = -2$.
The solutions are $x = 3, y = -3$ or $x = -2, y = 2$.

If $x = -y$, $x^2 - x^2 + x^2 = 0$
$$x^2 = 9$$
$$x = \pm 3$$

Thus, if $x = 3, y = -3$
and if $x = -3, y = 3$.

40. $\begin{cases} x^3 + y^3 = 26 \\ x + y = 2 \end{cases}$

$x = 2 - y$

$(2 - y)^3 + y^3 = 26$

$8 - 12y + 6y^2 - y^3 + y^3 = 26$

$6y^2 - 12y - 18 = 0$

$6(y^2 - 2y - 3) = 0$

$6(y - 3)(y + 1) = 0$

$y = 3$ or $y = -1$

If $y = 3, x = 2 - 3 = -1$.

If $y = -1, x = 2 - (-1) = 3$.

42. $\begin{cases} x^3 - 2x^2 + y^2 + 3y - 4 = 0 \\ x - 2 + \dfrac{y^2 - y}{x^2} = 0 \end{cases}$

$\begin{cases} x^3 - 2x^2 + y^2 + 3y - 4 = 0 \\ x^3 - 2x^2 + y^2 - y = 0 \end{cases}$

$4y - 4 = 0$

$y = 1$

If $y = 1, \quad x - 2 = 0$

$x = 2$

44. $\begin{cases} \log_x(2y) = 3 \\ \log_x(4y) = 2 \end{cases}$

$\begin{cases} 2y = x^3 \\ 4y = x^2 \end{cases}$

$4 \left[\dfrac{x^3}{2} \right] = x^2$

$2x^3 - x^2 = 0$

$x^2(2x - 1) = 0$

$x = 0, x = \dfrac{1}{2}$

0 is extraneous. Thus, $x = \dfrac{1}{2}, y = \dfrac{1}{16}$

46. Let x be one number and y be the other number. Then

$\begin{cases} x + y = 7 \\ x^2 - y^2 = 21 \end{cases}$

$x = 7 - y$

$(7 - y)^2 - y^2 = 21$

$49 - 14y + y^2 - y^2 = 21$

$-14y = -28$

$y = 2$

$x = 5, y = 2$

$x = 7 - 2 = 5$

Hence, the two numbers are 5 and 2.

48. Let x be one number and y be the other number. Then

$\begin{cases} xy = 10 \\ x^2 - y^2 = 21 \end{cases}$

$x = \dfrac{10}{y}$

$\left[\dfrac{10}{y} \right]^2 - y^2 = 21$

$100 - y^4 = 21y^2$

$y^4 + 21y^2 - 100 = 0$

Let $u = y^2$. Then

$u^2 + 21u - 100 = 0$

$(u + 25)(u - 4) = 0$

$u = -25$ or $u = 4$

Thus, $y^2 = -25$ (impossible) or $y^2 = 4$.

Hence, $y = \pm 2$

If $y = 2$, then $x = \dfrac{10}{2} = 5$

If $y = -2$, then $x = \dfrac{10}{-2} = -5$

The two numbers are 2 and 5 or -2 and -5.

50. Let x be one number and y be the other number. Then

$\begin{cases} x + y = xy \\ \dfrac{1}{x} - \dfrac{1}{y} = 3 \end{cases}$

$\begin{cases} x + y = xy \\ y - x = 3xy \end{cases}$

$y - x = 3(x + y)$

$y - x = 3x + 3y$

$y - 3y = 3x + x$

$-2y = 4x$

$y = -2x$

$x - 2x = x(-2x)$

$-x = -2x^2$

$2x^2 - x = 0$

$x(2x - 1) = 0$

$x = 0$ (impossible) or $x = \dfrac{1}{2}$

If $x = \dfrac{1}{2}, y = -1$

52. $\begin{cases} \dfrac{a}{b} = \dfrac{4}{3} \\ a + b = 14 \end{cases}$

$\dfrac{14 - a}{b} = \dfrac{4}{3}$

$3(14 - a) = 4b$

$42 = 7b$

$b = 6$

$a = 8$

$a - b = 2, a + b = 14$

Ratio of $a - b$ to $a + b$ is $\dfrac{2}{14} = \dfrac{1}{7}$

54. $x^2 + 1 = 4x + 1$

$x^2 - 4x = 0$

$x(x - 4) = 0$

$x = 0, x = 4$

$x = 0, y = 1; x = 4, y = 17$

$(0, 1); (4, 17)$

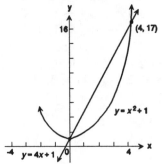

56. $\sqrt{4 - x^2} = 2x + 4$

$4 - x^2 = 4x^2 + 16x + 16$

$5x^2 + 16x + 12 = 0$

$(5x + 6)(x + 2) = 0$

$x = -\dfrac{6}{5} \qquad x = -2$

If $x = -\dfrac{6}{5}$, then $y = \dfrac{-12}{5} + 4 = \dfrac{8}{5}$

If $x = -2$, then $y = 0$

$\left(-\dfrac{6}{5}, \dfrac{8}{5}\right); (-2, 0)$

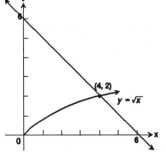

58. $\sqrt{x} = 6 - x$

$x = 36 - 12x + x^2$

$x^2 - 13x + 36 = 0$

$(x - 9)(x - 4) = 0$

$x = 9, x = 4$

$x = 9$ is extraneous

$x = 4, y = 2$

$(4, 2)$

60.
$$x - 1 = x^2 - 6x + 9$$
$$x^2 - 7x + 10 = 0$$
$$(x - 5)(x - 2) = 0$$
$$x = 5 \quad x = 2$$
$$x = 5, y = 4; x = 2, y = 1$$
$$(5, 4); (2, 1)$$

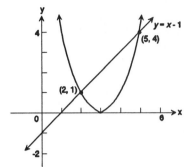

62.
$$\begin{cases} (x + 1)^2 + (y + 1)^2 = 5 \\ x + 2y + 6 = 0 \end{cases}$$
$$x = -2y - 6$$
$$(-2y - 5)^2 + (y + 1)^2 = 5$$
$$4y^2 + 20y + 25 + y^2 + 2y + 1 = 5$$
$$5y^2 + 22y + 21 = 0$$
$$(5y + 7)(y + 3) = 0$$
$$y = \frac{-7}{5} \text{ or } y = -3$$

When $y = \dfrac{-7}{5}$, $x = \dfrac{-16}{5}$

When $y = -3$, $x = 0$

Thus, the points of intersection are

$$(0, -3), \quad \left[-\frac{16}{5}, -\frac{7}{5}\right].$$

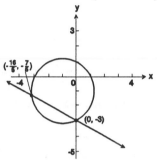

64.
$$\begin{cases} (x + 2)^2 + (y - 1)^2 = 4 \\ y^2 - 2y - x - 5 = 0 \end{cases}$$
$$(y - 1)^2 = x + 6$$
$$(x + 2)^2 + x + 6 = 4$$
$$x^2 + 4x + 4 + x + 6 - 4 = 0$$
$$x^2 + 5x + 6 = 0$$
$$(x + 2)(x + 3) = 0$$
$$x = -2 \text{ or } x = -3$$
$$(y - 1)^2 = 4 \text{ or } (y - 1)^2 = 3$$
$$y - 1 = \pm 2 \text{ or } y - 1 = \pm\sqrt{3}$$

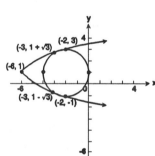

Thus, the points of intersection are $(-2, 3), (-2, -1), \left(-3, 1 + \sqrt{3}\right), \left(-3, 1 - \sqrt{3}\right).$

66.
$$\begin{cases} y = \dfrac{4}{x + 2} \\ x^2 + 4x + y^2 - 4 = 0 \end{cases}$$
$$y^2 = 4 - x^2 - 4x$$
$$\left[\frac{4}{x + 2}\right]^2 = 4 - x^2 - 4x$$
$$16 = (x + 2)^2(-x^2 - 4x + 4)$$
$$(x^2 + 4x + 4)(-x^2 - 4x + 4) - 16 = 0$$
$$-x^4 - 8x^3 - 16x^2 = 0$$
$$-x^2(x^2 + 8x + 16) = 0$$
$$-x^2(x + 4)^2 = 0$$
$$x = 0 \text{ or } x = -4$$

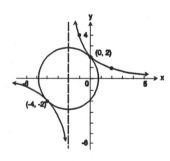

Thus, the points of intersection are $(0, 2), (-4, -2)$.

68. $\begin{cases} y = x^{3/2} \\ y = e^{-x} \end{cases}$

(0.64, 0.52)

70. $\begin{cases} x^3 + y^2 = 2 \\ x^2y = 4 \end{cases}$

(1.36, 2.13)

72. $\begin{cases} x^4 + y^4 = 6 \\ xy = 1 \end{cases}$

(0.64, 1.55)
(−0.64, −1.55)

74. $\begin{cases} x^2 + y^2 = 4 \\ y = \ln x \end{cases}$

(1.89, 0.63)
(0.13, −1.99)

76. Let $2x = $ side of the first square.
Let $3x = $ side of the second square.
$$(2x)^2 + (3x)^2 = 52$$
$$4x^2 + 9x^2 = 52$$
$$13x^2 = 52$$
$$x^2 = 4$$
$$x = 2$$
The sides of the first square measure 4 ft. The sides of the second square measure 6 ft.

78. Let x be the length of the two equal sides (an isosceles triangle has two sides equal); let y be the base, and let h be the altitude drawn to the base. We want to determine y, and we are given:
$$h = 3$$
and $\quad x + x + y = 18$
From the Pythagorean Theorem, applied to *half* of the triangle, we have:

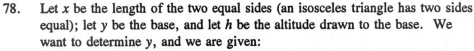

$$\left[\frac{y}{2}\right]^2 + h^2 = x^2$$
$$\frac{y^2}{4} + 9 = x^2 \quad (h = 3)$$
$$y^2 + 36 = 4x^2$$
This gives two equations in x and y:
$$\begin{cases} 2x + y = 18 & (1) \\ y^2 + 36 = 4x^2 & (2) \end{cases}$$
We are asked to find y, so let's eliminate x:
From (1): $2x + y = 18$
$$x = \frac{-y}{2} + 9$$
Then from (2):
$$y^2 + 36 = 4\left[\frac{-y}{2} + 9\right]^2$$
$$y^2 + 36 = 4\left[\frac{y^2}{4} - \frac{18}{2}y + 81\right]$$
$$y^2 + 36 = y^2 - 36y + 324$$
$$36y = 288$$
$$y = 8$$
Thus, the base of the triangle is $y = 8$ cm.

Chapter 12 Systems of Equations and Inequalities

80. Let v_1, v_2, v_3 = speeds of runners 1, 2, 3
t_1, t_2, t_3 = times of runners 1, 2, 3

$$\begin{cases} 5280 = v_1 t_1 \\ 5270 = v_2 t_1 \\ 5260 = v_3 t_1 \\ 5280 = v_2 t_2 \end{cases}$$

Distance between 2nd runner and 3rd runner after t_2 seconds is:

$$5280 - v_3 t_2 = 5280 - \left(\sqrt{3}\, t_1\right)\left[\frac{t_2}{t_1}\right]$$

$$= 5280 - 5260\left[\frac{5280}{5270}\right]$$

$$= 10.02 \text{ feet}$$

82. ℓ = length of cardboard, w = width;
$\ell w = 216$
Volume of tube = $V = \pi r^2 h$
where $2\pi r = w$ and $h = \ell$

$$224 = \pi \left[\frac{w}{2\pi}\right]^2 \cdot \ell$$

$$224 = \frac{w^2}{4\pi} \cdot \frac{216}{w}$$

$$224 = \frac{54w}{\pi}$$

$$w = \frac{224\pi}{54} = 13.03 \text{ cm}$$

$$\ell = \frac{216}{3} = 16.57 \text{ cm}$$

84. Let x = side of the square
Let r = radius of the circle
Then, Area of square = x^2; Area of circle = πr^2
Then, Perimeter of square = $4x$; Circumference of circle = $2\pi r$

$$\begin{cases} 4x + 2\pi r = 60 \\ x^2 + \pi r^2 = 100 \end{cases}$$

$$x = \frac{-1}{2}\pi r + 15$$

$$\left[\frac{-1}{2}\pi r + 15\right]^2 + \pi r^2 = 100$$

$$\frac{1}{4}\pi^2 r^2 - 15\pi r + 225 + \pi r^2 = 100$$

$$\left[\frac{\pi^2}{4} + \pi\right] r^2 - 15\pi r + 125 = 0$$

$$b^2 - 4ac = 225\pi^2 - 500\left[\frac{\pi^2}{4} + \pi\right] = 100\pi^2 - 500\pi < 0$$

No real solution. Therefore, it is not possible to bend 60 feet of wire into two pieces which have a total area of 100 square feet.

86.
$$P = b + 2\ell$$

$$h^2 + \frac{b^2}{4} = \ell^2$$

$$(4h^2 + P - 2\ell)^2 = 4\ell^2$$

$$4h^2 + P^2 - 4P\ell + 4\ell^2 = 4\ell^2$$

$$4P\ell = 4h^2 + P^2$$

$$\ell = \frac{4h^2 + P^2}{4P}$$

$$b = P - \frac{4h^2 + P^2}{2P}$$

88. $x^2 + y^2 = 10$ at $(1, 3)$

$$x^2 + y^2 = 10$$
$$y = mx + b$$
$$x^2 + (mx + b)^2 = 10$$
$$x^2 + m^2x^2 + 2mbx + b^2 - 10 = 0$$
$$(m^2 + 1)x^2 + 2mbx + b^2 - 10 = 0$$

This has one solution if

$$\begin{cases} (2mb)^2 - 4(m^2 + 1)(b^2 - 10) = 0 \\ \qquad\qquad\qquad\qquad 3 = m + b \end{cases}$$

$$4m^2b^2 - 4(m^2b^2 - 10m^2 + b^2 - 10) = 0$$
$$4m^2b^2 - 4m^2b^2 + 40m^2 - 4b^2 + 40 = 0$$

$$\begin{cases} 4(10m^2 - b^2 + 10) = 0 \\ \qquad\qquad m = 3 - b \end{cases}$$

$$10(3 - b)^2 - b^2 + 10 = 0$$
$$10(9 - 6b + b^2) - b^2 + 10 = 0$$
$$90 - 60b + 10b^2 - b^2 + 10 = 0$$
$$9b^2 - 60b + 100 = 0$$

$$b = \frac{60 \pm \sqrt{60^2 - 4(9)(100)}}{18}$$

$$b = \frac{60}{18} = \frac{10}{3}$$

$$m = 3 - \left[\frac{10}{3}\right] = \frac{-1}{3}$$

Thus, the equation of the tangent line is

$$y = \frac{-1}{3}x + \frac{10}{3}.$$

90. $x^2 + y = 5$ at $(-2, 1)$

$$\begin{cases} x^2 + y = 5 \\ y = mx + b \end{cases}$$

$$x^2 + mx + b - 5 = 0$$

This has one solution if
$$m^2 - 4(b - 5) = 0$$

$$\begin{cases} m^2 - 4b + 20 = 0 \\ 1 = -2m + b \Rightarrow b = 1 + 2m \end{cases}$$

$$m^2 - 4(1 + 2m) + 20 = 0$$
$$m^2 - 4 - 8m + 20 = 0$$
$$m^2 - 8m + 16 = 0$$
$$(m - 4)(m - 4) = 0$$
$$m = 4$$
$$b = 1 + 2(4) = 9$$

Thus, the equation of the tangent line is
$y = 4x + 9$.

92. $3x^2 + y^2 = 7$ at $(-1, 2)$

$$\begin{cases} 3x^2 + y^2 = 7 \\ \qquad y = mx + b \end{cases}$$

$$3x^2 + (mx + b)^2 = 7$$
$$3x^2 + m^2x^2 + 2bmx + b^2 = 7$$
$$(3 + m^2)x^2 + 2bmx + b^2 - 7 = 0$$

This has one solution if
$$4b^2m^2 - 4(3 + m^2)(b^2 - 7) = 0$$
$$4b^2m^2 - 4(3b^2 - 21 + m^2b^2 - 7m^2) = 0$$
$$-12b^2 + 28m^2 + 84 = 0$$

$$\begin{cases} -4(3b^2 - 7m^2 - 21) = 0 \\ 2 = -m + b \Rightarrow b = 2 + m \end{cases}$$

$$3(2 + m)^2 - 7m^2 - 21 = 0$$
$$3(4 + 4m + m^2) - 7m^2 - 21 = 0$$
$$12 + 12m + 3m^2 - 7m^2 - 21 = 0$$
$$4m^2 - 12m + 9 = 0$$
$$(2m - 3)(2m - 3) = 0$$

$$m = \frac{3}{2}$$

$$b = 2 + \frac{3}{2}$$

$$= \frac{7}{2}$$

Thus, the equation of the tangent line is

$$y = \frac{3}{2}x + \frac{7}{2}.$$

Chapter 12 Systems of Equations and Inequalities

94. $2y^2 - x^2 = 14$ at $(2, 3)$

$$\begin{cases} 2y^2 - x^2 = 14 \\ y = mx + b \end{cases}$$
$$2(mx + b)^2 - x^2 = 14$$
$$2(m^2x^2 + 2bmx + b^2) - x^2 = 14$$
$$(2m^2 - 1)x^2 + 4bmx + 2b^2 - 14 = 0$$

This has one solution if
$$16b^2m^2 - 4(2m^2 - 1)(2b^2 - 14) = 0$$
$$16b^2m^2 - 4(4m^2b^2 - 28m^2 - 2b^2 + 14) = 0$$
$$112m^2 + 8b^2 - 5b = 0$$
$$8(14m^2 + b^2 - 7) = 0$$

$$\begin{cases} 14m^2 + b^2 - 7 = 0 \\ 3 = 2m + b \Rightarrow b = 3 - 2m \end{cases}$$
$$14m^2 + (3 - 2m)^2 - 7 = 0$$
$$14m^2 + 9 - 12m + 4m^2 - 7 = 0$$
$$18m^2 - 12m + 2 = 0$$
$$2(9m^2 - 6m + 1) = 0$$
$$(3m - 1)(3m - 1) = 0$$

$$m = \frac{1}{3}$$

$$b = 3 - \frac{2}{3} \quad \frac{7}{3}$$

Thus, the equation of the tangent line is
$$y = \frac{1}{3}x + \frac{7}{3}.$$

12.5 Systems of Linear Inequalities

2. $y \geq 0$

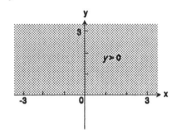

4. $y \leq 2$

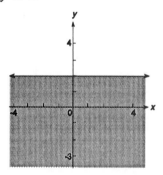

6. $3x + 2y \leq 6$

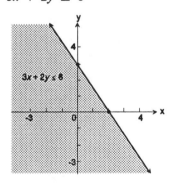

8. $x^2 + y^2 \leq 9$

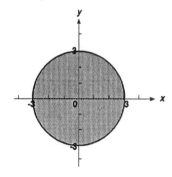

10. $y > x^2 + 2$

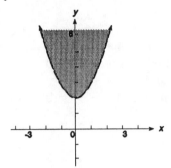

12. $xy \leq 1$

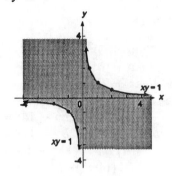

14. $\begin{cases} 3x - y \geq 6 \\ x + 2y \leq 2 \end{cases}$

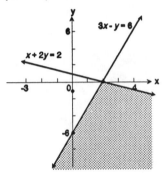

16. $\begin{cases} 4x - 5y \leq 0 \\ 2x - y \geq 2 \end{cases}$

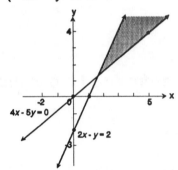

18. $\begin{cases} 4x - y \geq 2 \\ x + 2y \geq 2 \end{cases}$

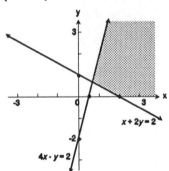

20. $\begin{cases} x^2 + y^2 \geq 9 \\ x + y \leq 3 \end{cases}$

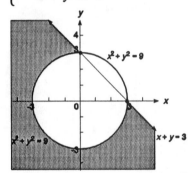

22. $\begin{cases} y^2 \leq x \\ y \geq x \end{cases}$

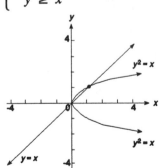

24. $\begin{cases} y + x^2 \leq 1 \\ y \geq x^2 - 1 \end{cases}$

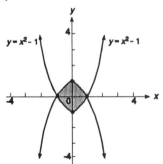

Chapter 12 Systems of Equations and Inequalities

26. $\begin{cases} x + 4y \leq 8 \\ x + 4y \geq 4 \end{cases}$

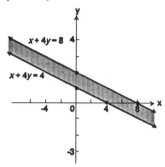

28. $\begin{cases} x - 4y \leq 4 \\ x - 4y \geq 0 \end{cases}$

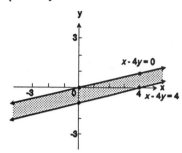

30. $\begin{cases} 2x + y \geq 0 \\ 2x + y \geq 2 \end{cases}$

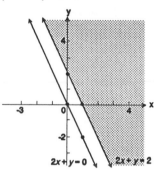

32. $\begin{cases} x \geq 0 \\ y \geq 0 \\ x + y \geq 4 \\ 2x + 3y \geq 6 \end{cases}$

Bounded; corner points $(0, 2)$, $(0, 4)$, $(4, 0)$, $(3, 0)$

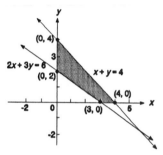

34. $\begin{cases} x \geq 0 \\ y \geq 0 \\ 3x + y \leq 6 \\ 2x + y \leq 2 \end{cases}$

Bounded; corner points $(0, 0)$, $(0, 2)$, $(1, 0)$

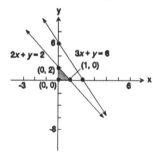

36. $\begin{cases} x \geq 0 \\ y \geq 0 \\ x + y \geq 2 \\ x + y \leq 10 \\ 2x + y \leq 3 \end{cases}$

Bounded; corner points $(0, 2)$, $(0, 3)$, $(1, 1)$

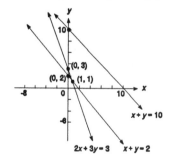

38. $\begin{cases} x \ge 0 \\ y \ge 0 \\ x + y \ge 2 \\ x + y \le 8 \\ x + 2y \ge 1 \end{cases}$

Bounded; corner points $(0, 2)$, $(2, 0)$, $(0, 8)$, $(8, 0)$

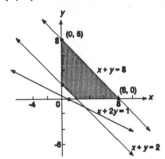

40. $\begin{cases} x \ge 0 \\ y \ge 0 \\ x + 2y \ge 1 \\ x + 2y \le 10 \\ x + y \ge 2 \\ x + y \le 8 \end{cases}$

Bounded; corner points $(0, 2)$, $(0, 5)$, $(6, 2)$, $(8, 0)$, $(2, 0)$

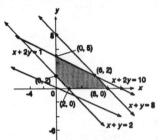

42. $\begin{cases} y \le 5 \\ x + y \ge 2 \\ x \le 6 \\ x \ge 0, y \ge 0 \end{cases}$

44. $\begin{cases} y \le 6 \\ x \le 5 \\ 3x + 4y \ge 12 \\ y \ge 2x - 8 \\ x \ge 0, y \ge 0 \end{cases}$

46. $\begin{cases} x \ge 0 \\ y \ge 0 \\ 3x + 4y \le 120 \quad (40 \text{ hours} \times 3 \text{ workers}) \\ 2x + 3y \le 80 \quad (40 \text{ hours} \times 2 \text{ workers}) \end{cases}$

Bounded; corner points $(0, 0)$, $\left(0, \dfrac{80}{3}\right)$, $(40, 0)$

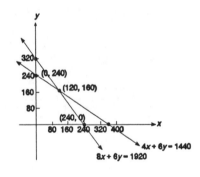

48. $\begin{cases} x \ge 0 \\ y \ge 0 \\ 8x + 6y \le 1920 \\ 4x + 6y \le 1440 \end{cases}$

There are 16 ounces in a pound, so
120 lbs = 1920 ounces and
90 lbs = 1440 ounces

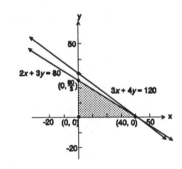

2. $z = 2x + 3y$ $z = 2(5) + 3(6)$ $z = 2(5) + 3(2)$
         $z = 28$ $z = 16$

    $z = 2(4) + 3(0)$ $z = 2(0) + 3(3)$ $z = 2(0) + 3(6)$
    $z = 8$ $z = 9$ $z = 18$
    Maximum value is 28, minimum value is 8.

4. $z = 10x + y$ $z = 10(5) + (6)$ $z = 10(5) + (2)$
         $z = 56$ $z = 52$

    $z = 10(4) + (0)$ $z = 10(0) + (3)$ $z = 10(0) + (6)$
    $z = 40$ $z = 3$ $z = 6$
    Maximum value is 56, minimum value is 3.

6. $z = 7x + 5y$ $z = 7(5) + 5(6)$ $z = 7(5) + 5(2)$
         $z = 65$ $z = 45$

    $z = 7(4) + 5(0)$ $z = 7(0) + 5(3)$ $z = 7(0) + 5(6)$
    $z = 28$ $z = 15$ $z = 30$
    Maximum value is 65, minimum value is 15.

8. Maximize $z = x + 3y$ subject to $x \geq 0$, $y \geq 0$, $x + y \geq 3$, $x \leq 5$, $y \leq 7$. The graph of the feasible points is shown as the outlined shaded region. The vertices are plotted.

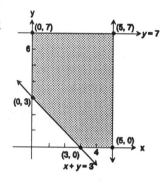

Corner Point	Value of the Objective Function
(x, y)	$z = x + 3y$
$(0, 3)$	$z = 0 + 3(3) = 9$
$(0, 7)$	$z = 0 + 3(7) = 21$
$(5, 7)$	$z = 5 + 3(7) = 26$
$(4, 0)$	$z = 5 + 3(0) = 5$
$(3, 0)$	$z = 3 + 3(0) = 3$

From the table, the maximum value of z is 26, and it occurs at the point $(5, 7)$.

10. Minimize $z = 3x + 4y$ subject to $x \geq 0$, $y \geq 0$, $2x + 3y \geq 6$, $x + y \leq 8$.

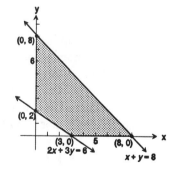

Corner Point	Value of the Objection Function
(x, y)	$z = 3x + 4y$
$(0, 2)$	$z = 3(0) + 4(2) = 8$
$(0, 8)$	$z = 3(0) + 4(8) = 32$
$(8, 0)$	$z = 3(8) + 4(0) = 24$
$(3, 0)$	$z = 3(3) + 4(0) = 9$

The minimum value of z is 8, and it occurs at the point $(0, 2)$.

12. Maximize $z = 5x + 3y$ subject to $x \geq 0$, $y \geq 0$, $x + y \geq 2$, $x + y \leq 8$, $2x + y \leq 10$.

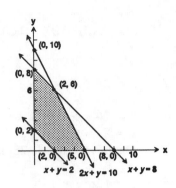

Corner Point	Value of Objection Function
(x, y)	$z = 5x + 3y$
$(0, 2)$	$z = 5(0) + 3(2) = 6$
$(0, 8)$	$z = 5(0) + 3(8) = 24$
$(2, 6)$	$z = 5(2) + 3(6) = 28$
$(5, 0)$	$z = 5(5) + 3(0) = 25$
$(2, 0)$	$z = 5(2) + 3(0) = 10$

The maximum value of z is 28, and it occurs at the point $(2, 6)$.

14. Minimize $z = 2x + 3y$ subject to $x \geq 0$, $y \geq 0$, $x + y \geq 3$, $x + y \leq 9$, $2x + 3y \geq 6$.

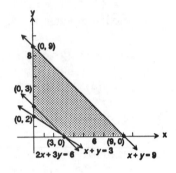

Corner Point	Value of the Objection Function
(x, y)	$z = 2x + 3y$
$(0, 2)$	$z = 2(0) + 3(2) = 6$
$(0, 9)$	$z = 2(0) + 3(9) = 27$
$(9, 0)$	$z = 2(9) + 3(0) = 18$
$(3, 0)$	$z = 2(3) + 3(0) = 6$

The minimum value of z is 6, and it occurs at the points $(0, 2)$ and $(3, 0)$.

16. Maximize $z = 2x + 4y$ subject to $x \geq 0$, $y \geq 0$, $2x + y \geq 4$, $x + y \leq 9$.

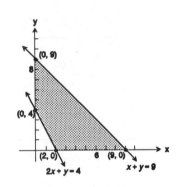

Corner Point	Value of Objective Function
(x, y)	$z = 2x + 4y$
$(0, 4)$	$z = 2(0) + 4(4) = 16$
$(0, 9)$	$z = 2(0) + 4(9) = 36$
$(9, 0)$	$z = 2(9) + 4(0) = 18$
$(2, 0)$	$z = 2(2) + 4(0) = 4$

The maximum value of z is 36, and it occurs at the point $(0, 9)$.

18. Let $x =$ number of downhill skis
  $y =$ number of cross$-$country skis
  If P denotes the profit, then
  $$P = 70x + 50y$$
  This expression is to be maximized (the objective function). The conditions imposed on the variables x and y are:
  $x \geq 0, y \geq 0$ Nonnegative constraints
  $2x + y \leq 48$ Time available to manufacture
  $x + y \leq 32$ Time available to finish skis

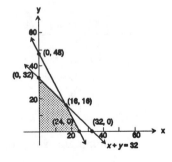

Chapter 12 Systems of Equations and Inequalities

The linear programming problem may be stated as:

Maximize $P = 70x + 50y$ subject to $x \geq 0$, $y \geq 0$, $2x + y \leq 48$, $x + y \leq 32$.

Vertex	Value of Profit
(0, 0)	$P = 70(0) + 50(0) = 0$
(0, 32)	$P = 70(0) + 50(32) = \$1600$
(16, 16)	$P = 70(16) + 50(16) = \$1920$
(24, 0)	$P = 70(24) + 50(0) = \$1680$

The maximum profit is \$1920 which occurs when producing 16 downhill skis and 16 cross-country skis.

20. Let x = number of ounces of supplement A
     y = number of ounces of supplement B
If C denotes the cost of the diet, then $C = 1.50x + 1.00y$.
This expression is to be minimized (the objective function). The conditions imposed on the variables x and y are:

$x \geq 0, y \geq 0$ Nonnegative constraints
$5x + 2y \geq 60$ Carbohydrates needed in diet
$3x + 2y \geq 45$ Protein needed in diet
$4x + y \geq 30$ Fat needed in diet

Minimize $C = 1.50x + 1.00y$ subject to $x \geq 0$, $y \geq 0$, $5x + 2y \geq 60$, $3x + 2y \geq 45$, $4x + y \geq 30$.

Vertex	Cost of Diet
(0, 30)	$C = 1.50(0) + 1.00(30) = \$30$
(7.5, 11.25)	$C = 1.50(7.5) + 1.00(11.25) = \$22.50$
(15, 0)	$C = 1.50(15) + 1.00(0) = \$22.50$

The minimum cost is \$22.50 which occurs when 7.5 ounces of supplement A and 11.25 ounces of supplement B are used in the diet or when 15 ounces of supplement A and 0 ounces of supplement B are used in the diet.

22. Let x = number of newer trees
     y = number of older trees
If C denotes the cost, then $C = 15x + 20y$
This expression is to be minimized.
The conditions imposed on the variables x and y are:

$x \geq 0, y \geq 0$ Nonnegative constraints
$x + y \geq 25$ (At least 25 trees to be pruned.)
$x + y \leq 50$ (Not more than 50 fruit trees.)
$x + 1.5y \geq 30$ (Crew works at least 30 hours.)

Minimize $C = 15x + 20y$ subject to $x \geq 0$, $y \geq 0$, $x + y \geq 25$, $x + y \leq 50$, $x + 1.5y \geq 30$.

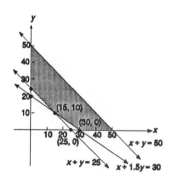

Vertex	Cost
(0, 50)	$15(0) + 20(50) = 1000$
(0, 25)	$15(0) + 20(25) = 500$
(15, 10)	$15(15) + 20(10) = 425$
(30, 0)	$15(30) + 20(0) = 450$
(50, 0)	$15(50) + 20(0) = 750$

The minimum cost is \$425 which occurs when pruning 15 newer trees and 10 older trees.

24. (a) Let x = amount invested in a junk bond
y = amount invested in Treasury bills
If I denotes the income, then
Maximize $I = .09x + .07y$ subject to $x \geq 0$, $y \geq 0$, $x + y \leq 20{,}000$, $y \geq 8{,}000$, $x \leq 12{,}000$, $y \geq x$. (Amount invested in treasury bills must equal or exceed the amount placed in junk bonds.)

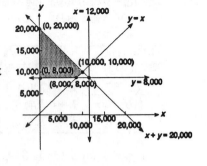

Vertex	Income
(0, 20,000)	$I = .09(0) + .07(20000) = \$1400$
(0, 8,000)	$I = .09(0) + .07(8000) = \$560$
(10,000, 10,000)	$I = .09(10000) + .07(10000) = \$1600$
(8,000, 8,000)	$I = .09(8000) + .07(8000) = \$1280$

The maximum income is $1600 which occurs when investing $10,000 in a junk bond and $10,000 in treasury bills.

(b) Maximize $I = .09x + .07y$ subject to $x \geq 0$, $y \geq 0$, $x + y \leq 20{,}000$, $y \geq 8{,}000$, $x \leq 12{,}000$, $y \leq x$. (Amount invested in treasury bills must not exceed the amount placed in junk bonds.)

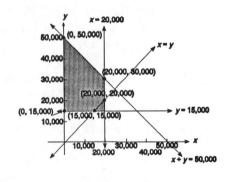

Vertex	Income
(8,000, 8,000)	$I = .09(8000) + .07(8000) = \$1280$
(10,000, 10,000)	$I = .09(10000) + .07(10000) = \$1600$
(12,000, 8,000)	$I = .09(12000) + .07(8000) = \$1640$

The maximum income is $1640 which occurs when investing $12,000 in a junk bond and $8,000 in treasury bills.

26. Let x = amount in a AAA bond
y = amount placed in a CD
I = Investment
Maximize $I = .08x + .04y$ subject to $x \geq 0$, $y \geq 0$, $x + y \leq 50{,}000$, $x \leq 20{,}000$, $y \geq 15{,}000$, $x \leq y$.

Vertex	Investment
(0, 15000)	$I = .08(0) + .04(15000) = \$600$
(15000, 15000)	$I = .08(15000) + .04(15000) = \$1800$
(20000, 20000)	$I = .08(20000) + .04(20000) = \$2400$
(20000, 30000)	$I = .08(20000) + .04(30000) = \$2800$
(0, 50000)	$I = .08(0) + .04(50000) = \$2000$

The maximum return on investment is $2800 which occurs when investing $20000 in a AAA bond and $30000 in a CD.

28. Let x = amount of "Gourmet Dog"
    y = amount of "Chow Hound"
    C = cost

Minimize $C = .40x + .32y$ subject to $x \geq 0$, $y \geq 0$, $x + y \leq 60$, $20x + 35y \geq 1175$, $75x + 50y \geq 2375$.

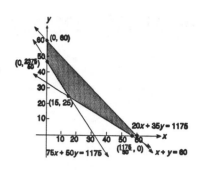

Vertex	Investment
$(0, 60)$	$C = .40(0) + .32(60) = \$19.20$
$\left[0, \dfrac{2375}{50}\right]$	$C = .40(0) + .32\left[\dfrac{2375}{50}\right] = \$15.20$
$(15, 25)$	$C = .40(15) + .32(25) = \$14.00$
$\left[\dfrac{1175}{20}, 0\right]$	$C = .40\left[\dfrac{1175}{20}\right] + .32(0) = \$23.50$
$(60, 0)$	$C = .40(60) + .32(0) = \$24.00$

The minimum cost is \$14.00 which occurs when buying 15 cans of "Gourmet Dog" and 25 cans of "Chow Hound."

30. Let x = amount of supplement A (per ounce)
    y = amount of supplement B (per ounce)
    C = cost

Minimize $C = .06x + .08y$ subject to $x \geq 0$, $y \geq 0$, $5x + 25y \geq 50$, $25x + 10y \geq 90$, $10x + 10y \geq 60$, $35x + 20y \geq 100$.

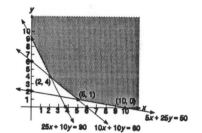

Vertex	Cost
$(0, 9)$	$C = .06(0) + .08(9) = .72$
$(2, 4)$	$C = .06(2) + .08(4) = .44$
$(5, 1)$	$C = .06(5) + .08(1) = .38$
$(10, 0)$	$C = .06(10) + .08(0) = .60$

The minimum cost per ounce of the supplements is \$.38 which occurs when buying 5 ounces of supplement A and 1 ounce of supplement B.

2.
$$\begin{cases} 2x + 3y = 2 \\ 7x - y = 3 \end{cases}$$
$$y = 7x - 3$$
$$2x + 3(7x - 3) = 2$$
$$2x + 21x - 9 = 2$$
$$23x = 11$$
$$x = \frac{11}{23}$$
$$y = 7\left[\frac{11}{23}\right] - 3$$
$$y = \frac{77}{23} - \frac{69}{23} = \frac{8}{23}$$

4.
$$\begin{cases} 2x + y = 0 \\ 5x - 4y = \dfrac{-13}{2} \end{cases}$$
$$y = -2x$$
$$5x - 4(-2x) = \frac{-13}{2}$$
$$5x + 8x = \frac{-13}{2}$$
$$13x = \frac{-13}{2}$$
$$x = -\frac{1}{2}$$
$$y = -2\left[-\frac{1}{2}\right] = 1$$

6.
$$\begin{cases} x - 3y + 5 = 0 \\ 2x + 3y - 5 = 0 \end{cases}$$
$$x = 3y - 5$$
$$2(3y - 5) + 3y - 5 = 0$$
$$6y - 10 + 3y - 5 = 0$$
$$9y = 15$$
$$y = \frac{15}{9}$$
$$= \frac{5}{3}$$
$$x = 3\left[\frac{5}{3}\right] - 5 = 0$$

8.
$$\begin{cases} x = 5y + 2 \\ y = 5x + 2 \end{cases}$$
$$y = 5(5y + 2) + 2$$
$$y = 25y + 12$$
$$24y = -12$$
$$y = -\frac{1}{2}$$
$$x = 5\left[-\frac{1}{2}\right] + 2 = -\frac{1}{2}$$

10.
$$\begin{cases} x + \dfrac{1}{4}y = 2 \\ y + 4x + 2 = 0 \end{cases}$$
$$y = -4x - 2$$
$$x + \frac{1}{4}(-4x - 2) = 2$$
$$x - x - \frac{1}{2} = 2$$
$$-\frac{1}{2} = 2$$
(impossible)
Thus, the system is inconsistent.

12.
$$\begin{cases} x - 3y + \dfrac{7}{2} = 0 \\ \dfrac{1}{2}x + 3y - 5 = 0 \end{cases}$$
$$\begin{cases} x - 3y + \dfrac{7}{2} = 0 \\ \dfrac{3}{2}x - \dfrac{3}{2} = 0 \end{cases}$$
$$\begin{cases} -3y + \dfrac{9}{2} = 0 \\ x = 1 \end{cases}$$
$$\begin{cases} y = \dfrac{3}{2} \\ x = 1 \end{cases}$$

14.
$$\begin{cases} 3x - 4y - 12 = 0 \\ 5x + 2y + 6 = 0 \end{cases}$$
$$\begin{cases} 3x - 4y - 12 = 0 \\ 10x + 4y + 12 = 0 \end{cases}$$
$$\begin{cases} 3x - 4y - 12 = 0 \\ \qquad\quad 13x = 0 \end{cases}$$
$$\begin{cases} -4y - 12 = 0 \\ \qquad\quad x = 0 \end{cases}$$
$$\begin{cases} y = -3 \\ x = 0 \end{cases}$$

16.
$$\begin{cases} 4x + 5y = 21 \\ 5x + 6y = 42 \end{cases}$$
$$\begin{cases} 20x + 25y = 105 \\ -20x - 24y = -168 \end{cases}$$
$$\begin{cases} 20x + 25y = 105 \\ \qquad\quad y = -63 \end{cases}$$
$$\begin{cases} 20x + 25(-63) = 105 \\ \qquad\qquad\quad y = -63 \end{cases}$$
$$\begin{cases} 20x = 1680 \\ \quad y = -63 \end{cases}$$
$$\begin{cases} x = 84 \\ y = -63 \end{cases}$$

18.
$$\begin{cases} 2x + 5y = 10 \\ 4x + 10y = 15 \end{cases}$$
$$\begin{cases} -4x - 10y = -20 \\ 4x + 10y = 15 \end{cases}$$
$$\begin{cases} -4x - 10y = -20 \\ \qquad\qquad 0 = -5 \end{cases} \quad \text{(impossible)}$$
Thus, the system is inconsistent.

20.
$$\begin{cases} x + 5y - z = 2 \\ 2x + y + z = 7 \\ x - y + 2z = 11 \end{cases}$$
$$\begin{cases} x + 5y - z = 2 \\ 3x + 6y = 9 \\ x - y + 2z = 11 \end{cases}$$
$$\begin{cases} 2x + 10y - 2z = 4 \\ 3x + 6y = 9 \\ x - y + 2z = 11 \end{cases}$$

$$\begin{cases} 2x + 10y - 2z = 4 \\ 3x + 6y = 9 \\ 3x + 9y = 15 \end{cases}$$
$$\begin{cases} 2x + 10y - 2z = 4 \\ -3x - 6y = -9 \\ 3x + 9y = 15 \end{cases}$$
$$\begin{cases} 2x + 10y - 2z = 4 \\ -3x - 6y = -9 \\ 3y = 6 \end{cases}$$

$$\begin{cases} -2x - 10y + 2z = -4 \\ -3x - 6y = -9 \\ y = 2 \end{cases}$$
$$\begin{cases} -2x - 10y + 2z = -4 \\ -3x = 3 \\ y = 2 \end{cases}$$
$$\begin{cases} -2(-1) - 10(2) + 2z = -4 \\ x = -1 \\ y = 2 \end{cases}$$
$$\begin{cases} x = -1 \\ y = 2 \\ z = 7 \end{cases}$$

22.
$$\begin{cases} 3x + 2y = 6 \\ x - y = -\dfrac{1}{2} \end{cases}$$

$$\begin{bmatrix} 3 & 2 & \bigg| & 6 \\ 1 & -1 & \bigg| & -\dfrac{1}{2} \end{bmatrix} \rightarrow \begin{bmatrix} 1 & -1 & \bigg| & -\dfrac{1}{2} \\ 3 & 2 & \bigg| & 6 \end{bmatrix} \rightarrow \begin{bmatrix} 1 & -1 & \bigg| & -\dfrac{1}{2} \\ 0 & 5 & \bigg| & \dfrac{15}{2} \end{bmatrix}$$

$$\rightarrow \begin{bmatrix} 1 & -1 & \bigg| & -\dfrac{1}{2} \\ 0 & 1 & \bigg| & \dfrac{3}{2} \end{bmatrix} \rightarrow \begin{bmatrix} 1 & 0 & \bigg| & 1 \\ 0 & 1 & \bigg| & \dfrac{3}{2} \end{bmatrix}, \; x = 1, \; y = \dfrac{3}{2}$$

24. $\begin{cases} 2x + y + z = 5 \\ 4x - y - 3z = 1 \\ 8x + y - z = 5 \end{cases}$

$$\begin{bmatrix} 2 & 1 & 1 & | & 5 \\ 4 & -1 & -3 & | & 1 \\ 8 & 1 & -1 & | & 5 \end{bmatrix} \rightarrow \begin{bmatrix} 1 & \frac{1}{2} & \frac{1}{2} & | & \frac{5}{2} \\ 4 & -1 & -3 & | & 1 \\ 8 & 1 & -1 & | & 5 \end{bmatrix} \rightarrow \begin{bmatrix} 1 & \frac{1}{2} & \frac{1}{2} & | & \frac{5}{2} \\ 0 & -3 & -5 & | & -9 \\ 0 & -3 & -5 & | & -15 \end{bmatrix}$$

$$\rightarrow \begin{bmatrix} 1 & \frac{1}{2} & \frac{1}{2} & | & \frac{5}{2} \\ 0 & 1 & \frac{5}{3} & | & 3 \\ 0 & -3 & -5 & | & -15 \end{bmatrix} \rightarrow \begin{bmatrix} 1 & 0 & -\frac{1}{3} & | & 1 \\ 0 & 1 & \frac{5}{3} & | & 3 \\ 0 & 0 & 0 & | & -6 \end{bmatrix}$$

Thus, the system is inconsistent.

26. $\begin{cases} x + 2y - z = 2 \\ 2x - 2y + z = -1 \\ 6x + 4y + 3z = 5 \end{cases}$

$$\begin{bmatrix} 1 & 2 & -1 & | & 2 \\ 2 & -2 & 1 & | & -1 \\ 6 & 4 & 3 & | & 5 \end{bmatrix} \rightarrow \begin{bmatrix} 1 & 2 & -1 & | & 2 \\ 0 & -6 & 3 & | & -5 \\ 0 & -8 & 9 & | & -7 \end{bmatrix} \rightarrow \begin{bmatrix} 1 & 2 & -1 & | & 2 \\ 0 & 1 & \frac{1}{2} & | & \frac{5}{6} \\ 0 & -8 & 9 & | & -7 \end{bmatrix}$$

$$\rightarrow \begin{bmatrix} 1 & 0 & 0 & | & \frac{1}{3} \\ 0 & 1 & \frac{-1}{2} & | & \frac{5}{6} \\ 0 & 0 & 5 & | & -\frac{1}{3} \end{bmatrix} \rightarrow \begin{bmatrix} 1 & 0 & 0 & | & \frac{1}{3} \\ 0 & 1 & 0 & | & \frac{4}{5} \\ 0 & 0 & 1 & | & -\frac{1}{15} \end{bmatrix}$$

$x = \frac{1}{3}$, $y = \frac{4}{5}$, $z = \frac{-1}{15}$

28. $\begin{cases} 4x - 3y + 5z = 0 \\ 2x + 4y - 3z = 0 \\ 6x + 2y + z = 0 \end{cases}$

$$\begin{bmatrix} 4 & -3 & 5 & | & 0 \\ 2 & 4 & -3 & | & 0 \\ 6 & 2 & 1 & | & 0 \end{bmatrix} \rightarrow \begin{bmatrix} 1 & -\frac{3}{4} & \frac{5}{4} & | & 0 \\ 0 & \frac{11}{2} & -\frac{11}{2} & | & 0 \\ 0 & \frac{13}{2} & \frac{17}{2} & | & 0 \end{bmatrix} \rightarrow \begin{bmatrix} 1 & 0 & \frac{1}{2} & | & 0 \\ 0 & 1 & -1 & | & 0 \\ 0 & 0 & 15 & | & 0 \end{bmatrix} \rightarrow \begin{bmatrix} 1 & 0 & 0 & | & 0 \\ 0 & 1 & 0 & | & 0 \\ 0 & 0 & 1 & | & 0 \end{bmatrix}$$

$x = 0, y = 0, z = 0$

30. $\begin{cases} x - 3y + 3z - t = 4 \\ x + 2y - z = -3 \\ x + 3z + 2t = 3 \\ x + y + 5z = 6 \end{cases}$

$$\begin{bmatrix} 1 & -3 & 3 & -1 & | & 4 \\ 1 & 2 & -1 & 0 & | & -3 \\ 1 & 0 & 3 & 2 & | & 3 \\ 1 & 1 & 5 & 0 & | & 6 \end{bmatrix} \rightarrow \begin{bmatrix} 1 & -3 & 3 & -1 & | & 4 \\ 0 & 5 & -4 & 1 & | & -7 \\ 0 & 3 & 2 & 1 & | & -1 \\ 0 & 4 & 2 & 1 & | & 2 \end{bmatrix} \rightarrow \begin{bmatrix} 1 & -3 & 3 & -1 & | & 4 \\ 0 & 1 & -\dfrac{4}{5} & \dfrac{1}{5} & | & -\dfrac{7}{5} \\ 0 & 3 & 0 & 3 & | & -1 \\ 0 & 4 & 2 & 1 & | & 2 \end{bmatrix} \rightarrow \begin{bmatrix} 1 & 0 & \dfrac{3}{5} & -\dfrac{2}{5} & | & -\dfrac{1}{5} \\ 0 & 1 & -\dfrac{4}{5} & \dfrac{1}{5} & | & -\dfrac{7}{5} \\ 0 & 0 & \dfrac{12}{5} & \dfrac{12}{5} & | & \dfrac{16}{5} \\ 0 & 0 & \dfrac{26}{5} & \dfrac{1}{5} & | & \dfrac{38}{5} \end{bmatrix}$$

$$\rightarrow \begin{bmatrix} 1 & 0 & \dfrac{3}{5} & -\dfrac{2}{5} & | & -\dfrac{1}{5} \\ 0 & 1 & -\dfrac{4}{5} & \dfrac{1}{5} & | & -\dfrac{7}{5} \\ 0 & 0 & 1 & 1 & | & \dfrac{4}{3} \\ 0 & 0 & \dfrac{26}{5} & \dfrac{1}{5} & | & \dfrac{38}{5} \end{bmatrix} \rightarrow \begin{bmatrix} 1 & 0 & 0 & -1 & | & -1 \\ 0 & 1 & 0 & 1 & | & -\dfrac{1}{3} \\ 0 & 0 & 1 & 1 & | & \dfrac{4}{3} \\ 0 & 0 & 0 & -5 & | & \dfrac{2}{3} \end{bmatrix} \rightarrow \begin{bmatrix} 1 & 0 & 0 & 0 & | & -\dfrac{17}{15} \\ 0 & 1 & 0 & 0 & | & -\dfrac{3}{15} \\ 0 & 0 & 1 & 0 & | & \dfrac{22}{15} \\ 0 & 0 & 0 & 1 & | & -\dfrac{2}{15} \end{bmatrix}$$

$x = -\dfrac{17}{15}, \ y = -\dfrac{3}{15}, \ z = \dfrac{22}{15}, \ t = -\dfrac{2}{15}$

32. $\begin{vmatrix} -4 & 0 \\ 1 & 3 \end{vmatrix} = -4(3) - 0(1) = -12$

34. $\begin{vmatrix} 2 & 3 & 10 \\ 0 & 1 & 5 \\ -1 & 2 & 3 \end{vmatrix} = 0 \begin{vmatrix} 3 & 10 \\ 2 & 3 \end{vmatrix} + 1 \begin{vmatrix} 2 & 10 \\ -1 & 3 \end{vmatrix} - 5 \begin{vmatrix} 2 & 3 \\ -1 & 2 \end{vmatrix}$

$\qquad\qquad = 6 + 10 - 5(4 + 3) = 16 - 35 = -19$

36. $\begin{vmatrix} -2 & 1 & 0 \\ 1 & 2 & 3 \\ -1 & 4 & 2 \end{vmatrix} = -2 \begin{vmatrix} 2 & 3 \\ 4 & 2 \end{vmatrix} - 1 \begin{vmatrix} 1 & 3 \\ -1 & 2 \end{vmatrix} + 0 \begin{vmatrix} 1 & 2 \\ -1 & 4 \end{vmatrix} = 2(-8) - 1(5) = 11$

38. $\begin{cases} x - 3y = -5 \\ 2x + 3y = 5 \end{cases}$

$D = \begin{vmatrix} 1 & -3 \\ 2 & 3 \end{vmatrix} = (1)(3) - (-3)(2) = 9$

$x = \dfrac{D_x}{D} = \dfrac{\begin{vmatrix} -5 & -3 \\ 5 & 3 \end{vmatrix}}{9} = \dfrac{0}{9} = 0, \ y = \dfrac{D_y}{D} = \dfrac{\begin{vmatrix} 1 & -5 \\ 2 & 5 \end{vmatrix}}{9} = \dfrac{15}{9}$

40. $\begin{cases} 3x - 4y - 12 = 0 \\ 5x + 2y + 6 = 0 \end{cases}$

$$D = \begin{vmatrix} 3 & -4 \\ 5 & 2 \end{vmatrix} = (3)(2) - (-4)(5) = 26$$

$$x = \frac{D_x}{D} = \frac{\begin{vmatrix} 12 & -4 \\ -6 & 2 \end{vmatrix}}{26} = \frac{0}{26} = 0, \quad y = \frac{D_y}{D} = \frac{\begin{vmatrix} 3 & 12 \\ 5 & -6 \end{vmatrix}}{26} = -\frac{78}{26} = -3$$

42. $\begin{cases} x - y + z = 8 \\ 2x + 3y - x = -2 \\ 3x - y - 9z = 9 \end{cases}$

$$D = \begin{vmatrix} 1 & -1 & 1 \\ 2 & 3 & -1 \\ 3 & -1 & -9 \end{vmatrix} = 1 \begin{vmatrix} 3 & -1 \\ -1 & -9 \end{vmatrix} - (-1) \begin{vmatrix} 2 & -1 \\ 3 & -9 \end{vmatrix} + 1 \begin{vmatrix} 2 & 3 \\ 3 & -1 \end{vmatrix}$$

$$= -27 - 1 + (-18) + 3 + (-2) - 9 = -54$$

$$x = \frac{D_x}{D} = \frac{\begin{vmatrix} 8 & -1 & 1 \\ -2 & 3 & -1 \\ 9 & -1 & -9 \end{vmatrix}}{-54} = \frac{8(-27 - 1) + 1(18 + 9) + 1(2 - 27)}{-54} = \frac{37}{9}$$

$$y = \frac{D_y}{D} = \frac{\begin{vmatrix} 1 & 8 & 1 \\ 2 & -2 & -1 \\ 3 & 9 & -9 \end{vmatrix}}{-54} = \frac{1(18 + 9) - 8(-18 + 3) + 1(18 + 6)}{-54} = -\frac{19}{6}$$

$$z = \frac{D_z}{D} = \frac{\begin{vmatrix} 1 & -1 & 8 \\ 2 & 3 & -2 \\ 3 & -1 & 9 \end{vmatrix}}{-54} = \frac{1(27 - 2) + 1(18 + 6) + 8(-2 - 9)}{-54} = \frac{39}{54}$$

44. $\begin{cases} x^2 + y^2 = 16 \\ 2x - y^2 = -8 \end{cases}$

$y^2 = 2x + 8$

$$x^2 + (2x + 8)^2 = 16$$
$$x^2 + 4x^2 + 32x + 64 - 16 = 0$$
$$5x^2 + 32x + 48 = 0$$
$$(5x + 12)(x + 4) = 0$$
$$x = -\frac{12}{5} \quad \text{or} \quad x = -4$$

$$y^2 = 16 - \left[-\frac{12}{5}\right]^2 \quad \text{or} \quad y^2 = 16 - (-4)^2$$

$$y^2 = \frac{400}{25} - \frac{144}{25} \quad \text{or} \quad y^2 = 16 - 16$$

$$y^2 = \frac{256}{25} \quad\quad\quad \text{or} \quad y^2 = 0$$

Thus, if $x = -\frac{12}{5}$, $y = \pm\frac{16}{5}$; if $x = -4$, $y = 0$.

46. $\begin{cases} 3x^2 - y^2 = 1 \\ 7x^2 - 2y^2 - 5 = 0 \end{cases}$

$y^2 = 3x^2 - 1$

$\quad 7x^2 - 2(3x^2 - 1) - 5 = 0$

$\quad\quad 7x^2 - 6x^2 + 2 - 5 = 0$

$\quad\quad\quad\quad\quad\quad\quad x^2 = 3$

$\quad\quad\quad\quad\quad\quad\quad\quad x = \pm\sqrt{3}$

If $x = \sqrt{3}$, $y^2 = 3\left(\sqrt{3}\right)^2 - 1 = 8$;

$\quad y = \pm\sqrt{8}$

If $x = -\sqrt{3}$, $y = \pm\sqrt{8}$

48. $\begin{cases} 2x^2 + y^2 = 9 \\ x^2 + y^2 = 9 \end{cases}$

$\quad x^2 = 0$

$\quad x = 0$

$\quad y^2 = 9 - x^2$

$\quad y = \pm 3$

50. $\begin{cases} 3x^2 + 2xy - 2y^2 = 6 \\ xy - 2y^2 + 4 = 0 \end{cases}$

$\quad \begin{cases} 12x^2 + 8xy - 8y^2 = 24 \\ 6xy - 12y^2 = -24 \end{cases}$

$\quad 12x^2 + 14xy - 20y^2 = 0$

$\quad 2(6x^2 + 7xy - 10y^2) = 0$

$\quad (6x - 5y)(x + 2y) = 0$

$\quad\quad\quad 6x - 5y = 0$

$\quad\quad\quad\quad\quad x = \dfrac{5}{6}y \quad\quad\quad \text{or} \quad x + 2y = 0$

$\quad\quad\quad\quad\quad\quad\quad\quad\quad\quad\quad\quad\quad\quad\quad x = -2y$

$\left[\dfrac{5y}{6}\right]y - 2y^2 + 4 = 0 \quad\quad \text{or} \quad (-2y)y - 2y^2 + 4 = 0$

$5y^2 - 12y^2 + 24 = 0 \quad\quad\quad\quad\quad\quad\quad\quad 4y^2 = 4$

$\quad\quad\quad\quad -7y^2 = -24 \quad\quad\quad\quad\quad\quad\quad\quad\quad y^2 = 1$

$\quad\quad\quad\quad\quad\quad y^2 = \dfrac{24}{7} \quad\quad\quad\quad\quad\quad\quad\quad\quad y = \pm 1$

$\quad\quad\quad\quad\quad\quad y = \pm\dfrac{\sqrt{168}}{7} \quad\quad\quad\quad\quad\quad\quad x = \pm 2$

If $x = \dfrac{\sqrt{168}}{7}$, $y = \dfrac{5\sqrt{168}}{42}$ $\quad\quad$ If $y = 1$, $x = -2$

If $x = \dfrac{-\sqrt{168}}{7}$, $y = \dfrac{-5\sqrt{168}}{42}$ $\quad\quad$ If $y = -1$, $x = 2$

52. $\begin{cases} x^2 + x + y^2 = y + 2 \\ \quad x + 1 = \dfrac{2 - y}{x} \end{cases}$

$\begin{cases} x^2 + x + y^2 = y + 2 \\ \quad x^2 + x = 2 - y \end{cases}$

$y^2 = 2y$

$y^2 - 2y = 0$

$y(y - 2) = 0$

$y = 0 \ \text{ or } \ y = 2$

If $y = 0$, $\quad x^2 + x - 2 = 0$

$(x + 2)(x - 1) = 0$

$x = -2 \ \text{ or } \ x = 1$

If $y = 2$, $\quad x^2 + x + 4 = 4$

$x^2 + x = 0$

$x(x + 1) = 0$

$x = 0 \ \text{ or } \ x = -1$

54. $\begin{cases} x - 2y \le 6 \\ 2x + y \ge 2 \end{cases}$
Unbounded; corner points $(2, -2)$

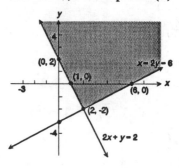

56. $\begin{cases} x \ge 0 \\ y \ge 0 \\ 3x + y \ge 6 \\ 2x + y \ge 2 \end{cases}$
Unbounded; corner points $(2, 0)$, $(0, 6)$

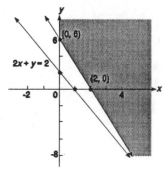

58. $\begin{cases} x \ge 0 \\ y \ge 0 \\ 3x + y \le 9 \\ 2x + 3y \ge 6 \end{cases}$
Bounded; corner points $(0, 2)$, $(0, 9)$, $(3, 0)$

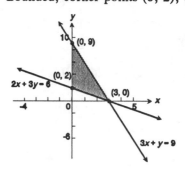

60. $\begin{cases} y^2 \le x - 1 \\ x - y \le 3 \end{cases}$

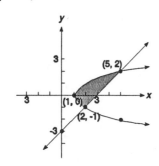

62. $\begin{cases} x^2 + y^2 \ge 1 \\ x^2 + y^2 \le 4 \end{cases}$

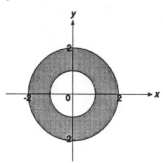

Chapter 12 Systems of Equations and Inequalities

64. Maximize $z = 2x + 4y$ subject to $x \geq 0, y \geq 0, x + y \leq 6, x \geq 2$

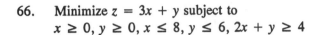

Corner Point	Value of the Objective Function
(2, 0)	$z = 2(2) + 4(0) = 4$
(2, 4)	$z = 2(2) + 4(4) = 20$
(6, 0)	$z = 2(6) + 4(0) = 12$

The maximum value of the objective function is 20 which occurs at (2, 4).

66. Minimize $z = 3x + y$ subject to
$x \geq 0, y \geq 0, x \leq 8, y \leq 6, 2x + y \geq 4$

Corner Point	Value of Objective Function
(0, 4)	$z = 3(0) + 4 = 4$
(0, 6)	$z = 3(0) + 6 = 6$
(8, 6)	$z = 3(8) + 6 = 30$
(8, 0)	$z = 3(8) + 0 = 24$
(2, 0)	$z = 3(2) + 0 = 6$

The minimum value is 4 occurring at (0, 4).

68. Maximize $z = 4x + 5y$ subject to $x \geq 0, y \geq 0, 2x + 3y \geq 6, x \geq y$, $2x + y \leq 12$.

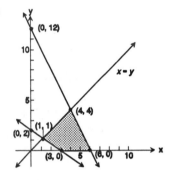

Corner Point	Value of Objective Function
(3, 0)	$z = 4(3) + 5(0) = 12$
(1, 1)	$z = 4(1) + 5(1) = 9$
(4, 4)	$z = 4(4) + 5(4) = 36$
(6, 0)	$z = 4(6) + 5(0) = 24$

The maximum value is 36 occurring at (4, 4).

70. If $A \neq 10$, the system is inconsistent.

72. $x^2 + y^2 + Dx + Ey + F = 0$, which passes through the points $(0, 1)$, $(1, 0)$, and $(-2, 0)$.

$$1 + E + F = 0$$
$$1 + D + F = 0$$
$$5 - 2D + E + F = 0$$

$$\begin{bmatrix} 0 & 1 & 1 & | & -1 \\ 1 & 0 & 1 & | & -1 \\ -2 & 1 & 1 & | & -5 \end{bmatrix} \rightarrow \begin{bmatrix} 1 & 0 & 1 & | & -1 \\ 0 & 1 & 1 & | & -1 \\ -2 & 1 & 1 & | & -5 \end{bmatrix} \rightarrow \begin{bmatrix} 1 & 0 & 1 & | & -1 \\ 0 & 1 & 1 & | & -1 \\ 0 & 1 & 3 & | & -7 \end{bmatrix}$$

$$\rightarrow \begin{bmatrix} 1 & 0 & 1 & | & -1 \\ 0 & 1 & 1 & | & -1 \\ 0 & 0 & 2 & | & -6 \end{bmatrix} \rightarrow \begin{bmatrix} 1 & 0 & 1 & | & -1 \\ 0 & 1 & 1 & | & -1 \\ 0 & 0 & 1 & | & -3 \end{bmatrix} \rightarrow \begin{bmatrix} 1 & 0 & 0 & | & 2 \\ 0 & 1 & 1 & | & 2 \\ 0 & 0 & 1 & | & -3 \end{bmatrix}$$

Thus, $D = 2, E = 2$, and $F = -3$, and the general equation of the circle is
$$x^2 + y^2 + 2x + 2y - 3 = 0$$

74. Let x = number of acres for corn
y = number of acres for soy beans

$$\begin{cases} x + y = 1000 \\ 65x + 45y = 54{,}325 \end{cases}$$

$y = 1000 - x$

$65x + 45(1000 - x) = 54325$

$65x + 45000 - 45x = 54325$

$20x = 9325$

$x = 466.25$

$y = 1000 - 466.25$

$y = 533.75$

Corn should be planted on 466.25 acres and soybeans should be allocated 533.75 acres.

76. (a) $\begin{cases} 8x + 6y \leq 120 \\ 4x + 6y \leq 80 \end{cases}$
$x \geq 0, y \geq 0$

(b)

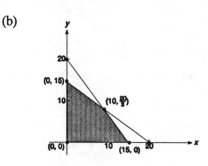

78. ℓ = length
w = width

$$\begin{cases} \ell w = 4 \\ \ell^2 + w^2 = 8 \end{cases}$$

$\left[\dfrac{4}{w}\right]^2 + w^2 = 8$

$16 + w^4 = 8w^2$

$w^4 - 8w^2 + 16 = 0$

$(w^2 - 4)^2 = v$

$w^2 = 4$

$w = 2$ feet

$\ell = 2$ feet

80. ℓ = length of equal sides
b = length of base

$$\begin{cases} 2\ell + b = 18 \\ \left[\dfrac{b}{2}\right]^2 + 36 = \ell^2 \end{cases}$$

$\left[\dfrac{b}{2}\right]^2 + 36 = \left[\dfrac{18 - b}{2}\right]^2$

$\dfrac{b^2}{4} + 36 = \dfrac{324 - 36b + b^2}{4}$

$36 = 81 - 9b$

$9b = 45$

$b = 5$ inches

82. x = amount of 20% HCl
y = amount of 25% HCl
z = amount of 40% HCl

$$\begin{cases} x + y + z = 100 \\ .10x + .25y + .40 = .30(100) \end{cases}$$

$\begin{bmatrix} 1 & 1 & 1 & | & 100 \\ .1 & .25 & .4 & | & 30 \end{bmatrix}$

$\begin{bmatrix} 1 & 1 & 1 & | & 100 \\ 0 & .15 & .3 & | & 20 \end{bmatrix}$

$\begin{bmatrix} 1 & 1 & 1 & | & 100 \\ 0 & 1 & 2 & | & 133.3 \end{bmatrix}$

$\begin{bmatrix} 1 & 1 & -1 & | & 33.3 \\ 0 & 1 & 2 & | & 133.3 \end{bmatrix}$

$x = z - 33.3$

$y = -2z + 133.3$

$33.3 \leq z \leq 66.7$

x	y	z
0	66.7	33.3
16.7	33.3	50.0
33.3	0	66.7

84. Let x = airspeed of the plane (i.e, the speed of the plane if it was flying in still air)

   y = speed of jet-stream

   d = distance from Chicago to Ft. Lauderdale.

We use the basic formula:

   rate × time = distance

We first need to determine if the jet-stream flows from Chicago to Ft. Lauderdale or vice-versa. The trip from Chicago to Ft. Lauderdale takes less time than the return trip, so the jet-stream flows from Chicago to Ft. Lauderdale.

From Chicago to Ft. Lauderdale, we have:

   rate: $x + y$ (going *with* the jet-stream)

   time: 2 hrs, 30 min = $2\frac{1}{2}$ = $\frac{5}{2}$ hr.

   distance: d

Therefore, $(x + y)\left[\dfrac{5}{2}\right] = d$

or $5x + 5y = 2d$ (1)   Multiply by 2.

From Ft. Lauderdale to Chicago we have:

   rate: $x - y$ (going *against* jet-stream)

   time: 2 hrs. 50 min = $2\frac{5}{6}$ = $\frac{17}{6}$ hr.

   distance: d

Therefore, $(x - y)\left[\dfrac{17}{6}\right] = d$

or $17x - 17y = 6d$ (1)   Multiply by 6.

We have two equations:

$$\begin{cases} 5x + 5y = 2d & (1) \\ 17x - 17y = 6d & (2) \end{cases}$$

But we know $x = 475$ (airspeed of plane), so we have:

$$\begin{cases} 2375 + 5y = 2d & (1) \\ 8075 - 17y = 6d & (2) \end{cases}$$

$$\begin{cases} -7125 - 15y = -6d & (1) \quad \text{Multiply by } -3. \\ 8075 - 17y = 6d & (2) \end{cases}$$

$$\begin{cases} -7125 - 15y = -6d & (1) \\ 950 - 32y = 0 & (2) \quad \text{Replace (2) by (1) + (2).} \end{cases}$$

From (2): $-32y = -950$

   $y = \dfrac{-950}{-32} = 29.69$

Therefore, the speed of the jet-stream is 29.69 mph.

86.

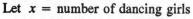

	Dancing Girl	Mermaid	Daily Labor
Molding	3	3	90
Painting	6	4	120
Glazing	2	3	60

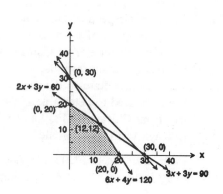

Let x = number of dancing girls
    y = number of mermaids
Maximize profit $P = 25x + 30y$ subject to $x \geq 0$, $y \geq 0$,
$3x + 3y \leq 90$, $6x + 4y \leq 120$, $2x + 3y \leq 60$.

Vertex	Value of Profit
(0, 0)	$P = 25(0) + 30(0) = 0$
(0, 20)	$P = 25(0) + 30(20) = \$600$
(12, 12)	$P = 25(12) + 30(12) = \$660$
(20, 0)	$P = 25(20) + 30(0) = \$500$

The maximum profit is \$660 which occurs when producing 12 dancing girls and 12 mermaids daily. The process of molding has excess work-hours assigned to it.

Mission Possible

1.

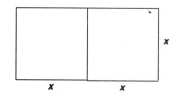

2. If students use ℓ and w, then the area, 4500, $= \ell \cdot w$ and the fencing, 300, $= 3w + 2\ell$ If the figure is labelled as above, then $4500 = x \cdot 2x$ and $300 = 7x$.

3. $2x^2 = 4500 \Rightarrow x^2 = 2250 \Rightarrow x = 47.43$ ft.; $300 = 7x \Rightarrow x = \dfrac{300}{7} = 42.86$ ft.

Obviously, the first problem is that the rancher does not have enough fence. She will need 332 feet of fence to provide 4500 square feet of pens in the configuration she imagines.

4. If she uses just 300 feet of fencing, then each square pen will contain $\left[\dfrac{300}{7}\right]^2$ sq. ft. or 1836.7 sq. ft., a total of 3673.5 sq. ft. altogether.

5. $4x + 3y = 300$. Area $= x^2 + \left[100 - \dfrac{4x}{3}\right]^2$. Area a maximum when $y = 0$ and $x = 75$. Area would equal 5625 sq. ft.

6. A perimeter of 300 might give you a rectangle with dimensions 100×50 for an area of 5000 sq. ft; many rectangles are possible. An equilateral triangle with 100 ft. sides would have an area of $2500\sqrt{3}$ or 4330.1 sq. ft. A square 75 feet per side would have an area of 5625 sq. ft. A circle with circumference 300 would have area of 7162.0 sq. ft.; this is the largest.

SEQUENCES; INDUCTION; COUNTING; PROBABILITY

13.1 Sequences

2. 2, 5, 10, 17, 26

4. $\dfrac{3}{2}, \dfrac{5}{4}, \dfrac{7}{6}, \dfrac{9}{8}, \dfrac{11}{10}$

6. $1, -\dfrac{2}{3}, \dfrac{3}{5}, -\dfrac{4}{7}, \dfrac{5}{9}$

8. $\dfrac{4}{3}, \dfrac{16}{9}, \dfrac{64}{27}, \dfrac{256}{81}, \dfrac{1024}{243}$

10. $3, \dfrac{9}{2}, 9, \dfrac{81}{4}, \dfrac{243}{5}$

12. $\dfrac{1}{2}, 1, \dfrac{9}{8}, 1, \dfrac{25}{32}$

14. $\dfrac{1}{n(n+1)}$

16. $\dfrac{2^n}{3^n} = \left(\dfrac{2}{3}\right)^n$

18. $\left(\dfrac{1}{n}\right)^{(-1)^n}$ or $n^{(-1)^{n+1}}$

20. $(-1)^{n+1}2n$

22. $a_1 = 3$
$a_2 = 1$
$a_3 = 3$
$a_4 = 1$
$a_5 = 3$

24. $a_1 = 1$
$a_2 = 0$
$a_3 = 2$
$a_4 = 1$
$a_5 = 3$

26. $a_1 = 2$
$a_2 = -2$
$a_3 = 2$
$a_4 = -2$
$a_5 = 2$

28. $a_1 = -2$
$a_2 = -5$
$a_3 = -13$
$a_4 = -36$
$a_5 = -104$

30. $a_1 = -1$
$a_2 = 1$
$a_3 = 0$
$a_4 = 2$
$a_5 = 2$

32. $a_1 = A$
$a_2 = rA$
$a_3 = r^2A$
$a_4 = r^3A$
$a_5 = r^4A$

34. $a_1 = \sqrt{2}$

$a_2 = \sqrt{\dfrac{\sqrt{2}}{2}} = \dfrac{1}{\sqrt[4]{2}} = \dfrac{1}{2^{1/4}}$

$a_3 = \sqrt{\dfrac{\sqrt{\dfrac{\sqrt{2}}{2}}}{2}} = \sqrt{\dfrac{1}{2\sqrt[4]{2}}} = \dfrac{1}{2^{5/8}}$

$a_4 = \sqrt{\dfrac{1}{2 \cdot 2^{5/8}}} = \dfrac{1}{2^{13/16}}$

$a_5 = \sqrt{\dfrac{1}{2 \cdot 2^{13/16}}} = \dfrac{1}{2^{29/32}}$

36. $\displaystyle\sum_{k=1}^{n} (2k + 1) = 3 + 5 + 7 + \cdots + (2n + 1)$

38. $\displaystyle\sum_{k=1}^{n} (k + 1)^2 = 4 + 9 + 16 + \cdots + (n + 1)^2$ **40.** $\displaystyle\sum_{k=0}^{n} \left[\frac{3}{2}\right]^k = 1 + \frac{3}{2} + \frac{9}{4} + \cdots + \left[\frac{3}{2}\right]^n$

42. $\displaystyle\sum_{k=0}^{n} (2k + 1) = 1 + 3 + 5 + \cdots + (2n - 1)$

44. $\displaystyle\sum_{k=3}^{n} (-1)^{k+1} 2^k = 2^3 - 2^4 + 2^5 - 2^6 + \cdots + (-1)^{n+1} 2^n$ **46.** $\displaystyle\sum_{k=1}^{n} k^3$

48. $\displaystyle\sum_{k=1}^{n} (2k - 1)$ **50.** $\displaystyle\sum_{k=1}^{n} (-1)^{k+1} \left[\frac{2}{3}\right]^k$ **52.** $\displaystyle\sum_{k=1}^{n} \frac{k}{e^k}$ **54.** $\displaystyle\sum_{k=1}^{n} ar^{k-1}$

56. (a) $u_1 = \dfrac{\left(1 + \sqrt{5}\right)^1 - \left(1 - \sqrt{5}\right)^1}{2^1 \sqrt{5}} = \dfrac{1 + \sqrt{5} - 1 + \sqrt{5}}{2\sqrt{5}} = \dfrac{2\sqrt{5}}{2\sqrt{5}} = 1$

$u_2 = \dfrac{\left(1 + \sqrt{5}\right)^2 - \left(1 - \sqrt{5}\right)^2}{2^2 \sqrt{5}} = \dfrac{1 + 2\sqrt{5} + 5 - \left(1 - 2\sqrt{5} + 5\right)}{4\sqrt{5}}$

$\qquad = \dfrac{1 + 2\sqrt{5} + 5 - 1 + 2\sqrt{5} - 5}{4\sqrt{5}} = \dfrac{4\sqrt{5}}{4\sqrt{5}} = 1$

(b) $u_{n+1} + u_n = \dfrac{\left(1 + \sqrt{5}\right)^{n+1} - \left(1 - \sqrt{5}\right)^{n+1}}{2^{n+1} \sqrt{5}} + \dfrac{\left(1 + \sqrt{5}\right)^n - \left(1 - \sqrt{5}\right)^n}{2^n \sqrt{5}}$

$\qquad = \dfrac{\left(1 + \sqrt{5}\right)^{n+1} - \left(1 - \sqrt{5}\right)^{n+1} + 2\left(1 + \sqrt{5}\right)^n - 2\left(1 - \sqrt{5}\right)^n}{2^{n+1} \sqrt{5}}$

$\qquad = \dfrac{\left(1 + \sqrt{5}\right)^n \left[1 + \sqrt{5} + 2\right] - \left(1 - \sqrt{5}\right)^n \left[1 - \sqrt{5} + 2\right]}{2^{n+1} \sqrt{5}}$

$\qquad = \dfrac{\left(1 + \sqrt{5}\right)^n \left(3 + \sqrt{5}\right) - \left(1 - \sqrt{5}\right)^n - \left[3 - \sqrt{5}\right]}{2^{n+1} \sqrt{5}}$

$\qquad = \dfrac{\left(1 + \sqrt{5}\right)^{n+2} \dfrac{\left(3 + \sqrt{5}\right)}{\left(1 + \sqrt{5}\right)^2} - \left(1 - \sqrt{5}\right)^{n+2} \dfrac{\left(3 - \sqrt{5}\right)}{\left(1 - \sqrt{5}\right)^2}}{2^{n+1} \sqrt{5}}$

$\qquad = \dfrac{\left(1 + \sqrt{5}\right)^{n+2} \dfrac{\left(3 + \sqrt{5}\right)}{\left(6 + \sqrt{5}\right)^2} - \left(1 - \sqrt{5}\right)^{n+2} \dfrac{\left(3 - \sqrt{5}\right)}{\left(6 - 2\sqrt{5}\right)^2}}{2^{n+1} \sqrt{5}}$

$\qquad = \dfrac{\left(1 + \sqrt{5}\right)^{n+2} \dfrac{1}{2} - \left(1 - \sqrt{5}\right)^{n+2} \left[\dfrac{1}{2}\right]}{2^{n+1} \sqrt{5}} = \dfrac{\left(1 + \sqrt{5}\right)^{n+2} - \left(1 - \sqrt{5}\right)^{n+2}}{2^{n+2} \sqrt{5}}$

$\qquad = u_{n+2}$

(c)   Since $u_1 = 1$, $u_2 = 1$, $u_{n+2} = u_n + u_{n+1}$, $\{u_n\}$ is the Fibonacci sequence.

2. $\{n - 5\}$
   $d = a_{n+1} - a_n$
   $d = (n - 4) - (n - 5) = 1$
   $\qquad -4, -3, -2, -1$

4. $\{3n + 1\}$
   $d = [3(n + 1) + 1] - (3n + 1)$
   $d = 3n + 4 - 3n - 1 = 3$
   $\qquad 4, 7, 10, 13$

6. $\{4 - 2n\}$
   $d = [4 - 2(n + 1)] - (4 - 2n)$
   $d = 4 - 2n - 2 - 4 + 2n = -2$
   $\qquad 2, 0, -2, -4$

8. $\left\{\dfrac{2}{3} + \dfrac{n}{4}\right\}$

   $d = \left[\dfrac{2}{3} + \dfrac{n + 1}{4}\right] - \left[\dfrac{2}{3} + \dfrac{n}{4}\right]$

   $\quad = \dfrac{n + 1 - n}{4} = \dfrac{1}{4}$

   $\quad \dfrac{11}{12}, \dfrac{7}{6}, \dfrac{17}{12}, \dfrac{5}{3}$

10. $\{e^{\ln n}\}$
    Since $e^{\ln n} = n$
    $\qquad d = n + 1 - n = 1$
    $\quad 1, 2, 3, 4$

12. $a_n = a + (n - 1)d$
    $a_n = -2 + 4(n - 1)$
    $\quad = -2 + 4n - 4 = 4n - 6$
    $a_5 = 4(5) - 6 = 14$

14. $a_n = a + (n - 1)d$
    $a_n = 6 - 2(n - 1)$
    $\quad = 6 - 2n + 2 = 8 - 2n$
    $a_5 = 8 - 2(5) = -2$

16. $a_n = a + (n - 1)d$

    $a_n = 1 - \dfrac{1}{3}(n - 1)$

    $\quad = 1 - \dfrac{n}{3} + \dfrac{1}{3} = \dfrac{4}{3} - \dfrac{n}{3} = \dfrac{4 - n}{3}$

    $a_5 = \dfrac{4 - 5}{3} = -\dfrac{1}{3}$

18. $a_n = a + (n - 1)d$
    $a_n = 0 + (n - 1)\pi = \pi n - \pi$
    $a_5 = 5\pi - \pi = 4\pi$

20. $a_n = a + (n - 1)d = -1 + (n - 1)(2)$
    $a_8 = -1 + 7(2) = 13$

22. $a_9 = 5 + 8(-5) = -35$

24. $a_7 = 2\sqrt{5} + 6\left(2\sqrt{5}\right) = 14\sqrt{5}$

26. $a_4 = a + 3d = 3$
    $a_{20} = a + 19d = 35$
    $\qquad\qquad -16d = -32$
    $\qquad\qquad\qquad d = 2$
    $a + 19d = 35$
    $a = 35 - 19(2)$
    $a = -3$
    $a_1 = -3, a_{n+1} = a_n + 2$

28. $a_8 = a + 7d = 4$
$a_{18} = a + 17d = -96$
$\phantom{a_{18} = a + 1}-10d = 100$
$\phantom{a_{18} = a + 17}d = -10$
$a = -96 - 17d$
$ = -96 - 17(-10)$
$ = 74$
$a_1 = 74, a_{n+1} = a_n - 10$

30. $a_5 = a + 4d = -2$
$a_{13} = a + 12d = 30$
$\phantom{a_{13} = a + 12}-8d = -32$
$\phantom{a_{13} = a + 12d}d = 4$
$a_1 = -2 - 4d = -2 - 4(4)$
$ = -18$
$a_1 = -18, a_{n+1} = a_n + 4$

32. $a_{12} = a + 11d = 4$
$a_{18} = a + 17d = 28$
$\phantom{a_{18} = a + 1}-6d = -24$
$\phantom{a_{18} = a + 17}d = 4$
$a_1 = 4 - 11d = 4 - 11(4) = -40$
$a_1 = -40, a_{n+1} = a_n + 4$

34. $S_n = \dfrac{n}{2}(2 + 2n) = n + n^2$

36. $S_n = \dfrac{n}{2}(-1 + 4n - 5) = \dfrac{n}{2}(4n - 6)$
$ = 2n^2 - 3n$

38. $S_{30} = \dfrac{30}{2}(1 + 59) = 15 \cdot 60 = 900$

40. $S_{14} = \dfrac{14}{2}(2 + 41) = 7 \cdot 43 = 301$

42.
$3x + 2 - 2x = d$
$5x + 3 - (3x + 2) = d$
$x + 2 = d$
$\underline{2x + 1 = d}$
$-x - 1 = 0$
$-x = 1$
$x = -1$

44. The total number of seats, S, is $S = 15 + 17 + 19 + \cdots$

$ n = 40$

This is the sum of an arithmetic sequence with $d = 2$ and $n = 40$.

$$S_n = \frac{n}{2}[2a + (n - 1)d]$$

$$S_{40} = \frac{40}{2}(30 + 39 \cdot 2) = 20(108) = 2160$$

Thus, there are 2160 seats in this section.

46. The bottom step requires 100 bricks and each successive step requires 2 less bricks than the prior step. This is an arithmetic sequence with $a_1 = 100$ and $d = -2$. Since there are 30 stairs, $n = 30$.
$$a_{30} = a + (n - 1)d = 100 + 29(-2) = 42$$
The total number of bricks required is
$$S = 100 + 98 + 96 + \cdots + 42$$
This is the sum of an arithmetic sequence. The number of terms to be added is 30.

$$S = \frac{30}{2}(100 + 42) = 2130$$

Thus, 2130 bricks will be needed.

Chapter 13 Sequences; Induction; Counting; Probability

2. $\{(-5)^n\}$

$$r = \frac{(-5)^{n+1}}{(-5)^n} = (-5)^{n+1-n} = -5$$

$-5, 25, -125, 625$

4. $\left\{\left(\frac{5}{2}\right)^n\right\}$

$$r = \frac{\left(\frac{5}{2}\right)^{n+1}}{\left(\frac{5}{2}\right)^n} = \left(\frac{5}{2}\right)^{n+1-n} = \frac{5}{2}$$

$\frac{5}{2}, \frac{25}{4}, \frac{125}{8}, \frac{625}{16}$

6. $\left\{\frac{3^n}{9}\right\}$

$$r = \frac{\frac{3^{n+1}}{9}}{\frac{3^n}{9}} = 3^{n+1-n} = 3$$

$\frac{1}{3}, 1, 3, 9$

8. $\{3^{2n}\}$

$$r = \frac{3^{2(n+1)}}{3^{2n}} = 3^{2n+2-2n} = 9$$

$9, 81, 729, 6561$

10. $\left\{\frac{2^n}{3^{n-1}}\right\}$

$$r = \frac{\frac{2^{n+1}}{3^n}}{\frac{2^n}{3^{n-1}}} = \frac{2^{n+1}}{3^n} \cdot \frac{3^{n-1}}{2^n} = \frac{2}{3}$$

$2, \frac{4}{3}, \frac{8}{9}, \frac{16}{27}$

12. $\{2n - 5\}$ Looking at $-3, -1, 1, 3, \ldots$
    Arithmetic:
$$d = [2(n + 1) - 5] - (2n - 5)$$
$$d = 2$$

14. $\{5n^2 + 1\}$ Looking at $6, 21, 46, 81, \ldots$
    Neither

16. $\left\{8 - \frac{3}{4}n\right\}$ Looking at $\frac{29}{4}, \frac{13}{2}, \frac{23}{4}, 5, \ldots$
    Arithmetic:
$$d = \left[8 - \frac{3}{4}(n + 1)\right] - \left(8 - \frac{3}{4}n\right)$$
$$d = \frac{-3}{4}$$

18. $2, 4, 6, 8, \ldots$
    Arithmetic: $d = 2$

20. $\left\{\left(\frac{5}{4}\right)^n\right\}$ Looking at $\frac{5}{4}, \frac{25}{16}, \frac{125}{64}, \frac{625}{256}, \ldots$

    Geometric: $r = \frac{\left(\frac{5}{4}\right)^{n+1}}{\left(\frac{5}{4}\right)^n} = \left(\frac{5}{4}\right)^{n+1-n} = \frac{5}{4}$

22. 1, 1, 2, 3, 5, 8, ... Fibonacci Sequence

Neither $\quad a_{n+2} = a_{n+1} + a_n$

24. $\{(-1)^n\}$ Looking at $-1, 1, -1, 1, \ldots$

Geometric: $r = \dfrac{(-1)^{n+1}}{(-1)^n} = -1$

26. $a_n = a \cdot r^{n-1}$

$a_5 = (-2)4^4 = -512$

$a_n = (-2)4^{n-1}$

28. $a_n = a \cdot r^{n-1}$

$a_5 = (6)(-2)^4 = 96$

$a_n = (6)(-2)^{n-1}$

30. $a_n = a \cdot r^{n-1}$

$a_5 = (1)\left[-\dfrac{1}{3}\right]^4 = \dfrac{1}{81}$

$a_n = (1)\left[-\dfrac{1}{3}\right]^{n-1} = \left[-\dfrac{1}{3}\right]^{n-1}$

32. $a_n = a \cdot r^{n-1}$

$a_5 = (0) \cdot \left[\dfrac{1}{\pi}\right]^4 = 0$

$a_n = 0$

34. $a_n = a_1 \cdot r^{n-1}$

$a_8 = (1) \cdot 3^7 = 2187$

36. $a_{10} = (-1) \cdot (-2)^9 = 512$

38. $a_7 = (.1) \cdot (10)^6 = 100{,}000$

40. The sequence $\left\{\dfrac{3^n}{9}\right\}$ is a geometric sequence with $a = \dfrac{3}{9}$ and $r = 3$.

$$S_n = \sum_{k=1}^{n} a_k = a_1 + a_2 + \cdots + a_n = a\left[\dfrac{1 - r^n}{1 - r}\right], \quad r \ne 1$$

In this sequence, $S_n = \displaystyle\sum_{k=1}^{n} \dfrac{3^k}{9} = \dfrac{3}{9} + \dfrac{3^2}{9} + \dfrac{3^3}{9} + \cdots + \dfrac{3^n}{9} = \dfrac{3}{9}\left[\dfrac{1 - 3^n}{1 - 3}\right] = \dfrac{-1}{6}(1 - 3^n)$

42. Geometric Sequence: $a = 4, r = 3$

$$S_n = \sum_{k=1}^{n}\left(4 \cdot 3^{k-1}\right) = 4 + 4 \cdot 3 + 4 \cdot 3^2 + \cdots + 4 \cdot 3^{n-1} = 4\left[\dfrac{1 - 3^n}{1 - 3}\right] = -2(1 - 3^n)$$

44. Geometric Sequence: $a = 2, r = \dfrac{3}{5}$

$$S_n = \sum_{k=0}^{n} 2\left[\dfrac{3}{5}\right]^k = 2 + \sum_{k=1}^{n} \dfrac{6}{5}\left[\dfrac{3}{5}\right]^{k-1} = 2 + \dfrac{6}{5}\cdot\dfrac{1 - \left[\dfrac{3}{5}\right]^n}{1 - \dfrac{3}{5}} = 2 + 3\left[1 - \left[\dfrac{3}{5}\right]^n\right]$$

46. This geometric series has first term $a = 2$ and common ratio $r = \dfrac{2}{3}$. Since $|r| < 1$, its sum is

$$2 + \dfrac{4}{3} + \dfrac{8}{9} + \cdots = \dfrac{2}{1 - \dfrac{2}{3}} = 6$$

48. Geometric Series: $a = 6$, $r = \dfrac{1}{3}$

$$6 + 2 + \frac{2}{3} + \cdots = \frac{6}{1 - \dfrac{1}{3}} = 9$$

50. Geometric Series: $a = 1$, $r = \dfrac{-3}{4}$

$$1 - \frac{3}{4} + \frac{9}{16} - \frac{27}{64} + \cdots = \frac{1}{1 - \dfrac{3}{4}} = \frac{4}{7}$$

52. Geometric Series: $a = 8$, $r = \dfrac{1}{3}$

$$\sum_{k=1}^{\infty} 8 \left[\frac{1}{3} \right]^{k-1} = \frac{8}{1 - \dfrac{1}{3}} = 12$$

54. Geometric Series: $a = 4$, $r = \dfrac{-1}{2}$

$$\sum_{k=1}^{\infty} 4 \left[\frac{-1}{2} \right]^{k-1} = \frac{4}{1 + \dfrac{1}{2}} = \frac{8}{3}$$

56. The ratio of successive terms must be the same.

Thus, $\dfrac{x}{x - 1} = \dfrac{x + 2}{x}$

$$x^2 = (x - 1)(x + 2)$$
$$x^2 = x^2 + x - 2$$
$$x = 2$$

58. This is an example of a geometric series, with $r = 0.8$.
   (a) After striking the 3rd time, the height is $30(.8)^3 = 15.36$ feet
   (b) After striking the nth time, the height is $30(.8)^n$
   (c) If the height is 6 inches (0.5 feet), then
   $$0.5 = 30(.8)^n$$
   $$.01667 = .8^n$$
   $$\log .01667 = n \log .8$$
   $$n = \frac{\log .01667}{\log .8} = 18.35$$
   The height is less than 6 inches after the 19th strike.
   (d) Distance $= 30 + 30(.8) + 30(.8) + 30(.8)^2 + 30(.8)^2 + \cdots + 30(.8)^n + 30(.8)^n + \cdots$
   $$= 30 + 60(.8) + 60(.8)^2 + \cdots + 60(.8)^n + \cdots$$
   $$= 30 + 60(.8)[1 + .8^5 + \cdots + .8^n + \cdots]$$
   $$= 30 + \frac{60(.8)}{1 - .8} = 30 + 240 = 270 \text{ feet}$$

60. Value of equipment after 5 years $= \$15000(.85)^5 = \$6655.60$

62. Geometric Sequence: $a = \$1000,$

$$r = \frac{9}{10}$$

$$a_n = \left[\frac{9}{10}\right]^n \$1,000$$

$$\left[\frac{9}{10}\right]^n \$1,000 = .01$$

$$\left[\frac{9}{10}\right]^n = 0.00001$$

$$n \log .9 = \log 10^{-5} = -5$$

$$n = 109.27$$

Thus, you will receive less than 1¢ in 110 days or on December 19, 1996.

$$S_n = 1000\frac{1 - .9^{110}}{1 - .9} = \$9,999.91$$

68. A: $S_n = \dfrac{1000}{2}(1000 + 0) = \$500,000$

B: $S_n = 1\left[\dfrac{1 - 2^{19}}{1 - 2}\right] = \$524,287$

Option B results in more money.

70. $\dfrac{1}{4}$ is shaded on each process, so eventually

$$A = \frac{1}{4} + \left[\frac{1}{4}\right]^2 + \cdots = \frac{1}{4}\left[1 + \frac{1}{4} + \cdots\right] = \frac{1}{4}\left[\frac{1}{1 - \frac{1}{4}}\right] = \frac{1}{3}$$

13.4 Mathematical Induction

2. (I) $n = 1$: $4 \cdot 1 - 3 = 1$ and $1(2 \cdot 1 - 1) = 1$

(II) If $1 + 5 + 9 \cdots + (4k - 3) = k(2k - 1)$, then

$1 + 5 + 9 + \cdots + (4k - 3) + (4k + 1)$

$= [1 + 5 + 9 + \cdots + (4k - 3)] + (4k + 1) = k(2k - 1) + (4k + 1)$

$= 2k^2 - k + 4k + 1 = 2k^2 + 3k + 1 = (k + 1)(2k + 1)$

Thus, by I and II, it is true for all natural numbers.

4. (I) $n = 1$: $2 \cdot 1 + 1 = 3$ and $1(1 + 2) = 3$

(II) If $3 + 5 + 7 \cdots + (2k + 1) = k(k + 2)$, then

$3 + 5 + 7 + \cdots + (2k + 1) + (2k + 3)$

$= [3 + 5 + 7 + \cdots + (2k + 1)] + (2k + 3) = k(k + 2) + (2k + 3)$

$= k^2 + 2k + 2k + 3 = k^2 + 4k + 3 = (k + 1)(k + 3)$

Thus, by I and II, it is true for all natural numbers.

6. (I) $n = 1$: $3 \cdot 1 - 2 = 1$ and $\frac{1}{2} \cdot 1(3 \cdot 1 - 1) = 1$

(II) If $1 + 4 + 7 \cdots + (3k - 2) = \frac{1}{2}k(3k - 1)$, then

$$1 + 4 + 7 + \cdots + (3k - 2) + (3k + 1)$$
$$= [1 + 4 + 7 + \cdots + (3k - 2)] + (3k + 1)$$
$$= \frac{1}{2}k(3k - 1) + (3k + 1) = \frac{3}{2}k^2 - \frac{1}{2}k + 3k + 1$$
$$= \frac{3}{2}k^2 + \frac{5}{2}k + 1 = \frac{1}{2}(3k^2 + 5k + 2) = \frac{1}{2}(k + 1)(3k + 2)$$

Thus, by I and II, it is true for all natural numbers.

8. (I) $n = 1$: $3^{1-1} = 1$ and $\frac{1}{2}(3^1 - 1) = 1$

(II) If $1 + 3 + 3^2 \cdots + 3^{k-1} = \frac{1}{2}(3^k - 1)$, then

$$1 + 3 + 3^2 + \cdots + 3^{k-1} + 3^k$$
$$= [1 + 3 + 3^2 + \cdots + 3^{k-1}] + 3^k$$
$$= \frac{1}{2}(3^k - 1) + 3^k = \frac{3^k}{2} - \frac{1}{2} + 3^k = \frac{3^k}{2} + 3^k - \frac{1}{2}$$
$$= 3^k\left[\frac{1}{2} + 1\right] - \frac{1}{2} = 3^k\left[\frac{3}{2}\right] - \frac{1}{2} = \frac{1}{2}(3^k(3) - 1) = \frac{1}{2}(3^{k+1} - 1)$$

Thus, by I and II, it is true for all natural numbers.

10. (I) $n = 1$: $5^{1-1} = 1$ and $\frac{1}{4}(5^1 - 1) = 1$

(II) If $1 + 5 + 5^2 \cdots + 5^{k-1} = \frac{1}{4}(5^k - 1)$, then

$$1 + 5 + 5^2 + \cdots + 5^{k-1} + 5^k$$
$$= [1 + 5 + 5^2 + \cdots + 5^{k-1}] + 5^k$$
$$= \frac{1}{4}(5^k - 1) + 5^k = \frac{5^k}{4} - \frac{1}{4} + 5^k = 5^k\left[\frac{1}{4} + 1\right] - \frac{1}{4}$$
$$= 5^k\left[\frac{5}{4}\right] - \frac{1}{4} = \frac{1}{4}(5^k \cdot 5 - 1) = \frac{1}{4}(5^{k+1} - 1)$$

12. (I) $n = 1$: $\dfrac{1}{(2 \cdot 1 - 1)(2 \cdot 1 + 1)} = \dfrac{1}{3}$ and $\dfrac{1}{2 \cdot 1 + 1} = \dfrac{1}{3}$

(II) If $\dfrac{1}{1 \cdot 3} + \dfrac{1}{3 \cdot 5} + \dfrac{1}{5 \cdot 7} + \cdots + \dfrac{1}{(2k - 1)(2k + 1)} = \dfrac{k}{2k + 1}$,

then $\left[\dfrac{1}{1 \cdot 3} + \dfrac{1}{3 \cdot 5} + \dfrac{1}{5 \cdot 7} + \cdots + \dfrac{1}{(2k - 1)(2k + 1)}\right] + \dfrac{1}{(2k + 1)(2k + 3)}$

$$= \dfrac{k}{2k + 1} + \dfrac{1}{(2k + 1)(2k + 3)} = \dfrac{k(2k + 3) + 1}{(2k + 1)(2k + 3)}$$
$$= \dfrac{2k^2 + 3k + 1}{(2k + 1)(2k + 3)} = \dfrac{(2k + 1)(k + 1)}{(2k + 1)(2k + 3)} = \dfrac{k + 1}{2k + 3}$$

Thus, by I and II, it is true for all natural numbers.

14. **(I)** $n = 1$: $1^3 = 1$ and $\frac{1}{4} \cdot 1^2(1 + 1)^2 = 1$

(II) If $1^3 + 2^3 + 3^3 \cdots + k^3 = \frac{1}{4}k^2(k + 1)^2$, then

$$1^3 + 2^3 + 3^3 + \cdots + k^3 + (k + 1)^3 = [1^3 + 2^3 + 3^3 + \cdots k^3] + (k + 1)^3$$

$$= \frac{1}{4}k^2(k + 1)^2 + (k + 1)^3 = (k + 1)^2\left[\frac{1}{4}k^2 + k + 1\right]$$

$$= (k + 1)^2 \cdot \frac{1}{4}(k^2 + 4k + 4) = (k + 1)^2 \cdot \frac{1}{4}(k + 2)^2 = \frac{1}{4}(k + 1)^2(k + 2)^2$$

Thus, by I and II, it is true for all natural numbers.

16. **(I)** $n = 1$: $-(1 + 1) = -2$ and $-\frac{1}{2} \cdot 1(1 + 3) = -2$

(II) If $-2 - 3 - 4 - \cdots - (k + 1) = -\frac{1}{2}k(k + 3)$, then

$$-2 - 3 - 4 - \cdots - (k + 1) - (k + 2)$$
$$= [-2 - 3 - 4 - \cdots - (k + 1)] - (k + 2)$$

$$= -\frac{1}{2}k(k + 3) - (k + 2) = -\frac{1}{2}k^2 - \frac{3}{2}k - k - 2$$

$$= -\frac{1}{2}k^2 - \frac{5}{2}k - 2 = -\frac{1}{2}(k^2 + 5k + 4) = -\frac{1}{2}(k + 1)(k + 4)$$

Thus, by I and II, it is true for all natural numbers.

18. **(I)** $n = 1$: $(2 \cdot 1 - 1)(2 \cdot 1) = 2$ and $\frac{1}{3} \cdot 1(1 + 1)(4 \cdot 1 - 1) = 2$

(II) If $1 \cdot 2 + 3 \cdot 4 + 5 \cdot 6 + \cdots + (2k - 1)(2k) = \frac{1}{3}k(k + 1)(4k - 1)$, then

$$1 \cdot 2 + 3 \cdot 4 + 5 \cdot 6 + \cdots + (2k - 1)(2k) + (2k + 1)(2k + 2)$$
$$= [1 \cdot 2 + 3 \cdot 4 + 5 \cdot 6 + \cdots + (2k - 1)(2k)] + (2k + 1)(2k + 2)$$

$$= \frac{1}{3}k(k + 1)(4k - 1) + (2k + 1)(2k + 2)$$

$$= \frac{1}{3}k(4k^2 + 3k - 1) + (2k + 1)(2k + 2) = \frac{4}{3}k^3 + k^2 - \frac{1}{3}k + 4k^2 + 6k + 2$$

$$= \frac{4}{3}k^3 + 5k^2 + \frac{17}{3}k + 2 = \frac{1}{3}(4k^3 + 15k^2 + 17k + 6)$$

$$= \frac{1}{3}(k + 1)(4k^2 + 11k + 6) = \frac{1}{3}(k + 1)(k + 2)(4k + 3)$$

Thus, by I and II, it is true for all natural numbers.

20. **(I)** $n = 1$: $1^3 + 2 \cdot 1 = 3$ is divisible by 3.
(II) If $k^3 + 2k$ is divisible by 3, then
$$(k + 1)^3 + 2(k + 1) = k^3 + 3k^2 + 3k + 1 + 2k + 2 = k^3 + 2k + 3k^2 + 3k + 3$$
Since $k^3 + 2k$ is divisible by 3 and $3k^2 + 3k + 3$ is divisible by 3, therefore, $(k + 1)^3 + 2(k + 1)$ is divisible by 3. Thus, by I and II, it is true for all natural numbers.

22. **(I)** $n = 1$: $1(1 + 2)(1 + 2) = 6$ is divisible by 6.
(II) If $k(k + 1)(k + 2)$ is divisible by 6, then
$$(k + 1)(k + 2)(k + 3) = k(k + 1)(k + 2) + 3(k + 1)(k + 2)$$
We know that $k(k + 1)(k + 2)$ is divisible by 6 and $3(k + 1)(k + 2)$ is also since either $k + 1$ or $k + 2$ must be even (divisible by 2). Thus, by I and II, it is true for all natural numbers.

24. **(I)** If $0 < x < 1$, then $0 < x^1 = x < 1$.

(II)   Assume that if $0 < x < 1$, then $0 < x^k < 1$ for any natural number k.
Show that, if $0 < x < 1$, then $0 < x^{k+1} < 1$:
$$0 < x^{k+1} = x^k \cdot x < 1 \cdot x = x < 1$$
$$\uparrow$$
$$x^k < 1$$
Thus, by I and II, it is true for all natural numbers.

26. (I) $n = 1$: $\ a + b$ is a factor of $a^{2 \cdot 1 + 1} + b^{2 \cdot 1 + 1} = a^3 + b^3$

      (II) Suppose for some natural number k that $a + b$ is a factor of $a^{2k+1} + b^{2k+1}$. Show that $a + b$ is a factor of $a^{2k+3} + b^{2k+3}$:
$$a^{2k+3} + b^{2k+3} = a^2 \cdot a^{2k+1} + a^2 b^{2k+1} - a^2 b^{2k+1} + b^{2k+3}$$
$$= a^2(a^{2k+1} + b^{2k+1}) - b^{2k+1}(a^2 - b^2)$$
Since $a + b$ divides $a^{2k+1} + b^{2k+1}$ and $a + b$ divides $a^2 - b^2$, then $a + b$ divides $a^{2k+3} + b^{2k+3}$.
Thus, by I and II, it is true for all natural numbers.

28. (II) If $2 + 4 + 6 + \cdots + 2k = k^2 + k + 2$,
then $[2 + 4 + 6 + \cdots + 2k] + 2k + 2 = k^2 + k + 2 + 2k + 2$
$$= k^2 + 3k + 4 = k^2 + 2k + k + 1 + 1 + 2 = (k^2 + 2k + 1) + k + 1 + 2$$
$$= (k + 1)^2 + (k + 1) + 2$$

      (I) $n = 1$: $\ 2 \cdot 1 = 2$ but $1^2 + 1 + 2 = 4 \neq 2$

30. (I) $n = 1$: $\ [a + (1 - 1)d] = a$ and $1a + d\dfrac{1(1 - 1)}{2} = a$

      (II) If $a + (a + d) + (a + 2d) + \cdots + [a + (k - 1)d] = ka + d\dfrac{k(k - 1)}{2}$, then
$$a + (a + d) + (a + 2d) + [a + (k - 1)d] + [a + kd]$$
$$= \{[a + (a + d) + (a + 2d) + \cdots + [a + (k - 1)d]\} + [a + kd]$$
$$= ka + d\frac{k(k - 1)}{2} + a + kd = ka + \frac{kd}{2}(k - 1) + a + kd$$
$$= ka + \frac{k^2 d}{2} - \frac{kd}{2} + a + kd = (k + 1)a + \frac{d}{2}k^2 - \frac{d}{2}k + dk$$
$$= (k + 1)a + \frac{d}{2}k^2 + \frac{d}{2}k = (k + 1)a + \frac{dk(k + 1)}{2}$$
Thus, by I and II, it is true for all natural numbers.

32. (I) $n = 4$: The number of diagonals of a quadrilateral is $\dfrac{1}{2} \cdot 4(4 - 3) = 2$

      (II) Assume that for any k the number of diagonals of a convex polygon of k sides
(k vertices) is $\dfrac{1}{2}k(k - 3)$. A convex polygon of $k + 1$ sides ($k + 1$ vertices)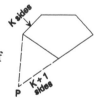
consists of a convex polygon of k sides (k vertices) plus a triangle for a total of
$k + 1$ vertices; see the illustration. The number of diagonals of this convex
polygon consists of all the original ones plus $k - 1$ additional ones, namely,
$$\frac{1}{2}k(k - 3) + (k - 1) = \frac{1}{2}k^2 - \frac{3}{2}k + k - 1 = \frac{1}{2}k^2 - \frac{1}{2}k - 1$$
$$= \frac{1}{2}(k + 1)(k - 2)$$
Since Conditions I and II have been met, the result follows.

2. $\begin{bmatrix} 7 \\ 3 \end{bmatrix} = \dfrac{7!}{3!4!} = \dfrac{7 \cdot 6 \cdot 5}{6} = 35$ 4. $\begin{bmatrix} 9 \\ 7 \end{bmatrix} = \dfrac{9!}{7!2!} = \dfrac{9 \cdot 8}{2} = 36$

6. $\begin{bmatrix} 100 \\ 98 \end{bmatrix} = \dfrac{100!}{98!2!} = \dfrac{100 \cdot 99}{2} = 4950$ 8. $\begin{bmatrix} 1000 \\ 0 \end{bmatrix} = \dfrac{1000!}{1000!} = 1$

10. $\begin{bmatrix} 60 \\ 20 \end{bmatrix} = 4.192 \times 10^{15}$ 12. $\begin{bmatrix} 37 \\ 19 \end{bmatrix} = 1.767 \times 10^{10}$

14. $(x-1)^5 = \begin{bmatrix} 5 \\ 0 \end{bmatrix} x^5 + \begin{bmatrix} 5 \\ 1 \end{bmatrix}(-1)x^4 + \begin{bmatrix} 5 \\ 2 \end{bmatrix}(-1)^2 x^3 + \begin{bmatrix} 5 \\ 3 \end{bmatrix}(-1)^3 x^2 + \begin{bmatrix} 5 \\ 4 \end{bmatrix} x(-1)^4 x + \begin{bmatrix} 5 \\ 5 \end{bmatrix}(-1)^5$

   $= x^5 - 5x^4 + 10x^3 - 10x^2 + 5x - 1$

16. $(x+3)^4 = \begin{bmatrix} 4 \\ 0 \end{bmatrix} x^4 + \begin{bmatrix} 4 \\ 1 \end{bmatrix} 3x^3 + \begin{bmatrix} 4 \\ 2 \end{bmatrix} 9x^2 + \begin{bmatrix} 4 \\ 3 \end{bmatrix} 27x + \begin{bmatrix} 4 \\ 4 \end{bmatrix} 81$

   $= x^4 + 12x^3 + 54x^2 + 108x + 81$

18. $(2x+3)^5 = \begin{bmatrix} 5 \\ 0 \end{bmatrix}(2x)^5 + \begin{bmatrix} 5 \\ 1 \end{bmatrix}(3)(2x)^4 + \begin{bmatrix} 5 \\ 2 \end{bmatrix}(3)^2(2x)^3 + \begin{bmatrix} 5 \\ 3 \end{bmatrix}(3)^3(2x)^2$

   $\qquad + \begin{bmatrix} 5 \\ 4 \end{bmatrix}(3)^4(2x) + \begin{bmatrix} 5 \\ 5 \end{bmatrix} 3^5$

   $= 32x^5 + 240x^4 + 720x^3 + 1080x^2 + 810x + 243$

20. $(x^2 - y^2)^6 = \begin{bmatrix} 6 \\ 0 \end{bmatrix}(x^2)^6 + \begin{bmatrix} 6 \\ 1 \end{bmatrix}(-y^2)(x^2)^5 + \begin{bmatrix} 6 \\ 2 \end{bmatrix}(-y^2)^2(x^2)^4 + \begin{bmatrix} 6 \\ 3 \end{bmatrix}(-y^2)^3(x^2)^3$

   $\qquad + \begin{bmatrix} 6 \\ 4 \end{bmatrix}(-y^2)^4(x^2)^2 + \begin{bmatrix} 6 \\ 5 \end{bmatrix}(-y^2)^5(x^2)^1 + \begin{bmatrix} 6 \\ 6 \end{bmatrix}(-y^2)^6$

   $= x^{12} - 6y^2 x^{10} + 15y^4 x^8 - 20y^6 x^6 + 15y^8 x^4 - 6y^{10} x^2 + y^{12}$

22. $\left(\sqrt{x} - \sqrt{3}\right)^4 = \begin{bmatrix} 4 \\ 0 \end{bmatrix}\left(\sqrt{x}\right)^4 + \begin{bmatrix} 4 \\ 1 \end{bmatrix}\left(-\sqrt{3}\right)\left(\sqrt{x^3}\right) + \begin{bmatrix} 4 \\ 2 \end{bmatrix}\left(-\sqrt{3}\right)^2\sqrt{x^2} + \begin{bmatrix} 4 \\ 3 \end{bmatrix}\left(-\sqrt{3}\right)^3\left(\sqrt{x}\right) + \begin{bmatrix} 4 \\ 4 \end{bmatrix}\left(-\sqrt{3}\right)^4$

   $= x^2 - 4\sqrt{3}\, x^{3/2} + 18x - 12\sqrt{3}\, x + 9$

24. $(ax - by)^4 = \begin{bmatrix} 4 \\ 0 \end{bmatrix}(ax)^4 + \begin{bmatrix} 4 \\ 1 \end{bmatrix}(-by)(ax)^3 + \begin{bmatrix} 4 \\ 2 \end{bmatrix}(-by)^2(ax)^2 + \begin{bmatrix} 4 \\ 3 \end{bmatrix}(-by)^3(ax) + \begin{bmatrix} 4 \\ 4 \end{bmatrix}(-by)^4$

   $= (ax)^4 - 4by(ax)^3 + 6(by)^2(ax)^2 - 4(by)^3(ax) + (by)^4$

26. $n = 10$, $a = -3$, $x = x$, and $j = 3$

   $\begin{bmatrix} 10 \\ 10-3 \end{bmatrix}(-3)^{10-3} x^3 = \begin{bmatrix} 10 \\ 7 \end{bmatrix}(-3)^7 x^3 = \dfrac{10!}{7!3!} \cdot (-3)^7 \cdot x^3 = 120 \cdot -2187 \cdot x^3 = -262{,}440 x^3$

   The coefficient of x^3 is $-262{,}440$.

28. $n = 12, a = 1, x = 2x,$ and $j = 3$

$$\begin{bmatrix} 12 \\ 12-3 \end{bmatrix} 1^{12-3}(2x)^3 = \begin{bmatrix} 12 \\ 9 \end{bmatrix} 2^3 \cdot x^3 = 220 \cdot 8 \cdot x^3 = 1760x^3$$

The coefficient of x^2 is 1760.

30. $n = 9, a = -3, x = 2x,$ and $j = 2$

$$\begin{bmatrix} 9 \\ 9-2 \end{bmatrix} (-3)^{9-2}(2x)^2 = \begin{bmatrix} 9 \\ 7 \end{bmatrix} (-3)^7 \cdot 2^2 \cdot x^2 = 36 \cdot -2187 \cdot 4 \cdot x^2 = -314{,}928x^2$$

The coefficient of x^2 is $-314{,}928$.

32. $n = 7, a = -3, x = x, j = 5$

$$\begin{bmatrix} 7 \\ 7-5 \end{bmatrix} (-3)^{7-5} \cdot x^5 = \begin{bmatrix} 7 \\ 2 \end{bmatrix} (-3)^2 x^5 = 21 \cdot 9 \cdot x^5 = 189x^5$$

34. $n = 8, a = 2, x = 3x,$ and $j = 3$

$$\begin{bmatrix} 8 \\ 8-3 \end{bmatrix} 2^{8-3}(3x)^3 = \begin{bmatrix} 8 \\ 5 \end{bmatrix} 2^5 \cdot 3^3 \cdot x^3 = 56 \cdot 32 \cdot 27 \cdot x^3 = 48{,}384x^3$$

36. $\left[x - \dfrac{1}{x^2} \right]^9$

Constant term $= \begin{bmatrix} 9 \\ 9-j \end{bmatrix} \left[\dfrac{-1}{x^2} \right]^{9-j} x^j$ where $j = 18 - 2j$ or $j = 6$

Constant term $= \begin{bmatrix} 9 \\ 3 \end{bmatrix} \left[\dfrac{-1}{x^2} \right]^3 x^6 = (-1) \begin{bmatrix} 9 \\ 3 \end{bmatrix} = -1 \cdot \dfrac{9 \cdot 8 \cdot 7}{3 \cdot 2} = -84$

38. $\left[\sqrt{x} + \dfrac{3}{\sqrt{x}} \right]^8$

x^2 term $= \begin{bmatrix} 8 \\ 8-j \end{bmatrix} \left[\dfrac{3}{\sqrt{x}} \right]^{8-j} \left(\sqrt{x} \right)^j$ where $\dfrac{j}{2} - \dfrac{8-j}{2} = 2$ or $2j - 8 = 4$ or $j = 6$

x^2 term $= \begin{bmatrix} 8 \\ 2 \end{bmatrix} \left[\dfrac{3}{\sqrt{x}} \right]^2 \left(\sqrt{x} \right)^6 = 9 \begin{bmatrix} 8 \\ 2 \end{bmatrix} \dfrac{x^3}{x} = \dfrac{9 \cdot 8 \cdot 7}{2} x^2 = 252x^2$

40. $(.998)^6 = (1 - .002)^6 = (1 - 2 \cdot 10^{-3})^6 = 1 + \begin{bmatrix} 6 \\ 1 \end{bmatrix} (2)(-10^{-3}) + \begin{bmatrix} 6 \\ 2 \end{bmatrix} (2)^2(-10^{-3})^2$

$+ \begin{bmatrix} 6 \\ 3 \end{bmatrix} (2)^3(-10^{-3})^3 + \cdots$

$= 1 - .012 + 15(.000002) + \cdots$

$= 1 - .006 + .000030$

$\approx .99403$ correct to five decimal places

42. To show that if n and j are integers with $0 \leq j \leq n$, then

$$\begin{pmatrix} n \\ j \end{pmatrix} = \begin{pmatrix} \dfrac{n!}{j!(n-j)!} \end{pmatrix} = \begin{pmatrix} \dfrac{n!}{(n-j)!(n-(n-j))!} \end{pmatrix} = \begin{pmatrix} n \\ n-j \end{pmatrix}$$

Therefore, we conclude that the Pascal triangle is symmetric with respect to a vertical line drawn from the topmost entry.

44. To show that $\begin{pmatrix} n \\ n-j \end{pmatrix} = \begin{pmatrix} n \\ 0 \end{pmatrix} - \begin{pmatrix} n \\ 1 \end{pmatrix} + \begin{pmatrix} n \\ 2 \end{pmatrix} - \dots + (-1)^n \begin{pmatrix} n \\ n \end{pmatrix} = 0$,

Let $0 = (1-1)^n = \begin{pmatrix} n \\ 0 \end{pmatrix} 1^n + \begin{pmatrix} n \\ 1 \end{pmatrix} (-1) 1^{n-1} + \begin{pmatrix} n \\ 2 \end{pmatrix} (-1)^2 1^{n-2} + \dots + \begin{pmatrix} n \\ n \end{pmatrix} (-1)^n$

Therefore, $\begin{pmatrix} n \\ 0 \end{pmatrix} - \begin{pmatrix} n \\ 1 \end{pmatrix} + \begin{pmatrix} n \\ 2 \end{pmatrix} + \dots + (-1)^n \begin{pmatrix} n \\ n \end{pmatrix} = 0$

46. $12! = 4.790016 \times 10^8$
$20! = 2.432902 \times 10^{18}$
$25! = 1.551121 \times 10^{25}$

$$12! \approx \sqrt{2 \cdot 12\pi} \left[\frac{12}{e} \right]^{12} \left[1 + \frac{1}{12 \cdot 12 - 1} \right]$$

$12! \approx \sqrt{24\pi}\, (54782414.5)(1.006993007)$

$12! \approx (8.683215055)(54782414.5)(1.006993007)$

$12! \approx 479013972.2$

$12! \approx 4.790139722 \times 10^8$

$$20! \approx \sqrt{2 \cdot 20\pi} \left[\frac{20}{e} \right]^{20} \left[1 + \frac{1}{12 \cdot 20 - 1} \right]$$

$20! \approx \sqrt{40\pi}\, (2.1612762 \times 10^{17})(1.0041841)$

$20! \approx (11.20998243)(2.1612762 \times 10^{17})(1.0041841)$

$20! \approx 2.432924 \times 10^{18}$

$$25! \approx \sqrt{2 \cdot 25\pi} \left[\frac{25}{e} \right]^{25} \left[1 + \frac{1}{12 \cdot 25 - 1} \right]$$

$25! \approx \sqrt{50\pi}\, (1.2334972 \times 10^{24})(1.003344481)$

$25! \approx (12.53314137)(1.2334972 \times 10^{24})(1.003344481)$

$25! \approx 1.5511299 \times 10^{25}$

13.6 Sets and Counting

In Problems 1–10, A = {1, 3, 5, 7, 9}, B = {1, 5, 6, 7}, and C = {1, 2, 4, 6, 8, 9}.

2. $A \cup C = \{1, 2, 3, 4, 5, 6, 7, 8, 9\}$ **4.** $A \cap B = \{1, 5, 7\}$

6. $(A \cap C) \cup (B \cap C) = \{1, 9\} \cup \{1, 6\} = \{1, 6, 9\}$

8. $(A \cup B) \cup C = \{1, 3, 5, 6, 7, 9\} \cup \{1, 2, 4, 6, 8, 9\} = \{1, 2, 3, 4, 5, 6, 7, 8, 9\}$

10. $(A \cap B) \cap C = \{1, 5, 7\} \cap \{1, 2, 4, 6, 8, 9\} = \{1\}$

In Problems 11–20, U = Universal set = {0, 1, 2, 3, 4, 5, 6, 7, 8, 9}, A = {1, 3, 4, 5, 9},
B = {2, 4, 6, 7, 8}, C = {1, 3, 4, 6}.

12. $C' = \{0, 2, 5, 7, 8, 9\}$ 14. $(B \cup C)' = \{0, 5, 9\}$

16. $B' \cap C' = \{0, 1, 3, 5, 9\} \cap \{0, 2, 5, 7, 8, 9\} = \{0, 5, 9\}$

18. $(B' \cup C)' = (\{0, 1, 3, 5, 9\} \cup \{1, 3, 4, 6\})' = (\{0, 1, 3, 4, 5, 6, 9\})' = \{2, 7, 8\}$

20. $(A \cap B \cap C)' = (\{4\})' = \{0, 1, 2, 3, 5, 6, 7, 8, 9\}$

22. The $2^5 = 32$ subset of $\{a, b, c, d, e\}$ are:
$\emptyset$, $\{a\}$, $\{b\}$, $\{c\}$, $\{d\}$, $\{e\}$, $\{a, b\}$, $\{a, c\}$, $\{a, d\}$, $\{a, e\}$, $\{b, c\}$, $\{b, d\}$, $\{b, e\}$, $\{c, d\}$,
$\{c, e\}$, $\{d, e\}$, $\{a, b, c\}$, $\{a, b, d\}$, $\{a, b, e\}$, $\{a, c, d\}$, $\{a, c, e\}$, $\{a, d, e\}$, $\{b, c, d\}$,
$\{b, c, e\}$, $\{b, d, e\}$, $\{c, d, e\}$, $\{a, b, c, d\}$, $\{a, b, c, e\}$, $\{a, b, d, e\}$, $\{a, c, d, e\}$, $\{b, c, d, e\}$,
$\{a, b, c, d, e\}$

24. $n(A) = 20$, $n(B) = 20$, and $n(A \cup B) = 35$, find $n(A \cap B)$.
By the counting formula, $n(A \cup B) = n(A) + n(B) - n(A \cap B)$
$$35 = 20 + 40 - n(A \cap B)$$
$$n(A \cap B) = 60 - 35 = 25$$

26. $n(A \cup B) = 60$, $n(A \cap B) = 40$, and $n(A) = n(B)$, find $n(A)$.
By the counting formula, $n(A \cup B) = n(A) + n(B) - n(A \cap B)$
$$60 = 2[n(A)] - 40$$
$$2[n(A)] = 100$$
$$n(A) = 50$$

In Problems 27–34, use figure shown here.

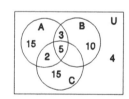

28. $n(B) = 10 + 2 + 5 + 3 = 20$ 30. $n(A \cap B) = 3 + 5 = 8$

32. $n(A') = 10 + 2 + 15 + 4 = 31$

34. $n(A \cup B \cup C) = 15 + 3 + 5 + 2 + 10 + 2 + 15 = 52$

36. Let A = Set of students in Summer Session I.
            B = Set of students in Summer Session II.
The number of students who participated in the survey is
$125 + 75 + 75 + 275 = 550$

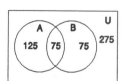

38. There are 8 different kinds of blood:
A−Rh+, B−Rh+, AB−Rh+, 0-Rh+, A−Rh−, B−Rh−,
AB-Rh−, 0−Rh−

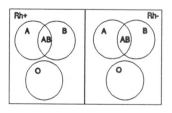

2. $P(7, 2) = 7 \cdot 6 = 42$

4. $P(4, 4) = 4 \cdot 3 \cdot 2 \cdot 1 = 24$

8. $P(8, 5) = 8 \cdot 7 \cdot 6 \cdot 5 \cdot 4 = 6720$

6. $P(9, 0) = \dfrac{9!}{(9 - 0)!} = \dfrac{9!}{9!} = 1$

10. $C(8, 6) = \dfrac{8!}{6!2!} = \dfrac{8 \cdot 7 \cdot 6!}{6!2!} = \dfrac{56}{2} = 28$

12. $C(6, 2) = \dfrac{6!}{4!2!} = \dfrac{6 \cdot 5 \cdot 4!}{4! \cdot 2!} = \dfrac{30}{2} = 15$

14. $C(18, 1) = \dfrac{18!}{17!1!} = \dfrac{18 \cdot 17!}{17! \cdot 1} = 18$

16. $C(18, 9) = \dfrac{18!}{9!9!} = \dfrac{18 \cdot 17 \cdot 16 \cdot 15 \cdot 14 \cdot 13 \cdot 12 \cdot 11 \cdot 10 \cdot 9!}{9 \cdot 8 \cdot 7 \cdot 6 \cdot 5 \cdot 4 \cdot 3 \cdot 2 \cdot 1 \cdot 9} = 48620$

18. {*ab, ac, ad, ae, ba, bc, bd, be, ca, cb, cd, ce, da, db, dc, de, ea, eb, ec, ed*}

$P(5, 2) = \dfrac{5!}{(5 - 2)!} = \dfrac{5!}{3!} = 5 \cdot 4 = 20$

20. {123, 124, 125, 126, 132, 134, 135, 136, 142, 143, 145, 146, 152, 153, 154, 156, 162, 163, 164, 165, 213, 214, 215, 216, 231, 234, 235, 236, 241, 243, 245, 246, 251, 253, 254, 256, 261, 263, 264, 265, 312, 314, 315, 316, 321, 324, 325, 326, 341, 342, 345, 346, 351, 352, 354, 356, 361, 362, 364, 365, 412, 413, 415, 416, 421, 423, 425, 426, 431, 432, 435, 436, 451, 452, 453, 456, 461, 462, 463, 465, 512, 513, 514, 516, 521, 523, 524, 526, 531, 532, 534, 536, 541, 542, 543, 546, 561, 562, 563, 564, 612, 613, 614, 615, 621, 623, 624, 625, 631, 632, 634, 635, 641, 642, 643, 645, 651, 652, 653, 654}

$P(6, 3) = \dfrac{6!}{(6 - 3)!} = \dfrac{6!}{3!} = 6 \cdot 5 \cdot 4 = 120$

22. {*ab, ac, ad, ae, bc, bd, be, cd, ce, de*}

$C(5, 2) = \dfrac{5!}{3!2!} = \dfrac{5 \cdot 4 \cdot 3!}{3! \cdot 2 \cdot 1} = \dfrac{20}{2} = 10$

24. {123, 124, 125, 126, 234, 235, 236, 345, 346, 456, 134, 135, 136, 245, 246, 356, 145, 146, 256, 156}

$C(6, 3) = \dfrac{6!}{3!3!} = \dfrac{6 \cdot 5 \cdot 4 \cdot 3!}{3!3!} = \dfrac{6 \cdot 5 \cdot 4}{3 \cdot 2 \cdot 1} = 20$

26. $3 \cdot 5 = 15$ different outfits

28. $5 \cdot 5 = 25$ 2-letter codes

30. $10 \cdot 10 \cdot 10 = 1000$ 3-digit numbers

32. $5 \cdot 4 \cdot 3 \cdot 2 \cdot 1 = 120$ ways to arrange 5 books

34. $6 \cdot 5 \cdot 4 \cdot 3 = 360$ different 4-letter codes

36. $5 \cdot 4 \cdot 3 \cdot 2 \cdot 1 = 120$ arrangements of the digits in 51,342.

38. $C(8, 3) = \dfrac{8!}{5!3!} = \dfrac{8 \cdot 7 \cdot 6 \cdot 5!}{5! \cdot 3 \cdot 2 \cdot 1} = 56$
ways of forming the committee.

40. $4 \cdot 4 \cdot 4 \cdot 4 \cdot 4 = 1024$ possible arrangements

42. $8 \cdot 10 \cdot 10 \cdot 10 \cdot 10 = 80,000$ 5-digit numbers

44. (a) $26 \cdot 26 \cdot 10 \cdot 10 \cdot 10 \cdot 10 = 6,760,000$ different license plate numbers

    (b) $26 \cdot 26 \cdot 10 \cdot 9 \cdot 8 \cdot 7 = 3,407,040$ different license numbers.

    (c) $26 \cdot 25 \cdot 10 \cdot 9 \cdot 8 \cdot 7 = 3,276,000$ different license numbers.

46. $50 \cdot 50 \cdot 50 = 125,000$ different lock combinations

48. $C(4, 1) \cdot C(11, 8) = \dfrac{4!}{3!1!} \cdot \dfrac{11!}{3!8!} = \dfrac{4 \cdot 3!}{3! \cdot 1!} = \dfrac{11 \cdot 10 \cdot 9 \cdot 8!}{3! \cdot 8!}$

$= \dfrac{4}{1} \cdot \dfrac{990}{6} = 660$ different teams

50. $C(10, 5) \cdot C(10, 3) \cdot C(5, 3) = \dfrac{10!}{5!5!} \cdot \dfrac{10!}{7!3!} \cdot \dfrac{5!}{2!3!}$

$= \dfrac{10 \cdot 9 \cdot 8 \cdot 7 \cdot 6 \cdot 5!}{5! \cdot 5!} \cdot \dfrac{10 \cdot 9 \cdot 8 \cdot 7!}{7! \cdot 3!} \cdot \dfrac{5 \cdot 4 \cdot 3!}{2! \cdot 3!}$

$= 302,400$ different teams

52. $8 \cdot 7 \cdot 6 \cdot 5 \cdot 4 \cdot 3 \cdot 2 \cdot 1 = 40,320$ different batting orders.

54. $\dfrac{11!}{2!2!2!} = 4,989,600$ different words

56. Let's look at the number of sequences possible for the American League to win the World Series. Once we find this result, we'll multiply it by 2 and we'll have our answer. To win the World Series, you must win the last game played. To win in 4 games, there is only 1 possibility, that is to win all 4 games. To win in 5 games, you must win 3 of the first 4 and then win the 5th game. There are

$\begin{bmatrix} 4 \\ 3 \end{bmatrix} = 4$ ways to win 3 of the first 4 games. Therefore, there are 4 ways to win in 5 games.

Similarly, there are $\begin{bmatrix} 5 \\ 3 \end{bmatrix} = 10$ ways to win in 6 games and $\begin{bmatrix} 6 \\ 3 \end{bmatrix} = 20$ ways to win in 7 games. Therefore, there are $1 + 4 + 10 + 20 = 35$ ways for the American League to win the World Series. Similarly, there are 35 ways for the National League to win the World Series. There is a total of 70 different sequences possible.

58. $\begin{bmatrix} 2 \\ 1 \end{bmatrix} \cdot \begin{bmatrix} 3 \\ 2 \end{bmatrix} \cdot \begin{bmatrix} 7 \\ 2 \end{bmatrix} = 126$ different teams.

60. (a) $\begin{bmatrix} 15 \\ 5 \end{bmatrix} \begin{bmatrix} 10 \\ 0 \end{bmatrix} = \dfrac{15!}{10!5!} \cdot \dfrac{10!}{10!0!} = \dfrac{15 \cdot 14 \cdot 13 \cdot 12 \cdot 11 \cdot 10!}{10! \cdot 5 \cdot 4 \cdot 3 \cdot 2 \cdot 1} = 3003$

    (b) $\begin{bmatrix} 15 \\ 3 \end{bmatrix} \begin{bmatrix} 10 \\ 2 \end{bmatrix} = \dfrac{15!}{12!3!} \cdot \dfrac{10!}{8!2!} = \dfrac{15 \cdot 14 \cdot 13 \cdot 12!}{12! \cdot 3 \cdot 2 \cdot 1} \cdot \dfrac{10 \cdot 9 \cdot 8!}{8! \cdot 2 \cdot 1} = 20475$

    (c) $\begin{bmatrix} 15 \\ 4 \end{bmatrix} \begin{bmatrix} 10 \\ 1 \end{bmatrix} + \begin{bmatrix} 15 \\ 5 \end{bmatrix} \begin{bmatrix} 10 \\ 0 \end{bmatrix} = \dfrac{15!}{11!4!} \cdot \dfrac{10!}{9!1!} + 3003$

$= \dfrac{15 \cdot 14 \cdot 13 \cdot 12 \cdot 11!}{11! \cdot 4 \cdot 3 \cdot 2 \cdot 1} \cdot \dfrac{10 \cdot 9!}{9!} + 3003 = 16653$

2. $S = \{HH, HT, TH, TT\}$

$$P(HH) = \frac{1}{4}, \; P(HT) = \frac{1}{4}, \; P(TH) = \frac{1}{4}, \; P(TT) = \frac{1}{4}$$

4. $S = \{$H1H, H2H, H3H, H4H, H5H, H6H, H1T, H2T, H3T, H4T, H5T, H6T, T1H, T2H, T3H, T4H, T5H, T6H, T1T, T2T, T3T, T4T, T5T, T6T$\}$

Each outcome has the probability of $\frac{1}{24}$

6. $S = \{$HHH, HHT, HTH, THH, THT, TTH, HTT, TTT$\}$

Each outcome has the probability of $\frac{1}{8}$

8. $S = \{$Forward Yellow, Forward Green, Forward Red, Backward Yellow, Backward Green, Backward Red$\}$

Each outcome has the probability of $\frac{1}{6}$; thus

$$P(\text{Forward Yellow}) + P(\text{Forward Red}) = \frac{1}{6} + \frac{1}{6} = \frac{1}{3}$$

10. $S = \{$Yellow 1 Forward, Yellow 1 Backward, Yellow 2 Forward, Yellow 2 Backward, Yellow 3 Forward, Yellow 3 Backward, Yellow 4 Forward, Yellow 4 Backward, Green 1 Forward, Green 1 Backward, Green 2 Forward, Green 2 Backward, Green 3 Forward, Green 3 Backward, Green 4 Forward, Green 4 Backward, Red 1 Forward, Red 1 Backward, Red 2 Forward, Red 2 Backward, Red 3 Forward, Red 3 Backward, Red 4 Forward, Red 4 Backward$\}$

Each outcome has the probability $\frac{1}{24}$; thus,

$$P(\text{Yellow 2 Forward}) + P(\text{Yellow 4 Forward}) = \frac{1}{24} + \frac{1}{24} = \frac{1}{12}$$

12. $S = \{$Forward 11, Forward 12, Forward 13, Forward 14, Forward 21, Forward 22, Forward 23, Forward 24, Forward 31, Forward 32, Forward 33, Forward 34, Forward 41, Forward 42, Forward 43, Forward 44, Backward 11, Backward 12, Backward 13, Backward 14, Backward 21, Backward 22, Backward 23, Backward 24, Backward 31, Backward 32, Backward 33, Backward 34, Backward 41, Backward 42, Backward 43, Backward 44$\}$

Each outcome has the probability $\frac{1}{32}$; thus,

$$P(\text{Forward 12}) + P(\text{Forward 14}) + P(\text{Forward 32}) + P(\text{Forward 34})$$
$$\frac{1}{32} + \frac{1}{32} + \frac{1}{32} + \frac{1}{32} = \frac{1}{8}$$

14. A

16. F

18. $P(H) = \dfrac{1}{2}P(T)$

Since $P(H) + P(T) = 1$

$\dfrac{1}{2}P(T) + P(T) = 1$

$\dfrac{3}{2}P(T) = 1$

$P(T) = \dfrac{2}{3}$, so $P(H) = \dfrac{1}{3}$

20. There are 5 outcomes, with each outcome equally likely to occur; therefore, the probability assigned to each face is $\dfrac{1}{5}$.

22. $P(A \cap B) = 0$

24. $P(A \cup B) = P(A) + P(B) - P(A \cap B)$; therefore, $P(A \cap B) = P(A) + P(B) - P(A \cup B) = .3 + .4 - .6 = 0.1$

26. $P(\text{Green}) = \dfrac{n(\text{Green})}{n(S)} = \dfrac{8}{20} = \dfrac{2}{5}$

28. $P(\text{not white}) = P(\text{Green} \cup \text{Orange}) = \dfrac{n(\text{Green}) + n(\text{Orange})}{n(S)} = \dfrac{11}{20}$

30. $P(5 \text{ or } 9) = P(5) + P(9) - P(5 \text{ and } 9) = P(5) + P(9) = \dfrac{n(5) + n(9)}{n(S)} = \dfrac{4+4}{36} = \dfrac{8}{36} = \dfrac{2}{9}$

32. $n(S) = 5 + 35 + 30 + 20 + 10 = 100$
$n(E) = 35 + 30 = 65$
$P(E) = \dfrac{n(E)}{n(S)} = \dfrac{65}{100} = 0.65$

34. $n(S) = 100$ (from Problem 32)
$n(E) = 30 + 20 + 10 = 60$
$P(E) = \dfrac{n(E)}{n(S)} = \dfrac{60}{100} = 0.6$

36. (a) $P(\text{at most } 2) = P(0) + P(1) + P(2) = .1 + .15 + .20 = 0.45$
     (b) $P(\text{at least } 2) = P(2) + P(3) + P(4 \text{ or more}) = .2 + .24 + .31 = 0.75$
     (c) $P(\text{at least } 1) = 1 - P(0) = 1 - .10 = 0.9$

38. The number of ways 6 people can be chosen from 14 is
$C(14, 6) = \dfrac{14!}{8!6!} = 3003$ ways.
The number of ways 0 supervisors can be chosen from 2 is
$C(2, 0) = \dfrac{2!}{2!0!} = 1$ way.
The number of ways 2 skilled workers can be chosen from 5 is
$C(5, 2) = \dfrac{5!}{3!2!} = 10$
The number of ways 4 unskilled workers can be chosen from 7 is
$C(7, 4) = 35$
Using the multiplication principle, $P(2 \text{ skilled and } 4 \text{ unskilled})$
$= \dfrac{1 \cdot 10 \cdot 35}{3003} \approx 0.1166$

40. (a) Construct a tree diagram to list all the possible outcomes:

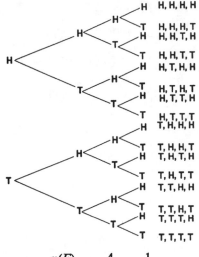

$$P(E) = \frac{n(E)}{n(S)} = \frac{4}{16} = \frac{1}{4}$$

(b) $P(\text{no more than } 1) = P(0 \text{ tails}) + P(1 \text{ tail}) = \frac{1}{16} + \frac{4}{16} = \frac{5}{16}$

42. (a) $P(\text{sum} \neq 2) = 1 - P(\text{sum} = 2) = 1 - \frac{1}{36} = \frac{35}{36}$

$$P(\text{sum} \neq 2 \text{ on } 5 \text{ tosses}) = \left[\frac{35}{36}\right]^5 = .869$$

(b) $P(\text{sum} \neq 7) = 1 - P(\text{sum} = 7) = 1 - \frac{6}{36} = \frac{30}{36} = \frac{5}{6}$

$$P(\text{sum} \neq 7 \text{ on } 5 \text{ tosses}) = \left[\frac{5}{6}\right]^5 = .402$$

44. $P(30 \text{ nondefective}) = \dfrac{\text{Number of ways 30 nondefectives can be chosen 40 nondefectives}}{\text{Number of ways 30 transformers can be chosen out of 50}}$

$$= \frac{C(40, 30)}{C(50, 30)} = \frac{\dfrac{40!}{30!10!}}{\dfrac{50!}{30!20!}} \approx 1.80 \times 10^{-5}$$

13 Chapter Review

2. $6! = 6 \cdot 5 \cdot 4 \cdot 3 \cdot 2 \cdot 1 = 720$

4. $\begin{bmatrix} 8 \\ 6 \end{bmatrix} = \dfrac{8!}{(8-6)!6!} = \dfrac{8 \cdot 7 \cdot 6!}{2 \cdot 1 \cdot 6!} = 28$

6. $P(7, 3) = \dfrac{7!}{(7-3)!} = \dfrac{7!}{4!} = \dfrac{7 \cdot 6 \cdot 5 \cdot 4!}{4!}$
$= 210$

8. $C(7, 3) = \dfrac{7!}{(7-3)!3!} = \dfrac{7 \cdot 6 \cdot 5 \cdot 4!}{4! \cdot 3 \cdot 2 \cdot 1}$
$= 35$

10. $\{(-1)^{n+1}(2n + 3)\}$
    $5, -7, 9, -11, 13$

12. $\left\{\dfrac{e^n}{n}\right\}$

    $e, \dfrac{e^2}{2}, \dfrac{e^3}{3}, \dfrac{e^4}{4}, \dfrac{e^5}{5}$

14. $a_1 = 4, a_{n+1} = \dfrac{-1}{4}a_n$

    $4, -1, \dfrac{1}{4}, \dfrac{-1}{16}, \dfrac{1}{64}$

16. $a = -3, a_{n+1} = 4 + a_n$
    $-3, 1, 5, 9, 13$

18. $\{4n + 3\}$ Arithmetic
    $d = [4(n + 1) + 3] - (4n + 3) = 4$
    $a_1 = 7, a_n = (4n + 3)$
    $S_n = \dfrac{n}{2}[7 + (4n + 3)] = \dfrac{n}{2}(4n + 10) = 2n^2 + 5n$

20. $\{2n^2 - 1\}$ Looking at 1, 7, 17, 31
    Neither

22. $\{3^{2n}\}$ Looking at 9, 81, 729, 6561
    Geometric, $r = 9$

    $S_n = \displaystyle\sum_{k=1}^{n} (3^{2n}) = 9\left[\dfrac{1 - 9^n}{1 - 9}\right] = \dfrac{-9}{8}(1 - 9^n)$

24. $1, -3, -7, -11,$ Arithmetic
    $d = -4$
    $a = 1, a_n = 1 + (n - 1) \cdot -4 = -4n + 5$
    $S_n = \dfrac{n}{2}[1 + (-4n + 5)] = \dfrac{n}{2}(-4n + 6) = 3n - 2n^2$

26. $5, \dfrac{-5}{3}, \dfrac{5}{9}, \dfrac{-5}{27}, \dfrac{5}{81}, \ldots$

    Geometric: $r = \dfrac{-1}{3}$

    $S_n = 5\left[\dfrac{1 + \left[\dfrac{1}{3}\right]^n}{1 + \dfrac{1}{3}}\right] = \dfrac{15}{4}\left[1 + \left[\dfrac{1}{3}\right]^n\right]$

28. $\dfrac{3}{2}, \dfrac{5}{4}, \dfrac{7}{6}, \dfrac{9}{8}, \dfrac{11}{10}, \ldots$
    Neither.

30. $a_8 = a + 7d, d = -2$
    $a_8 = 1 + 7(-2) = -13$

32. Geometric sequence, $r = 2$
    $a_{11} = 2^{10}(1) = 1024$

34. Geometric sequence, $r = \sqrt{2}$

    $a_9 = \sqrt{2^8}\left(\sqrt{2}\right) = 16\sqrt{2}$

36. $a_8 = a + 7d = -20$
    $a_{17} = a + 16d = -47$
    $\phantom{a_{17} = a + 1} -9d = 27$
    $\phantom{a_{17} = a + 16d = -4} d = \dfrac{-27}{9} = -3$
    $a = -20 - 7\left[\dfrac{-327}{9}\right] = -20 + \dfrac{189}{9} = 1$
    $a_n = 1 - \dfrac{27}{9}(n - 1) = 1 - \dfrac{27}{9}n + \dfrac{27}{9} = 4 - 3n$

38.
$$a_{12} = a + 11d = 30$$
$$a_{22} = a + 21d = 50$$
$$\underline{\quad\quad\quad -10d = -20}$$
$$d = 2$$
$$a = 30 - 11(2) = 8$$
$$a_n = 8 + 2(n - 1) = 6 + 2n$$

40. $\quad a = 2, r = \dfrac{1}{2}$
$$2 + 1 + \frac{1}{2} + \frac{1}{4} + \ldots = \frac{2}{1 - \dfrac{1}{2}} = 4$$

42. $\quad a = 6, r = \dfrac{-2}{3}$
$$6 - 4 + \frac{8}{3} - \frac{16}{9} + \cdots = \frac{6}{1 + \dfrac{2}{3}} = \frac{18}{5}$$

44. $\quad a = 3, r = \dfrac{-3}{4}$
$$\sum_{k=1}^{\infty} 3\left[\frac{-3}{4}\right]^{k-1} = \frac{3}{1 + \dfrac{3}{4}} = \frac{12}{7}$$

46. (I) $\quad n = 1:\ 4 \cdot 1 - 2 = 2$ and $2 \cdot 1^2 = 2$

(II) $\quad$ If $2 + 6 + 10 + \cdots + (4k - 2) = 2k^2$, then
$$2 + 6 + 10 + \cdots + (4k - 2) + (4k + 2)$$
$$= [2 + 6 + 10 + \cdots + (4k - 2)] + 4(k + 2)$$
$$= 2k^2 + (4k + 2) = 2(k^2 + 2k + 1) = 2(k + 1)^2$$

48. (I) $\quad n = 1:\ 3 \cdot 2^{1-1} = 3$ and $3(2^1 - 1) = 3$

(II) $\quad$ If $3 + 6 + 12 + \cdots + 3 \cdot 2^{k-1} = 3(2^k - 1)$, then
$$3 + 6 + 12 + \cdots + 3 \cdot 2^{k-1} + 3 \cdot 2k$$
$$= [3 + 6 + 12 + \cdots + 3 \cdot 2^{k-1}] + 3 \cdot 2^k$$
$$= 3(2^k - 1) + 3 \cdot 2^k = 3 \cdot 2^k - 3 + 3 \cdot 2^k = 3(2^{k+1} - 1)$$

50. (I) $\quad n = 1:\ 1(1 + 2) = 3$ and $\dfrac{1}{6}(1 + 1)(2 \cdot 1 + 7) = 3$

(II) $\quad$ If $1 \cdot 3 + 2 \cdot 4 + 3 \cdot 5 + \cdots + k(k + 2) = \dfrac{k}{6}(k + 1)(2k + 7)$, then
$$1 \cdot 3 + 2 \cdot 4 + 3 \cdot 5 + \cdots + k(k + 2) + (k + 1)(k + 3)$$
$$= [1 \cdot 3 + 2 \cdot 4 + 3 \cdot 5 + \cdots + k(k + 2)] + (k + 1)(k + 3)$$
$$= \frac{k}{6}(k + 1)(2k + 7) + (k + 1)(k + 3)$$
$$= (k + 1)\left[\frac{k}{6}(2k + 7) + (k + 3)\right]$$
$$= (k + 1)\left[\frac{2k^2}{6} + \frac{7k}{6} + k + 3\right] = (k + 1)\left[\frac{2k^2 + 7k + 6k + 18}{6}\right]$$
$$= \frac{k + 1}{6}(2k^2 + 13k + 18) = \frac{k + 1}{6}(k + 2)(2k + 9)$$

52. $\quad (x - 3)^4 = \begin{bmatrix} 4 \\ 0 \end{bmatrix} x^4 + \begin{bmatrix} 4 \\ 1 \end{bmatrix}(-3)x^3 + \begin{bmatrix} 4 \\ 2 \end{bmatrix}(-3)^2 x^2 + \begin{bmatrix} 4 \\ 3 \end{bmatrix}(-3)^3 x + \begin{bmatrix} 4 \\ 4 \end{bmatrix}(-3)^4$

$$= x^4 + 4 \cdot (-3x^3) + 6 \cdot (9x^2) + 4(-27x) + 81$$
$$= x^4 - 12x^3 + 54x^2 - 108x + 81$$

54. $\quad (3x - 4)^4 = \begin{bmatrix} 4 \\ 0 \end{bmatrix}(3x)^4 + \begin{bmatrix} 4 \\ 1 \end{bmatrix}(-4)(3x)^3 + \begin{bmatrix} 4 \\ 2 \end{bmatrix}(-4)^2(3x)^2 + \begin{bmatrix} 4 \\ 3 \end{bmatrix}(-4)^3(3x) + \begin{bmatrix} 4 \\ 4 \end{bmatrix}(-4)^4$

$$= 81x^4 + 4(-108x^3) + 6(144x^2) + 4(-192x) + 256$$
$$= 81x^4 - 432x^3 + 864x^2 - 768x + 256$$

56. $n = 8$, $a = -3$, $x = x$, and $j = 3$

$$\begin{bmatrix} 8 \\ 8-3 \end{bmatrix} (-3)^{8-3} x^3 = \begin{bmatrix} 8 \\ 5 \end{bmatrix} \cdot -243 x^3 = \frac{8!}{5!3!} \cdot -243 \cdot x^3 = -13,608 x^3$$

The coefficient is $-13,608$.

58. $n = 8$, $a = 1$, $x = 2x^6$, and $j = 6$

$$\begin{bmatrix} 8 \\ 8-6 \end{bmatrix} 1^{8-6} (2x)^6 = \begin{bmatrix} 8 \\ 2 \end{bmatrix} (2x)^6 = 28 \cdot 64 \cdot x^7 = 1792 x^6$$

The coefficient of x^6 is 1792.

In Problems 60–68, $U = \{1, 2, 3, 4, 5, 6, 7, 8, 9\}$, $A = \{1, 3, 5, 7\}$, $B = \{3, 5, 6, 7, 8\}$, and
$C = \{2, 3, 7, 8, 9\}$.

60. $B \cup C = \{2, 3, 5, 6, 7, 8, 9\}$ 62. $A \cap B = \{3, 5, 7\}$

64. $B' \cap C' = \{1, 2, 4, 9\} \cap \{1, 4, 5, 6\} = \{1, 4\}$

66. $(A \cup B)' = \{1, 3, 5, 6, 7, 8\}' = \{2, 4, 9\}$

68. Using the formula, $n(A \cup B) = n(A) + n(B) - n(A \cap B)$
$$30 = 12 + n(B) - 6$$
$$n(B) = 24$$

70. $20 + 1 + 6 + 2 + 0 + 5 = 34$ in A or B.   72. $20 + 1 + 4 + 20 = 45$ not in B.

74. $2 + 5 = 7$ in B but not C.   76. $P(5, 5) = 5! = 120$

78. $2 \cdot 2 \cdot 2 \cdot 2 \cdot 2 \cdot 2 = 2^6 = 64$ different   80. $P(4, 4) = 4! = 12$
outcomes are possible.

82. $C(10, 3) = \dfrac{10!}{7!3!} = \dfrac{10 \cdot 9 \cdot 8 \cdot 7!}{7! \cdot 3!} = 120$ different tests.

84. (a) $5! \cdot 5! = 14,400$ different arrangements
    (b) $5 \cdot 5 \cdot 4 \cdot 4 \cdot 3 \cdot 3 \cdot 2 \cdot 2 \cdot 1 \cdot 1 = 5! \cdot 5! = 14,400$ different arrangements.

86. $5 \cdot 3 \cdot 4 = 60$ different home designs   88. $2^8 = 256$ different numbers

90. $\dfrac{10!}{4!3!2!1!} = 12,600$ different vertical arrangements.

92. (a) $\begin{bmatrix} 5 \\ 1 \end{bmatrix} \begin{bmatrix} 8 \\ 3 \end{bmatrix} = \dfrac{5!}{4!1!} \cdot \dfrac{8!}{5!3!} = 280$ committees containing exactly 1 man

(b) $\begin{bmatrix} 5 \\ 2 \end{bmatrix} \begin{bmatrix} 8 \\ 2 \end{bmatrix} = \dfrac{5!}{3!2!} \cdot \dfrac{8!}{6!2!} = 280$ committees containing exactly 2 women

(c) $\begin{bmatrix} 5 \\ 1 \end{bmatrix} \begin{bmatrix} 8 \\ 3 \end{bmatrix} + \begin{bmatrix} 5 \\ 2 \end{bmatrix} \begin{bmatrix} 8 \\ 2 \end{bmatrix} + \begin{bmatrix} 5 \\ 3 \end{bmatrix} \begin{bmatrix} 8 \\ 1 \end{bmatrix} + \begin{bmatrix} 5 \\ 4 \end{bmatrix} \begin{bmatrix} 8 \\ 0 \end{bmatrix} = 280 + 280 + 80 + 5$
$$= 645 \text{ committees containing at}$$
$$\text{least 1 man}$$

94. The number of all logical possibilities is

$$n(S) = C(9, 1) = \frac{9!}{8!1!} = 9$$

$$n(E) = \text{Number of ways of choosing a \$1 bill}$$

$$= c(4, 1) = \frac{4!}{3!1!} = 4$$

$$P(E) = \frac{4}{9} = 0.44 \ldots$$

96. Let S be all possible selections, let D be the card is divisible by 5, and let PN be the card is 1 or a prime number, then

$$N(S) = 100$$

$$n(D) = 20 \ \text{(There are 20 numbers divisible by 5 between 1 and 100).}$$

$$n(PN) = 26 \ \text{(There are 26 prime numbers or 1 less than or equal to 100).}$$

$$P(D) = \frac{20}{100} = 0.2$$

$$P(PN) = \frac{26}{100} = 0.26$$

98. (a) $$P(5 \text{ heads}) = \frac{n(5 \text{ heads})}{n(S)} = \frac{\begin{bmatrix} 10 \\ 5 \end{bmatrix}}{2^{10}} = \frac{\frac{10!}{5!5!}}{1024} = \frac{252}{1024} \approx 0.246$$

(b) $$P(\text{all heads}) = \frac{n(\text{all heads})}{n(S)} = \frac{1}{2^{10}} = \frac{1}{1024} \approx 9.8 \times 10^{-4}$$

100. The total number of tiles required is

$$S = 30 + 29 + 28 + \cdots + 15$$

This is the sum of an arithmetic sequence; the common difference is -1. The number of terms to be added is 16, with the first term $a = 30$ and the last term $a_{16} = 15$. The sum S is

$$S = \frac{16}{2}(30 + 15) = 360$$

Thus, 360 tiles will be required.

102. Since the salary increases are 4%, $r = 1.04$. To find a_5, we use the formula

$$a_n = a_1 \cdot r^{n-1}$$

Therefore, $a_5 = \$20,000(1.04)^{5-1} = \$23.397.17$

Mission Possible

1. (a) \$210,000; \$210,000; \$210,000; \$210,000; \$210,000; \$210,000
   (b) \$200,000; \$209,000; \$218,405; \$228,233; \$238,504; \$249,236
   (c) \$200,000; \$200,000; \$212,000; \$224,720; \$238,203; \$252,495
   (d) \$200,000; \$209,500; \$219,000; \$228,500; \$238,000; \$247,500

2. Options (b) and (c) are geometric, (d) is arithmetic, and (a) can be classified as either geometric with $r = 1$ or arithmetic with $d = 0$.

3. (a) \$1,260,000 (b) \$1,343,378 (c) \$1,327,419 (d) \$1,342,500

4.   Option (b) appears to be best because it pays the most over all. However, contracts are made to be broken, so you may be better off to take option (d), for example, which pays the most over the first three years. Then if things are going well, you can demand a new contract. If things are not going well and the recording studio tears up your contract, you will have made the most of those three years. Another factor to bear in mind is that the money made earlier can be invested which will (should) earn more money, increasing its "worth."

Chapter 14

MISCELLANEOUS TOPICS

14.1 Matrix Algebra

2. $A - B = \begin{bmatrix} 0 & 3 & -5 \\ 1 & 2 & 6 \end{bmatrix} - \begin{bmatrix} 4 & 1 & 0 \\ -2 & 3 & -2 \end{bmatrix} = \begin{bmatrix} -4 & 2 & -5 \\ 3 & -1 & 8 \end{bmatrix}$

4. $-3B = -3\begin{bmatrix} 4 & 1 & 0 \\ -2 & 3 & -2 \end{bmatrix} = \begin{bmatrix} -12 & -3 & 0 \\ 6 & -9 & 6 \end{bmatrix}$

6. $2A + 4B = 2\begin{bmatrix} 0 & 3 & -5 \\ 1 & 2 & 6 \end{bmatrix} + 4\begin{bmatrix} 4 & 1 & 0 \\ -2 & 3 & -2 \end{bmatrix} = \begin{bmatrix} 0 & 6 & -10 \\ 2 & 4 & 12 \end{bmatrix} + \begin{bmatrix} 16 & 4 & 0 \\ -8 & 12 & -8 \end{bmatrix} = \begin{bmatrix} 16 & 10 & -10 \\ -6 & 16 & 4 \end{bmatrix}$

8. $BC = \begin{bmatrix} 4 & 1 & 0 \\ -2 & 3 & -2 \end{bmatrix}\begin{bmatrix} 4 & 1 \\ 6 & 2 \\ -2 & 3 \end{bmatrix} = \begin{bmatrix} 16 + 6 + 0 & 4 + 2 + 0 \\ -8 + 18 + 4 & -2 + 6 - 6 \end{bmatrix} = \begin{bmatrix} 22 & 6 \\ 14 & -2 \end{bmatrix}$

10. $CB = \begin{bmatrix} 4 & 1 \\ 6 & 2 \\ -2 & 3 \end{bmatrix}\begin{bmatrix} 4 & 1 & 0 \\ -2 & 3 & -2 \end{bmatrix} = \begin{bmatrix} 16 - 2 & 4 + 3 & 0 - 2 \\ 24 - 4 & 6 + 6 & 0 - 4 \\ -8 - 6 & -2 + 9 & 0 - 6 \end{bmatrix} = \begin{bmatrix} 14 & 7 & -2 \\ 20 & 12 & -4 \\ -14 & 7 & -6 \end{bmatrix}$

12. $(A + B)C = \left(\begin{bmatrix} 0 & 3 & -5 \\ 1 & 2 & 6 \end{bmatrix} + \begin{bmatrix} 4 & 1 & 0 \\ -2 & 3 & -2 \end{bmatrix} \right)\begin{bmatrix} 4 & 1 \\ 6 & 2 \\ -2 & 3 \end{bmatrix} = \begin{bmatrix} 4 & 4 & -5 \\ -1 & 5 & 4 \end{bmatrix}\begin{bmatrix} 4 & 1 \\ 6 & 2 \\ -2 & 3 \end{bmatrix}$

$= \begin{bmatrix} 16 + 24 + 10 & 4 + 8 - 15 \\ -4 + 30 - 8 & -1 + 10 + 12 \end{bmatrix} = \begin{bmatrix} 50 & -3 \\ 18 & 21 \end{bmatrix}$

14. $CA + 5I_3 = \begin{bmatrix} 4 & 1 \\ 6 & 2 \\ -2 & 3 \end{bmatrix} \begin{bmatrix} 0 & 3 & -5 \\ 1 & 2 & 6 \end{bmatrix} + 5 \begin{bmatrix} 1 & 0 & 0 \\ 0 & 1 & 0 \\ 0 & 0 & 1 \end{bmatrix} = \begin{bmatrix} 0+1 & 12+2 & -20+6 \\ 0+2 & 18+4 & -30+12 \\ 0+3 & -6+6 & 10+18 \end{bmatrix} \begin{bmatrix} 5 & 0 & 0 \\ 0 & 5 & 0 \\ 0 & 0 & 5 \end{bmatrix}$

$= \begin{bmatrix} 1 & 14 & -14 \\ 2 & 22 & -18 \\ 3 & 0 & 28 \end{bmatrix} + \begin{bmatrix} 5 & 0 & 0 \\ 0 & 5 & 0 \\ 0 & 0 & 5 \end{bmatrix} = \begin{bmatrix} 6 & 14 & -14 \\ 2 & 27 & -18 \\ 3 & 0 & 33 \end{bmatrix}$

16. $AC + BC = \begin{bmatrix} 0 & 3 & -5 \\ 1 & 2 & 6 \end{bmatrix} \begin{bmatrix} 4 & 1 \\ 6 & 2 \\ -2 & 3 \end{bmatrix} + \begin{bmatrix} 4 & 1 & 0 \\ -2 & 3 & -2 \end{bmatrix} \begin{bmatrix} 4 & 1 \\ 6 & 2 \\ -2 & 3 \end{bmatrix}$

$= \begin{bmatrix} 0+18+10 & 0+6-15 \\ 4+12-12 & 1+4+18 \end{bmatrix} + \begin{bmatrix} 16+6+0 & 4+2+0 \\ -8+18+4 & -2+6-6 \end{bmatrix}$

$= \begin{bmatrix} 28 & -9 \\ 4 & 23 \end{bmatrix} + \begin{bmatrix} 22 & 6 \\ 14 & -2 \end{bmatrix} = \begin{bmatrix} 50 & -3 \\ 18 & 21 \end{bmatrix}$

18. $\begin{bmatrix} 4 & 1 \\ 2 & 1 \end{bmatrix} \begin{bmatrix} -6 & 6 & 1 & 0 \\ 2 & 5 & 4 & -1 \end{bmatrix} = \begin{bmatrix} -24+2 & 24+5 & 4+4 & 0-1 \\ -12+2 & 12+5 & 2+4 & 0-1 \end{bmatrix} = \begin{bmatrix} -22 & 29 & 8 & -1 \\ -10 & 17 & 6 & -1 \end{bmatrix}$

20. $\begin{bmatrix} 4 & -2 & 3 \\ 0 & 1 & 2 \\ -1 & 0 & 1 \end{bmatrix} \begin{bmatrix} 2 & 6 \\ 1 & -1 \\ 0 & 2 \end{bmatrix} = \begin{bmatrix} 8-2+0 & 24+2+6 \\ 0+1+0 & 0-1+4 \\ -2+0+0 & -6+0+2 \end{bmatrix} = \begin{bmatrix} 6 & 32 \\ 1 & 3 \\ -2 & -4 \end{bmatrix}$

22. $\begin{bmatrix} 3 & -1 & | & 1 & 0 \\ -2 & 1 & | & 0 & 1 \end{bmatrix} \rightarrow \begin{bmatrix} 1 & 0 & | & 1 & 1 \\ -2 & 1 & | & 0 & 1 \end{bmatrix} \rightarrow \begin{bmatrix} 1 & 0 & | & 1 & 1 \\ 0 & 1 & | & 2 & 3 \end{bmatrix}$

$\qquad\qquad\qquad\uparrow \qquad\qquad\quad \uparrow$

$\qquad\qquad R_1 = r_2 + r_1 \quad R_2 = 2r_1 + r_2$

$A^{-1} = \begin{bmatrix} 1 & 1 \\ 2 & 3 \end{bmatrix}$

24. $\begin{bmatrix} -4 & 1 & | & 1 & 0 \\ 6 & -2 & | & 0 & 1 \end{bmatrix} \rightarrow \begin{bmatrix} 1 & -\frac{1}{4} & | & -\frac{1}{4} & 0 \\ 6 & -2 & | & 0 & 1 \end{bmatrix} \rightarrow \begin{bmatrix} 1 & -\frac{1}{4} & | & -\frac{1}{4} & 0 \\ 0 & -\frac{1}{2} & | & \frac{3}{2} & 1 \end{bmatrix} \rightarrow \begin{bmatrix} 1 & 0 & | & -1 & -\frac{1}{2} \\ 0 & -\frac{1}{2} & | & \frac{3}{2} & 1 \end{bmatrix} \rightarrow \begin{bmatrix} 1 & 0 & | & -1 & -\frac{1}{2} \\ 0 & 1 & | & -3 & -2 \end{bmatrix}$

$\qquad\quad\uparrow \qquad\qquad\qquad \uparrow \qquad\qquad\qquad\quad \uparrow \qquad\qquad\qquad \uparrow$

$\qquad R_1 = -\frac{1}{4}r_1 \qquad R_2 = -6r_1 + r_2 \quad R_1 = -\frac{1}{2}r_2 + r_1 \quad R_2 = -2r_2$

$A^{-1} = \begin{bmatrix} -1 & -\frac{1}{2} \\ -3 & -2 \end{bmatrix}$

26.
$$\begin{bmatrix} b & 3 & | & 1 & 0 \\ b & 2 & | & 0 & 1 \end{bmatrix} \rightarrow \begin{bmatrix} 1 & \frac{3}{b} & | & \frac{1}{b} & 0 \\ b & 2 & | & 0 & 1 \end{bmatrix} \rightarrow \begin{bmatrix} 1 & \frac{3}{b} & | & \frac{1}{b} & 0 \\ 0 & -1 & | & -1 & 1 \end{bmatrix} \rightarrow \begin{bmatrix} 1 & \frac{3}{b} & | & \frac{1}{b} & 0 \\ 0 & 1 & | & 1 & -1 \end{bmatrix} \rightarrow \begin{bmatrix} 1 & 0 & | & -\frac{2}{b} & \frac{3}{b} \\ 0 & 1 & | & 1 & -1 \end{bmatrix}$$

$$R_1 = \frac{1}{b}r_1 \qquad R_2 = -br_1 + r_2 \quad R_2 = -r_2 \qquad\qquad R_1 = -\frac{3}{b}r_2 + r_1$$

$$A^{-1} = \begin{bmatrix} -\frac{2}{b} & \frac{3}{b} \\ 1 & -1 \end{bmatrix}$$

28.
$$\begin{bmatrix} 1 & 0 & 2 & | & 1 & 0 & 0 \\ -1 & 2 & 3 & | & 0 & 1 & 0 \\ 1 & -1 & 0 & | & 0 & 0 & 1 \end{bmatrix} \rightarrow \begin{bmatrix} 1 & 0 & 2 & | & 1 & 0 & 0 \\ 0 & 2 & 5 & | & 1 & 1 & 0 \\ 0 & -1 & -2 & | & -1 & 0 & 1 \end{bmatrix} \rightarrow \begin{bmatrix} 1 & 0 & 2 & | & 1 & 0 & 0 \\ 0 & 1 & \frac{5}{2} & | & \frac{1}{2} & \frac{1}{2} & 0 \\ 0 & -1 & -2 & | & -1 & 0 & 1 \end{bmatrix}$$

$$R_2 = r_1 + r_2 \qquad\qquad R_2 = \frac{1}{2}r_2$$
$$R_3 = -r_1 + r_3$$

$$\rightarrow \begin{bmatrix} 1 & 0 & 2 & | & 1 & 0 & 0 \\ 0 & 1 & \frac{5}{2} & | & \frac{1}{2} & \frac{1}{2} & 0 \\ 0 & 0 & 1 & | & -1 & 1 & 2 \end{bmatrix} \rightarrow \begin{bmatrix} 1 & 0 & 0 & | & 3 & -2 & -4 \\ 0 & 1 & 0 & | & 3 & -2 & -5 \\ 0 & 0 & 1 & | & -1 & 1 & 2 \end{bmatrix}$$

$$R_3 = 2(r_2 + r_3) \qquad R_1 = -2r_3 + r_1$$
$$R_2 = -\frac{5}{2}r_3 + r_2$$

$$A^{-1} = \begin{bmatrix} 3 & -2 & -4 \\ 3 & -2 & -5 \\ -1 & 1 & 2 \end{bmatrix}$$

30.
$$\begin{bmatrix} 3 & 3 & 1 & | & 1 & 0 & 0 \\ 1 & 2 & 1 & | & 0 & 1 & 0 \\ 2 & -1 & 1 & | & 0 & 0 & 1 \end{bmatrix} \rightarrow \begin{bmatrix} 1 & 1 & \frac{1}{3} & | & \frac{1}{3} & 0 & 0 \\ 1 & 2 & 1 & | & 0 & 1 & 0 \\ 2 & -1 & 1 & | & 0 & 0 & 1 \end{bmatrix} \rightarrow \begin{bmatrix} 1 & 1 & \frac{1}{3} & | & \frac{1}{3} & 0 & 0 \\ 0 & 1 & \frac{2}{3} & | & -\frac{1}{3} & 1 & 0 \\ 0 & -3 & \frac{1}{3} & | & -\frac{2}{3} & 0 & 1 \end{bmatrix}$$

$$R_1 = \frac{1}{3}r_1 \qquad\qquad R_2 = -r_1 + r_2$$
$$R_3 = -2r_1 + r_3$$

$$\rightarrow \begin{bmatrix} 1 & 0 & -\frac{1}{3} & \frac{2}{3} & -1 & 0 \\ 0 & 1 & \frac{2}{3} & -\frac{1}{3} & 1 & 0 \\ 0 & 0 & \frac{7}{3} & -\frac{5}{3} & 3 & 1 \end{bmatrix} \rightarrow \begin{bmatrix} 1 & 0 & -\frac{1}{3} & \frac{2}{3} & -1 & 0 \\ 0 & 1 & \frac{2}{3} & -\frac{1}{3} & 1 & 0 \\ 0 & 0 & 1 & -\frac{5}{7} & \frac{9}{7} & \frac{3}{7} \end{bmatrix} \rightarrow \begin{bmatrix} 1 & 0 & 0 & \frac{3}{7} & -\frac{4}{7} & \frac{1}{7} \\ 0 & 1 & 0 & \frac{1}{7} & \frac{1}{7} & -\frac{2}{7} \\ 0 & 0 & 1 & -\frac{5}{7} & \frac{9}{7} & \frac{3}{7} \end{bmatrix}$$

$\uparrow$ $\qquad\qquad\qquad\qquad\quad \uparrow$ $\qquad\qquad\qquad\qquad\quad \uparrow$

$R_1 = -r_2 + r_1 \qquad\qquad R_3 = \frac{3}{7}r_3 \qquad\qquad\qquad R_2 = -\frac{2}{3}r_3 + r_2$

$R_3 = 3r_2 + r_3 \qquad\qquad\qquad\qquad\qquad\qquad R_1 = \frac{1}{3}r_3 + r_1$

$$A^{-1} = \begin{bmatrix} \frac{3}{7} & -\frac{4}{7} & \frac{1}{7} \\ \frac{1}{7} & \frac{1}{7} & -\frac{2}{7} \\ -\frac{5}{7} & \frac{9}{7} & \frac{3}{7} \end{bmatrix}$$

32. $A = \begin{bmatrix} 3 & -1 \\ -2 & 1 \end{bmatrix}$, $X = \begin{bmatrix} x \\ y \end{bmatrix}$, $B = \begin{bmatrix} 8 \\ 4 \end{bmatrix}$

We know that $A^{-1} = \begin{bmatrix} 1 & 1 \\ 2 & 3 \end{bmatrix}$ from Problem 22.

$X = \begin{bmatrix} x \\ y \end{bmatrix} = A^{-1}B = \begin{bmatrix} 1 & 1 \\ 2 & 3 \end{bmatrix} \begin{bmatrix} 8 \\ 4 \end{bmatrix} = \begin{bmatrix} 12 \\ 28 \end{bmatrix}$ Thus, $x = 12$, $y = 28$.

34. $X = \begin{bmatrix} x \\ y \end{bmatrix} = A^{-1}B = \begin{bmatrix} 1 & 1 \\ 2 & 3 \end{bmatrix} \begin{bmatrix} 4 \\ 5 \end{bmatrix} = \begin{bmatrix} 9 \\ 23 \end{bmatrix}$ Thus, $x = 9$, $y = 23$.

36. $X = \begin{bmatrix} x \\ y \end{bmatrix} = A^{-1}B = \begin{bmatrix} -1 & -\frac{1}{2} \\ -3 & -2 \end{bmatrix} \begin{bmatrix} 0 \\ 14 \end{bmatrix} = \begin{bmatrix} -7 \\ -28 \end{bmatrix}$ Thus, $x = -7$, $y = -28$

38. $X = \begin{bmatrix} x \\ y \end{bmatrix} = A^{-1}B = \begin{bmatrix} -1 & -\frac{1}{2} \\ -3 & -2 \end{bmatrix} \begin{bmatrix} 5 \\ -9 \end{bmatrix} = \begin{bmatrix} -\frac{1}{2} \\ 3 \end{bmatrix}$ Thus, $x = -\frac{1}{2}$, $y = 3$

40. $X = \begin{bmatrix} x \\ y \end{bmatrix} = A^{-1}B = \begin{bmatrix} -\frac{2}{b} & \frac{3}{b} \\ 1 & -1 \end{bmatrix} \begin{bmatrix} 2b + 3 \\ 2b + 2 \end{bmatrix} = \begin{bmatrix} \frac{-2}{b}(2b + 3) + \frac{3}{b}(2b + 2) \\ 2b + 3 - (2b + 2) \end{bmatrix}$

$\qquad\qquad\qquad\qquad\qquad\qquad = \begin{bmatrix} -4 - \frac{6}{b} + 6 + \frac{6}{b} \\ 1 \end{bmatrix} = \begin{bmatrix} 2 \\ 1 \end{bmatrix}$

Thus, $x = 2$, and $y = 1$

42. $X = \begin{bmatrix} x \\ y \end{bmatrix} = A^{-1}B = \begin{bmatrix} -\dfrac{2}{b} & \dfrac{3}{b} \\ 1 & -1 \end{bmatrix} \begin{bmatrix} 14 \\ 10 \end{bmatrix} = \begin{bmatrix} \dfrac{2}{b} \\ 4 \end{bmatrix}$

Thus, $x = \dfrac{2}{b}$, and $y = 4$

44. $X = \begin{bmatrix} x \\ y \\ z \end{bmatrix} = A^{-1}B = \begin{bmatrix} 3 & -2 & -4 \\ 3 & -2 & -5 \\ -1 & 1 & 2 \end{bmatrix} \begin{bmatrix} 6 \\ -5 \\ 6 \end{bmatrix} = \begin{bmatrix} 18 + 10 - 24 \\ 18 + 10 - 30 \\ -6 - 5 + 12 \end{bmatrix} = \begin{bmatrix} 4 \\ -2 \\ 1 \end{bmatrix}$

Thus, $x = 4$, $y = -2$, and $z = 1$

46. $X = \begin{bmatrix} x \\ y \\ z \end{bmatrix} = A^{-1}B = \begin{bmatrix} 3 & -2 & -4 \\ 3 & -2 & -5 \\ -1 & 1 & 2 \end{bmatrix} \begin{bmatrix} 2 \\ -\dfrac{3}{2} \\ 2 \end{bmatrix} = \begin{bmatrix} 6 + 3 - 8 \\ 6 + 3 - 10 \\ -2 - \dfrac{3}{2} + 4 \end{bmatrix} = \begin{bmatrix} 1 \\ -1 \\ \dfrac{1}{2} \end{bmatrix}$

Thus, $x = 1$, $y = -1$, and $z = \dfrac{1}{2}$

48. $X = \begin{bmatrix} x \\ y \\ z \end{bmatrix} = A^{-1}B = \begin{bmatrix} \dfrac{3}{7} & -\dfrac{4}{7} & \dfrac{1}{7} \\ \dfrac{1}{7} & \dfrac{1}{7} & -\dfrac{2}{7} \\ -\dfrac{5}{7} & \dfrac{9}{7} & \dfrac{3}{7} \end{bmatrix} \begin{bmatrix} 8 \\ 5 \\ 4 \end{bmatrix} = \begin{bmatrix} \dfrac{24}{7} - \dfrac{20}{7} + \dfrac{4}{7} \\ \dfrac{8}{7} + \dfrac{5}{7} - \dfrac{8}{7} \\ -\dfrac{40}{7} + \dfrac{45}{7} + \dfrac{12}{7} \end{bmatrix} = \begin{bmatrix} \dfrac{8}{7} \\ \dfrac{5}{7} \\ \dfrac{17}{7} \end{bmatrix}$

Thus, $x = \dfrac{8}{7}$, $y = \dfrac{5}{7}$, and $z = \dfrac{17}{7}$

50. $X = \begin{bmatrix} x \\ y \\ z \end{bmatrix} = A^{-1}B = \begin{bmatrix} \dfrac{3}{7} & -\dfrac{4}{7} & \dfrac{1}{7} \\ \dfrac{1}{7} & \dfrac{1}{7} & -\dfrac{2}{7} \\ -\dfrac{5}{7} & \dfrac{9}{7} & \dfrac{3}{7} \end{bmatrix} \begin{bmatrix} 1 \\ 0 \\ 4 \end{bmatrix} = \begin{bmatrix} \dfrac{3}{7} + 0 + \dfrac{4}{7} \\ \dfrac{1}{7} + 0 - \dfrac{8}{7} \\ -\dfrac{5}{7} + 0 + \dfrac{12}{7} \end{bmatrix} = \begin{bmatrix} 1 \\ -1 \\ 1 \end{bmatrix}$

Thus, $x = 1$, $y = -1$, and $z = 1$

52. $\begin{bmatrix} -3 & \dfrac{1}{2} & | & 1 & 0 \\ 6 & -1 & | & 0 & -1 \end{bmatrix} \rightarrow \begin{bmatrix} 1 & -\dfrac{1}{6} & | & -\dfrac{1}{3} & 0 \\ 6 & -1 & | & 0 & 1 \end{bmatrix} \rightarrow \begin{bmatrix} 1 & -\dfrac{1}{6} & | & -\dfrac{1}{3} & 0 \\ 0 & 0 & | & 2 & 1 \end{bmatrix}$

$\uparrow \qquad\qquad\qquad\qquad \uparrow$

$R_1 = -\dfrac{1}{3}r_1 \qquad R_2 = -6r_1 + r_2$

No inverse.

54.

$$\begin{bmatrix} -3 & 0 & | & 1 & 0 \\ 4 & 0 & | & 0 & 1 \end{bmatrix} \rightarrow \begin{bmatrix} 1 & 0 & | & -\dfrac{1}{3} & 0 \\ 4 & 0 & | & 0 & 1 \end{bmatrix} \rightarrow \begin{bmatrix} 1 & 0 & | & -\dfrac{1}{3} & 0 \\ 0 & 0 & | & \dfrac{4}{3} & 1 \end{bmatrix}$$

$$\uparrow \qquad\qquad\qquad \uparrow$$
$$R_1 = -\dfrac{1}{3}r_1 \qquad R_2 = -4r_1 + r_2$$

No inverse.

56.

$$\begin{bmatrix} 1 & 1 & -3 & | & 1 & 0 & 0 \\ 2 & -4 & 1 & | & 0 & 1 & 0 \\ -5 & 7 & 1 & | & 0 & 0 & 1 \end{bmatrix} \rightarrow \begin{bmatrix} 1 & 1 & -3 & | & 1 & 0 & 0 \\ 0 & -6 & 7 & | & -2 & 1 & 0 \\ 0 & 12 & -14 & | & 5 & 0 & 1 \end{bmatrix}$$

$$\uparrow$$
$$R_2 = -2r_1 + r_2$$
$$R_3 = 5r_1 + r_3$$

$$\rightarrow \begin{bmatrix} 1 & 1 & -3 & | & 1 & 0 & 0 \\ 0 & 1 & -\dfrac{7}{6} & | & \dfrac{1}{3} & -\dfrac{1}{6} & 0 \\ 0 & 12 & -14 & | & 5 & 0 & 1 \end{bmatrix} \rightarrow \begin{bmatrix} 1 & 0 & -\dfrac{11}{6} & | & \dfrac{2}{3} & \dfrac{1}{6} & 0 \\ 0 & 1 & -\dfrac{7}{6} & | & \dfrac{1}{3} & -\dfrac{1}{6} & 0 \\ 0 & 0 & 0 & | & 1 & 2 & 1 \end{bmatrix}$$

$$\uparrow \qquad\qquad\qquad\qquad \uparrow$$
$$R_2 = -\dfrac{1}{6}r_2 \qquad\qquad R_1 = -r_2 + r_1$$
$$R_3 = -12r_2 + r_3$$

No inverse.

58.
$$\begin{bmatrix} .26 & -.28 & -.2 \\ -1.21 & 1.63 & 1.2 \\ -1.84 & 2.52 & 1.8 \end{bmatrix}$$

60.
$$\begin{bmatrix} .00 & .03 & -.00 & .02 \\ .02 & -.01 & .00 & .00 \\ -.04 & .01 & .03 & .06 \\ .05 & -.01 & -.00 & -.09 \end{bmatrix}$$

62. $x = 4.79, y = -6.63, z = -24.13$

64. $x = -2.05, y - 3.87, z = 13.35$

66. (a)

January

	Subcompacts	Intermediate	Station Wagons
City	400	250	50
Suburbs	450	200	140

February

	Subcompacts	Intermediate	Station Wagons
City	350	100	30
Suburbs	350	300	100

(b) **Total Sales**

$$\begin{bmatrix} 400 & 250 & 50 \\ 450 & 200 & 140 \end{bmatrix} + \begin{bmatrix} 350 & 100 & 30 \\ 350 & 300 & 100 \end{bmatrix} = \begin{bmatrix} 750 & 350 & 80 \\ 800 & 500 & 240 \end{bmatrix}$$

(c)

Profit

$$\begin{matrix} \text{Subcompact} \\ \text{Intermediate} \\ \text{Station Wagon} \end{matrix} \begin{bmatrix} \$100 \\ \$150 \\ \$200 \end{bmatrix}$$

(d) $\begin{bmatrix} 750 & 350 & 80 \\ 800 & 500 & 240 \end{bmatrix} \begin{bmatrix} \$100 \\ \$150 \\ \$200 \end{bmatrix} = \begin{bmatrix} 750 + 52{,}500 + 16{,}000 \\ 800 + 75{,}000 + 48{,}000 \end{bmatrix} = \begin{bmatrix} \$\ 69{,}250 \\ \$123{,}800 \end{bmatrix}$

14.2 Partial Fraction Decomposition

2. $\dfrac{5x + 2}{x^3 - 1}$, Proper

4. $\dfrac{3x^2 - 2}{x^2 - 1}$
$$x^2 - 1 \overline{\smash{\big)}\ 3x^2 - 2} \quad \overset{3}{}$$
$$\underline{3x^2 - 3}$$
$$1$$
Improper, $3 + \dfrac{1}{x^2 - 1}$

6. $\dfrac{3x^4 + x^2 - 2}{x^3 + 8}$

$$x^3 + 8 \overline{\smash{\big)}\ 3x^4 + x^2 \qquad - 2} \quad \overset{3x}{}$$
$$\underline{3x^4 \qquad + 24x}$$
$$x^2 - 24x - 2$$

Improper, $3x + \dfrac{x^2 - 24x - 2}{x^3 + 8}$

8. $\dfrac{2x^3 + 8x}{x^2 + 1}$
$$x^2 + 1 \overline{\smash{\big)}\ 2x^3 + 8x} \quad \overset{2x}{}$$
$$\underline{2x^3 + 2x}$$
$$6x$$
Improper, $2x + \dfrac{6x}{x^2 + 1}$

10. $\dfrac{3x}{(x + 2)(x - 1)} = \dfrac{A}{x + 2} + \dfrac{B}{x - 1}$
$3x = A(x - 1) + B(x + 2)$
Let $x = 1$: $\quad 3 = 3B$, or $B = 1$
Let $x = -2$: $\quad -6 = -3A$, or $A = 2$
$$\dfrac{3x}{(x + 2)(x - 1)} = \dfrac{2}{x + 2} + \dfrac{1}{x - 1}$$

12.
$$\frac{1}{(x + 1)(x^2 + 4)} = \frac{A}{x + 1} + \frac{Bx + C}{x^2 + 4}$$
$$1 = A(x^2 + 4) + (Bx + C)(x + 1)$$
$$= Ax^2 + 4A + Bx^2 + (B + C)x + C$$
$$0 = A + B$$
$$0 = B + C$$
$$1 = 4A + C$$
$$-A = B \text{ so} \qquad 0 = -A + C$$
$$\underline{-1 = -4A - C}$$
$$-1 = -5A$$
$$\frac{1}{5} = A$$
$$B = -\frac{1}{5} \text{ and } C = -B$$
$$C = \frac{1}{5}$$
$$\frac{1}{(x + 1)(x^2 + 4)} = \frac{\frac{1}{5}}{x + 1} + \frac{-\frac{1}{5}x + \frac{1}{5}}{x^2 + 4}$$

14.
$$\frac{3x}{(x + 2)(x - 4)} = \frac{A}{x + 2} + \frac{B}{x - 4}$$
$$3x = A(x - 4) + B(x + 2)$$
Let $x = 4$: $\qquad 12 = 6B$, or $B = 2$
Let $x = -2$: $\qquad -6 = -6A$, or $A = 1$
$$\frac{3x}{(x + 2)(x - 4)} = \frac{1}{x + 2} + \frac{2}{x - 4}$$

16.
$$\frac{x + 1}{x^2(x - 2)} = \frac{A}{x} + \frac{B}{x^2} + \frac{C}{x - 2}$$
$$x + 1 = Ax(x - 2) + B(x - 2) + Cx^2$$
Let $x = 0$: $\quad 1 = -2B$, or $B = -\frac{1}{2}$
Let $x = 2$: $\quad 3 = 4C$, or $C = \frac{3}{4}$
Let $x = 1$: $\quad 2 = -A - B + C$
$$2 = -A + \frac{1}{2} + \frac{3}{4}$$
$$A = \frac{-8}{4} + \frac{2}{4} + \frac{3}{4} = \frac{-3}{4}$$
$$\frac{x + 1}{x^2(x - 2)} = \frac{-\frac{3}{4}}{x} + \frac{-\frac{1}{2}}{x^2} + \frac{\frac{3}{4}}{x - 2}$$

18.
$$\frac{2x + 4}{x^3 - 1} = \frac{A}{x - 1} + \frac{Bx + C}{x^2 + x + 1}$$
$$2x + 4 = A(x^2 + x + 1) + (Bx + C)(x - 1)$$
Let $x = 1$: $\qquad 6 = 3A$, or $A = 2$
Let $x = 0$: $\qquad 4 = A - C$, or $C = -2$
Let $x = -1$: $\qquad 2 = A + 2B - 2C$
$$2 = 2 + 2B + 4$$
$$-4 = 2B$$
$$-2 = B$$
$$\frac{2x + 4}{x^3 - 1} = \frac{2}{x - 1} + \frac{-2x - 2}{x^2 + x + 1}$$

20.
$$\frac{x + 1}{x^2(x - 2)^2} = \frac{A}{x} + \frac{B}{x^2} + \frac{C}{x - 2} + \frac{D}{(x - 2)^2}$$
$$x + 1 = Ax(x - 2)^2 + B(x - 2)^2 + Cx^2(x - 2) + Dx^2$$
Let $x = 0$: $\qquad 1 = 4B$, or $B = \frac{1}{4}$
Let $x = 2$: $\qquad 3 = 4D$, or $D = \frac{3}{4}$
Let $x = 1$: $\qquad 2 = A + B - C + D$
$$2 = A + \frac{1}{4} - C + \frac{3}{4}$$
$$1 = A - C$$

$$\text{Let } x = 3: \qquad 4 = 3A + B + 9C + 9D$$

$$4 = 3A + \frac{1}{4} + 9C + 9\left(\frac{3}{4}\right)$$

$$4 = 3A + 9C + 7$$

$$-3 = 3A + 9C$$

$$-1 = A + 3C$$

$$1 = A - C$$

$$-2 = 4C \text{ or } C = -\frac{1}{2}$$

$$A = \frac{1}{2}$$

$$\frac{x + 1}{x^2(x - 2)^2} = \frac{\frac{1}{2}}{x} + \frac{\frac{1}{4}}{x^2} - \frac{\frac{1}{2}}{x - 2} + \frac{\frac{3}{4}}{(x - 2)^2}$$

22. $\dfrac{x^2 + x}{(x + 2)(x - 1)^2} = \dfrac{A}{x + 2} + \dfrac{B}{x - 1} + \dfrac{C}{(x - 1)^2}$

$$x^2 + x = A(x - 1)^2 + B(x - 1)(x + 2) + C(x + 2)$$

Let $x = 1$: $\qquad 2 = 3C$, or $C = \dfrac{2}{3}$

Let $x = -2$: $\qquad 2 = 9A$, or $A = \dfrac{2}{9}$

Let $x = 0$: $\qquad 0 = A - 2B + 2C$

$$0 = \frac{2}{9} - 2B + \frac{4}{3}$$

$$\frac{-14}{9} = -2B, \text{ or } B = \frac{7}{9}$$

$$\frac{x^2 + x}{(x + 2)(x - 1)^2} = \frac{\frac{2}{9}}{x + 2} + \frac{\frac{7}{9}}{x - 1} + \frac{\frac{2}{3}}{(x - 1)^2}$$

24. $\dfrac{10x^2 + 2x}{(x - 1)^2(x^2 + 2)} = \dfrac{A}{x - 1} + \dfrac{B}{(x - 1)^2} + \dfrac{Cx + D}{x^2 + 2}$

$10x^2 + 2x = A(x - 1)(x^2 + 2) + B(x^2 + 2) + (Cx + D)(x - 1)^2$
Let $x = 1$: $12 = 3B$, or $B = 4$

$$10x^2 + 2x - 4x^2 - 8 = A(x - 1)(x^2 + 2) + (Cx + D)(x - 1)^2$$

$$6x^2 + 2x - 8 = (x - 1)[A(x^2 + 2) + (Cx + D)(x - 1)]$$

$$2(3x^2 + x + 4) = (x - 1)[A(x^2 + 2) + (Cx + D)(x - 1)]$$

$$2(x - 1)(3x + 4) = (x - 1)[A(x^2 + 2) + (Cx + D)(x - 1)]$$

$$2(3x + 4) = A(x^2 + 2) + (Cx + D)(x - 1)$$

$x = 1$: $2(7) = A(3)$

$$A = \frac{14}{3}$$

Let $x = 0$:
$$0 = -2A + 2B + D$$
$$0 = -2\left[\frac{14}{3}\right] + 2(4) + D$$
$$0 = -\frac{4}{3} + D$$
$$D = \frac{4}{3}$$

Let $x = -1$:
$$8 = -6A + 3B - 4C + 4D$$
$$8 = -6\left[\frac{14}{3}\right] + 3(4) - 4C + 4\left[\frac{4}{3}\right]$$
$$8 = \frac{-84}{3} + 12 - 4C + \frac{16}{3}$$
$$8 = \frac{-32}{3} - 4C$$
$$4C = \frac{-56}{3}$$
$$C = \frac{-14}{3}$$

$$\frac{10x^2 + 2x}{(x-1)^2(x^2+2)} = \frac{\frac{14}{3}}{x-1} + \frac{4}{(x-1)^2} + \frac{\frac{-14}{3}x + \frac{4}{3}}{x^2+2}$$

26. $\dfrac{x^2 - 11x - 18}{x(x^2 + 3x + 3)} = \dfrac{A}{x} + \dfrac{Bx + C}{x^2 + 3x + 3}$

$x^2 - 11x - 18 = A(x^2 + 3x + 3) + (Bx + C)x$
Let $x = 0$: $-18 = 3A$, or $A = -6$
Let $x = 1$: $-28 = 7A + B + C$
$\qquad\qquad -28 = -42 + B + C$
$\qquad\qquad 14 = B + C$
Let $x = -1$: $-6 = A + B - C$
$\qquad\qquad\quad -6 = -6 + B - C$
$\qquad\qquad\quad 0 = B - C$
$\qquad\qquad\quad 14 = B + C$
$\qquad\qquad\quad 14 = 2B$ or $B = 7$
$\qquad\qquad\quad C = 14 - B = 14 - 7 = 7$

$$\frac{x^2 - 11x - 18}{x(x^2 + 3x + 3)} = \frac{-6}{x} + \frac{7x + 7}{x^2 + 3x + 3}$$

28. $\dfrac{1}{(2x + 3)(4x - 1)} = \dfrac{A}{2x + 3} + \dfrac{B}{4x - 1}$
$1 = A(4x - 1) + B(2x + 3)$

Let $x = \dfrac{1}{4}$: $\quad 1 = \dfrac{7}{2}B$, or $B = \dfrac{2}{7}$

Let $x = \dfrac{-3}{2}$: $\quad 1 = -7A$, or $A = \dfrac{-1}{7}$

$$\frac{1}{(2x + 3)(4x - 1)} = \frac{-\frac{1}{7}}{2x + 3} + \frac{\frac{2}{7}}{4x - 1}$$

30. $\dfrac{x^2 - x - 8}{(x + 1)(x^2 + 5x + 6)} = \dfrac{A}{x + 1} + \dfrac{B}{x + 2} + \dfrac{C}{x + 3}$

$x^2 - x - 8 = A(x + 2)(x + 3) + B(x + 1)(x + 3) + C(x + 1)(x + 2)$
Let $x = -2$: $-2 = -B$, or $B = 2$
Let $x = -3$: $\quad 4 = 2C$, or $C = 2$
Let $x = 0$: $\quad -8 = 6A + 3B + 2C$
$\qquad\qquad\quad -8 = 6A + 6 + 4$
$\qquad\qquad\quad -18 = 6A$, or $A = -3$

$$\frac{x^2 - x - 8}{(x + 1)(x^2 + 5x + 6)} = \frac{-3}{x + 1} + \frac{2}{x + 2} + \frac{2}{x + 3}$$

32. $$\frac{x^3 + 1}{(x^2 + 16)^2} = \frac{Ax + B}{x^2 + 16} + \frac{Cx + D}{(x^2 + 16)^2}$$

$x^3 + 1 = (Ax + B)(x^2 + 16) + (Cx + D)$
$x^3 + 1 = Ax^3 + Bx^2 + (16A + C)x + (16B + D)$
Equating coefficients,
$\qquad A = 1, B = 0, 16A + C = 0, 16B + D = 1$
$\qquad A = 1, B = 0, C = -16, D = 1$

$$\frac{x^3 + 1}{(x^2 + 16)^2} = \frac{x}{x^2 + 16} + \frac{-16x + 1}{(x^2 + 16)^2}$$

34. $$\frac{x^5 + 1}{x^6 - x^4} = \frac{(x + 1)(x^4 - x^3 + x^2 - x + 1)}{x^4(x + 1)(x - 1)}$$

$$\frac{x^4 - x^3 + x^2 - x + 1}{x^4(x - 1)} = \frac{A}{x} + \frac{B}{x^2} + \frac{C}{x^3} + \frac{D}{x^4} + \frac{E}{x - 1}$$

$x^4 - x^3 + x^2 - x + 1 = Ax^3(x - 1) + Bx^2(x - 1) + Cx(x - 1) + D(x - 1) + Ex^4$

Let $x = 0$: $\qquad\qquad 1 = -D$ or $D = -1$
Let $x = 1$: $\qquad\qquad 1 = E$
Let $x = -1$: $\qquad\quad 5 = 2A - 2B + 2C - 2D + E$
$\qquad\qquad\qquad\quad 5 = 2A - 2B + 2C + 2 + 1$
$\qquad\qquad\qquad\quad 2 = 2A - 2B + 2C$
Let $x = 2$: $\qquad\quad 11 = 8A + 4B + 2C + D + 16E$
$\qquad\qquad\qquad\quad 11 = 8A + 4B + 2C - 1 + 16$
$\qquad\qquad\qquad\; -4 = 8A + 4B + 2C$
Let $x = -2$: $\quad 31 = 24A - 12B + 6C - 3D + 16E$
$\qquad\qquad\qquad 31 = 24A - 12B + 6C + 3 + 16$
$\qquad\qquad\qquad 12 = 24A - 12B + 6C$
$\qquad\qquad\qquad\; 2 = 2A - 2B + 2C$
$\qquad\qquad\quad -2 = 4A + 2B + C$
$\qquad\qquad\qquad\; 2 = 4A - 2B + C$
$\qquad\qquad\qquad\; 0 = 8A + 2C \rightarrow 0 = -24A - 6C$
$\qquad\qquad\qquad\; 0 = 6A + 3C \rightarrow 0 = 12A + 6C$
$\qquad\qquad\qquad\qquad\quad 0 = -12A$ or $A = 0$
$\qquad\qquad\qquad\qquad\quad C = 0$
$\qquad\qquad\qquad\qquad\quad 2 = 4(0) - 2B + 0$
$\qquad\qquad\qquad\qquad\quad B = -1$

$$\frac{x^5 + 1}{x^6 - x^4} = \frac{x^4 - x^3 + x^2 - x + 1}{x^4(x - 1)} = \frac{-1}{x^2} - \frac{1}{x^4} + \frac{1}{x - 1}$$

36. $$\frac{x^2 + 1}{x^3 + x^2 - 5x + 3} = \frac{A}{x - 1} + \frac{B}{(x - 1)^2} + \frac{C}{x + 3}$$

$x^2 + 1 = A(x - 1)(x + 3) + B(x + 3) + C(x - 1)^2$

Let $x = 1$: $\qquad 2 = 4B$, or $B = \dfrac{1}{2}$

Let $x = -3$: $\quad 10 = 16C$, or $C = \dfrac{5}{8}$

Let $x = 0$: $\quad\quad 1 = -3A + 3B + C$

$$1 = -3A + 3\left[\frac{1}{2}\right] + \frac{5}{8}$$

$$\frac{-9}{8} = -3A$$

$$A = \frac{3}{8}$$

$$\frac{x^2 + 1}{x^3 + x^2 - 5x + 3} = \frac{\frac{3}{8}}{x - 1} + \frac{\frac{1}{2}}{(x - 1)^2} + \frac{\frac{5}{8}}{x + 3}$$

38. $\quad \dfrac{x^2}{(x^2 + 4)^3} = \dfrac{Ax + B}{x^2 + 4} + \dfrac{Cx + D}{(x^2 + 4)^2} + \dfrac{Ex + F}{(x^2 + 4)^3}$

$x^2 = (Ax + B)(x^2 + 4)^2 + (Cx + D)(x^2 + 4) + Ex + F$

$x^2 = Ax^5 + Bx^4 + 8Ax^3 + 8Bx^2 + 16Ax + 16B + Cx^3 + Dx^2 + 4Cx + 4D + Ex + F$

$0 = A$

$0 = B$

$0 = 8A + C$ or $C = 0$

$1 = 8B + D$ or $D = 1$

$0 = 16A + 4C + E$ or $E = 0$

$0 = 16B + 4D + F$ or $F = -4$

$$\frac{x^2}{(x^2 + 4)^3} = \frac{1}{(x^2 + 4)^2} - \frac{4}{(x^2 + 4)^3}$$

40. $\quad \dfrac{4x}{2x^2 + 3x - 2} = \dfrac{4x}{(2x - 1)(x + 2)} = \dfrac{A}{2x - 1} + \dfrac{B}{x + 2}$

$4x = A(x + 2) + B(2x - 1)$

Let $x = -2$: $\quad\quad -8 = -5B$, or $B = \dfrac{8}{5}$

Let $x = \dfrac{1}{2}$: $\quad\quad 2 = \dfrac{5}{2}A$, or $A = \dfrac{4}{5}$

$$\frac{4x}{2x^2 + 3x - 2} = \frac{\frac{4}{5}}{2x - 1} + \frac{\frac{8}{5}}{x + 2}$$

42. $\quad \dfrac{x^2 + 9}{x^4 - 2x^2 - 8} = \dfrac{x^2 + 9}{(x^2 - 4)(x^2 + 2)} = \dfrac{x^2 + 9}{(x - 2)(x + 2)(x^2 + 2)} = \dfrac{A}{x - 2} + \dfrac{B}{x + 2} + \dfrac{Cx + D}{x^2 + 2}$

$x^2 + 9 = A(x + 2)(x^2 + 2) + B(x - 2)(x^2 + 2) + (Cx + D)(x - 2)(x + 2)$

Let $x = 2$: $\quad\quad 13 = A(4)(6)$, or $A = \dfrac{13}{24}$

Let $x = -2$: $\quad\quad 13 = B(-4)(6)$, or $B = \dfrac{-13}{24}$

Let $x = 0$: $9 = 4A - 4B - 4D$

$$9 = 4\left[\frac{13}{24}\right] - 4\left[\frac{-13}{24}\right] - 4D$$

$$9 = \frac{26}{6} - 4D$$

$$4D = \frac{26}{6} - \frac{54}{6} = \frac{-28}{6} = \frac{-14}{3}$$

$$D = \frac{-7}{6}$$

Let $x = 1$: $10 = 9A - 3B - 3C - 3D$

$$10 = 9\left[\frac{13}{24}\right] - 3\left[\frac{-13}{24}\right] - 3C - 3\left[\frac{-7}{6}\right]$$

$$10 = \frac{39}{8} + \frac{13}{8} - 3C + \frac{21}{6}$$

$$3C = -\frac{240}{24} + \frac{117}{24} + \frac{39}{24} + \frac{84}{24}$$

$$3C = 0$$

$$C = 0$$

$$\frac{x^2 + 9}{x^4 - 2x^2 - 8} = \frac{\frac{13}{24}}{x - 2} + \frac{\frac{-13}{24}}{x + 2} + \frac{\frac{-7}{6}}{x^2 + 2}$$

14.3 Vectors

2.

4.

6.

8.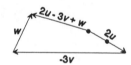

10. If **x** + D = E, **x** = −G − H

12. G = −K + C + D − E

14. E = A + B + C + D

16. If **x** = A + B + C + H + G, **x** = 0

18. If $\|\mathbf{v}\| = 2$, then
$\|-4\mathbf{v}\| = |-4| \, \|\mathbf{v}\| = 8.$

20. $P = (0, 0)$, $Q = (-3, -5)$
$\mathbf{v} = (-3 - 0)\mathbf{i} + (-5 - 0)\mathbf{j} = -3\mathbf{i} - 5\mathbf{j}$

22. $P = (-3, 2)$, $Q = (6, 5)$
$\mathbf{v} = 9\mathbf{i} + 3\mathbf{j}$

24. $P = (-1, 4)$, $Q = (6, 2)$
$\mathbf{v} = 7\mathbf{i} - 2\mathbf{j}$

26. $P = (1, 1)$, $Q = (2, 2)$
$\mathbf{v} = \mathbf{i} + \mathbf{j}5$

28. For $\mathbf{v} = -5\mathbf{i} + 12\mathbf{j}$, $\|\mathbf{v}\| = 13$

30. For $\mathbf{v} = -\mathbf{i} - \mathbf{j}$, $\|\mathbf{v}\| = \sqrt{2}$

32. For $\mathbf{v} = 6\mathbf{i} + 2\mathbf{j}$,
$$\|\mathbf{v}\| = \sqrt{36 + 4} = 2\sqrt{10}$$

34. $3\mathbf{v} + 2\mathbf{w} = 3(3\mathbf{i} - 5\mathbf{j}) - 2(-2\mathbf{i} + 3\mathbf{j}) = 9\mathbf{i} - 15\mathbf{j} + 4\mathbf{i} - 6\mathbf{j} = 13\mathbf{i} - 21\mathbf{j}$

36. $\|\mathbf{v} + \mathbf{w}\| = \|3\mathbf{i} - 5\mathbf{j} + -2\mathbf{i} + 3\mathbf{j}\| = \|\mathbf{i} - 2\mathbf{j}\|$ so $\|\mathbf{v} + \mathbf{w}\| = \sqrt{1 + 4} = \sqrt{5}$

38. $\|\mathbf{v}\| + \|\mathbf{w}\| = \sqrt{34} + \sqrt{13}$

40. For $\mathbf{v} = -3\mathbf{j}$, $\|\mathbf{v}\| = 9$
$$\mathbf{u} = \frac{\mathbf{v}}{\|\mathbf{v}\|} = \frac{-3\mathbf{j}}{9} = \frac{-1}{3}\mathbf{j}$$

42. For $\mathbf{v} = -5\mathbf{i} + 12\mathbf{j}$,
$$\|\mathbf{v}\| = \sqrt{25 + 144} = 13$$
$$\mathbf{u} = \frac{\mathbf{v}}{\|\mathbf{v}\|} = \frac{-5\mathbf{i} + 12\mathbf{j}}{13} = \frac{-5}{13}\mathbf{i} + \frac{12}{13}\mathbf{j}$$

44. For $\mathbf{v} = 2\mathbf{i} - \mathbf{j}$, $\|\mathbf{v}\| = \sqrt{4 + 1} = \sqrt{5}$
$$\mathbf{u} = \frac{\mathbf{v}}{\|\mathbf{v}\|} = \frac{2\mathbf{i} - \mathbf{j}}{\sqrt{5}} = \frac{2\sqrt{5}}{5}\mathbf{i} - \frac{\sqrt{5}}{5}\mathbf{j}$$

46. $\|\mathbf{v}\| = 3 = \sqrt{a^2 + a^2} = \sqrt{2a^2}$
$$3 = \sqrt{2a^2}$$
$$9 = 2a^2$$
$$\frac{9}{2} = a^2$$
$$\frac{3\sqrt{2}}{2} = a$$
$$\mathbf{v} = \frac{3\sqrt{2}}{2}\mathbf{i} + \frac{3\sqrt{2}}{2}\mathbf{j}$$

48.
$$5 = \sqrt{(x + 3)^2 + (3)^2} = \sqrt{(x + 3)^2 + 9}$$
$$25 = (x + 3)^2 + 9$$
$$16 = (x + 3)^2$$
$$x + 3 = \pm 4$$
$$x = 1 \text{ or } x = -7$$

50. From Example 6, $\mathbf{v_g} = \mathbf{v_a} + \mathbf{v_w}$ where
$\mathbf{v_g}$ = velocity of airplane relative to the ground = $-200\mathbf{j}$

$\mathbf{v_w}$ = velocity of wind (30 mph from northwest) = $30\left[\dfrac{1}{\sqrt{2}}\right](\mathbf{i} - \mathbf{j}) = 15\sqrt{2}\,\mathbf{i} - 15\sqrt{2}\,\mathbf{j}$

$\mathbf{v_a}$ = velocity of airplane in the air = unknown

$\mathbf{v_a} = \mathbf{v_g} - \mathbf{v_w} = -200\mathbf{j} - \left(15\sqrt{2}\,\mathbf{i} - 15\sqrt{2}\,\mathbf{j}\right) = -15\sqrt{2}\,\mathbf{i} + \left(-200 + 15\sqrt{2}\right)\mathbf{j}$

$\|\mathbf{v_a}\| = \sqrt{\left(-15\sqrt{2}\right)^2 + \left(-200 + 15\sqrt{2}\right)^2} = 180$ miles/hour

52. Let $\quad \mathbf{v_c}$ = velocity of the current
$\quad\quad\quad \mathbf{v_m}$ = velocity of the motorboat relative to the water
and $\quad \mathbf{v_\ell}$ = velocity of the motorboat relative to the land
We are given *two* of the three vectors:
$\quad\quad \mathbf{v_c} = 4\mathbf{i}$, if we let $\mathbf{i}$ point directly downstream
and $\quad \mathbf{v_m} = 10\mathbf{j}$, since the motorboat is heading perpendicular to the current, at 10 miles per hour.
Then, $\quad \mathbf{v_\ell} = \mathbf{v_c} + \mathbf{v_m} = 4\mathbf{i} + 10\mathbf{j}$
The motorboat's actual speed is:
$$\|\mathbf{v_\ell}\| = \sqrt{16 + 100} = \sqrt{116} = 2\sqrt{29} \approx 10.8 \text{ miles per hour}$$

2. $\mathbf{v} \cdot \mathbf{w} = (\mathbf{i} + \mathbf{j}) \cdot (-\mathbf{i} + \mathbf{j}) = 1(-1) + 1(1) = 0$

$\|\mathbf{v}\| = \sqrt{(1)^2 + (1)^2} = \sqrt{2}\,; \|\mathbf{w}\| = \sqrt{(-1)^2 + (1)^2} = \sqrt{2}$

$\cos \theta = \dfrac{\mathbf{v} \cdot \mathbf{w}}{\|\mathbf{v}\|\,\|\mathbf{w}\|} = 0$

4. $\mathbf{v} \cdot \mathbf{w} = (2\mathbf{i} + 2\mathbf{j}) \cdot (\mathbf{i} + 2\mathbf{j}) = 2(1) + 2(2) = 6$

$\|\mathbf{v}\| = \sqrt{(2)^2 + (2)^2} = 2\sqrt{2}\,; \|\mathbf{w}\| = \sqrt{(1)^2 + (2)^2} = \sqrt{5}$

$\cos \theta = \dfrac{6}{2\sqrt{2}\left(\sqrt{5}\right)} = \dfrac{3\sqrt{10}}{10}$

6. $\mathbf{v} \cdot \mathbf{w} = (\mathbf{i} + \sqrt{3}\,\mathbf{j}) \cdot (\mathbf{i} - \mathbf{j}) = 1(1) + \sqrt{3}\,(-1) = 1 - \sqrt{3}$

$\|\mathbf{v}\| = \sqrt{(1)^2 + \left(\sqrt{3}\right)^2} = 2$

$\|\mathbf{w}\| = \sqrt{(1)^2 + (-1)^2} = \sqrt{2}$

$\cos \theta = \dfrac{1 - \sqrt{3}}{2\sqrt{2}\left(\sqrt{2}\right)} = \dfrac{\sqrt{2} - \sqrt{6}}{4}$

8. $\mathbf{v} \cdot \mathbf{w} = (3\mathbf{i} - 4\mathbf{j}) \cdot (4\mathbf{i} - 3\mathbf{j}) = 3(4) + (-4)(-3) = 24$

$\|\mathbf{v}\| = \sqrt{(3)^2 + (-4)^2} = 5\,; \|\mathbf{w}\| = \sqrt{4^2 + (-3)^2} = 5$

$\cos \theta = \dfrac{24}{5(5)} = \dfrac{24}{25}$

10. $\mathbf{v} \cdot \mathbf{w} = (\mathbf{i}) \cdot (-3\mathbf{j}) = 1(0) + 0(-3) = 0$

$\|\mathbf{v}\| = \sqrt{1^2} = 1\,; \|\mathbf{w}\| = \sqrt{(-3)^2} = 3$
$\cos \theta = 0$

12. $\mathbf{v} \cdot \mathbf{w} = (\mathbf{i} + \mathbf{j}) \cdot (\mathbf{i} + b\mathbf{j}) = 1(1) + 1(b) = 1 + b$

$\|\mathbf{v}\| = \sqrt{(1)^2 + (1)^2} = \sqrt{2}\,; \|\mathbf{w}\| = \sqrt{(1)^2 + (b)^2} = \sqrt{1 + b^2}$

$\cos \dfrac{\pi}{2} = \dfrac{1 + b}{\sqrt{2}\left(\sqrt{1 + b^2}\right)}$

$0 = \dfrac{1 + b}{\sqrt{2}\left(\sqrt{1 + b^2}\right)}$

$0 = 1 + b$

$-1 = b$

14. $\mathbf{v} = -3\mathbf{i} + 2\mathbf{j}, \ \mathbf{w} = 2\mathbf{i} + \mathbf{j}$

$\mathbf{v}_1 = \text{proj}_{\mathbf{w}}\mathbf{v} = \dfrac{\mathbf{v} \cdot \mathbf{w}}{\|\mathbf{w}\|^2}\mathbf{w} = \dfrac{-4}{\left(\sqrt{5}\right)^2}\mathbf{w} = \dfrac{-4}{5}(2\mathbf{i} + \mathbf{j})$

$\mathbf{v}_2 = \mathbf{v} - \mathbf{v}_1 = (-3\mathbf{i} + 2\mathbf{j}) + \dfrac{4}{5}(2\mathbf{i} + \mathbf{j}) = -3\mathbf{i} + 2\mathbf{j} + \dfrac{8}{5}\mathbf{i} + \dfrac{4}{5}\mathbf{j} = \dfrac{-7}{5}\mathbf{i} + \dfrac{14}{5}\mathbf{j}$

16. $\mathbf{v} = 2\mathbf{i} - \mathbf{j}$, $\mathbf{w} = \mathbf{i} - 2\mathbf{j}$

$$\mathbf{v}_1 = \text{proj}_\mathbf{w}\mathbf{v} = \frac{\mathbf{v} \cdot \mathbf{w}}{\|\mathbf{w}\|^2}\mathbf{w} = \frac{4}{5}(\mathbf{i} - 2\mathbf{j})$$

$$\mathbf{v}_2 = \mathbf{v} - \mathbf{v}_1 = (2\mathbf{i} - \mathbf{j}) - \frac{4}{5}(\mathbf{i} - 2\mathbf{j}) = -2\mathbf{i} - \frac{4}{5}\mathbf{i} - \mathbf{j} + \frac{8}{5}\mathbf{j} = \frac{6}{5}\mathbf{i} + \frac{3}{5}\mathbf{j}$$

18. $\mathbf{v} = \mathbf{i} - 3\mathbf{j}$, $\mathbf{w} = 4\mathbf{i} - \mathbf{j}$

$$\mathbf{v}_1 = \text{proj}_\mathbf{w}\mathbf{v} = \frac{\mathbf{v} \cdot \mathbf{w}}{\|\mathbf{w}\|^2}\mathbf{w} = \frac{7}{17}(4\mathbf{i} - \mathbf{j})$$

$$\mathbf{v}_2 = \mathbf{v} - \mathbf{v}_1 = (\mathbf{i} - 3\mathbf{j}) - \frac{7}{17}(4\mathbf{i} - \mathbf{j}) = -\mathbf{i} - 3\mathbf{j} - \frac{28}{17}\mathbf{i} + \frac{7}{17}\mathbf{j} = \frac{-11}{17}\mathbf{i} - \frac{44}{17}\mathbf{j}$$

20. We set up a coordinate system in which north (N) is along the positive y-axis. Let

$\mathbf{v}_a$ = velocity of the aircraft = $250\mathbf{i}$

$\mathbf{v}_w$ = velocity of the wind = $40\left[\dfrac{\mathbf{i} - \mathbf{j}}{\|\mathbf{i} - \mathbf{j}\|}\right] = 40\left[\dfrac{\mathbf{i} - \mathbf{j}}{\sqrt{1 + 1}}\right] = \dfrac{40}{\sqrt{2}}(\mathbf{i} - \mathbf{j})$

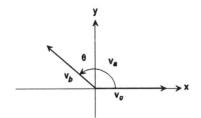

$\mathbf{v}_g$ = actual velocity of the aircraft relative to the ground

$$\mathbf{v}_g + \mathbf{v}_w = \mathbf{v}_a$$

$$\mathbf{v}_g + \frac{40}{\sqrt{2}}(\mathbf{i} + \mathbf{j}) = 250\mathbf{i}$$

$$\mathbf{v}_g = \left(250 - \frac{40}{\sqrt{2}}\right)\mathbf{i} - \frac{40}{\sqrt{2}}\mathbf{j}$$

$$\|\mathbf{v}_g\| = \sqrt{\left(250 - \frac{40}{\sqrt{2}}\right)^2 + \left(\frac{40}{\sqrt{2}}\right)^2} \approx 233.51 \text{ mph}$$

$$\cos\theta = \frac{\mathbf{v}_g \cdot \mathbf{v}_a}{\|\mathbf{v}_g\|\ \|\mathbf{v}_a\|} = \frac{250\left(250 - \frac{40}{\sqrt{2}}\right)}{250 \cdot 223.51} = .99197$$

$$\theta = 7.265°$$

A compass heading of 7.265° should be maintained and the actual speed of the aircraft is 223.51 mph.

22. Let $\mathbf{v}_a$ = velocity to get across
$\mathbf{v}_b$ = velocity of the boat
$\mathbf{v}_c$ = velocity of the current

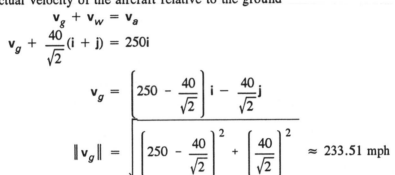

$$\mathbf{v}_a\mathbf{j} = \mathbf{v}_b\cos\theta\mathbf{i} + \mathbf{v}_b\sin\theta\mathbf{j} + \mathbf{v}_c$$
$$\mathbf{v}_a\mathbf{j} = 20\cos\theta\mathbf{i} + 20\sin\theta\mathbf{j} + 5\mathbf{i}$$
$$0 = 20\cos\theta + 5$$
$$\cos\theta = \frac{-1}{4}$$
$$\theta = 104.5°$$

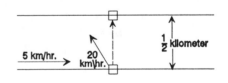

The boat should head 14.5° into the current.

$$\mathbf{v}_a \cdot \mathbf{v}_a = 400 \sin^2 \theta = 400 \left[1 - \frac{1}{16} \right] = 375$$

$$\mathbf{v}_a \approx 19.36 \text{ km/hr}$$

Time = .0258 hr. ≈ 1.55 minutes

24. The airplane travels 200 miles due west, represented by $-200\mathbf{i}$, and 150 miles 60° north of west, represented by:

$$-150(\cos 60°)\mathbf{i} + 150(\sin 60°)\mathbf{j} = -150 \left[\frac{1}{2} \right] \mathbf{i} + 150 \left[\frac{\sqrt{3}}{2} \right] \mathbf{j} = -75\mathbf{i} + 75\sqrt{3}\,\mathbf{j}$$

We have $-200\mathbf{i} + (-75\mathbf{i} + 75\sqrt{3}\,\mathbf{j}) = (-200) + (-75)\mathbf{i} = -275\mathbf{i}$

The resultant vector is $\sqrt{(-275)^2 + (75\sqrt{3})^2}$ = 304 miles.

26. $\mathbf{F} = \dfrac{1(2\mathbf{i} + 2\mathbf{j})}{4} \qquad \overrightarrow{AB} = 3\mathbf{i} + 4\mathbf{j}$

$$\mathbf{W} = \mathbf{F} \cdot \overrightarrow{AB} = \frac{1}{4}(2\mathbf{i} + 2\mathbf{j}) \cdot (3\mathbf{i} + 4\mathbf{j}) = \left[\frac{1}{2}\mathbf{i} + \frac{1}{2}\mathbf{j} \right] \cdot (3\mathbf{i} + 4\mathbf{j}) = \frac{3}{2}\mathbf{i} + 2\mathbf{j}$$

$$= \frac{5}{2} = 2.5 \text{ foot-pounds}$$

28. $\mathbf{W} = \mathbf{F} \cdot \overrightarrow{AB}$

$\mathbf{W} = 2, \ \overrightarrow{AB} = 4\mathbf{i}$

$\mathbf{F} = \cos x°\mathbf{i} - \sin x°\mathbf{j}$

$2 = (\cos x°\mathbf{i} - \sin x°\mathbf{j}) \cdot 4\mathbf{i}$

$2 = 4 \cos x°\mathbf{i}$

$\dfrac{1}{2} = \cos x°$

$x° = 60°$

30. Since $\phi = 0\mathbf{i}$ and $0\mathbf{j}$ and $\mathbf{v} = a_1\mathbf{i} + b_1\mathbf{j}$, then $\phi \cdot \mathbf{v} = 0 \cdot a_1 + 0b_1 = 0$

32.

$$\mathbf{v} = \cos \alpha \mathbf{i} + \sin \alpha \mathbf{j}$$

$$\mathbf{w} = \cos \beta \mathbf{i} + \sin \beta \mathbf{j}$$

$$\mathbf{v} \cdot \mathbf{w} = \|\mathbf{v}\| \, \|\mathbf{w}\| = 1 \cdot 1 \cdot \cos(\alpha - \beta) = \cos(\alpha - \beta)$$

$$\cos(\alpha - \beta) = (\cos \alpha \mathbf{i} + \sin \alpha \mathbf{j}) + (\cos \beta \mathbf{i} + \sin \beta \mathbf{j})$$

$$= \cos \alpha \cos \beta \mathbf{i} \cdot \mathbf{i} + \cos \alpha \sin \beta \mathbf{j} \cdot \mathbf{j} + \cos \alpha \sin \beta \mathbf{i} \cdot \mathbf{j} + \sin \alpha \sin \beta \mathbf{j} \cdot \mathbf{j}$$

$$\cos(\alpha - \beta) = \cos \alpha \cos \beta + \sin \alpha \sin \beta$$

34. (a) Let $\mathbf{u} = a_1\mathbf{i} + b_1\mathbf{j}$ and $\mathbf{v} = a_2\mathbf{i} + b_2\mathbf{j}$.

If $\mathbf{u}$ and $\mathbf{v}$ have the same magnitude, then $\sqrt{a_1^2 + b_1^2} = \sqrt{a_2^2 + b_2^2}$

To prove that $u + v$ and $u - v$ are orthogonal, we need to show that

$$0 = \frac{(u + v) \cdot (u - v)}{\|u + v\| \ \|u - v\|}$$

$$0 = \frac{[(a_1 + a_2)i + (b_1 + b_2)j] \cdot [(a_1 - a_2)i + (b_1 - b_2)j]}{\sqrt{(a_1 + a_2)^2 + (b_1 + b_2)^2} \cdot \sqrt{(a_1 - a_2)^2 + (b_1 - b_2)^2}}$$

$$0 = (a_1 + a_2)(a_1 - a_2)i + (b_1 + b_2)(b_1 - b_2)j$$
$$0 = (a_1^2 - a_2^2)i + (b_1^2 - b_2^2)j$$

Because $a_1^2 = a_2^2$ and $b_1^2 = b_2^2$, we have shown that $u + v$ and $u - v$ are orthogonal.

(b) Because the vectors u and v are radii of the circle, we know they have the same magnitude. Therefore, $u + v$ and $u - v$ are orthagonal and form a right angle.

36. Let $v = a_1 i + b_1 j$ and $w = a_2 i + b_2 j$.

$$\|v\| = \sqrt{a_1^2 + b_1^2} \ \text{ and } \ \|w\| = \sqrt{a_2^2 + b_2^2}$$

To show that $\|w\| v + \|v\| w$ and $\|w\| v - \|v\| w$ are orthogonal, we need to show that

$$0 = \frac{(\|w\| v + \|v\| w)(\|w\| v - \|v\| w)}{\|(\|w\| v + \|v\| w)\| \|(\|w\| v - \|v\| w)\|}$$

38. In order to prove the polarization identity, we want to show that
$$\|u + v\|^2 = \|u - v\|^2 = 4(u \cdot v)$$
$$(u + v)(u + v) - (u - v)(u - v) = 4(u \cdot v)$$
$$u^2 + 2(u \cdot v) + v^2 - (u^2 - 2(u \cdot v) + v^2) = 4(u \cdot v)$$
$$u^2 + 2(u \cdot v) + v^2 - u^2 + 2(u \cdot v) - v^2 = 4(u \cdot v)$$
$$2(u \cdot v) + 2(u \cdot v) = 4(u \cdot v)$$
$$4(u \cdot v) = 4(u \cdot v)$$

14 Chapter Review

2. $A - C = \begin{bmatrix} 1 & 0 \\ 2 & 4 \\ -1 & 2 \end{bmatrix} - \begin{bmatrix} 3 & -4 \\ 1 & 5 \\ 5 & -2 \end{bmatrix} = \begin{bmatrix} -2 & 4 \\ 1 & -1 \\ -6 & 4 \end{bmatrix}$

4. $-4B = -4 \begin{bmatrix} 4 & -3 & 0 \\ 1 & 1 & -2 \end{bmatrix} = \begin{bmatrix} -16 & 12 & 0 \\ -4 & -4 & 8 \end{bmatrix}$

6. $BA = \begin{bmatrix} 4 & -3 & 0 \\ 1 & 1 & -2 \end{bmatrix} \begin{bmatrix} 1 & 0 \\ 2 & 4 \\ -1 & 2 \end{bmatrix} \begin{bmatrix} 4 - 6 + 0 & 0 - 12 + 0 \\ 1 + 2 + 2 & 0 + 4 - 4 \end{bmatrix} = \begin{bmatrix} -2 & -12 \\ 5 & 0 \end{bmatrix}$

8. $BC = \begin{bmatrix} 4 & -3 & 0 \\ 1 & 1 & -2 \end{bmatrix} \begin{bmatrix} 3 & -4 \\ 1 & 5 \\ 5 & -2 \end{bmatrix} \begin{bmatrix} 12 - 3 + 0 & -16 - 15 + 0 \\ 3 + 1 - 10 & -4 + 5 + 4 \end{bmatrix} = \begin{bmatrix} 9 & -31 \\ -6 & 5 \end{bmatrix}$

10. $\begin{bmatrix} -3 & 2 & | & 1 & 0 \\ 1 & -2 & | & 0 & 1 \end{bmatrix} \rightarrow \begin{bmatrix} 1 & -2 & | & 0 & 1 \\ -3 & 2 & | & 1 & 0 \end{bmatrix} \rightarrow \begin{bmatrix} 1 & -2 & | & 0 & 1 \\ 0 & -4 & | & 1 & 3 \end{bmatrix} \rightarrow \begin{bmatrix} 1 & -2 & | & 0 & 1 \\ 0 & 1 & | & \frac{-1}{4} & \frac{-3}{4} \end{bmatrix} \rightarrow \begin{bmatrix} 1 & 0 & | & \frac{-1}{2} & \frac{-1}{2} \\ 0 & 1 & | & \frac{-1}{4} & \frac{-3}{4} \end{bmatrix}$

$\qquad\qquad\uparrow \qquad\qquad\qquad \uparrow \qquad\qquad\qquad \uparrow \qquad\qquad\qquad\qquad \uparrow$

Interchange rows $\qquad R_2 = 3r_1 + r_2 \quad R_2 = \frac{-1}{4}r_2 \qquad\qquad R_1 = 2r_2 + r_1$

Therefore, $A^{-1} = \begin{bmatrix} \frac{-1}{2} & \frac{-1}{2} \\ \frac{-1}{4} & \frac{-3}{4} \end{bmatrix}$

12. $\begin{bmatrix} 3 & 1 & 2 & | & 1 & 0 & 0 \\ 3 & 2 & -1 & | & 0 & 1 & 0 \\ 1 & 1 & 1 & | & 0 & 0 & 1 \end{bmatrix} \rightarrow \begin{bmatrix} 1 & 1 & 1 & | & 0 & 0 & 1 \\ 3 & 2 & -1 & | & 0 & 1 & 0 \\ 3 & 1 & 2 & | & 1 & 0 & 0 \end{bmatrix} \rightarrow \begin{bmatrix} 1 & 1 & 1 & | & 0 & 0 & 1 \\ 0 & -1 & -4 & | & 0 & 1 & -3 \\ 0 & -2 & -1 & | & 1 & 0 & -3 \end{bmatrix} \rightarrow \begin{bmatrix} 1 & 1 & 1 & | & 0 & 0 & 1 \\ 0 & 1 & 4 & | & 0 & -1 & 3 \\ 0 & -2 & -1 & | & 1 & 0 & -3 \end{bmatrix}$

$\qquad\qquad\uparrow \qquad\qquad\qquad\qquad\qquad\qquad \uparrow \qquad\qquad\qquad\qquad \uparrow$

Interchange r_1 and r_3 $\qquad\quad R_2 = -3r_1 + r_2 \qquad\qquad R_2 = -r_2$
$\qquad\qquad\qquad\qquad\qquad\qquad\quad R_3 = -3r_1 + r_3$

$\rightarrow \begin{bmatrix} 1 & 0 & -3 & | & 0 & 1 & -2 \\ 0 & 1 & 4 & | & 0 & -1 & 3 \\ 0 & 0 & 7 & | & 1 & -2 & 3 \end{bmatrix} \rightarrow \begin{bmatrix} 1 & 0 & -3 & | & 0 & 1 & -2 \\ 0 & 1 & 4 & | & 0 & -1 & 3 \\ 0 & 0 & 1 & | & \frac{1}{7} & \frac{-2}{7} & \frac{3}{7} \end{bmatrix} \rightarrow \begin{bmatrix} 1 & 0 & 0 & | & \frac{3}{7} & \frac{1}{7} & \frac{-5}{7} \\ 0 & 1 & 0 & | & \frac{-4}{7} & \frac{1}{7} & \frac{9}{7} \\ 0 & 0 & 1 & | & \frac{1}{7} & \frac{-2}{7} & \frac{3}{7} \end{bmatrix}$

$\qquad\quad\uparrow \qquad\qquad\qquad\qquad\qquad \uparrow \qquad\qquad\qquad\qquad \uparrow$

$R_1 = -r_2 + r_1 \qquad\qquad R_3 = \frac{1}{7}r_3 \qquad\qquad R_1 = 3r_3 + r_1$
$R_3 = 2r_2 + r_3 \qquad\qquad\qquad\qquad\qquad\qquad R_2 = -4r_3 + r_2$

Therefore, $A^{-1} = \begin{bmatrix} \frac{3}{7} & \frac{1}{7} & \frac{-5}{7} \\ \frac{-4}{7} & \frac{1}{7} & \frac{9}{7} \\ \frac{1}{7} & \frac{-2}{7} & \frac{3}{7} \end{bmatrix}$

14. $\begin{bmatrix} -3 & 1 & | & 1 & 0 \\ -6 & 2 & | & 0 & 1 \end{bmatrix} \rightarrow \begin{bmatrix} 1 & \frac{-1}{3} & | & \frac{-1}{3} & 0 \\ -6 & 2 & | & 0 & 1 \end{bmatrix} \rightarrow \begin{bmatrix} 1 & \frac{-1}{3} & | & \frac{-1}{3} & 0 \\ 0 & 0 & | & -2 & 1 \end{bmatrix}$

$\qquad\qquad\uparrow \qquad\qquad\qquad \uparrow$

$R_1 = \frac{-1}{3}r_1 \qquad\quad R_2 = 6r_1 + r_2$

There is no inverse. Therefore, the matrix is singular.

16. $\dfrac{x}{(x+2)(x-3)} = \dfrac{A}{x+2} + \dfrac{B}{x-3}$

$x = A(x-3) + B(x+2)$

Let $x = 3$: $\quad 3 = 5B$, or $B = \dfrac{3}{5}$

Let $x = -2$: $-2 = -5A$, or $A = \dfrac{2}{5}$

$\dfrac{x}{(x+2)(x-3)} = \dfrac{\frac{2}{5}}{x+2} + \dfrac{\frac{3}{5}}{x-3}$

18. $\dfrac{2x-6}{(x-2)^2(x-1)} = \dfrac{A}{x-2} + \dfrac{B}{(x-2)^2} + \dfrac{C}{x-1}$

$2x - 6 = A(x-2)(x-1) + B(x-1) + C(x-2)^2$

Let $x = 2$: $\quad -2 = B$

Let $x = 1$: $\quad -4 = C$

Let $x = 0$: $\quad -6 = 2A - B + 4C$

$\qquad\qquad\quad -6 = 2A + 2 - 16$

$\qquad\qquad\quad\ 8 = 2A$, or $A = 4$

$\dfrac{2x-6}{(x-2)^2(x-1)} = \dfrac{4}{x-2} + \dfrac{-2}{(x-2)^2} + \dfrac{-4}{x-1}$

20. $\dfrac{3x}{(x-2)(x^2+1)} = \dfrac{A}{x-2} + \dfrac{Bx+C}{x^2+1}$

$3x = A(x^2+1) + (Bx+C)(x-2)$

$3x = Ax^2 + A + Bx^2 - 2Bx + Cx - 2C$

$3x = (A+B)x^2 + (-2B+C)x + (A-2C)$

For x^2: $\quad 0 = A + B$

For x: $\quad 3 = -2B + C$

For the Constant: $0 = A - 2C$

Since $A = -B$, we have

$\qquad 3 = -2B + C \rightarrow 6 = -4B + 2C$

$\qquad 0 = -B - 2C \rightarrow 0 = -B - 2C$

$\qquad\qquad\qquad\quad\ 6 = -5B \text{ or } B = \dfrac{-6}{5}$

$\qquad\quad 3 = -2\left[\dfrac{-6}{5}\right] + C$

$\qquad\quad \dfrac{3}{5} = C$

$\qquad\quad A = \dfrac{6}{5}$

$\dfrac{3x}{(x-2)(x^2+1)} = \dfrac{\frac{6}{5}}{x-2} + \dfrac{\frac{-6}{5}x + \frac{3}{5}}{x^2+1}$

22. $$\frac{x^3 + 1}{(x^2 + 16)^2} = \frac{Ax + B}{(x^2 + 16)} + \frac{Cx + D}{(x^2 + 16)^2}$$

$x^3 + 1 = (Ax + B)(x^2 + 16) + Cx + D$

$x^3 + 1 = Ax^3 + Bx^2 + 16Ax + 16B + Cx + D$

$x^3 + 1 = Ax^3 + Bx^2 + (16A + C)x + (16B + D)$

$A = 1$

$B = 0$

$16A + C = 0$ or $16 + C = 0$ or $C = -16$

$16B + D = 1$ or $D = 1$

$$\frac{x^3 + 1}{(x^2 + 16)^2} = \frac{x}{x^2 + 16} + \frac{-16x + 1}{(x^2 + 16)^2}$$

24. $$\frac{4}{(x^2 + 4)(x^2 - 1)} = \frac{Ax + B}{x^2 + 4} + \frac{C}{x - 1} + \frac{D}{x + 1}$$

$4 = (Ax + B)(x^2 - 1) + C(x^2 + 4)(x + 1) + D(x^2 + 4)(x - 1)$

Let $x = -1$: $\quad 4 = -10D$, or $D = \dfrac{-2}{5}$

Let $x = 1$: $\quad 4 = 10C$, or $C = \dfrac{2}{5}$

Let $x = 0$: $\quad 4 = -B + 4C - 4D$

$\qquad\qquad\qquad 4 = -B + \dfrac{8}{5} + \dfrac{8}{5}$

$\qquad\qquad\qquad \dfrac{4}{5} = -B$, or $B = \dfrac{-4}{5}$

Let $x = 2$: $\quad 4 = 6A + 3B + 24C + 8D$

$\qquad\qquad\qquad 4 = 6A + 3\left[\dfrac{-4}{5}\right] + 24\left[\dfrac{2}{5}\right] + 8\left[\dfrac{-2}{5}\right]$

$\qquad\qquad\qquad 0 = 6A$, or $A = 0$

$$\frac{4}{(x^2 + 4)(x^2 - 1)} = \frac{\frac{-4}{5}}{x^2 + 4} + \frac{\frac{2}{5}}{x - 1} + \frac{\frac{-2}{5}}{x + 1}$$

26. $P = (-3, 1), Q = (4, -2)$

$\mathbf{v} = (4 + 3)\mathbf{i} + (-2 - 1)\mathbf{j} = 7\mathbf{i} - 3\mathbf{j}$

$\|\mathbf{v}\| = \sqrt{7^2 + (-3)^2} = \sqrt{58}$

28. $P = (3, -4), Q = (-2, 0)$

$\mathbf{v} = (-2 - 3)\mathbf{i} + (0 + 4)\mathbf{j} = -5\mathbf{i} + 4\mathbf{j}$

$\|\mathbf{v}\| = \sqrt{(-5)^2 + 4^2} = \sqrt{41}$

30. $-\mathbf{v} + 2\mathbf{w} = -(-2\mathbf{i} + \mathbf{j}) + 2(4\mathbf{i} - 3\mathbf{j}) = 2\mathbf{i} - \mathbf{j} + 8\mathbf{i} - 6\mathbf{j} = 10\mathbf{i} - 7\mathbf{j}$

32. $\|\mathbf{v} + \mathbf{w}\| = \|-2\mathbf{i} + \mathbf{j} + 4\mathbf{i} - 3\mathbf{j}\| = \|2\mathbf{i} - 2\mathbf{j}\|$

$\|\mathbf{v} + \mathbf{w}\| = \sqrt{2^2 + (-2)^2} = \sqrt{8} = 2\sqrt{2}$

34. $\|2\mathbf{v}\| - 3\|\mathbf{w}\| = \|2(-2\mathbf{i} + \mathbf{j})\| - 3\|4\mathbf{i} - 3\mathbf{j}\| = \|-4\mathbf{i} + 2\mathbf{j}\| - 3\left(\sqrt{4^2 + (-3)^2}\right)$

$\qquad\qquad\qquad = \sqrt{(-4)^2 + 2^2} - 3(5) = 2\sqrt{5} - 15$

36. $\|\mathbf{w}\| = \sqrt{4^2 + (-3)^2} = 5$

$\mathbf{u} = \dfrac{-\mathbf{w}}{\|\mathbf{w}\|} \quad \dfrac{-(4\mathbf{i} - 3\mathbf{j})}{5}$

$\mathbf{u} = -\dfrac{4\mathbf{i}}{5} + \dfrac{3\mathbf{j}}{5}$

Check: $\|\mathbf{u}\| = \sqrt{\left(\dfrac{-4}{5}\right)^2 + \left(\dfrac{3}{5}\right)^2} = \sqrt{\dfrac{16 + 9}{25}} = 1$

38. $\mathbf{v} = 3\mathbf{i} - \mathbf{j}, \; \mathbf{w} = \mathbf{i} + \mathbf{j}$

$\mathbf{v} \cdot \mathbf{w} = (3)(1) + (-1)(1) = 2$

$\cos\theta = \dfrac{2}{\sqrt{10} \cdot \sqrt{2}} = \dfrac{2}{2\sqrt{5}} = \dfrac{\sqrt{5}}{5}$

40. $\mathbf{v} = \mathbf{i} + 4\mathbf{j}, \; \mathbf{w} = 3\mathbf{i} - 2\mathbf{j}$

$\mathbf{v} \cdot \mathbf{w} = (1)(3) + (4)(-2) = -5$

$\cos\theta = \dfrac{-5}{\sqrt{17} \cdot \sqrt{13}} = \dfrac{-5}{\sqrt{221}} = \dfrac{-5\sqrt{221}}{221}$

42. $\mathbf{v} = -\mathbf{i} + 2\mathbf{j}, \; \mathbf{w} = 3\mathbf{i} - \mathbf{j}$

$\text{proj}_{\mathbf{w}} = \dfrac{-5}{\left(\sqrt{10}\right)^2}(3\mathbf{i} - \mathbf{j}) = \dfrac{-1}{2}(3\mathbf{i} - \mathbf{j})$

44. $\mathbf{v} = \mathbf{i} - \mathbf{j}$ and $\mathbf{w} = 2\mathbf{i} + \mathbf{j}$

$\cos\theta = \dfrac{\mathbf{v} \cdot \mathbf{w}}{\|\mathbf{v}\| \, \|\mathbf{w}\|} = \dfrac{1}{\sqrt{2} \cdot \sqrt{5}} = \dfrac{1}{\sqrt{10}}$

$\theta = 72°$

46. Let $\mathbf{v_w}$ = velocity of the motorboat relative to the water

$\mathbf{v_c}$ = velocity of the current

and $\mathbf{v_m}$ = true velocity of the motorboat

Then, $\mathbf{v_m} = \mathbf{v_w} + \mathbf{v_c}$ or $\mathbf{v_w} = \mathbf{v_m} - \mathbf{v_c}$

$\mathbf{v_m} = -11\mathbf{j}$

From the northeast: $-\mathbf{i} - \mathbf{j}$

The unit vector in this direction is $= \dfrac{-\mathbf{i} - \mathbf{j}}{\|\mathbf{i} - \mathbf{j}\|} = \dfrac{-\mathbf{i} - \mathbf{j}}{\sqrt{2}} = \dfrac{-1}{\sqrt{2}}(\mathbf{i} + \mathbf{j})$

$\mathbf{v_c} = \dfrac{-3\sqrt{2}}{2}\mathbf{i} - \dfrac{3\sqrt{2}}{2}\mathbf{j}$

$\mathbf{v_w} = \mathbf{v_w} = -11\mathbf{j} - \left[\dfrac{-3\sqrt{2}}{2}\mathbf{i} - \dfrac{3\sqrt{2}}{2}\mathbf{j}\right] = \dfrac{3\sqrt{2}}{2}\mathbf{i} + \dfrac{3\sqrt{2} - 22}{2}\mathbf{j}$

The speed of the motorboat relative to the water is:

$$\|v_w\| = \sqrt{\left[\frac{3\sqrt{2}}{2}\right]^2 + \left[\frac{3\sqrt{2} - 22}{2}\right]^2}$$

$$\|v_w\| = \sqrt{83.33} \approx 9.13 \text{ mph}$$

$$\cos\theta = \frac{v \cdot v_b}{\|v\| \cdot \|v_b\|} = \frac{-11\left[\frac{3\sqrt{2}}{2} - 11\right]}{(9.13)(11)} = 0.97$$

$$\theta = 13.5°$$

48. Let $v_a = 500j$, $v_w = \frac{60 \cdot (i - j)}{\sqrt{2}} = 30\sqrt{2}\,i - j$

Then, $v_g = v_a + v_w$

$$v_g = 500j + 30\sqrt{2}\,i - 30\sqrt{2}\,j$$

$$v_g = 30\sqrt{2}\,i + (500 - 30\sqrt{2})j$$

Speed $= \|v_g\| = \sqrt{\left(30\sqrt{2}\right)^2 + \left(500 - 30\sqrt{2}\right)^2} = \sqrt{1800 + 2099{,}373.6}$
$= 459.5$ kilometers per hour

$$\cos\theta = \frac{v_g v_a}{\|v_g\|\|v_a\|} = \frac{500(500 - 30\sqrt{2})}{500(459.5)} = 0.996$$

$$\theta = 5.2°$$

Mission Possible

1. Graph

2-7. Answers will vary. I would use points #4, 5, 7 or 4, 6, 7 since I think the fuel efficiency story was affected permanently by the oil crisis of the 70s. Of the two, I would choose the one with the steepest increase for 2000, partly to be safe, partly because fewer and fewer of the old gas guzzlers are in existence, which will affect the average upward.

Example:
$$13.5 = a \cdot 30^2 + b \cdot 30 + c$$
$$15.5 = a \cdot 40^2 + b \cdot 40 + c$$
$$21.7 = a \cdot 51^2 + b \cdot 51 + c$$

$$900a + 30b + c = 13.5$$
$$1600a + 40b + c = 15.5$$
$$2601a + 51b + c = 21.7$$

According to my TI-82, $a = 0.0173$, $b = -1.012$, $c = 28.279$, which yields the equation $y = 0.0173x^2 + -1.012x + 28.279$. This curve is very accurate for the last four points, not very good for the first three points, but according to my reasoning, that is okay. Others might not agree.

Using my equation, in 2000 the value of average miles per gallon in the U.S. would be predicted to be 29.8; in 2010 it will be 12.2.

Exercise 1

2. xmin = −3
    xmax = 7
    xscl = 1
    ymin = −4
    ymax = 9
    yscl = 1

4. xmin = −90
    xmax = 30
    xscl = 10
    ymin = −50
    ymax = 70
    yscl = 10

6. xmin = −20
    xmax = 110
    xscl = 10
    ymin = −10
    ymax = 60
    yscl = 10

8. xmin = −3
    xmax = 3
    xscl = 1
    ymin = −2
    ymax = 2
    yscl = 1

10. xmin = −3
    xmax = 3
    xscl = 1
    ymin = −10
    ymax = 10
    yscl = 5

12. xmin = −12
    xmax = 12
    xscl = 2
    ymin = −4
    ymax = 4
    yscl = 1

14. xmin = −9
    xmax = 9
    xscl = 3
    ymin = −12
    ymax = 4
    yscl = 4

16. xmin = −22
    xmax = −10
    xscl = 2
    ymin = 4
    ymax = 8
    yscl = 1

Exercise 2

2. (a) 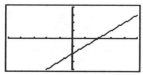 (b) 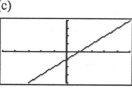 (c) 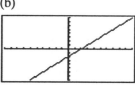 (d)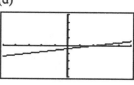

4. (a) 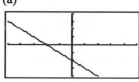 (b) 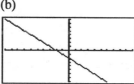 (c) 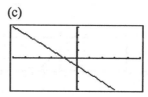 (d)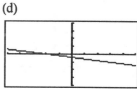

6. (a)　　(b)　　(c)　　(d)

8. (a)　　(b)　　(c)　　(d)

10. (a)　　(b)　　(c)　　(d)

12. (a)　　(b)　　(c)　　(d)

14. (a)　　(b)　　(c)　　(d)

16. (a)　　(b)　　(c)　　(d)

18. (a)　　(b)　　(c)　　(d)

20. (a)　　(b)　　(c)　　(d)

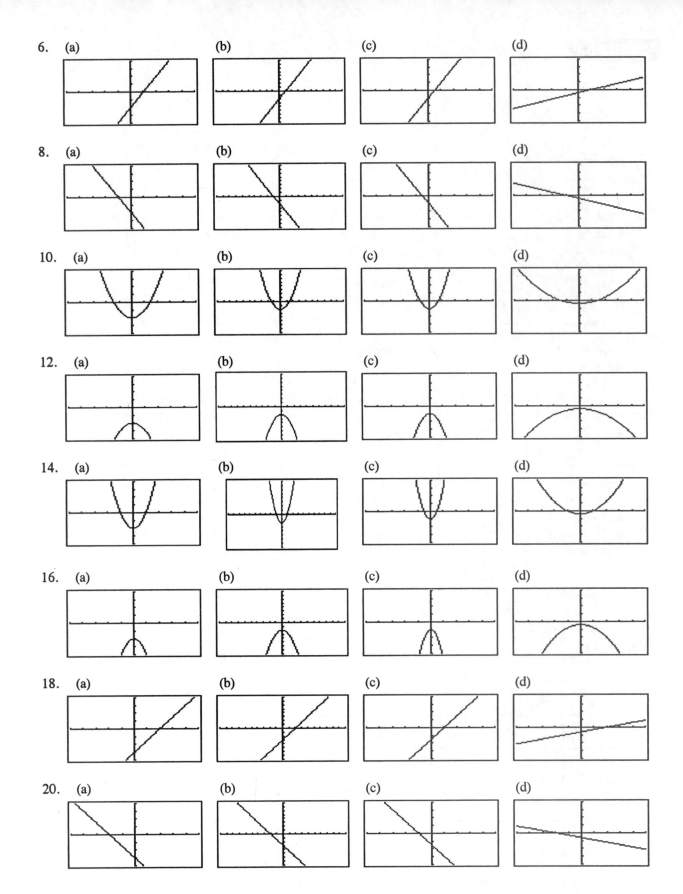

22.

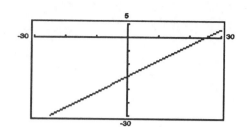

24.

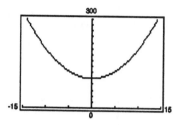

26.

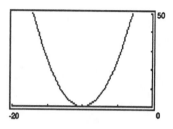

28.

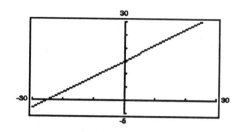

30.

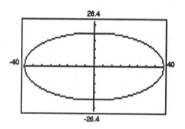

32.

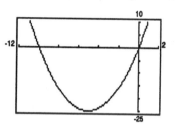

34.

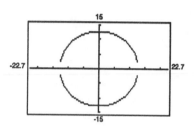

36.

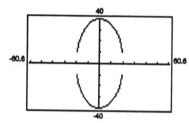

38.

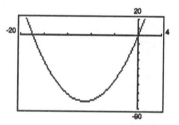

40.

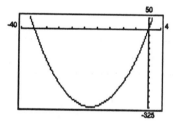

Exercise 4

2.   No   4.   Yes   6.   Yes   8.   Yes

10. $ymin = -2$
      $ymax = 10$
      $yscl = 2$

   Note: Other answers are possible ($ymax - ymin$ must equal 12).

Exercise 5

2. 0.181   4. 2.449   6. 1.442   8. −4.64

10. −1.43   12. −0.22   14. 2.00   16. 1.70

18. 0.63, 13.59   20. 1.07